Lecture Notes in Physics

The Editorial Policy for Edited Volumes

The series *Lecture Notes in Physics* (LNP), founded in 1969, reports new developments in physics research and teaching - quickly, informally but with a high degree of quality. Manuscripts to be considered for publication are topical volumes consisting of a limited number of contributions, carefully edited and closely related to each other. Each contribution should contain at least partly original and previously unpublished material, be written in a clear, pedagogical style and aimed at a broader readership, especially graduate students and nonspecialist researchers wishing to familiarize themselves with the topic concerned. For this reason, traditional proceedings cannot be considered for this series though volumes to appear in this series are often based on material presented at conferences, workshops and schools.

Acceptance

A project can only be accepted tentatively for publication, by both the editorial board and the publisher, following thorough examination of the material submitted. The book proposal sent to the publisher should consist at least of a preliminary table of contents outlining the structure of the book together with abstracts of all contributions to be included. Final acceptance is issued by the series editor in charge, in consultation with the publisher, only after receiving the complete manuscript. Final acceptance, possibly requiring minor corrections, usually follows the tentative acceptance unless the final manuscript differs significantly from expectations (project outline). In particular, the series editors are entitled to reject individual contributions if they do not meet the high quality standards of this series. The final manuscript must be ready to print, and should include both an informative introduction and a sufficiently detailed subject index.

Contractual Aspects

Publication in LNP is free of charge. There is no formal contract, no royalties are paid, and no bulk orders are required, although special discounts are offered in this case. The volume editors receive jointly 30 free copies for their personal use and are entitled, as are the contributing authors, to purchase Springer books at a reduced rate. The publisher secures the copyright for each volume. As a rule, no reprints of individual contributions can be supplied.

Manuscript Submission

The manuscript in its final and approved version must be submitted in ready to print form. The corresponding electronic source files are also required for the production process, in particular the online version. Technical assistance in compiling the final manuscript can be provided by the publisher's production editor(s), especially with regard to the publisher's own LaTeX macro package which has been specially designed for this series.

LNP Homepage (springerlink.com)

On the LNP homepage you will find:

–The LNP online archive. It contains the full texts (PDF) of all volumes published since 2000. Abstracts, table of contents and prefaces are accessible free of charge to everyone. Information about the availability of printed volumes can be obtained.

–The subscription information. The online archive is free of charge to all subscribers of the printed volumes.

–The editorial contacts, with respect to both scientific and technical matters.

–The author's / editor's instructions.

G. Amelino-Camelia J. Kowalski-Glikman (Eds.)

Planck Scale Effects in Astrophysics and Cosmology

Springer

Editors

Jurek Kowalski-Glikman
University Wroclaw
Institute of Theoretical Physics
Pl. Maxa Borna 90
50-204 Wroclaw, Poland
Email: jurekk@ift.uni.wroc.pl

Giovanni Amelino-Camelia
Universitá di Roma La Sapienza
Dipto. di Matematica
Piazzale Aldo Moro, 2
00185 Roma, Italy
Email: amelino@roma1.infn.it

J. Kowalski-Glikman, G. Amelino-Camelia (Eds.), *Planck Scale Effects in Astrophysics and Cosmology*, Lect. Notes Phys. 669 (Springer, Berlin Heidelberg 2005), DOI 10.1007/b105189

ISBN 978-3-642-06441-8 e-ISBN 978-3-540-31527-8
ISSN 0075-8450

Springer is a part of Springer Science+Business Media
springeronline.com

Softcover reprint of the hardcover 1st edition 2005

Cover production: *design &production* GmbH, Heidelberg

Preface

This volume is composed of notes from lectures given at the 40th Karpacz Winter School, which focused on "Quantum Gravity Phenomenology" particularly its applications in astrophysics and cosmology.

After several decades in which the quantum-gravity problem was studied in a way that did not involve at all the confrontation with experiments [1, 2], over the last few years the idea of testing quantum-gravity ideas using experimental data has attracted significant interest. It was just around the time of the 35th Karpacz Winter School [3] that this change of attitude materialized in a part of the quantum-gravity community. And discussions that got started at the time of the 35th Karpacz Winter School finally led to the choice of topic of this 40th Karpacz Winter School.

The idea was to give students attending the school an opportunity for a short introduction to the heavily mathematical subjects that compose pure-theory quantum-gravity research, and then expose them to the core ideas that allow us to test some Planck-scale effects, especially in astrophysics and cosmology.

The lectures by Alvarez provide a brief introduction and review of string theory and loop quantum gravity, the two most popular approaches to the quantum-gravity problem. His lectures of course do not provide a detailed account of all of the technical developments in these heavily technical fields, but they strike a nice balance, combining an elementary technical introduction to the subjects with a perspective which emphasizes some key strengths and some key weaknesses of each of these two approaches. Readers interested in a more detailed technical introduction to string theory and loop quantum gravity will find useful the [4, 5] and [6, 7, 8, 9], respectively.

On the loop-quantum-gravity side the lectures by Smolin nicely complement Alvarez's lectures. In fact, Smolin provides a pedagogical introduction to some advanced aspects of the loop-quantum-gravity research program which have recently taken center stage. In particular, in Smolin's lecture notes the reader is exposed to the idea of recovering Minkowski space, in quantum gravity, only through a procedure which requires, as an intermediate step, the (quantum) description of deSitter spacetime.

Some advanced topics in "loop quantum gravity" were also introduced in the invited seminars by Pullin, which are not covered in this volume. Following the line of analysis of [10] and references therein, he presented the "consistent discretization" approach to general relativity, showing that this leads to a theory that has as its physical space what is usually considered the kinematical space of loop quantum gravity.

A first introduction to the ideas and to the most fundamental techniques used in quantum-gravity phenomenology is given in the lecture notes by Amelino-Camelia, who also stresses the importance of relying on some suitable test theories in developing this phenomenology.

The theme of working with test theories and pushing forward the experimental bounds on some commonly-adopted reference test theories was further explored in the lectures by Laemmerzahl. His lectures focus on the use of interferometry in various areas of interest for the quantum-gravity problem, including tests of the equivalence principle and tests of Lorentz symmetry.

In addition to parts of the lectures by Amelino-Camelia and Laemmerzahl, several other lectures also focused or at least touched upon the subject of the fate of Lorentz symmetry in quantum-gravity theories. The fact that in various approaches to the quantum-gravity problem there is some evidence of departures from Lorentz symmetry, and the fact that several observatories are preparing to provide us with a gigantic leap forward in the quality of Lorentz-symmetry tests, combined to bring this subject to the top of the list of priorities for the School. On the theory side a key issue here is the one of establishing whether Lorentz symmetry is "broken", in the sense commonly encountered in the analysis of particle physics in the presence of external media, or "deformed", in the sense of the "doubly-special relativity" proposal of [11, 12]. While the various scenarios for broken Lorentz symmetry were discussed briefly when appropriate in various lectures, the concept of deformation of Lorentz symmetry was introduced pedagogically in the dedicated lectures by Kowalski-Glikman, since this familiar concept of broken symmetries did not require a significant tutoring effort. His lectures emphasize in particular some delicate issues that have emerged in doubly-special-relativity research, including the role that, at least to some extent, could be played by the mathematics of κ-Poincaré Hopf algebras.

From a more phenomenological perspective the possibility of Planck-scale modifications of Lorentz symmetry was the main focus of the lectures by Jacobson, Grillo and Piran. The lectures delivered by Jacobson gave an overview of several opportunities that modern astrophysics provides for testing Lorentz symmetry with Planck-scale sensitivity. Grillo focused on the study of the cosmic-ray spectrum, especially as it will soon be studied by the Pierre Auger Observatory, which should be the best opportunity for dramatic improvement in the quality of our tests of Planck-scale modifications of Lorentz symmetry. Piran gave detailed pedagogical lectures on the research line that intends to constrain Planck-scale departures of Lorentz symmetry using data on gamma-ray bursters, and in particular he stressed some features of gamma-ray bursters which could effectively could the act as troublesome background for the quantum-gravity studies.

The lectures by Mavromatos and Ng focused on some examples of phenomenological programs which can be primarily motivated by some descriptions of "spacetime foam". Both lectures provided further encouragement for the idea of Planck-scale departures from Lorentz symmetry. Mavromatos emphasized even more strongly the possibility of Planck-scale departures from CPT symmetry, and discussed a rich CPT phenomenology. Ng also discussed some other spacetime-foam effects which could be investigated with modern interferometers.

Martin's lectures gave a pedagogical introduction to the research area that investigates the possibility of quantum-gravity effects in cosmology.

This "quantum-gravity cosmology" was also the subject of invited seminars by de Bernardis, which are not covered in this volume. He gave a detailed description of the BOOMERANG and WMAP experiments following roughly the line of analysis presented in [13].

The invited seminars by Lipari, which were based on some of his works in preparation, provided a perspective on several aspects of gamma-ray and cosmic-ray physics, which are relevant for the topics covered by other lecturers.

The invited seminar by Urrutia presented yet another intriguing perspective on the phenomenology of Planck-scale departures from Lorentz symmetry, following roughly the line of analysis presented in [14].

Also the seminars contributed by several participants were very important for the overall balance of the school. In particular, lively discussions were generated by the seminars by Arzano [15], Bruno [16], Doplicher [17], Hinterleitner [18], Liberati [19], Mandanici [20], Martinetti, Mattingly [19], Mendez [21], Oriti, Penna-Firme, Rembielinski, Rychkov [22], Sudarsky [23] and Turko.

We owe special thanks to all lecturers and all other speakers, and we are particularly grateful for the lucky assortment of students and senior participants who attended the school. The enthusiasm of the students for all lectures was a major source of energy for the school. We were amazed to see students requesting on several occasions additional hours of lecture by some lecturers, which were often scheduled after dinner (the feared "8pm–10pm extra lectures"). Perhaps the unfriendly weather outside the hotel that hosted the school had something to do with all this enthusiasm for lectures, but it nonetheless contributed to a wonderful 10 days of physics.

Finally we would like to thank the Rector of the University of Wroclaw, the Polish Ministry of Education, the Foundation for Karpacz Winter Schools in Theoretical Physics, and the European Physical Society for their generous financial support. Thanks are due Professor Jerzy Lukierski, Director of the Institute for Theoretical Physics of the University of Wroclaw, for his encouragement support and help, to Mrs. Katarzyna Imilkowska, who did a great job as the School secretary, and to Dr. Wojciech Cegla, who oversaw the School finances.

Giovanni Amelino-Camelia
April, 2005 *Jerzy Kowalski-Glikman*

References

1. J. Stachel: Early History of Quantum Gravity. In *Black Holes, Gravitational Radiation and the Universe*, ed by B.R. Iyer, B. Bhawal (Kluwer Academic Publisher, Netherlands, 1999).
2. C.J. Isham: *Structural issues in quantum gravity.* In *Proceedings of General relativity and gravitation 1995* (World Scientic, Singapore 1997).
3. *Towards Quantum Gravity*, ed by J. Kowalski-Glikman, Lect. Notes Phys **541** (2000).
4. M.B. Green, J.H. Schwarz and E. Witten: *Superstring theory* (Cambridge Univ. Press, Cambridge, 1987)
5. J. Polchinski: *Superstring Theory and Beyond*, (Cambridge University Press, Cambridge, 1998).
6. C. Rovelli: gr-qc/9710008, Living Rev. Rel. **1**, 1 (1998).
7. A. Ashtekar: gr-qc/0112038.
8. L. Smolin: hep-th/0303185.
9. T. Thiemann: Lect. Notes Phys. **631**, 41 (2003).
10. R. Gambini and J. Pullin, gr-qc/0409057.
11. G. Amelino-Camelia: Int. J. Mod. Phys. D **11**, 35 (2002); Phys. Lett. B **510**, 255 (2001).
12. J. Kowalski-Glikman: Phys. Lett. A **286**, 391 (2001); R. Bruno, G. Amelino-Camelia and J. Kowalski-Glikman: hep-th/0107039, Phys. Lett. B **522**, 133 (2001).
13. P. De Bernardis et al, Nucl. Phys. Proc. Suppl. **113**, 18 (2002).
14. L.F. Urrutia, hep-ph/0402271.
15. G. Amelino-Camelia, M. Arzano, Y.J. Ng, T. Piran and H. Van Dam, hep-ph/0307027, JCAP **0402**, 009 (2004).
16. A. Ballesteros, N.R. Bruno and F.J. Herranz, hep-th/0401244.
17. Doplicher Giovanni Amelino-Camelia, L. Doplicher, hep-th/0312313, Class. Quant. Grav. **21**, 4927 (2004).
18. D. Heyman, F. Hinterleitner, and S. Major, gr-qc/0312089, Phys. Rev. D **69**, 105016 (2004).
19. T. Jacobson, S. Liberati and D. Mattingly, gr-qc/0404067.
20. G. Amelino-Camelia, J. Kowalski-Glikman, G. Mandanici and A. Procaccini, gr-qc/0312124.
21. R. Aloisio, J.M. Carmona, J.L. Cortes, A. Galante, A.F. Grillo and F. Mendez, hep-th/0404111, JHEP **0405**, 028 (2004).
22. V.S. Rychkov, hep-th/0305187, JCAP **0307**, 006 (2003).
23. J. Collins, A. Perez, D. Sudarsky, L.F. Urrutia, H. Vucetich, gr-qc/0403053.

Contents

List of Contributors

Roberto Aloisio
INFN, Laboratori Nazionali del Gran Sasso, SS 17bis
67010 Assergi, L' Aquila Italy

Enrique Alvarez
Instituto de Fisica Teorica
UAM/CSIC, 28049 Madrid Spain
enrique.alvarez@uam.es

Giovanni Amelino-Camelia
Dipartimento di Fisica
Universitá di Roma "La Sapienza" and Sez. Roma1 INFN
P.le Moro 2, I-00185 Roma, Italy
amelino@roma1.infn.it

Pasquale Blasi
INAF, Osservatorio Astrofisico di Arcetri, Largo E. Fermi5, 50125
Firenze Italy

Angelo Galante
INFN, Laboratori Nazionali del Gran Sasso, SS 17bis 67010
Assergi, L' Aquila Italy
and
Dipartimento di Fisica
Universita' di L'Aquila
Via Vetoio, 67100
Coppito (AQ) Italy

Aurelio F. Grillo
INFN, Laboratori Nazionali del Gran Sasso, SS 17bis
67010 Assergi, L' Aquila (Italy)
and
Speaker
Aurelio.Grillo@lngs.infn.it

Ted Jacobson
Institut d'Astrophysique de Paris
98bis bd Arago, 75014 Paris
France
and
Department of Physics
University of Maryland
College Park, MD 20742
USA
jacobson@umd.edu

Jerzy Kowalski-Glikman
Institute for Theoretical Physics
University of Wrocław
Pl. Maxa Borna 9, 50204 Wrocław
Poland
jurekk@ift.uni.wroc.pl

Claus Lämmerzahl
ZARM, University of Bremen
Am Fallturm 28359 Bremen
Germany
laemmerzahl@zarm.uni-bremen.de

Stefano Liberati
SISSA, Via Beirut 2-4, 34014 Trieste
Italy
and
INFN, Trieste, Italy
liberati@sissa.it

Jérôme Martin
Institut d'Astrophysique
de Paris
GreCO, FRE 2435-CNRS
98bis boulevard Arago
75014 Paris
France
jmartin@iap.fr

David Mattingly
Department of Physics
University of California at Davis
USA
mattingly@physics.ucdavis.edu

Nick E. Mavromatos
King's College London
Department of Physics, Strand
London WC2R 2LS, U.K.
Nikolaos.Mavromatos@kcl.ac.uk

Y. Jack Ng
Institute of Field Physics
Department of Physics
and Astronomy
University of North Carolina
Chapel Hill
NC 27599-3255, USA
yjng@physics.unc.edu

Tsvi Piran
Racah Institute for Physics
The Hebrew University
Jerusalem, Israel
tsvi@phys.huji.ac.il

Lee Smolin
Perimeter Institute for Theoretical
Physics
Waterloo, Canada N2J 2W9
and
Department of Physics
University of Waterloo
lsmolin@perimeterinstitute.ca

Planck Scale Kinematics and the Pierre Auger Observatory

R. Aloisio[1], P. Blasi[2], A. Galante[1,3] and A.F. Grillo[1,4]

[1] INFN, Laboratori Nazionali del Gran Sasso, SS 17bis, 67010 Assergi, L' Aquila Italy
[2] INAF, Osservatorio Astrofisico di Arcetri, Largo E. Fermi5, 50125 Firenze Italy
[3] Dipartimento di Fisica, Universita' di L'Aquila, Via Vetoio, 67100 Coppito (AQ) Italy
[4] Presented by A.F. Grillo
`Aurelio.Grillo@lngs.infn.it`

Abstract. Quite unexpectedly, to many of us at least, Planck scale physics has in last years made irruption in present experimental physics. In these lectures I try to describe why this happened particularly in relation to Ultra High Energy Cosmic Ray Physics, and will discuss the potentialities of experiments in this field, in particular of the Pierre Auger Observatory. I will also present some (more theoretical) speculations.

1 Motivations

In this report I will mix theoretical and phenomenological considerations. However, being a theorist who has spent a large part of his activity as a Cosmic Ray experimentalist, I will always try to keep experimental verification/falsification as the main guide in approaching theories, and more so in this field connected to the possibly most complex entities (Quantum Gravity effects) for the experimentalist (and even for most theorists). In this connection I think that a description of how I entered in this field might be of some interest.

In fact I remember exactly how this happened: I was attending a seminar from a colleague of mine (experimentalist, we were both part of the MACRO experiment at Gran Sasso Laboratory) reporting about the so called Greisen-Zatsepin-Kuzmin break [1] and the first experimental data from AGASA [2] which did not show it. While I was mentally searching for possible explanations, I realized that the processes giving rise to the expected break are in fact *low energy* processes, since it is always possible to boost back the UHE proton to a frame where it is at rest, and there the photon must have an energy only larger than $\approx 100\,\mathrm{MeV}$ to photoproduce a pion. So, no particle physics explanation for its possible absence was at hand.

R. Aloisio et al.: *Planck Scale Kinematics and the Pierre Auger Observatory*, Lect. Notes Phys. **669**, 1–30 (2005)
www.springerlink.com

On the other hand I appreciated that in fact the expected presence of the break is entirely based on an extrapolation of the validity of the Lorentz transformations up to Lorentz factors $\gamma_L \approx 10^{11}$ or velocities $1 - \beta \approx 10^{-22}$. A verification of the presence of the break would imply a *direct* comparison of physics in two frames moving at extreme relative speed.

Unfortunately (and contrary to what is often said in the literature) the absence of the break *does not* point by itself to a violation of Lorentz invariance: since we ultimately do not know the origin of the highest energy particles in the Cosmic Radiation, there are at least an handful of more mundane explanations for its absence. I will discuss later under which conditions Cosmic Ray experiments would imply such a radical departure, but from what I said it is clear that this is tied to the recognition of the sources of these particles.

In March 1997 I presented under the title "The Auger Observatory as a test ground for very fundamental physics" a short report at the European Auger meeting that was held in Gran Sasso, and (in collaboration with P. Blasi) we sent an abstract to the 1997 ICRC conference. However, since we were not able at the moment to make a more quantitative description of possible departures from Lorentz Invariance, we withdraw the paper.

Seven years have passed, and I have discovered a very interesting theoretical physics that was largely unknown to me. We also discovered that some related (and prophetic) ideas were presented in a paper by D.A. Kirzhnits and V.A. Chechin in 1971, shortly after the prediction of the GZK break [3] .

What has really dramatically changed in the last few years has been the recognition that the effects of Planck scale physics need not to be confined to the Planck regime. There are several ways to understand why this happens at an intuitive level. First, violations of Lorentz invariance should be parameterized by some indicator of "relativisticity" so to say; for instance at an energy of 10^{20} eV a proton, on a logarithmic scale is nearer to the Planck scale (10^{28} eV) than to its rest mass.

Second, in the Cosmic Microwave Background Radiation (CMBR) frame, i.e. the frame in which the CMBR is isotropic with a photon energy distribution corresponding to a Planckian of temperature $\approx 2.7°$K, a 10^{20} eV proton only needs a fractional gain in energy $\approx 10^{-22}$ to perform the transition to the final πp state. Of course there is nothing miraculous in this since Lorentz invariance guarantees that this is exactly equivalent to what happens in a frame in which the proton is at rest and the photon has an energy larger than $\approx$100 MeV. But this also displays the fact that even very tiny violations of relativistic invariance are bound to give, in some selected reactions at least, observable effects.

And, third, an example of reactions very sensitive to even small departures from L.I. is given by particle production thresholds, which typically sensitively depend on the *rest masses* of the (massive) involved particles. The simplest way of parameterizing departures from relativistic invariance is to change the form of the LI dispersion relation $E^2 - p^2 = m^2$ by rewriting it as $E^2 - p^2 = \mu^2(E, p)$. This may be seen as the introduction of an (energy and momentum dependent) effective mass that in turn will affect the threshold energy-momenta, in principle differently in different reference frames.

2 Introduction

The hunt for possible minuscule violations of the fundamental Lorentz invariance (LI) has been object of renewed interest, in particular because it has been understood that cosmic ray physics has an unprecedented potential for investigation in this field [3, 4, 5, 6, 7, 8]. Some authors [5, 6, 9] have even invoked possible violations of LI as a plausible explanation to some puzzling observations related to the detection of ultra high energy cosmic rays (UHECRs) with energy above the so-called GZK feature [1], and to the unexpected shape of the spectrum of photons with super-TeV energy from sources at cosmological distances.

Both types of observations have in fact many uncertainties, that will be diffusely discussed in the following, either coming from limited statistics of very rare events, or from accuracy issues in the energy determination of the detected particles, and it is very possible that the solution to the alleged puzzles will come from more accurate observations rather than by a violation of fundamental symmetries.

For this reason, from the very beginning we proposed [7] that cosmic ray observations should be used as an ideal tool to constrain the minuscule violations of LI, rather than as evidence for the need to violate LI. The reason why the cases of UHECRs and TeV gamma rays represent such good test sites for LI is that both are related to physical processes with a kinematical energy threshold, which is in turn very sensitive to the smallest violations of LI. UHECRs are expected to suffer severe energy losses due to photopion production off the photons of the cosmic microwave background (CMB), and this should suppress the flux of particles at the Earth at energies above $\sim 10^{20}$ eV, the so called GZK feature.

Present largest operating (or just ended) experiments are AGASA [10] and HiRes [11], and they do not provide strong evidence either in favor or against the detection of the GZK feature [12]. A substantial increase in the statistics of events, as expected with the Auger project [13] and with EUSO [14], should dramatically change the situation and allow to detect the presence or lack of the GZK feature in the spectrum of UHECRs (see next section for more detail). Moreover they should in principle be able to perform rather sensitive anisotropy studies, in particular to search for possible correlations with distant sources.

These are the observations that will provide the right ground for imposing a strong limit on violations of LI.In this report I will direct myself only to Ultra High Energy hadronic Cosmic rays. It is perhaps worth remembering that a potentially very interesting arena for detecting violations of LI is also the study of TeV γ sources. For the case of TeV sources, the process involved is pair production [15] of high energy gamma rays on the photons of the infrared background. Also in this case, a small violation of LI can move the threshold to energies which are smaller than the classical ones, or move them to infinity, making the reactions impossible. The detection of the GZK suppression or the cutoff in the gamma ray spectra of gamma ray sources at cosmological distances will prove that LI is preserved to correspondingly high accuracy [7].

The recipes for the violations of LI generally consist of requiring an *explicit* modification of the dispersion relation of high energy particles, due to their propagation in the "vacuum", now affected by quantum gravity (QG). This effect is generally parameterized by introducing a typical mass, expected to be of the order of the Planck mass (M_P), that sets the scale for QG to become effective.

This approach has been extensively discussed in the literature (and in several reports in these proceedings) so it will be presented here for completeness, and to set the ground for comparison with possible experimental outcomes of the new experiments, in particular of the Pierre Auger Observatory.

In particular we discuss at some length the possible outcomes of the next experiments, and their relevance for the detection of (Quantum Gravity inspired) modifications of special relativity.

We next pass to a more speculative level. Explicit modifications of the dispersion relation are not really necessary in order to produce detectable effects, as was recently pointed out in [16, 17, 18, 19] for the case of propagation of UHECRs. It is in fact generally believed that coordinate measurements cannot be performed with precision better than the Planck distance (time) $\delta x \geq l_P$, namely the distance where the metric of space-time must feature quantum fluctuations.

A similar line of thought implies that an uncertainty in the measurement of energy and momentum of particles can be expected, according with the relation $\delta p \simeq \delta E \simeq p^2/M_P$. As discussed also in [17, 18] the apparent problem of super-GZK particles might find a solution also in the context of this uncertainty approach.

In the second part of these lectures we discuss this approach in some detail, by taking into account the effects of the propagation of CRs in the QG vacuum in the presence of the universal microwave background radiation. A fluctuating metric implies that different measurements of the particle energy or momentum may result in different outcomes.

A consequence of this approach is that particles with classical energy below the standard Lorentz invariant threshold have a certain probability of interacting. In the same way, particles above the classical threshold have a finite probability of evading interaction. We show here that the most striking consequences of the approach described above derive from low energy particles rather than from particles otherwise above the threshold for photopion production.

However, the possibility of a fluctuating energy and momentum is mainly constrained by other processes that could arise. The fluctuations of energy and momentum are responsible, in fact, for decaying processes otherwise impossible, typically prevented by energy and momentum conservation. These decaying processes represent the most stringent test of the proposed model. From a general point of view a particle propagating in a fluctuating vacuum acquires an energy dependent fluctuating effective mass (the fluctuating dispersion relations introduced in [20]) which may be responsible for kinematically forbidden decay reactions to become kinematically allowed.

If this happens, particles that are known to be stable would decay, provided no other fundamental conservation law is violated (e.g.: baryon number conservation, charge conservation).

The fact that particles that would be otherwise stable could decay has been known for some time now [21, 22] and in fact it rules out a class of non-fluctuating modifications of the dispersion relations for some choices of the sign of the modification: the new point here is that it does not appear to be possible to fix the sign of the fluctuations, so that the conclusions illustrated above seem unavoidable. This result represents the most striking test of the fluctuating picture discussed in this paper and could in principle invalidate the basis of the proposed model itself.

The plan of the paper is the following: in Sect. 3 we discuss the experimental results on the UHE Cosmic Ray spectrum and their (marginal) disagreement. This will allow us to give a brief presentation of the Auger experiment and of its performances.

Next (Sect. 4), we discuss on general grounds the modification of thresholds coming from violations of LI, mainly to present the level of sensitivity of the experiments, whose possible outcomes are discussed in Sect. 5.

On a more speculative side, we then discuss (in Sect. 6) the effect of fluctuations on the propagation of high energy particles.

3 The Experimental Data and the Pierre Auger Observatory

Cosmic Rays particles reach the highest energies known in the present Universe (see Fig. 1).

It is an open problem the mechanism through which particles are accelerated to such extreme (in fact macroscopic) energies.

Since the discovery of the Cosmic Microwave Background Radiation it has been clear that the Universe could become opaque to extremely high energy particles propagating in it via inelastic interactions, in case of protons through the reaction $p_{UHE}\gamma_{CMB} \rightarrow N\pi$ in which the proton loses part of its energy. The threshold energy for this reaction is $\approx 5\ 10^{19}$ eV.

This was appreciated independently by Greisen and Kuzmin and Zatsepin [1] in 1966. The interaction length can be computed from the data on pion photoproduction and the number density of target photons and is of the order of 6 Mparsec; a proton having a production energy much larger than the

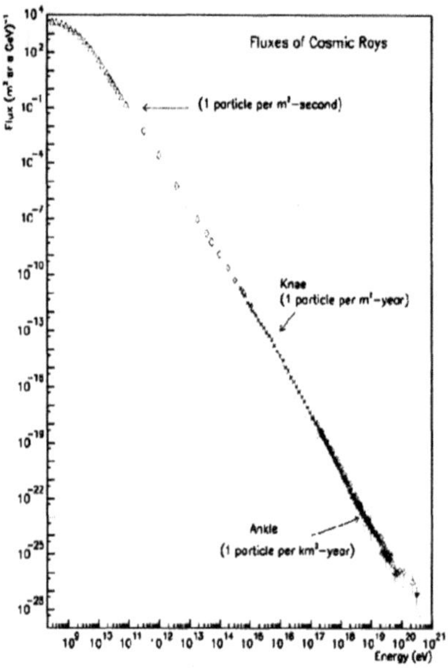

Fig. 1. The overall flux of Cosmic Rays [23] up to 10^{21} eV

threshold would be brought below threshold after approximately 100 Mparsec. These numbers are only representative since we are not interested in numerical details here: it is only necessary to remark that they depend on many details like for instance phenomenology and cosmological evolution of the possible sources of UHECRs.

One point that is not often appreciated is that one should not expect a dramatic change in the CR spectrum at these energies. For instance, assume that one could associate CRs with their sources[1], if the source is more distant than (approximately) 100 Mparsec, then particle emitted with energy lower than the threshold travel essentially undisturbed from any distance. On the contrary, essentially all the particles produced by distant sources above the threshold are degraded in energy until they go below threshold and propagate (almost) without further energy change. So the spectrum from a single ($D >$ 100 Mparsec) source is expected to decrease exponentially above the GZK threshold. However if only the (directions) integrated spectrum is measured the expected decrease in the flux is

$$\Delta\phi \approx \frac{\int^{<100Mp} N_s/d^2}{\int^{D_{max}} N_s/d^2} \approx \frac{D_{max}}{100Mpc} \tag{1}$$

[1] Although charged CR are deflected at least in the Galactic magnetic field, at these energies the deflection is only of few degrees if the extragalactic M.F is not unexpectedly large, so selecting particles approximately pointing to their origin is in principle possible

where N_s is the (unknown) density of UHECR sources, and D_{max} is the maximum distance from which CR can reach detectors in absence of absorption, this depending on the evolution of sources and cosmological parameters. More details can be found for instance in [25], but in any case, if UHECR are of astrophysical origin a decrease around one order of magnitude is expected.

It is interesting to note that in a few years after 1966 Cosmic Ray experiments (Volcano Ranch, Haverah Park, Yakutsk) were able to claim events at estimated energies above 10^{20} eV, inspiring a prophetic article by Russian physicists D.A. Kirzhnits and V.A. Chechin [3] who discussed the absence of the GZK structure in view of possible modifications of relativistic invariance. It is also interesting to notice that this article passed essentially unnoticed, to be rediscovered from time to time.

The point is that, already in the seventies there appeared to be a problem at the end of the observed CR spectrum. The same problem may be present nowadays: when putting together the results of the experiments operating above the nominal value of the GZK threshold, the data are (marginally) conflicting, and may (or not) imply the presence of events above it (Fig. 2).

Notice that in the above figure the scale is logaritmic on both axes. Often the CR spectrum is presented in an amplified way, by multiplying it by an appropriate (positive) power of energy as in the following figure where the (reanalyzed) results also of the Yakutsk experiment are presented (Fig. 3).

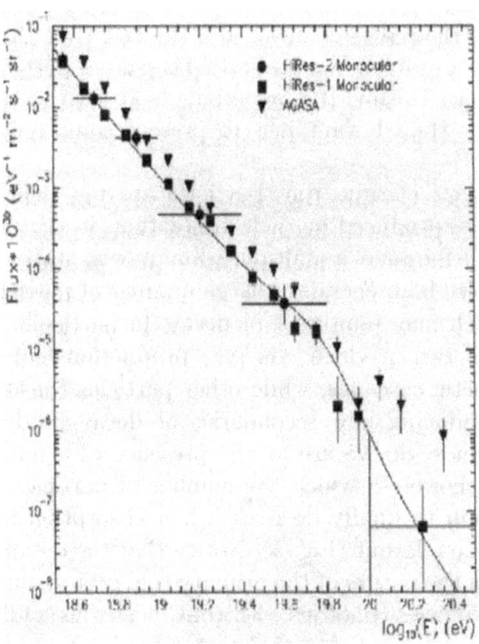

Fig. 2. The most recent data (Agasa, HiRes) in the highest energy region

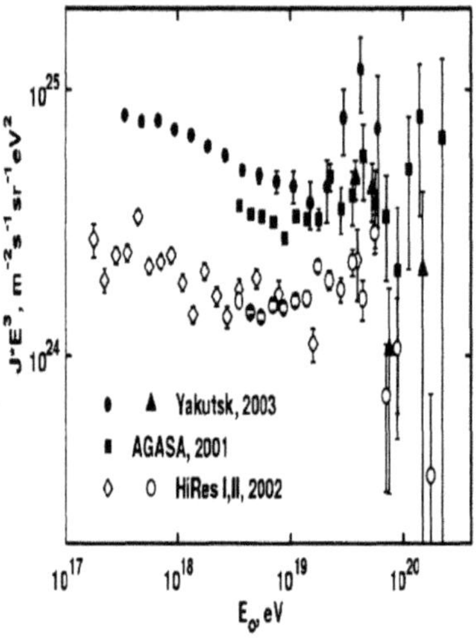

Fig. 3. Same data as above, with Yakutsk data added and flux multiplied by E^3 [24]

In this last way of presenting data the discrepancy between experiments and with the theoretical prediction of the GZK feature is clearly amplified: however, taking into account possible systematics, it is in fact of marginal statistical significance [12].

The reason for these discrepancies and the fact that already 30 years ago there seemed to be a problem not yet settled is related to the intrinsic difficulty of performing precise Cosmic Ray experiments at least at the energies we are here discussing of. Here I want just to present some points useful for our discussion.

At these energies Cosmic Ray Particles are not detected directly, but through secondaries produced by their interactions in the atmosphere. In fact when a CR particle interacts a multiplication process starts: in the interaction in particular at very high energies a large number of particles (mostly pions) are produced which may reinteract or decay. In particular π^o almost immediately decay into two γ which, via pair production and brehmsttrahlung, start electromagnetic cascades, while other particles (mostly charged pions) can reinteract producing new secondaries or decay producing for instance muons and neutrinos. So we are at the presence of a mixed (hadronic and electromagnetic) shower in which the number of particles quickly increases, reaches a maximum to finally decrease when absorption interactions in the atmosphere become relevant (Fig. 5). Notice that the details of shower development depend on the nature of the primary (i.e. proton, nucleus, photon...). Ultimately e.m. particles (photons, electrons, positrons) and muons are by far the most numerous ones arriving at detector level.

Produced particles move essentially at the speed of light, in a approximately thin disk (with some curvature far from the center). Due to transverse momentum distribution, the density of particles in the front rapidly decreases away from the direction of the primary. In the Extensive Air Showers (EAS) arrays the thin disk containing the shower front and when it intersects the detector level is sampled by detectors placed on the surface (Fig. 4). Clearly the largest area sampled gives better sensitivity and ultimately allows to reach higher energies, and the smallest possible distance gives better sampling but at an increased cost.

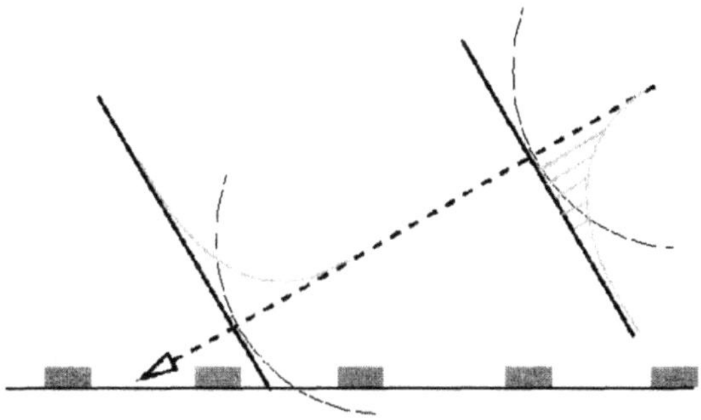

Fig. 4. A shower impinging detector level. Both the lateral distribution and the curvature of the front are unrealistic

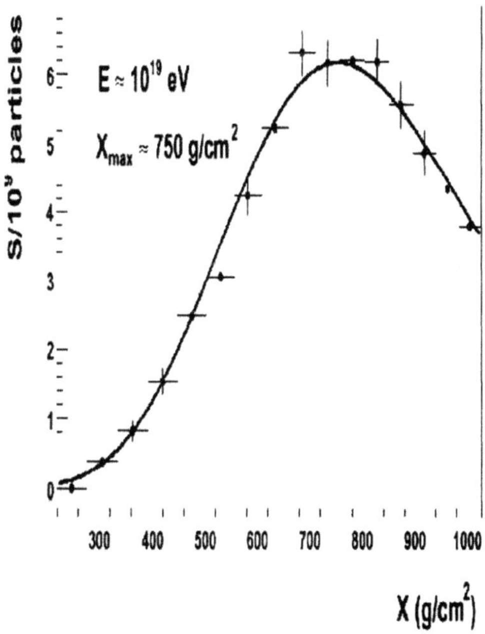

Fig. 5. Longitudinal profile of a high energy shower in the atmosphere

The time delay between detectors is used to deduce the direction of the shower axis; the number of particles depositing energy in each sampling detector, when supplemented by a (theoretical) lateral distribution function of their number with respect to the shower axis allows to compute an estimator of the primary energy. This seems a simple enough procedure to implement, but apart from experimental difficulties, is in principle model dependent since the shower development must be estimated from simulations, which on the other hand depend on not so well known quantities: for instance the center-of-mass energy corresponding to the interaction of a 10^{20} eV proton is $\approx$100 TeV, much higher than that attainable in terrestrial accelerators. Therefore cross sections used are in fact (theory guided) extrapolations from known hadronic physics, and in principle introduce difficult to evaluate systematic errors.

Several different experimental approaches can be followed. In fact the charged particles which make up the shower emit radiation in the atmosphere in two ways: they move faster than light in the atmosphere, so they emit Cherenkov radiation, and excite nitrogen molecules, which de-excite emitting (UV) light; both these forms of radiation can be detected and give information on Cosmic Ray properties. Cherenkov radiation is strongly directional, and its detection through dedicated telescopes has opened the field of γ astronomy.

Fluorescence radiation is instead isotropic, and, since it is emitted during the whole development of the shower, in principle its detection allows a *calorimetric* study of the showers. Moreover, if the telescopes have enough angular/time resolution, it would allow to follow the details of shower development (Fig. 5) and to have hints on the nature of the primary.

Even this detection procedure however has its problems: first, it can only be exploited when the sky is clear and (in part at least) moonless. Second, it is relatively easy to measure the integral of the *light* emitted by the shower, and it is clear that it is proportional to the total energy. However the proportionality constant depends critically on the atmospheric transparency, and this has to be monitored and some aspects might be poorly known. This also introduces daily variations and systematic errors.

The two most significant experiments reporting data in the GZK region are the largest representative of these two approaches to UHECR detection. AGASA [2] in Japan has been the largest operating extensive air shower (EAS) sampling detector array, having a sensitive area of $\approx$100 km^2 with single detectors placed at a distance of the order of hundreds of metres. It has been now superseded by the Auger observatory, although still in construction.

The measured events, when compared with expectations (see Fig. 6) were the confirmation that a puzzle might exist at these extreme energies.

HiRes in the US is the largest experiment exploiting the atmospheric fluorescence detection mechanism. Its released flux does not appear to contain events above expectations (Fig. 7).

It is clear that the two experimental methods are totally independent; this is certainly an advantage since they address complementary features of the shower development, but it implies that their systematics is completely

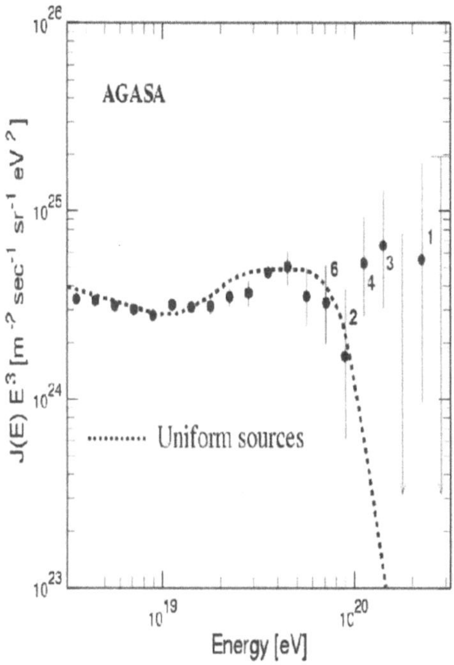

Fig. 6. AGASA spectrum [2]

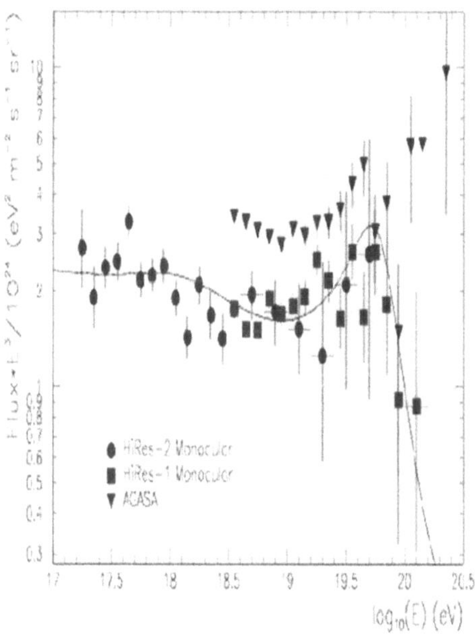

Fig. 7. HiRes spectrum compared with AGASA [11]

different. So, unknown (relatively small) systematic errors have been invoked to explain the difference: for instance the fluxes differ even below the GZK threshold. This can be corrected but a residual, marginally significant difference remains,

It is clear that new data are badly needed. The new generation of experiment has started with the design, construction and start of operations of the Pierre Auger Observatory (PAO) [13]. The main aim of the experiment is to reach a sensitive area more than one order of magnitude larger than previous experiments, giving a unprecedented statistics, some thousands of events around the GZK threshold. But also of fundamental importance is the fact that the PAO employs both detection techniques (particle sampling and fluorescence) of the previous conflicting experiments in its "hybrid" design (Fig. 8).

The Observatory in its final form will have a sensitive area of 3500 square km (comparable e.g. with the extension of the Rome province) with 1600 single detectors at a distance of 1.5 Km each other composing the Surface Detector (SD). Moreover it will have 4 fluorescence telescopes ("eyes" of the fluorescence detector, FD) that will cover the sensitive area of the SD (Fig. 9). The coincident operation of the SD and FD, possible during clear, almost moonless nights for an estimated fraction ≈15% of events will allow cross check and cross correlation of both types of detection reducing their systematic errors.

The experiment is in its5 building phase near Malargue (Argentina) in the pampa, and, being of modular nature, has already started taking data: in fact in its present form is already the largest area Cosmic Ray experiment ever built. The final configuration should be reached in summer 2005, when the first physics result will be presented.

Fig. 8. Conceptual idea of a hybrid detector

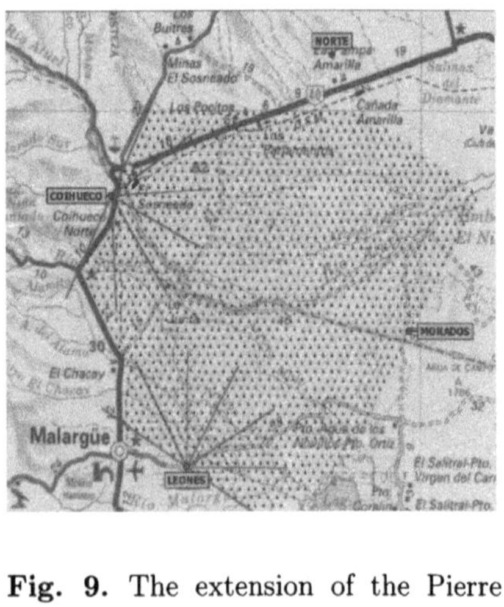

Fig. 9. The extension of the Pierre Auger experiment

To have such a large number of single detectors and telescopes many technological problems had to be solved: for instance single surface detectors have to be completely autonomous and are sun powered and no cables can be put over such large distance, so data communication is wireless.

The distance between detectors and between telescopes also constrains the sensitivity range of the Observatory: in fact to be reliably reconstructed a shower must hit several SDs implying a relatively large low energy threshold (above 10^{18} eV, although it might be possible to decrease it); the high energy limit is dictated by statistics and possibly physics (the GZK threshold). A last very important feature is the presence of various systems to monitor the atmospheric transparency, fundamental for an exact reconstruction of energy in FD detection.

The statistic per year of the events detected by this experiment are reported in Table 1. It is important to notice that in principle one year of operation of full Auger wold correspond to more than 10–30 years of earlier experiments, allowing a clear definition of possible features of the flux, and a study of possible correlations with astrophysical sources. More on the expected results from Auger will be discussed in the conclusions.

Table 1. Number of expected events vs. energy

$> 10^{19}$	$> 510^{19}$	$> 10^{20}$ eV
5000/y	500	50–100

4 The Particle Production Thresholds

Since the subject of these proceedings concerns possible phenomenological effects of violations of Lorentz invariance we will briefly discuss here how these violations can affect measurable quantities and what is their sensitivity. This subject is described in a more systematic way in other reports, so we will be very brief on this, while in Sect. 6 we will indulge in more radical speculations. Also, we will restrict our considerations to processes connected to relevant thresholds for UHECR physics, so will not discuss e.g. effects related to variation of speed of light with energy.

In general, a departure from L.I. will introduce modifications in the invariants, for instance in the (dispersion) relation between energy and momentum, which we will assume in the general form (assuming rotational invariance)

$$E^2 - \mathbf{p}^2 = \mu^2 = f(\mathbf{p}, E, \Lambda) + m^2 \qquad (2)$$

where μ can be thought as an effective mass and Λ is a mass scale parameterizing departures from strict LI[2], and m, possibly zero, is the rest mass of the particle we are considering. It is clear that we expect strong effects when $f \approx m^2$ and in particular in reactions in which the rest mass of involved particles as are generally threshold relations. For instance the LI value for the GZK threshold is $E \approx m_p m_\pi / \omega_{CMB}$.

The modified dispersion relation can be thought as the norm of four-momentum computed with metric experienced by a particle propagating in the QG vacuum (which might depend on the 4-momentum of the particle):

$$g_{\mu\nu}(E, \mathbf{p}^2, \Lambda) p^\mu p^\nu = g(m^2) \tag{3}$$

Clearly, $g_{\mu\nu}(E, \mathbf{p}^2, \Lambda)$ should be derived from the vacuum metric of Quantum Gravity in some appropriate limit. It is not clear how to do it, however some intuitive example can be given: for instance, if Planck mass (virtual) black holes are relevant it is easy to derive that a particle of wave-length λ would satisfy a modified dispersion relation

$$m^2 \approx (1 - \frac{l_P}{\lambda}) E^2 - p^2 \approx E^2 - p^2 - l_P |p| E^2$$

which is of te form used above.

It is in general assumed that $\Lambda \propto M_{Planck}$ so that, at the energies relevant for UHECR propagation $\ll M_{Planck}$ the precise form of f is not important, only the first term in the Taylor expansion will affect kinematics, i.e.

$$E^2 - p^2 \approx \pm \frac{p^3}{\Lambda} + m^2$$

$$E^2 - p^2 \approx \pm \frac{p^4}{\Lambda^2} + m^2$$

In this approach reaction thresholds are then computed in a single frame (assumed the one with respect to which the CMBR is isotropic) assuming energy momentum conservation. The results have been exposed in several papers. For our purposes here it is only important to note that:

- Modifications of dispersion relations with the positive sign tend to move the thresholds to lower energies. However they also in general produce the decay of otherwise stable particles, and there are strong limits on them.
- On the contrary, the negative sign pushes the threshold to larger energies and quickly to infinity.

  It is interesting to estimate the sensitivity of CR experiments to these modifications. In the case of the GZK feature, if experiments will detect it at

[2] This modification of the invariant dispersion relation is the most useful for the purpose of computing physical effects on absorption thresholds. We expect that also space-time invariants (i.e ds^2) are modified, but it is not clear how these would affect experimentally measurable quantities.

an energy (say) within a factor 2 from the theoretical LI prediction very stringent bounds on Λ will follow:

$$\Lambda > 10^3(10^{13})M_{Pl} \tag{4}$$

where the figure in parentheses refers to the larger (cubic) modifications of the dispersion relation.

5 Discussion: Phenomenology, and: Will CR Experiments Detect Lorentz Violations?

Before passing to more speculative arguments it is useful to stop here and discuss in these quite general grounds what will be the impact of the future experimental results from e.g. the PAO. It is clear that we have two possible cases:

1. the GZK feature will be found more or less at the expected energy. It is tempting to claim that in this case LI would be verified at the levels quoted above. There are however two *caveats* to take into account:
   - It is very difficult to invent mechanisms to bring particles to such high energies. It is not inconceivable that the number of possible sources drastically decrease just around the GZK feature, although it does not seem that there could be a physical connection between the GZK threshold and the acceleration mechanisms.
   - Lorentz invariance might not be violated but deformed [26]. This would imply the possibility of (non-linear) frame transformations, modifications of energy-momentum conservation and (unless in relatively contrived models) very little effects on thresholds. This point is discussed in many other reports. From an experimental point of view, at least for what concerns UHECR propagation, it is in general difficult to distinguish deformations for instance a la DSR, from strict LI.

   It is however important to notice that the statistic collected by PAO will allow to reconstruct in detail the form of the HECR spectrum at energies *below* the GZK threshold, and this could change under different hypotheses: in fact in general in the LI case a pile-up of events is expected just below the threshold, produced by particles generated at high energies but degraded from propagation in the CMBR, and this feature might be different in different scenarios.
2. The GZK break will not be present in the experimental data. Clearly this is the possibility that would more stimulate speculations, however one should be very cautious in conclusions for a fundamental reason:
   - sources of UHECR are largely unknown. For instance it is entirely possible (although maybe unlikely) that sources of CR are local, or galactic: young neutron stars can do it [27] if primaries are iron nuclei, or a nearby extragalactic source might be responsible for the whole budget of events

above GZK threshold with the conspiracy of a specific form of extra-galactic magnetic field [28]. There is also the possibility that UHCRs are not prodced in astrophysical sources, as in the so-called Top-Down models. In any way, in the case of strict LI, if UHECRs are known particles, their origin, whatever might be, must be local.

However the perspectives are not so dark. It is important to remind that the large statistic of PAO will allow a detailed analysis of the distribution in sky of the events possibly detected above the GZK threshold. At these energies deflection in the galactic magnetic field is quite small if primaries are protons so in these case the events have to point to some extent to the sources. Even in case of nuclei, a galactic origin should be discernible[3].
A clear association with distant sources would be a unambiguous indication that the propagation of particles in the Universe is not properly understood, and this will imply violations of LI as a possible (strong) candidate for explanation.

6 A More Speculative View: Space-Time Indetermination

6.1 The Effect of Space-Time Fluctuations on the Propagation of High Energy Particles

While electroweak and strong interactions propagate through space-time, gravity turns out to be a property of the space-time itself. This simple statement has profound implications in the quantization of gravity. Our belief that gravity can be turned into a quantum theory immediately implies that the structure of space-time has quantum fluctuations itself. Another way of rephrasing this concept is that space-time is expected to have a granular (or foamy) structure, where however the size of space-time cells fluctuates stochastically, thereby causing an intrinsic uncertainty in the measurements of space-time lengths, and indirectly of energy and momentum of a particle moving through space-time. The uncertainty appears on scales comparable with the Planck scale (the quantization scale of gravity).

It is generally argued that measurements of distances (times) smaller than the Planck length (time) are conceptually unfeasible, since the process of measurement collects in a Planck size cell an energy in excess of the Planck mass, hence forming a black hole, in which information is lost. This can be translated in different ways into an uncertainty on energy-momentum measurements [17, 18]. The Planck length is a good estimate of the uncertainty in the De Broglie wave-length λ of a particle with momentum p. Therefore $\delta\lambda \approx l_P$, and $\delta p = \delta(1/\lambda) \approx (p^2 l_P) = (p^2/M_P)$.

Speculating on the exact characteristics of the fluctuations induced by QG is beyond the scope of the present paper, and it would probably be useless

[3] PAO we should have some capability of distinguishing the nature of UHECRs.

anyway, since the current status of QG approaches does not allow such a kind of knowledge. We decided then to adopt a purely phenomenological approach, in which some reasonable assumptions are made concerning the fluctuations in the fabric of space-time, and their consequences for the propagation of high energy particles are inferred. Comparison with experimental data then possibly constrains QG models.

Following [17], we assume that in each measurement:

- the values of energy (momentum) fluctuate around their average values (assumed to be the result theoretically recoverable for an infinite number of measurements of the same observable):

$$E \approx \bar{E} + \alpha \frac{\bar{E}^2}{M_P} \tag{5}$$

$$p \approx \bar{p} + \beta \frac{\bar{p}^2}{M_P} \tag{6}$$

  with α, β normally distributed variables and p the modulus of the 3-momentum (for simplicity we assume rotationally invariant fluctuations);
- the dispersion relation fluctuates as follows:

$$P_\mu g^{\mu\nu} P_\nu = E^2 - p^2 + \gamma \frac{p^3}{M_P} = m^2 \tag{7}$$

  and γ is again a normally distributed variable.

Ideally, QG should predict the type of fluctuations introduced above, but, as already stressed, this is currently out of reach, therefore we assume here that the fluctuations are gaussian. Our conclusions are however not sensitive to this assumption: essentially any symmetrical distribution with variance $\approx$1, within a large factor, would give essentially the same results. Furthermore we *assume* that α, β and γ are uncorrelated random variables; again, this assumption reflects our ignorance in the dynamics of QG

The fluctuations described above will in general derive from metric fluctuations of magnitude $\delta g^{\mu\nu} \sim h^{\mu\nu} \frac{l_P}{l}$ [5, 18]. Our assumption reflects the fact that, while the magnitude of the fluctuation can be guessed, we do not make any assumption on its tensorial structure $h^{\mu\nu}$.

Our interest will be now concentrated upon processes of the type

$$a + b \rightarrow c + d$$

where we assume that a kinematic threshold is present; in the realm of UHECR physics (a,b) is either (γ, γ_{3K}) or (p, γ_{3K}) and (c,d) is (e^+, e^-) or (N, π).

To find the value of initial momenta for which the reaction occurs we write down energy-momentum conservation equations and solve them with the help of the dispersion relations, as discussed in detail in [7].

The energy momentum conservation relations are (in the laboratory frame, and specializing to the case in which the target (b) is a low energy background photon for which fluctuations can be entirely neglected)

$$E_a + \alpha_a \frac{E_a^2}{M_P} + \omega = E_c + \alpha_c \frac{E_c^2}{M_P} + E_d + \alpha_d \frac{E_d^2}{M_P} \tag{8}$$

$$p_a + \beta_a \frac{p_a^2}{M_P} - \omega = p_c + \beta_c \frac{p_c^2}{M_P} + p_d + \beta_d \frac{p_d^2}{M_P} \ . \tag{9}$$

These equations refer to head-on collisions and collinear reaction products, which is appropriate for threshold computations. Together with the modified dispersion relations, these equations, after some manipulations, lead to a cubic equation for the initial momentum as a function of the momentum of one of products, and, after minimization, they define the threshold for the process considered. In Fig. 10 we report the distribution of thresholds in the $\approx$70% of cases in which the solution is physical; in the other cases the kinematics does not allow the reaction.

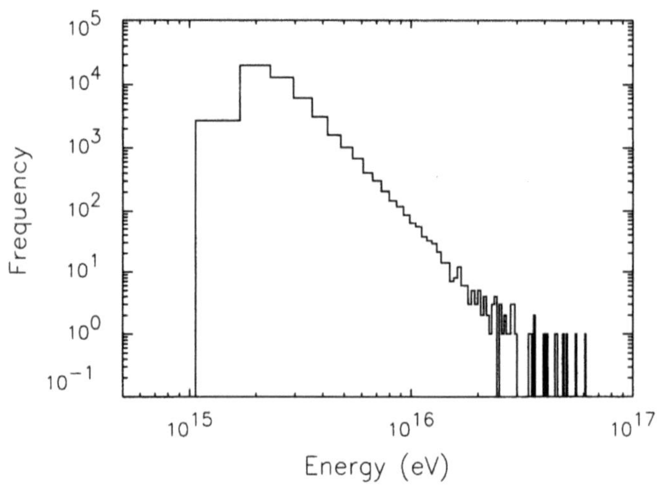

Fig. 10. A shower impinging detector level. Both the lateral distribution and the curvature of the front are not realistic

This threshold distribution can be interpreted in the following way: a particle with energy above $\sim 10^{15}$ eV has essentially 70% probability of being above threshold, and therefore to be absorbed. In the other 30% of the cases the protons do not interact.

In (8,9) the fluctuations are taken independently for each particle, which is justified as long as the energies are appreciably smaller than the Planck energy. At that point it becomes plausible that different particles experience the same fluctuations, or more precisely fluctuations of the same region of space-time.

It is instructive to consider this case in some more detail: we introduce then the four-momenta (and dispersion relations) of *all* particles fluctuating in the same way. Specializing to proton interaction on CMBR, the equation which defines the threshold p_{th} is [7]:

$$\eta \frac{2p_0^3}{(m_\pi^2 + 2m_\pi m_p)M_P} \frac{m_\pi m_p}{(m_\pi + m_p)^2} \left(\frac{p_{th}}{p_0}\right)^3 + \left(\frac{p_{th}}{p_0}\right) - 1 = 0 \qquad (10)$$

where η is a gaussian variable with zero average and variance of the order of (but not exactly equal to) one, and p_0 is the L.I. threshold (GZK). The threshold is the positive solution of this equation.

The coefficient of the cubic term is very large, of the order of 10^{13} in this case, so that unless η is $O(10^{-13})$, we can write, neglecting pion mass

$$p_{th} \approx p_0 \left(\frac{m_p^2 M_P}{\eta p_0^3}\right)^{\frac{1}{3}} . \qquad (11)$$

When η becomes negative, the above equation has no positive root; this happens essentially in 50% of the cases. Since the gaussian distribution is flat in a small interval around zero, the distribution of thresholds for positive η peaks around the value for $\eta \approx 1$, meaning that the threshold moves almost always down to a value of $\approx 10^{15}$ eV [7]; essentially the same result holds for fluctuations affecting only the incident (highest energy) particle. For *independent* fluctuations of final momenta, the asymmetry in the probability distribution of allowed thresholds arises from the fact that even exceedingly small negative values of the fluctuations lead to unphysical solutions.

Building upon our findings, we now apply the same calculations to the case of UHECR protons propagating on cosmological distances. An additional ingredient is needed to complete the dynamics of the process of photopion production, namely the cross section. The rather strong assumption adopted here is that the cross section remains the same as the Lorentz invariant one, provided the reaction is *kinematically* allowed. This implies that the interaction lengths remain unchanged.

In order to assess the situation of UHECRs, we first consider the case of particles above the threshold for photopion production in a Lorentz invariant world. According with (8, 9), in this case particles have a probability of $\approx$30% of being not kinematically allowed to interact inelastically with a photon in the CMBR. Therefore, if our assumption on the invariance of the interaction length is correct, then each proton is still expected to make photopion production, although with a slightly larger path length.

The situation is however even more interesting for particles that are below the Lorentz invariant threshold for the process of photopion production. If the energy is below a few 10^{18} eV, a galactic origin seems to be in good agreement with measurements of the anisotropy of cosmic ray arrival directions [29, 30]. We will not consider these energies any longer. On the other hand, at energies in excess of 10^{19} eV, cosmic rays are believed to be extragalactic protons, mainly on the ground of the comparison of the size of the magnetized region of our Galaxy and the Larmor radius of these particles. We take these pieces of information as the basis for our line of thought. If the cosmic rays observed in the energy range $E > 10^{19}$ eV are extragalactic protons, then our previous calculations apply and we may expect that these particles have a ~70% probability of suffering photopion production at each interaction with the CMB photons, even if their energy is below the classical threshold for this process. Note that the path length associated with the process is of the order of the typical path length for photopion production (a few tens of Mpc), therefore we are here discussing a dramatic process in which the absorption length of particles drops from Gpc, which would be pertinent to particles with energy below $\sim 10^{20}$ eV in a Lorentz invariant world, to several Mpc, with a corresponding suppression of the flux. What are the consequences for the observed fluxes of cosmic rays? The above result implies that *all* protons with $E > 10^{15}$ eV are produced within a radius of several tens of Mpc, and above this energy there is no dramatic change of path length with energy. There is no longer anything special about $E \sim 10^{20}$ eV, and any mechanism invoked to explain the flux of super-GZK particles must be at work also at lower energies.

The basic situation remains the same in the case of pair production as the physical process under consideration. For a source at cosmological distance, a cutoff is expected due to pair production off the far infrared background (FIR) or the microwave background. Using the results in [7] we expect that the modified thresholds are a factor 0.06 (0.73) lower than the Lorentz invariant ones for the case of interaction on the CMBR (FIR). There is also a small increase in the path lengths above the threshold, which would appear exponentially in the expression for the flux. Therefore there are two effects that go in opposite directions: the first moves the threshold to even lower energies, and the second increases the flux of radiation at Earth because of the increase of the path length. It seems that geometry fluctuations do not provide an immediate explanation of the possible detection of particles in excess of the expected ones from distance sources in the TeV region. In any case the experimental evidence for such an excess seems at present all but established.

6.2 Astrophysical Observations

As discussed in the previous section fluctuations in the space-time metric may induce a violation of Lorentz invariance that changes the thresholds for the photopion production of a very high energy proton off the photons of the CMBR, or for the pair production of a high energy gamma ray in the bath of the FIR or CMBR photons.

For the case of UHECRs interacting with the CMBR, we obtained a picture that changes radically our view of the effect of QG on this phenomenon, as introduced in previous papers: not only particles with energy above $\sim 10^{20}$ eV are affected by the fluctuations in space-time, but also particles with lower energy, down to $\sim 10^{15}$ eV seem to be affected by such fluctuations. In fact the latter, as a result of a fluctuating space-time, may end up being above the threshold for photopion production, so that particles may suffer significant absorption. Our conclusion is that all particles with energy in excess of $\sim 10^{15}$ eV eventually detected at Earth would be generated at distances comparable with the path length for photopion production ($\sim$100 Mpc). A consequence of this is that there is no longer anything special characterizing the energy $\sim 10^{20}$ eV.

Since the conclusion reached in the previous section is quite strong, it is important to summarize in detail some tests that may allow to understand whether the current or future astrophysical observations are compatible with the scenario discussed in this paper.

(a) Future experiments [13, 14] dedicated to the detection of UHECRs will provide a substantial increase in the statistics, so that the spectral features of the UHECRs in the energy region $E > 10^{19}$ eV can be resolved, and further indications on the nature of primaries and their possible extragalactic origin will be obtained. In particular the present possible disagreement between AGASA [31] and HiRes [32] will be clarified.

One should also keep in mind that an evaluation of the expected flux in terms of sources distributed as normal galaxies is in contradiction with AGASA data by an amount ranging from 2 to 6σ depending on the assumed source spectrum [25]. Since the nature of the sources is not known, it is not clear if their abundance within the absorption path length is sufficient to explain the observed flux in presence of space-time fluctuations, nor if they can induce observable anisotropies.

In any case, in a Lorentz invariant framework a suppression in the flux at $\sim 10^{20}$ eV is expected. If such a feature is unambiguously detected in the UHECR spectrum, no much room would be left for the fluctuations of space-time discussed in this paper, since in this scenario nothing special happens around 10^{20} eV. In quantitative terms [7] this would imply a phenomenological bound on l_P now interpreted as a parameter: $l_P < 10^{-46}$ cm instead of $l_P \approx 10^{-33}$ cm; in other words, only fluctuations with variance $\approx 10^{-13}$, instead of 1, would be allowed[4]

(b) According with our findings, all particles with energy in excess of $\sim 10^{15}$ eV lose their energy by photopion production on cosmological spatial scales, as a result of the metric fluctuations. This energy ends up mainly in gamma rays, neutrinos and protons. The protons pile up in the energy region right below $\sim 10^{15}$ eV. The gamma ray component actually generates an electromagnetic cascade that ends up contributing low energy gamma rays, in the energy band accessible to instruments like EGRET [35] and GLAST [36]. This cascade flux cannot be larger than the measured electromagnetic energy density in the same band $\omega_{cas}^{\rm exp} = 10^{-6}$ eV/cm^3 [35]. The cascade flux

[4] Alternatively, one can assume a more general form of fluctuations, i.e. $\delta E \approx E(E/M_P)^\alpha$ and similar for momentum and dispersion relations [34]. In this case the basic conclusions reached here remain unchanged.

in our scenario can be estimated as follows. Let $\Phi(E) = \Phi_0(E/E_0)^{-\gamma}$ be the emissivity in UHECRs (particles/cm^3/s/GeV). Let us choose the energy $E_0 = 10^{10}$ GeV and let us normalize the flux to the observations at the energy E_0. The total energy going into the cascade can be shown to be

$$\omega_{cas} \approx \frac{5 \times 10^{-4}}{\gamma - 2} \; x_{min}^{2-\gamma} \; \xi \, \mathrm{eV} \; cm^{-3} \, ,$$

where ξ is the fraction of energy going into gamma rays in each photopion production, and $x_{min} = (E_{th}/E_0) = 10^{-4}$ for $E_{th} = 10^{15}$ eV. It is easy to see that, for $\gamma = 2.7$, the cascade bound is violated unless $\xi \ll 10^{-3}$.

One note of warning has to be sent concerning the development of the electromagnetic cascade: the same violations of LI discussed here affect other processes, as stressed in the paper. For instance pair production and pion decay are also affected by violations of LI [26]. Therefore the possibility that the cascade limit is exceeded concerns only those scenarios of violations of LI that do not inhibit appreciably pair production and the decay of neutral pions.

The protons piled up at energies right below 10^{15} eV, would be a nice signature of this scenario, but it seems difficult to envision a way of detecting these remnants. In fact, even a tiny magnetic field on cosmological scales would make the arrival time of these particles to Earth larger than the age of the universe. Moreover, even assuming an exactly zero extragalactic magnetic field, these particles need to penetrate the magnetic field of our own Galaxy and mix with the galactic cosmic rays, making their detection extremely problematic if not impossible.

Clearly a more detailed flux computation, taking into account propagation of primaries as well as generation and propagation of the secondaries is needed in order to assess in a more quantitative way observable effects of possible metric fluctuations on UHECRs.

Let us conclude this section sending a note of warning concerning (11), in this expression the dependence on the CMB photon energy is washed out by the approximation done (we have neglected the pion mass). From the physical point of view this corresponds to the appearance of an effective mass (momentum dependent) of the proton due to the effect of fluctuations. The effective mass of the proton may be responsible for the decay of this particle. As we will discuss in the next section the possibility of a decaying proton is a very stringent test for the fluctuations picture much powerful than the astrophysical observations discussed in the present section.

6.3 A Fluctuating Space-Time can Make the World Unstable

Let us discuss in this section the most striking test of the models that predict energy and momentum fluctuations. We will discuss here the possibility that these fluctuations may induce particles decays otherwise impossible. This possibility, already discussed in the framework of non-fluctuating modifications of the dispersion relation [21, 22], could in principle rule out the models with fluctuations. In this section we will discuss the basic features of the decays, leaving a detailed discussion of the implications and possible way out to the next section.

We will consider three specific decay channels, that illustrate well, in our opinion, the consequences of the quantum fluctuations introduced above. We start with the reaction

$$p \to p + \pi^0$$

and we denote with p (p') the momentum of the initial (final) proton, and with k the momentum of the pion. Clearly this reaction cannot take place in the reality as we know it, due to energy conservation. However, since fluctuations have the effect of emulating an effective mass of the particles, it may happen that for some realizations, the effective mass induced to the final proton is smaller than the mass of the proton in the initial state, therefore allowing the decay from the kinematical point of view. Since no conservation law or discrete symmetry is violated in this reaction, it may potentially take place. For the sake of clarity, it may be useful to invoke as an example the decay of the Δ^+ resonance, which is structurally identical to a proton, but may decay to a proton and a pion according to the reaction $\Delta^+ \to p + \pi^0$, since its mass is larger than that of a proton. From the physical point of view, the effect of the quantum fluctuations may be imagined as that of *exciting* the proton, inducing a mass slightly larger than its own (average) physical mass.

Following the discussion of the previous sections we expect to find that for momenta above a given threshold, depending on the value of the random variables, the decay may become kinematically allowed. In general, the probability for this to happen has to be calculated numerically from the conservation equations supplemented by the dispersion relations [37].

Although a full calculation is possible, it is probably more instructive to proceed in a simplified way, in which only the fluctuations in the dispersion relation of the particle in the initial state are taken into account. Neglecting the corresponding fluctuations in the final state should not affect the conclusions in any appreciable way, unless the fluctuations in the initial and final states are correlated (we will return to this possibility at the end of Sect. 5).

In this approximation, the threshold for the process of proton decay to a proton and a neutral pion can be written as follows (neglecting corrections to order higher than p/M_P):

$$\gamma \frac{2p_{th}^3}{M_P} - 2m_\pi m_p - m_\pi^2 = 0 \, , \tag{12}$$

with solution

$$p_{th} = \left(\frac{(2m_p m_\pi + m_\pi^2) M_P}{2\gamma} \right)^{\frac{1}{3}} \, . \tag{13}$$

For negative values of γ, the above equation has no positive root; this happens in 50% of the cases. Since the gaussian distribution is essentially flat in a small interval around zero, the distribution of thresholds for positive γ (i.e. in the remaining 50% of the cases) peaks around the value for $\gamma \approx 1$, meaning that the threshold moves almost always down to a value of $\approx 10^{15}$ eV [7, 20]; essentially the same result holds for generic fluctuations (i.e. not confined to the dispersion relations) affecting only the incident particle, namely the one with the highest energy [37].

The reason why the effects of fluctuations are expected to occur at such low energies is that starting from that energy region the fluctuation term becomes comparable with the rest mass of the particle. In fact the same concept of rest mass of a particle may lose its traditional meaning at sufficiently high energies [19].

It can be numerically confirmed that *independent* fluctuations of momenta (and/or of the dispersion relations) of the decay products are more likely to make the decay easier rather than more difficult, due to the non linear dependence of the threshold on the strength of fluctuations: the probability that the decay does not take place is in fact $\approx$30%. In the remaining cases, the decay will occur if the momentum of the initial proton is larger than p_{th}[37]. The distribution of p_{th} is essentially identical to the one reported in Sect. 2 for the photopion production.

All the discussion reported so far remains basically unchanged if similar reactions are considered. For instance the reaction $p \to \pi^+ n$ is kinematically identical to the one discussed above. For all these reactions, we expect that once they become kinematically allowed, the energy loss of the parent baryon is fast. For the case of nuclei, all the decays that do not change the nature of the nucleon leave (A,Z) unchanged, so we do not expect any substantial blocking effect in nuclei.

Another reaction that may be instructive to investigate is the spontaneous pair production from a single photon, namely [37]

$$\gamma \to e^+ e^- .$$

In this case, following the calculations described above, we obtain the following expression for the threshold:

$$p'_{th} = \left(\frac{4 m_e^2 M_P}{2\gamma'} \right)^{\frac{1}{3}} , \tag{14}$$

and p'_{th} is of the order of 10^{13} eV. Again, if the reaction becomes kinematically allowed, there does not seem to be any reason why the reaction should not take place with a rate dictated by the typical cross section of electromagnetic interactions.

Finally, we propose a third reaction that in its simplicity may represent the clearest example of reactions that should occur in a world in which quantum fluctuations behave in the way described above. Let us consider a proton that moves in the vacuum with constant velocity, and let us consider the elementary reaction of spontaneous photon emission. In the Lorentz invariant world the process of photon emission is known to happen only in the presence of an external field that may provide the conditions for energy and momentum conservation. However, in the presence of quantum fluctuations, one can think of the gravitational fluctuating field as such an external field, so that the particle can in fact radiate a photon without being in the presence of a nucleus or some other external recognizable field. The threshold for this process, calculated following the above procedure, is

$$p''_{th} \approx \left(\frac{m^2 M_P \omega}{\gamma''}\right)^{\frac{1}{4}}, \tag{15}$$

where ω is the energy of the photon. This threshold approaches zero when $\omega \to 0$: for instance, if $\omega = 1\,\mathrm{eV}$, then $p_{th} \approx 300\,\mathrm{GeV}$ for protons and $p_{th} \approx 45\,\mathrm{GeV}$ for electrons. In other words there should be a sizable energy loss of a particle in terms of soft photons. This process can be viewed as a sort of bremsstrahlung emission of a charged particle in the presence of the (fluctuating) vacuum gravitational potential.

Based on the arguments provided in this section, it appears that all particles that we do know are stable in our world, should instead be unstable at sufficiently high energy, due to the quantum fluctuations described above. In the next section we will take a closer look at the implications of the existence of these quantum fluctuations, and possibly propose some plausible avenues to avoid these dramatic conclusions.

6.4 Discussion

If the decays discussed in the previous section could take place, our universe, at energies above a few PeV or even at much lower energies might be unstable, nothing like what we actually see. The decays

$$nucleon \to nucleon + \pi$$

would start to be kinematically allowed at energies that are of typical concern for cosmic ray physics, while the spontaneous emission of photons in vacuum might even start playing a role at much lower energies, testable in laboratory experiments. Without detailed calculations of energy loss rates it is difficult to assess the experimental consequences of this process.

For the nucleon decay, the situation is slightly simpler if we assume that the quantum fluctuations affect only the kinematics but not the dynamics, an assumption also used in in the photopion production study [20]. In this case one would expect the proton to suffer the decay to a proton and a pion on a time scale of the same order of magnitude of typical decays mediated by strong interactions. This would basically cause no cosmic ray with energy above

$\sim 10^{15}$ eV to be around, something that appears to be in evident contradiction with observations[5].

In the following we will try to provide a plausible answer to these three very delicate questions:

1. If the particles were kinematically allowed to decay, and there were no fundamental symmetries able to prevent the decay, would it take place?
2. Is the form adopted for the quantum fluctuations correct and if so, how general is it?
3. If in fact the form adopted for the fluctuations is correct, how general and unavoidable is the consequence that (experimentally) unobserved decays should take place?

Although the result that particles are kinematically allowed to decay is fairly general, the (approximate) lack of relativistic invariance forbids the computation of life-times[6]. Two comments are in order: first, the phase space for the decays described above, as calculated in the laboratory frame, is non zero and in fact it increases with the momentum of the parent particle. The effect of fluctuations can be seen as the generation of an effective $(\text{mass})^2 \propto p^3/M_P$. A similar effect, although in a slightly different context, was noted in [6].

Second, we do not expect dynamics to forbid the reactions: one must keep in mind that we are considering very small effects, at momenta much smaller than the Planck scale. For instance the gravitational potential of the vacuum fluctuations is expected to move quarks in a proton to excited levels, not to change its content, nor the properties of strong interactions.

There is a subtler possibility, which must be taken very seriously in our opinion, since it might invalidate completely the line of thought illustrated above, namely that the quantum fluctuations of the momenta of the particles involved in a reaction occur on time scales that are enormously smaller than the typical interaction/decay times. This situation might resemble the so called Quantum Zeno paradox, where continuously checking for the decay of an unstable particle effectively impedes its decay. This possibility is certainly worth a detailed study, that would however force one to handle the intricacies of matter in a Quantum Gravity regime. We regard this possibility as the most serious threat to the validity of the arguments in favor of quantum fluctuations discussed in this paper and in many others before it.

Let us turn out attention toward the question about the correctness and generality of the form adopted for the momentum fluctuations. It is generally accepted that the geometry of space-time suffers profound modifications at

[5] From a phenomenological point of view, consistency with experiments would require either that the variance of the fluctuations considered above is ridiculously small ($<10^{-24}$) or, allowing more generic fluctuations $\Delta l \propto l_P(l_P/l)^\alpha$, that a fairly large value for α should be adopted [20].

[6] In fact life-times can be in principle estimated in approaches in which it is possible to make transformations between frames [19, 38, 39], despite the lack of LI.

length (time) scales of the order of the Planck length (time), and that this leads to the emergence of a minimum measurable length. This may be reflected in a non commutativity of space-time and in a generalized form of the uncertainty principle.

The transition from uncertainty in the length or time scales to uncertainty in momenta of particles is undoubtedly more contrived and deserves some attention. The expressions in (5, 6) and (7) have been motivated in various ways [16, 17, 19, 20, 41] in previous papers. For instance, the condition $\Delta l \geq l_P$ seems to imply the following constraint on wavelengths $\Delta\lambda \geq l_P$, otherwise it would be possible to design an experimental set-up capable of measuring distances with precision higher than l_P. Therefore $\Delta p \propto \Delta(\lambda^{-1}) \propto l_P p^2$. Similar arguments have been proposed, all based to some extent on the de Broglie relation $p \propto \lambda^{-1}$.

There is certainly no guarantee that the de Broglie relation continues to keep its meaning in the extreme conditions we are discussing, in particular in models in which the coordinates and coordinate-momentum commutators are modified with respect to standard quantum mechanics and the representation of momentum in terms of coordinate derivatives generally fails. For instance in a specific (although non-relativistic) example [40] the existence of a minimum length is shown to imply that

$$p = \frac{2}{\pi l_P} \tan\left(\frac{\pi l_P}{2\lambda}\right) . \tag{16}$$

In other words, the de Broglie relation may be modified in such a way that a minimum wavelength corresponds to an unbound momentum. Notice, however, that we are considering here the effects of these modifications at length scales much larger than the Planck scale, where the correction is likely to be negligible. In general, if $p \propto \lambda^{-1} g(l_P/\lambda)$ then $\Delta p \propto l_P p^2 + p\, O(l_P^2 p^2)$. Hence, we do not expect that the result shown in the previous Section is appreciably modified.

Last but not least we notice that the fluctuations in the dispersion relations can be easily derived from fluctuations of the (vacuum) metric in the form given in [41]:

$$ds^2 = (1+\phi)dt^2 - (1+\psi)d\mathbf{r}^2 \tag{17}$$

where ϕ, ψ are functions of the position in space-time.

The fluctuations of the dispersion relation, (7), follow if $\phi \neq \psi$ (i.e. non conformal fluctuations), assuming at least approximate validity of the de Broglie relation; if $\phi = \psi$ a much milder modification $(\mathrm{O}(pm^2/M_P))$ follows.

Having given plausibility arguments in favor of the form adopted for the fluctuations, at least for the case of non conformal fluctuations, we are left with the goal of proving an answer to the last question listed above, namely does a decay actually occur once it is kinematically allowed? Certainly the answer is positive if one continues to assume momentum and energy conservation, and modifications of these conservation laws with random terms of order $O(p^2/M_P)$ do not change this conclusion. The question then is whether we are justified in assuming energy and momentum conservation in the form used above. For instance, in the so-called Doubly Special Relativity (DSR) [38], theories and in general in models with deformed Poincaré invariance, the conservation relations may be modified in a non trivial way.

This certainly makes the probability of being above threshold smaller, but not zero if fluctuations are uncorrelated. However in order to qualitatively modify our results this probability should be in fact vanishingly small. For the case of *low* energy cosmic rays, this probability should be of the order of a typical decay time divided by the residence time of cosmic rays (mostly galactic at these energies) in our Galaxy.

We are led to conclude that allowing for modifications of the conservation relations does not appear to improve the situation to the point that the strong conclusions derived in the previous section can be avoided. In the same perspective, cancellation between fixed modifications of the dispersion relation and fluctuations (of the same order of magnitude) does not seem a viable way to proceed.

It is important however to notice that we have considered the above fluctuations as independent. In a full theory one should take into account possible correlations between fluctuations. The effect of correlations is very important because it pushes to higher energies the fluctuation scale of the particle momentum (energy). Let us discuss in more detail this point. Quantum fluctuations of the momenta of the particles involved in a reaction occur on time scales that are much smaller than the typical interaction time. Particles during the interaction time experience a large number of fluctuations, typically

$$N = \frac{\tau}{\tau_P} = \frac{1}{p\tau_P} = \frac{M_P}{p} ,$$

where we have used $\tau \sim 1/p$ for the interaction time scale and $\tau_P \sim 1/M_P$ for the fluctuation time scale. Assuming independent fluctuations of energy and momentum the fluctuation variance σ will be

$$\sigma^2 = \frac{p^3}{M_P\sqrt{N}} = \frac{p^3}{M_P}\left(\frac{p}{M_P}\right)^{1/2} ,$$

and the fluctuation variance becomes of the order of the proton mass $\sigma \simeq m_p$ already at momentum $p \simeq 10^{17}$ eV. In this case the situation resembles as discussed above and, for instance, the decaying of the proton arises already at lower energies. Let us consider now the case in which there is some degree of correlation in the momentum (energy) fluctuations. In this case the fluctuation variance σ will be

$$\sigma^2 = \frac{p^3}{M_P N^\alpha} = \frac{p^3}{M_P}\left(\frac{p}{M_P}\right)^\alpha ,$$

where we have introduced the exponent $\alpha > 1/2$ that parameterizes the effect of correlations. In this case the fluctuation variance becomes of the order of the proton mass at larger energies, namely $\sigma \simeq m_p$ at momentum of the order of

$$p \simeq M_P \left(\frac{m_p}{M_P}\right)^{\frac{2}{3+\alpha}} .$$

A detailed analysis of possible correlations between fluctuations, namely an analytic determination of α, is impossible at this stage because it implies a better knowledge of the theory, and in particular of the dynamics of the QG regime.

Finally, a separate discussion is needed for those theories that include the relativity principle (exemplified by DSR models). The DSR theories are characterized by an extended Lorentz invariance [38] with two separate invariant scales: the light velocity and the Planck length. Moreover, in the low energy limit of DSR, or for distances much larger than the Planck length, the usual Lorentz invariance is recovered.

Using these two characteristics of the DSR theories it is easy to prove that particle kinematics in DSR is the same as in the usual Lorentz invariant theories. This result holds in the case in which there are no fluctuations of energy and momentum. In the most general case in which fluctuations of energy and momentum are taken into account it is difficult to prove that the situation remains unchanged. Nevertheless, if in DSR the relativity principle remains at work also in the fluctuating case the DSR approach seems the most promising in order to escape the particles decays discussed in this paper that seems to invalidate all the other models.

References

1. K. Greisen, Phys. Rev. Lett. **16**, 748 (1966); G.T. Zatsepin and V.A. Kuzmin, Pis'ma Zh. Ekps. Teor. Fiz. **4**, 114 (1966) [JETP Lett. **4**, 78 (1966)].
2. N. Hagashida et al., Astrophys. J. **522** 224 (1999).
3. D.A. Kirzhnits and V.A. Chechin, Sov. Jour. Nucl. Phys. **15**, 585 (1971).
4. L. Gonzalez-Mestres, Proc. 26th ICRC (Salt Lake City, USA), **1**, 179 (1999).
5. G. Amelino Camelia, J. Ellis, N.E. Mavromatos and S. Sarkar Nature **393** (1998) 763.
6. S. Coleman and S.L. Glashow, Phys. Rev. **D59**, 116008 (1999).
7. R. Aloisio, P. Blasi, P.L. Ghia and A.F. Grillo, Phys. Rev. **D62**, 053010 (2000).
8. J.M. Carmona, J.L. Cortes, J. Gamboa and F. Mendez, *preprint* hep-th/0301248.
9. O. Bertolami and C.S. Carvalho, Phys. Rev. **D61**, 103002 (2000); O. Bertolami, *preprint* astro-ph/0012462.
10. N. Hayashida et al. [AGASA collaboration], Phys. Rev. Lett. **73**, 3491 (1994).
11. T. Abu-Zayyad et al. [Hires collaboration], Phys. Rev. Lett. **92**, 15110 (2004)
12. D. De Marco, P. Blasi and A.V. Olinto, Astrop. Phys. **20**, 53 (2003).
13. J. Blümer et al. [Auger Collaboration], J. Phys. **G29** 867 (2003).
14. EUSO space Observatory, http://www.euso-mission.org.

15. A.I. Nikishov, Sov. Phys. – JETP **14**, 393 (1962); P. Goldreich and P. Morrison, Sov. Phys. – JETP **18**, 239 (1964); R.J. Gould and G.P. Schreder, Phys. Rev. Lett. **16**, 252 (1966).
16. L.H. Ford Int. J. Theor. Phys. **38** 2941 (1999).
17. Y.J. Ng, D.S. Lee, M.C. Oh and H. van Dam, Phys. Lett. **B507** 236 (2001).
18. Y.J. Ng, Int. J. Mod. Phys. **D11** 1585 (2002) (*preprint* gr-qc/0201022).
19. R. Lieu, Astrophys. J. **568** L67 (2002).
20. R. Aloisio, P. Blasi, A. Galante, P.L. Ghia and A.F. Grillo, Astrop. Phys. **19**, 127 (2003).
21. L. Gonzalez-Mestres, *preprint* hep-ph/9905430.
22. T. Jacobson, S. Liberati and D. Mattingly, Phys. Rev. **D66**, 081302 (2002).
23. P. Bhattacharjee, G. Sigl, Phys. Rept. **327**, 109 (2000)
24. A.V. Glushkov et al. [Yakutsk collaboration], 28th ICRC
25. M. Blanton, P. Blasi and A.V. Olinto, Astrop. Phys., **15**, 275 (2001)
26. G. Amelino-Camelia, Phys. Lett. **B528** (2002) 181.
27. A.V. Olinto, R.I. Epstein and P. Blasi, Astrophys. J. **533**, L123 (2000)
28. P.L. Biermann, E.-J. Ann, G. Medina Tanco and T. Stanev, Nucl. Phys. (Proc. Supp.) **87**, 417 (2000)
29. N.Hagashida et al., Astrop. Phys. **10** 303 (1999).
30. D.J. Bird et al., Ap. J. **511** 739 (1999).
31. N. Sakay et al. Proceedings of 2001 ICRC.
32. C.C.H. Jui et al., Proceedings of 2001 ICRC.
33. M. Blanton, P. Blasi and A.V. Olinto, Astrop. Phys. **15**, 275 (2001).
34. M. Jankiewicz, T.W. Kephart and T.J. Weiler, *preprint* hep-ph/0312221
35. P. Sreekumar et al. [EGRET collaboration], Astroph. J. **494** (1998) 523.
36. GLAST Telescope, http://www-glast.stanford.edu.
37. R. Aloisio, P. Blasi, A. Galante and A.F. Grillo, Astrop. Phys. **20**, 369 (2003).
38. N.R. Bruno, G. Amelino-Camelia and J. Kowalski-Glikman, Phys. Lett. **B522** 133 (2001). G. Amelino-Camelia, Int. J. Mod. Phys. **D11** 1643 (2002). G. Amelino-Camelia Nature **418** 34 (2002).
39. T. Tamaki, T. Harada, U. Miyamoto and T. Torii, Phys. Rev. **D65** 083003 (2002).
40. A. Kempf, G. Mangano and R.B. Mann, Phys. Rev. **D52** 1108 (1995).
41. A. Camacho, Gen. Rel. Grav. **35** 319 (2003).

Quantum Gravity

E. Alvarez

Instituto de Fisica Teorica UAM/CSIC, 28049 Madrid, Spain
enrique.alvarez@uam.es

1 General Questions on Quantum Gravity

It is not clear at all what is the problem in quantum gravity (cf. [3] or [8] for general reviews, written in the same spirit as the present one). The answers to the following questions are not known, and I believe it can do no harm to think about them before embarking in a more technical discussion.

To begin, it has been proposed that gravity should not be quantized, owing to its special properties as determining the background on which all other fields propagate. There is a whole line of thought on the possibility that gravity is not a fundamental theory, and this is certainly an alternative one has to bear in mind. Indeed, even the *holographic principle* of G. 't Hooft, to be discussed later, can be interpreted in this sense.

Granting that, the next question is whether it does make any sense to consider gravitons propagating in some background; that is, whether there is some useful approximation in which there is a particle physics approach to the physics of gravitons as quanta of the gravitational field. A related question is whether semiclassical gravity, i.e., the approximation in which the source of the classical Einstein equations is replaced by the expectation value of the energy momentum tensor of some quantum theory has some physical [22] validity in some limit. We shall say more on this problems towards the end.

At any rate, even if it is possible at all, the at first sight easy problem of graviton interactions in an otherwise flat background has withstood analysis of several generations of physicists. The reason is that the coupling constant has mass dimension -1, so that the structure of the perturbative counterterms involve higher and higher orders in the curvature invariants (powers of the Riemann tensor in all possible independent contractions), schematically,

$$S = \frac{1}{2\kappa_R^2}\int R + \int R^2 + \kappa_R^2 \int R^4 + \dots \tag{1}$$

Nobody knows how to make sense of this approach, except in one case, to be mentioned later on.

E. Alvarez: *Quantum Gravity*, Lect. Notes Phys. **669**, 31–58 (2005)
www.springerlink.com

It could be possible, *sensu stricto* to stop here. But if we believe that quantum gravity should give answers to such questions as to the fate of the initial cosmological singularity, its is almost unavoidable to speak of the wave function of the universe. This brings its own set of problems, such as to whether it is possible to do quantum mechanics without classical observers or whether the wave function of the Universe has a probabilistic interpretation. Paraphrasing C. Isham [38], one would not known when to qualify a probabilistic prediction on the whole Universe as a successful one.

The aim of the present paper is to discuss in some detail established results on the field. In some strong sense, the review could be finished at once, because there are none. There are, nevertheless, some interesting attempts, which look promising from certain points of view. Perhaps the two approaches that have attracted more attention have been the loop approach, on the one hand and strings on the other. We shall try to critically assess prospects in both. Interesting related papers are [34, 59].

Even if for the time being there is not (by far) consensus on the scientific community of any quantum gravity physical picture, many great physicist have not been able to resist the temptation of working (usually only for a while) on it. This has produced a huge spinoff in quantum field theory; to name only a few, constrained quantization, compensating ghosts, background field expansion and topological theories are concepts or techniques first developed in thinking about these problems, and associated to the names of Dirac, Pauli, Weinberg, Feynman, De Witt, Witten etc. In many cases, more or less surprising relationships have been found with quantum gauge theories. There are probably more in store, if one is to judge from the success of the partial implementation of holographic ideas in Maldacena's conjecture (more on this later).

This should be kept into account when reading the references. We have not attempted to be comprehensive, and we have used only the references familiar to us; but in some of the references, in particular in our own review article of 1989 ([3] there are more entry points into the vast literature. After all, paraphrasing Feynman, we still do not know what could be relevant in a field until the main problems are solved.

2 The Issue of Background Independence

One of the main differences between both attacks to the quantum gravity problem is the issue of background independence, by which it is understood that no particular background should enter into the definition of the theory itself. Any other approach is purportedly at variance with diffeomorphism invariance.

Work in particle physics in the second half of last century led to some understanding of ordinary gauge theories. Can we draw some lessons from there?

Gauge theories can be formulated in the *background field approach*, as introduced by B. de Witt and others (cf. [20]). In this approach, the quantum field theory depends on a background field, but not on any one in particular, and the theory enjoys background gauge invariance.

Is it enough to have quantum gravity formulated in such a way?[1]

It can be argued that the only vacuum expectation value consistent with diffeomorphisms invariance is

$$\langle 0|g_{\alpha\beta}|0\rangle = 0 \tag{2}$$

in which case the answer to the above question ought to be in the negative, because this is a singular background and curvature invariants do not make sense. It all boils down as to whether the ground state of the theory is diffeomorphism invariant or not. There is an example, namely three-dimensional gravity in which invariant quantization can be performed [70]. In this case at least, the ensuing theory is almost topological.

In all attempts of a canonical quantization of the gravitational field, one always ends up with an (constraint) equation corresponding physically to the fact that the total hamiltonian of a parametrization invariant theory should vanish. When expressed in the Schrödinger picture, this equation is often dubbed the *Wheeler-de Witt equation*. This equation is plagued by operator ordering and all other sorts of ambiguities. It is curious to notice that in ordinary quantum field theory there also exists a Schrödinger representation, which came recently to be controlled well enough as to be able to perform lattice computations [42].

Gauge theories can be expressed in terms of gauge invariant operators, such as Wilson loops . They obey a complicated set of equations, the loop equations, which close in the large N limit as has been shown by Makeenko and Migdal [43]. These equations can be properly regularized, e.g. in the lattice. Their explicit solution is one of the outstanding challenges in theoretical physics. Although many conjectures have been advanced in this direction, no definitive result is available.

3 The Canonical Approach

It is widely acknowledged that there is a certain tension between a $(3+1)$ decomposition implicit in any canonical approach, privileging a particular notion of time, and the beautiful geometrical structure of general relativity, with its invariance under general coordinate transformations.

Let us now nevertheless explore how far we can go on this road, following the still very much worth reading work of De Witt [20].

[1] This was, incidentally, the way G. Hooft and M. Veltman did the first complete one-loop calculation [65]

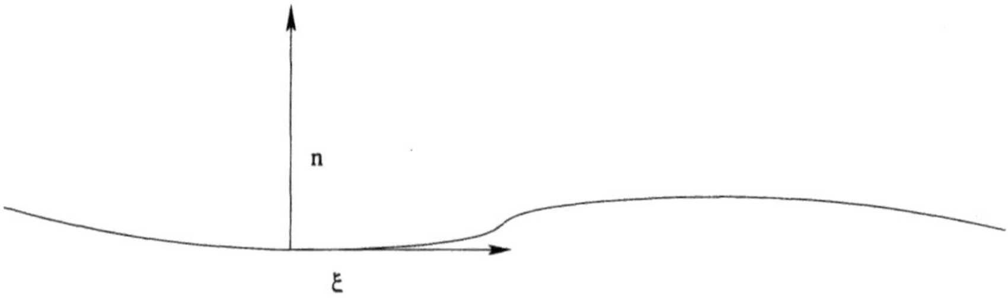

Fig. 1. Spacelike surface of codimension one

If (Fig. 1) we are given a spacelike surface (which will represent physically all spacetime events to which it will be assigned a fixed time), say

$$y^\alpha = f^\alpha(x^i)$$

The tangent vector to the surface are

$$\xi_i \equiv \partial_i f^\alpha \partial_\alpha$$

and the induced metric (that is, the pull-back to the surface of the spacetime metric) is

$$h_{ij} \equiv g_{\alpha\beta}\xi_i^\alpha \xi_j^\beta$$

The unit normal is then defined as;

$$g_{\alpha\beta} n^\alpha \xi_i^\beta = 0$$
$$n^2 \equiv g_{\alpha\beta} n^\alpha n^\beta = 1$$

We are interested now in a set of such surfaces which covers all spacetime; that is, a foliation of (a portion of) the spacetime; namely a one-parameter family of spacelike disjoint hypersurfaces

$$\Sigma_t \equiv \{y^\alpha = f^\alpha(x^i, t)\}$$

In a classical analysis Arnowitt, Deser and Misner (ADM) [1] characterized the embedding via two functions: the *lapse* and the *shift* : we first define the vector (see Fig. 2)

$$N^\alpha \equiv \frac{\partial f^\alpha}{\partial t}$$

in terms of which the lapse, N, is just the projection in the direction of the normal, and the *shift*, N_i the (three) projections tangent to the hypersurface.

$$N^\alpha = N n^\alpha + N^i \xi_i^\alpha$$

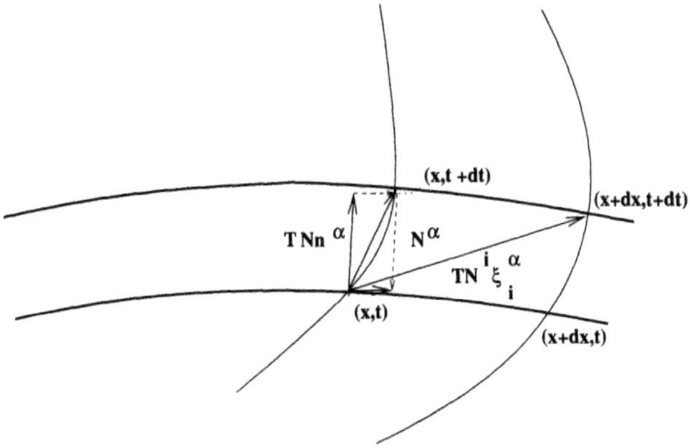

Fig. 2. ADM lapse and shift variables

All this amounts to a particular splitting of the full spacetime metric:

$$\begin{aligned} ds^2 &= g_{\alpha\beta}dx^\alpha dx^\beta = g_{\alpha\beta}df^\alpha df^\beta \\ &= g_{\alpha\beta}(N^\alpha dt + \xi_i^\alpha dx^i)(N^\beta dt + \xi_j^\beta dx^j) \\ &= N^2 dt^2 + h_{jk}(N^j dt + dx^j)(N^k dt + dx^k) \end{aligned}$$

or, what is the same,

$$g_{\mu\nu} = h^{ij}\xi_{i\mu}\xi_{j\nu} + n_\mu n_\nu$$

All surfaces which are equivalent from the intrinsic point of view, can be however embedded differently; the extrinsic curvature discriminates between them:

$$K_{ij} = -\xi_i^\alpha \nabla_\rho n_\alpha \xi_j^\rho$$

The Gauss-Codazzi equations relate intrinsic curvatures associated with the intrinsic geometry in the hypersurface with spacetime curvatures precisely through the extrinsic curvature:

$$R[h]_{lijk} = R[g]_{\alpha\beta\sigma\rho}\xi_l^\alpha\xi_i^\beta\xi_j^\sigma\xi_k^\rho - K_{ij}K_{lk} + K_{ik}K_{lj} \tag{3}$$

and

$$\nabla[h]_k K_{ij} - \nabla[h]_j K_{ik} = R_{\alpha\beta\sigma\rho}n^\alpha\xi_i^\beta\xi_j^\rho\xi_k^\sigma \tag{4}$$

whereas the curvature scalar is given by

$$R = R^{\alpha\beta}{}_{\alpha\beta} = 2R^{ni}{}_{ni} + R^{ij}{}_{ij} \tag{5}$$

In terms of this splitting, the Einstein-Hilbert action reads:

$$L_{EH} \equiv \sqrt{g}R[g] = N\sqrt{h}(R[h] + K_{ij}K^{ij} - K^2) - \partial_\alpha V^\alpha \tag{6}$$

with

$$V^{\alpha} = 2\sqrt{g}(n^{\beta}\nabla_{\beta}n^{\alpha} - n^{\alpha}\nabla_{\beta}n^{\beta})$$

Primary constraints appear when defining canonical momenta:

$$p^{\mu} \equiv \frac{\partial L}{\partial \dot{N}_{\mu}} \sim 0$$

the momenta conjugate to the spatial part of the metric is:

$$\pi^{ij} \equiv \frac{\delta L}{\delta \dot{h}_{ij}} = -h^{1/2}(K^{ij} - Kh^{ij})$$

The canonical commutation relations yield:

$$\{\pi^{ij}(\boldsymbol{x}), h_{kl}(\boldsymbol{x}')\} = -\delta(\boldsymbol{x}, \boldsymbol{x}')\frac{1}{2}(\delta^i_k\delta^j_l + \delta^j_k\delta^i_l)$$

The total Hamiltonian reads

$$H \equiv \int d^3x(\pi_{\mu}\dot{N}^{\mu} + \pi^{ij}\dot{h}_{ij} - L) = \int d^3x(N\mathcal{H} + N^i\mathcal{H}^i)$$

where

$$\mathcal{H}(h,\pi) = h^{-1/2}\left(\pi_{ij}\pi^{ij} - \frac{1}{2}\pi^2\right) - h^{1/2}R[h] \tag{7}$$

and

$$\mathcal{H}_i(h,\pi) = -2h_{ik}\partial_j\pi^{kj} - (2\partial_j h_{ki} - \partial_i h_{kj})\pi^{kj} = -2\nabla[h]_j{\pi_i}^j \tag{8}$$

The system of constraints is now consistent (that is, that the classical time evolution of the constraints is still a linear combination of constraints):

$$\dot{p}^{\mu} = \{p^{\mu}, H\} = (\mathcal{H}, \mathcal{H}_i) \sim 0$$

Second class constraints

$$N^{\mu} = f^{\mu}$$

can now be imposed. The whole hamiltonian analysis boils down to the two constraint equations

$$\mathcal{H} = 0$$
$$\mathcal{H}_i = 0$$

Much of the preceding analysis is actually quite generic for generally covariant systems. The full set of constraints obeys the Dirac-Schwinger algebra

$$\begin{aligned}
\{\mathcal{H}(\boldsymbol{x}), \mathcal{H}(\boldsymbol{y})\} &= [\mathcal{H}^i(\boldsymbol{x}) + \mathcal{H}^i(\boldsymbol{y})]\partial_i\delta(\boldsymbol{x},\boldsymbol{y}) \\
\{\mathcal{H}_i(\boldsymbol{x}), \mathcal{H}(\boldsymbol{y})\} &= \mathcal{H}(\boldsymbol{x})\partial_i\delta(\boldsymbol{x},\boldsymbol{y}) \\
\{\mathcal{H}_i(\boldsymbol{x}), \mathcal{H}_j(\boldsymbol{y})\} &= \mathcal{H}_i(\boldsymbol{y})\partial_j\delta(\boldsymbol{x},\boldsymbol{y}) + \mathcal{H}_j(\boldsymbol{x})\partial_i\delta(\boldsymbol{x},\boldsymbol{y})
\end{aligned} \tag{9}$$

which is nothing else than the Σ-projected algebra of the Diff(M) group.

Usually no reduction is made on the dynamical variables of the system, which amounts to keep h_{ij}, π^{ij} as (redundant) quantum variables. It is not clear how singular metrics can be avoided, because it is not easy to impose the condition that the metric is a positive definite operator.

Physical states in the Hilbert space are provisionally defined à la Dirac

$$\hat{\mathcal{H}}|\psi\rangle = 0$$
$$\hat{\mathcal{H}}_i|\psi\rangle = 0$$

It has been realized since long that this whole approach suffers from the *frozen time problem*, i.e., the Hamiltonian reads

$$H \equiv \int d^3x (N\mathcal{H} + N^i\mathcal{H}^i)$$

so that acting on physical states

$$\hat{H}|\psi\rangle = 0 \tag{10}$$

in such a way that Schrödinger's equation

$$i\frac{\partial}{\partial t}|\psi\rangle = \hat{H}|\psi\rangle \tag{11}$$

seemingly forbids any time dependence.

There are many unsolved problems in this approach, which has been kept at a formal level. The first one is an obvious operator ordering ambiguity owing to the nonlinearity. In the same vein, it is not clear whether it is possible to make the constraints hermitian. Besides, it is not clear that one recovers the full Diff invariance from the Dirac-Schwinger algebra. Actually, it is not known whether this is necessary; that is, what is the full symmetry of the quantum theory.

We can proceed further, still formally[2], using the Schrödinger representation defined in such a way that

$$(\hat{h}_{ij}\psi)[h] \equiv h_{ij}(x)\psi[h] \tag{12}$$

and

$$(\hat{\pi}^{ij}\psi)[h] \equiv -i\hbar\frac{\delta\psi}{\delta h_{ij}(x)}[h] \tag{13}$$

If we assume that diffeomorphisms act on wave functionals as:

$$\psi[f^*h] = \psi[h] \tag{14}$$

then the whole setup for the quantum dynamics of the gravitational field lies in Wheeler's *superspace* (nothing to do with supersymmetry) which is the

[2] It is bound to be formal as long as the problem of the infinities is not fully addressed. We know from the analysis of this representation for gauge theories in the lattice that those are the most difficult problems to solve.

set of three-dimensional metrics modulo three-dimensional diffs: $Riem(\Sigma)/Diff(\Sigma)$.

The Hamiltonian constraint then implies the famous *Wheeler-De Witt equation.*

$$- \hbar^2 2\kappa^2 G_{ijkl} \frac{\delta^2 \psi}{\delta h_{ik} \delta h_{jl}}[h] - \frac{h}{2\kappa^2} R^{(3)}[h]\psi[h] = 0 \tag{15}$$

where the De Witt tensor is:

$$G_{ijkl} \equiv \frac{1}{\sqrt{h}} \left(h_{ij} h_{kl} - \frac{1}{2} h_{ik} h_{jl} \right) \tag{16}$$

Needless to say, this equation, suggestive as it is, is plagued with ambiguities. The manifold of positive definite metrics has been studied by De Witt. He showed that it has signature $(-1, +1^5)$, where the timelike coordinate is given by the breathing mode of the metric:

$$\zeta = \sqrt{\frac{32}{3}} h^{1/4} \tag{17}$$

and in terms of other five coordinates ζ^a orthogonal to the timelike coordinate, the full metric reads

$$ds^2 = -d\zeta^2 + \frac{3}{32} \zeta^2 g_{ab} d\zeta^a d\zeta^b \tag{18}$$

with

$$g_{ab} = tr\, h^{-1} \partial_a h h^{-1} \partial h \tag{19}$$

The five dimensional submanifold with metric g_{ab} is the coset space

$$SL(3, \mathbb{R})/SO(3) \tag{20}$$

It has been much speculated whether the timelike character of the dilatations lies at the root of the concept of time. The Wheeler-De Witt equation can be written in a form quite similar to the Klein-Gordon equation:

$$\left(- \frac{\partial^2}{\partial \zeta^2} + \frac{32}{3\zeta^2} g^{ab} \partial_a \partial_b + \frac{3}{32} \zeta^2 R^{(3)} \right) \Psi = 0 \tag{21}$$

The analogy goes further in the sense that also here there is a naturally defined scalar product which is not positive definite:

$$(\psi, \chi) \equiv \int_\Sigma \psi^* d\Sigma^{ij} G_{ijkl} \frac{\delta \chi}{i \delta h_{kl}} - \chi^* d\Sigma^{ij} G_{ijkl} \frac{\delta \psi}{i \delta h_{kl}} \tag{22}$$

4 Using Ashtekar and Related Variables

The whole philosophy of this approach is canonical, i.e., an analysis of the evolution of variables defined classically through a foliation of spacetime by a family of spacelike three-surfaces Σ_t. The standard choice in this case as we have just reviewed, is the three-dimensional metric, g_{ij}, and its canonical conjugate, related to the extrinsic curvature.

Here, as in any canonical approach the way one chooses the canonical variables is fundamental.

Ashtekar's clever insight started from the definition of an original set of variables [10] stemming from the Einstein-Hilbert lagrangian written in the form[3]

$$S = \int e^a \wedge e^b \wedge R^{cd} \epsilon_{abcd} \tag{23}$$

where e^a are the one-forms associated to the tetrad,

$$e^a \equiv e^a_\mu dx^\mu . \tag{24}$$

Tetrads are defined up to a local Lorentz transformation

$$(e^a)' \equiv L^a{}_b(x) e^b \tag{25}$$

The associated $SO(1,3)$ connection one-form $\omega^a{}_b$ is usually called the *spin connection*. Its field strength is the curvature expressed as a two form:

$$R^a{}_b \equiv d\omega^a{}_b + \omega^a{}_c \wedge \omega^c{}_b \, . \tag{26}$$

Ashtekar's variables are actually based on the $SU(2)$ self-dual connection

$$A = \omega - i * \omega \tag{27}$$

Its field strength is

$$F \equiv dA + A \wedge A \tag{28}$$

The dynamical variables are then $(A_i, E^i \equiv F^{0i})$. The main virtue of these variables is that constraints are then linearized. One of them is exactly analogous to Gauss'law:

$$D_i E^i = 0 \, . \tag{29}$$

There is another one related to three-dimensional diffeomorphisms invariance,

$$tr\, F_{ij} E^i = 0 \tag{30}$$

and, finally, there is the Hamiltonian constraint,

$$tr F_{ij} E^i E^j = 0 \tag{31}$$

[3] Boundary terms have to be considered as well. We refer to the references for details.

On a purely mathematical basis, there is no doubt that Astekhar's variables are of a great ingenuity. As a physical tool to describe the metric of space, they are not real in general. This forces a reality condition to be imposed, which is awkward. For this reason it is usually preferred to use the Barbero-Immirzi [13, 37] formalism in which the connection depends on a free parameter, γ,

$$A_a^i = \omega_a^i + \gamma K_a^i \tag{32}$$

ω being the spin connection and K the extrinsic curvature. When $\gamma = i$ Astekhar's formalism is recovered; for other values of γ the explicit form of the constraints is more complicated. Thiemann [67] has proposed a form for the Hamiltonian constraint which seems promising, although it is not clear whether the quantum constraint algebra is isomorphic to the classical algebra (cf. [54]). A comprehensive reference is [66].

Some states which satisfy the Astekhar constraints are given by the loop representation, which can be introduced from the construct (depending both on the gauge field A and on a parameterized loop γ)

$$W(\gamma, A) \equiv tr\, P e^{\oint_\gamma A} \tag{33}$$

and a functional transform mapping functionals of the gauge field $\psi(A)$ into functionals of loops, $\psi(\gamma)$:

$$\psi(\gamma) \equiv \int \mathcal{D}A\, W(\gamma, A)\psi(A) \tag{34}$$

When one divides by diffeomorphisms, it is found that functions of knot classes (diffeomorphisms classes of smooth, non self-intersecting loops) satisfy all the constraints.

Some particular states sought to reproduce smooth spaces at coarse graining are the *weaves*. It is not clear to what extent they also approach the conjugate variables (that is, the extrinsic curvature) as well.

In the presence of a cosmological constant the hamiltonian constraint reads:

$$\epsilon_{ijk} E^{ai} E^{bj} \left(F_{ab}^k + \frac{\lambda}{3} \epsilon_{abc} E^{ck} \right) = 0 \tag{35}$$

A particular class of solutions of the constraint [60] are self-dual solutions of the form

$$F_{ab}^i = -\frac{\lambda}{3} \epsilon_{abc} E^{ci} \tag{36}$$

Kodama [41] has shown that the Chern-Simons state

$$\psi_{CS}(A) \equiv e^{\frac{3}{2\lambda} S_{CS}(A)} \tag{37}$$

is a solution of the hamiltonian constraint. He even suggested that the *sign* of the coarse grained, classical cosmological constant was always positive, irrespectively of the sign of the quantum parameter λ, but it is not clear whether this result is general enough. There is some concern [71] that this state as such is not normalizable with the usual norm. It has been argued that this is only natural, because the physical relevant norm must be very different from the naïve one (cf. [59]) and indeed normalizability of the Kodama state has been suggested as a criterion for the correctness of the physical scalar product (cf. for example the discussion in [24]) or else that a Euclidean interpretation could be given to it.

Loop states in general (suitable symmetrized) can be represented as spin network [56] states: colored lines (carrying some $SU(2)$ representation) meeting at nodes where intertwining $SU(2)$ operators act. A beautiful graphical representation of the group theory has been successfully adapted for this purpose. There is a clear relationship between this representation and the Turaev-Viro [68] invariants. Many of these ideas have been foresighted by Penrose (cf. [48]).

There is also a path integral representation, known as *spin foam* (cf. [12]), a topological theory of colored surfaces representing the evolution of a spin network. These are closely related to topological BF theories, and many independent generalizations have been proposed. Spin foams can also be considered as an independent approach to the quantization of the gravitational field [14].

In addition to its specific problems, this approach shares with all canonical approaches to covariant systems the problem of time. It is not clear its definition, at least in the absence of matter. Dynamics remains somewhat mysterious; the hamiltonian constraint does not say in what sense (with respect to what) the three-dimensional dynamics evolve.

4.1 Big Results of this Approach

One of the main successes of the loop approach is that the area (as well as the volume) operator is discrete. This allows, assuming that a black hole has been formed (which is a process that no one knows how to represent in this setting), to explain the formula for the black hole entropy . The result is expressed in terms of the Barbero-Immirzi parameter [57]. The physical meaning of this dependence is not well understood.

It has been pointed out [15] that there is a potential drawback in all theories in which the area (or mass) spectrum is discrete with eigenvalues A_n if the level spacing between eigenvalues δA_n is uniform because of the predicted thermal character of Hawking's radiation. The explicit computations in the present setting, however, lead to a space between (dimensionless) eigenvalues

$$\delta\, A_n \sim e^{-\sqrt{A_n}}, \tag{38}$$

which seemingly avoids this set of problems.

It has also been pointed out that [23] not only the spin foam, but almost all other theories of gravity can be expressed as topological BF theories with constraints. While this is undoubtedly an interesting and potentially useful remark, it is important to remember that the difference between the linear sigma model (a free field theory) and the nonlinear sigma models is just a matter of constraints. This is enough to produce a mass gap and asymptotic freedom in appropriate circumstances.

5 Euclidean Quantum Gravity

It can be boldly asserted that just by analogy with ordinary quantum field theory, the wave functional of quantum gravity must be given by:

$$\psi[h] \equiv \int_{g(\partial M)=h} \mathcal{D}g e^{-S_E[g]} \tag{39}$$

where we integrate over all riemannian metrics that obey the relevant boundary conditions, and the Einstein-Hilbert action has to be supplemented with boundary terms. This approach is problematic from the very beginning, due to the fact that the Wick analytic continuation of a lorentzian space-time is not riemannian in general (not even real), so that the whole setup seems to demand the study of real sections in a complex formulation. The point of view put forward by Hawking and collaborators [32] is that the needed analytical continuations could be hopefully made after Green's functions are evaluated.

There however is a well-known mathematical theorem of Markov (explained for physicists in [5]) asserting that there is no algorithmic way of predicting when two arbitrary manifolds are homeomorphic: $M_1 \sim M_2$. The problem stems essentially from the fundamental group: any finitely presented group can be the $\pi_1(M)$ of a four-dimensional manifold, M. So one proof of the result is to simply write down a family of groups G_k such that we cannot algorithmically recognisee when $G = \{e\}$. There are, in addition, further subtleties with the diffeomorphism class in $d = 4$: there is a uncountable set of non equivalent differentiable structures in $\mathbb{R}^4$: the so-called *exotic* $\mathbb{R}^4$ (cf. [18] for a physical approach; a relevant recent reference is [50]).

Working in lorentzian signature, a Hamiltonian path integral can be dreamt of, where a functional integral is performed over three-dimensional geometries (cf. [6]) only. Here the situation is slightly better: it seems that there is recent progress in the proof of Thurston's geometrization conjecture, which implies in particular Poincaré's conjecture, and which explains all three-dimensional manifolds in terms of eight different *geometries*. Incidentally, the work of Perelman [49] uses what mathematicians call the Ricci flow, which is *exactly* the flow of the renormalization group of the sigma model associated to the bosonic string in a curved background.

Let us finally comment that even if the basic theory of Nature is topological one needs to *enumerate* topologies to discriminate between different ones. Besides, topological symmetry has to be broken al low energies.

In order to reach a probabilistic interpretation, a scalar product ought to be defined. The one which is naturally associated to the Wheeler-de Witt equation is not positive-definite, so this remains as an open problem in this approach.

Were somebody apply these ideas to the whole Universe (the so called Quantum Cosmology) there are other problems in store. It is not clear what is the physical interpretation of probabilities associated to a single event. A related problem is the one of the physical interpretation of Quantum Mechanics without classical observers. Many people have related this to the decoherence mechanisms (cf. for example [30]) but it seems to me that the situation is still to be clarified.

6 Perturbative (Graviton) Approach

A much more modest approch is to study gravitons as ordinary (massless, spin two) particles in Minkowski space-time.

$$g_{\alpha\beta} = \bar{g}_{\alpha\beta} + \kappa h_{\alpha\beta} \tag{40}$$

It seems to many people (including the author) that this is at least a preliminary step before embarking in more complicated adventures. As a quantum field theory, quantum general relativity has got a dimensionful coupling : $d(\kappa) = -1$, which means that it is not renormalizable in the usual sense of the word.

In spite of this, the theory is one loop finite on shell , as was shown in a brilliant calculation by G. 't Hooft and M. Veltman [33]. They computed the counterterm:

$$\Delta L^{(1)} = \frac{\sqrt{\bar{g}}}{\epsilon}\frac{203}{80}\bar{R}^2 \tag{41}$$

No more miracles are expected for higher loops, and none happen. Goroff and Sagnotti [27] were the first to show that to two loops,

$$\Delta L^{(2)} = \frac{209}{2880(4\pi)^4}\frac{1}{\epsilon}\bar{R}^{\alpha\beta}{}_{\gamma\delta}\bar{R}^{\gamma\delta}{}_{\rho\sigma}\bar{R}^{\rho\sigma}{}_{\alpha\beta} \tag{42}$$

The general structure of perturbation theory is governed by the fact we have just mentioned that the coupling constant is dimensionful. A general diagram will then behave in the s-channel as $\kappa^n s^n$ and counterterms as:

$$\Delta L \sim \sum \int \kappa^n R^{(2+n/2)} \tag{43}$$

(where a symbolic notation has been used), packing all invariants with the same dimension; for example, R^2 stands for an arbitrary combination of R^2, $R_{\alpha\beta}R^{\alpha\beta}$ and $R_{\alpha\beta\gamma\delta}R^{\alpha\beta\gamma\delta}$ conveying the fact that that the theory is non-renormalizable.

It may however be pondered whether effective lagrangians are really useful for $E \ll m_P$. This possibility has been forcefully explored by Donoghue and collaborators (cf. [21]). There are some caveats: for example, when horizons are present, it seems necessary in order to be able to apply these ideas, to use some particular foliations, the so called *nice slices*). The mere fact that we are unable to predict the cosmological constant (which is the mother of all infrared problems) means that our understanding has ample room for improvement.

Could it be that in spite of the fact that general relativity is not renormalizable, there is a non perturbative sector in which the theory makes sense as a quantum theory? First of all, were that true, it would be most remarkable: there are no known QFT which are defined only nonperturbatively. Besides, at the classical level, perturbation theory works wonderfully, and there is indeed a whole framework, the parameterized post-newtonian (PPN) formalism to discriminate between alternate theories of gravity. It is then most unclear why at the quantum (and only there) level perturbation theory should fail.

We want to mention in closing this chapter, some fascinating relationships uncovered by Z. Berm and collaborators (cf. [16]) between purely field-theoretical S matrix elements in (super)gravity and gauge theories: the so called *Gravity* = *Gauge* × *Gauge* conjecture. In spite of several attempts, it is not clear how this can be understood from the Einstein-Hilbert action. The relationship is of course automatic in strings, because closed string amplitudes (which include the graviton) are given by products of open string amplitudes (which contain the gauge fields). The KLT relations [40] are a quantitative formulation of this fact.

In view of all this, one can try to study particular extensions of the Einstein-Hilbert actions. Modifications quadratic in the curvature improve renormalizability [39], but have problems with unitarity at a very fundamental level [61].

Local supersymmetry is expected to improve the ultraviolet behavior through cancellations between fermionic and bosonic degrees of freedom. In spite of that, some infinities are allowed by the symmetries of the problem, and are thus expected to appear; for example in extended supergravities this is expected to happen at loop order $L > \frac{10}{D-2}$ in the maximally supersymmetric case in which there are 32 supercharges.

The (sad) conclusion of all this is that ordinary QFT (with a finite number of fields) does not work, even for describing small (quantum) ripples in Minkowski space.

7 Strings

It should be clear by now that we probably still do not know what is exactly the problem to which string theories are the answer. At any rate, the starting point is that all elementary particles are viewed as quantized excitations of a one dimensional object, the string, which can be either open (free ends) or closed (a loop). Excellent books are avaliable, such as [29, 52].

String theories enjoyed a convoluted history. Their origin can be traced to the Veneziano model of strong interactions. A crucial step was the reinterpretation by Scherk and Schwarz [58] of the massless spin two state in the closed sector (previously thought to be related to the Pomeron) as the graviton and consequently of the whole string theory as a potential theory of quantum gravity, and potential unified theories of all interactions. Now the wheel has made a complete turn, and we are perhaps back through the Maldacena conjecture [44] to a closer relationship than previously thought with ordinary gauge theories.

Fig. 3. String theorist at work

From a certain point of view, their dynamics is determined by a two-dimensional non-linear sigma model, which geometrically is a theory of imbeddings of a two-dimensional surface (the world sheet of the string) to a (usually ten-dimensional) target space:

$$x^{\mu}(\xi) : \Sigma_2 \rightarrow M_n \tag{44}$$

There are two types of interactions to consider. Sigma model interactions (in a given two-dimensional surface) are defined as an expansion in powers of momentum, where a new dimensionful parameter, $\alpha' \equiv l_s^2$ sets the scalc. This scale is a priori believed to be of the order of the Planck length. The first terms in the action always include a coupling to the massless backgrounds: the spacetime metric, the two-index Maxwell like field known as the Kalb-Ramond or b-field, and the dilaton. To be specific,

$$S = \frac{1}{l_s^2}\int_{\Sigma_2} g_{\mu\nu}(x(\xi))\partial_a x^\mu(\xi)\partial_b x^\nu(\xi)\gamma^{ab}(\xi) + \ldots \qquad (45)$$

There are also string interactions, (changing the two-dimensional surface) proportional to the string coupling constant, g_s, whose variations are related to the logarithmic variations of the dilaton field. Open strings (which have gluons in their spectrum) *always* contain closed strings (which have gravitons in their spectrum) as intermediate states in higher string order (g_s) corrections. This interplay open/closed is one of the most fascinating aspects of the whole string theory.

It has been discovered by Friedan (cf. [25]) that in order for the quantum theory to be consistent with all classical symmetries (diffeomorphisms and conformal invariance), the beta function of the generalized couplings[4] must vanish:

$$\beta(g_{\mu\nu}) = R_{\mu\nu} = 0 \qquad (46)$$

This result remains until now one of the most important in string theory, hinting at a deep relationship between Einstein's equations and the renormalization group.

Polyakov [53] introduced the so called *non-critical strings* which have in general a two-dimensional cosmological constant (forbidden otherwise by Weyl invariance). The dynamics of the conformal mode (often called Liouville in this context) is, however, poorly understood.

Fundamental strings live in $D = 10$ spacetime dimensions, and so a Kaluza-Klein mecanism of sorts must be at work in order to explain why we only see four non-compact dimensions at low energies. Strings have in general tachyons in their spectrum, and the only way to construct seemingly consistent string theories (cf. [26]) is to project out those states, which leads to supersymmetry. This means in turn that all low energy predictions heavily depend on the supersymmetry breaking mechanisms.

String perturbation theory is probably well defined although a full proof is not available.

Several stringy symmetries are believed to be exact: T-duality, relating large and small compactification volumes, and S-duality, relating the strong coupling regime with the weak coupling one. Besides, extended configurations (*D branes*); topological defects in which open strings can end are known to be important [51]. They couple to Maxwell-like fields which are p-forms called Ramond-Ramond (RR) fields. These dualities [36] relate all five string theories (namely, Heterotic $E(8)\times E(8)$, Heterotic $SO(32)$, type I, IIA and IIB) and

[4] There are corrections coming from both dilaton and Kalb-Ramond fields. The quoted result is the first term in an expansion in derivatives, with expansion parameter $\alpha' \equiv l_s^2$.

it is conjectured that there is an unified eleven-dimensional theory, dubbed M-theory of which $\mathcal{N}=1$ supergravity in $d=11$ dimensions is the low energy limit.

7.1 Big Results

Perhaps the main result is that graviton physics in flat space is well defined for the first time, and this is no minor accomplishment.

Besides, there is evidence that at least some geometric singularities are harmless in the sense that strings do not feel them. Topology change amplitudes do not vanish in string theory.

The other Big Result [62] is that one can correctly count states of extremal black holes as a function of charges. This is at the same time astonishing and disappointing. It clearly depends strongly on the objects being BPS states (that is, on supersymmetry), and the result has not been extended to non-supersymmetric configurations. On the other hand, as we have said, it *exactly* reproduces the entropy as a function of a sometimes large number of charges, without any adjustable parameter.

7.2 The Maldacena Conjecture

Maldacena [44] proposed as a conjecture that IIB string theories in a background $AdS_5 \times S_5$ with common radius $l \sim l_s(g_s N)^{1/4}$ and N units of RR flux that is, $\int_{S_5} F_5 = N$ (which implies that $F_5 \sim \frac{N}{r^5}$) is equivalent to a four dimensional ordinary gauge theory in flat four-dimensional Minkowski space, namely $\mathcal{N}=4$ super Yang-Mills with gauge group $SU(N)$ and coupling constant $g = g_s^{1/2}$.

Although there is much supersymmetry in the problem and the kinematics largely determine correlators, (in particular, the symmetry group $SO(2,4) \times SO(6)$ is realized as an isometry group on the gravity side and as an R-symmetry group as well as conformal invariance on the gauge theory side) this is not fully so[5] and the conjecture has passed many tests in the semiclassical approximation to string theory.

The action of the RR field, given schematically by $\int F_5^2$, scales as N^2, whereas the ten-dimensional Einstein-Hilbert $\int R$, depends on the overall geometric scale as the eighth power of the common radius, l^8. The 't Hooft coupling is $\lambda = g^2 N \sim \frac{l^4}{l_s^4}$ and the tenth dimensional Newton's constant is $\kappa_{10}^2 \sim G_{10} \sim l_p^8 = g_s^2 l_s^8 \sim \frac{l^8}{N^2}$.

If we consider the effective five dimensional theory after compactifying on a five sphere of radius r, the RR term yields a negative contribution $\sim -(\frac{N}{r^5})^2 r^5$, whereas the positive curvature of the five sphere S^5 gives a positive contribution, $\sim \frac{1}{r^2} r^5$. The competition between these two terms in the

[5] The only correlators that are completely determined through symmetry are the two and three-point functions.

effective potential is responsible for the minimum with negative cosmological constant.

The way the dictionary works in detail [69] is that the supergravity action corresponding to fields with prescribed boundary values is related to gauge theory correlators of certain gauge invariant operators corresponding to the particular field studied:

$$e^{-S_{sugra}[\Phi_i]}\Big|_{\Phi_i|_{\partial AdS}=\phi_i} = \langle e^{\int \mathcal{O}_i\phi_i}\rangle_{CFT} \tag{47}$$

This is the first time that a precise holographic description of spacetime in terms of a (boundary) gauge theory is proposed and, as such it is of enormous potential interest. It has been conjectured by 't Hooft [64] and further developed by Susskind [63] that there should be much fewer degrees of freedom in quantum gravity than previously thought. The conjecture claims that it should be enough with one degree of freedom per unit Planck surface in the two-dimensional boundary of the three-dimensional volume under study. The reason for that stems from an analysis of the Bekenstein-Hawking [15, 31] entropy associated to a black hole, given in terms of the two-dimensional area A[6] of the horizon by

$$S = \frac{c^3}{4G\hbar}A\,. \tag{49}$$

This is a deep result indeed, still not fully understood.

It is true on the other hand that the Maldacena conjecture has only been checked for the time being in some corners of parameter space, namely when strings can be approximated by supergravity in the appropriate background.

8 Dualities and Branes

The so- called *T-duality* is the simplest of all dualities and the only one which can be shown to be true, at least in some contexts. At the same time it is a very stringy characteristic, and depends in an essential way on strings being extended objects. In a sense, the web of dualities rests on this foundation, so that it is important to understand clearly the basic physics involved. Let us consider strings living on an external space with one compact dimension, which we shall call y, with topology S^1 and radius R. The corresponding field in the imbedding of the string, which we shall call y (i.e. we are dividing the target-spacetime dimensions as (x^μ, y), where y parameterizes the circle), has then the possibility of winding around it:

[6] The area of the horizon for a Schwarzschild black hole is given by:

$$A = \frac{8\pi G^2}{c^4}M^2 \tag{48}$$

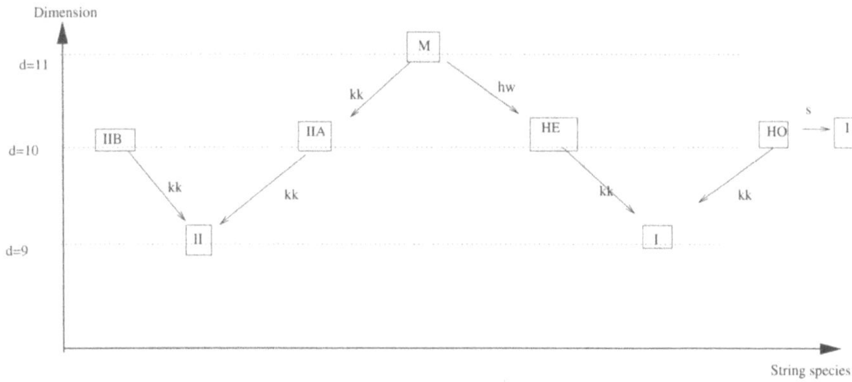

Fig. 4. Conjectured relationships between string theories in different dimensions

$$y(\sigma + 2\pi, \tau) = y(\sigma, \tau) + 2\pi R m \, . \tag{50}$$

A closed string can close in general up to an isometry of the external spacetime.

The zero mode expansion of this coordinate (that is, forgetting about oscillators) would then be

$$y = y_c + 2p_c\tau + mR\sigma \, . \tag{51}$$

Canonical quantization leads to $[y_c, p_c] = i$, and single-valuedness of the plane wave $e^{iy_c p_c}$ enforces as usual $p_c \in \mathbb{Z}/R$, so that $p_c = \frac{n}{R}$.

The zero mode expansion can then be organized into left and right movers in the following way

$$\begin{aligned} y_L(\tau + \sigma) &= y_c/2 + \left(\frac{n}{R} + \frac{mR}{2}\right)(\tau + \sigma) \, , \\ y_R(\tau - \sigma) &= y_c/2 + \left(\frac{n}{R} - \frac{mR}{2}\right)(\tau - \sigma) \, . \end{aligned} \tag{52}$$

The mass shell conditions reduce to

$$\begin{aligned} m_L^2 &= \frac{1}{2}\left(\frac{n}{R} + \frac{mR}{2}\right)^2 + N_L - 1 \, , \\ m_R^2 &= \frac{1}{2}\left(\frac{n}{R} - \frac{mR}{2}\right)^2 + N_R - 1 \, . \end{aligned} \tag{53}$$

Level matching, $m_L = m_R$, implies that there is a relationship between momentum and winding numbers on the one hand, and the oscillator excess on the other

$$N_R - N_L = nm \, . \tag{54}$$

At this point it is already evident that the mass formula is invariant under

$$R \to R^* \equiv 2/R\,, \tag{55}$$

and exchanging momentum and winding numbers. This is the simplest instance of *T-Duality*.

On the other hand, it is an old observation (which apparently originated in Schrödinger) that Maxwell's equations are almost symmetrical with respect to interchange between electric and magnetic degrees of freedom. This idea was explored by Dirac and eventually lead to the discovery of the consistency conditions that have to be fulfilled if there are magnetic monopoles in nature. The fact that nonsingular magnetic monopoles appear as classical solutions in some gauge theories led further support to this duality viewpoint. In order to be able to make a consisting conjecture, first put forward by Montonen and Olive [45], supersymmetry is needed, as first remarked by Osborn [47].

Now in strings there are the so-called Ramond-Ramond (RR) fields, which are p-forms of different degrees. In the same way that one forms (i.e., the Maxwell field) couples to charged particles that is, from the spacetime point of view, to objects of dimension 0 with one-dimensional trajectories, a p-form

$$A_{\mu_1 \ldots \mu_p} \tag{56}$$

would couple to a $(p-1)$-dimensional object, whose world history is described by a p-dimensional hypersurface

$$x^\mu = x^\mu(\xi_1 \ldots \xi_p) \tag{57}$$

These objects are traditionally denoted by the name p-branes (it all originated in a dubious joke). That is, ordinary particles are 0-branes, a string is a 1-brane, a membrane is a 2-brane, and so on.

Dualities relate branes of different dimensions in different theories; this means that if one is to take this symmetry seriously, it is not clear at all that strings are the more fundamental objects: in the so called M-theory branes appear as fundamental as strings.

If we are willing to make the hypothesis that supersymmetry is not going to be broken whilst increasing the coupling constant, g_s, some astonishing conclusions can be drawn. Assuming this, massless quanta can become massive as g_s grows only if their number, charges and spins are such that they can combine into massive multiplets (which are all larger than the irreducible massless ones). The only remaining issue, then, is whether any other massless quanta can appear at strong coupling.

Now, in the IIA string theory there are states associated to the Ramond-Ramond (RR) one form, A_1, namely the D-0-branes, whose tension goes as $m \sim \frac{1}{g_s}$. This clearly gives new massless states in the strong coupling limit.

There are reasons[7] to think that this new massless states are the first level of a Kaluza-Klein tower associated to compactification on a circle of an

[7] In particular: The fact that there is the possibility of a central extension in the IIA algebra, related to the Kaluza-Klein compactification of the $d = 11$ Supergravity algebra.

11-dimensional theory. Actually, assuming an 11-dimensional spacetime with an isometry $k = \frac{\partial}{\partial y}$, an Ansatz which exactly reproduces the dilaton factors of the IIA string is

$$ds^2_{(11)} = e^{\frac{4}{3}\phi}(dy - A^{(1)}_{\mu} dx^{\mu})^2 + e^{-\frac{2}{3}\phi} g_{\mu\nu} dx^{\mu} dx^{\nu} \ . \tag{58}$$

Equating the two expressions for the D0 mass,

$$\frac{1}{g_s} = \frac{1}{R_{11}} \ , \tag{59}$$

leads to $R_{11} = e^{\frac{2}{3}\phi} = g_A^{2/3}$.

This means that a new dimension appears at strong coupling, and this dimension is related to the dilaton. The only reason why we do not see it at low energies is precisely because of the smallness of the string coupling, related directly to the dilaton field. The other side of this is that this eleven dimensional theory, dubbed *M-theory* does not have any weak coupling limit; it is always strongly coupled. Consequently, not much is known on this theory, except for the fact that its field theory, low curvature limit is $\mathcal{N} = 1$ supergravity in $d = 11$ dimensions.

All supermultiplets of massive one-particle states of the IIB string supersymmetry algebra contain states of at least spin 4. This means that under the previous set of hypothesis, the set of massless states at weak coupling must be exactly the same as the corresponding set at strong coupling. This means that there must be a symmetry mapping weak coupling into strong coupling.

There is a well-known candidate for this symmetry: Let us call, as usual, l the RR scalar and ϕ the dilaton (NSNS). We can pack them together into complex scalar

$$S = l + ie^{-\frac{\phi}{2}} \ . \tag{60}$$

The IIB supergravity action in $d = 10$ is invariant under the $SL(2, \mathbb{R})$ transformations

$$S \to \frac{aS + b}{cS + d} \ , \tag{61}$$

if at the same time the two two-forms, $B_{\mu\nu}$ (the usual, ever-present, NS field), and $A^{(2)}$, the RR field transform as

$$\begin{pmatrix} B \\ A^{(2)} \end{pmatrix} \to \begin{pmatrix} d & -c \\ -b & a \end{pmatrix} \begin{pmatrix} B \\ A^{(2)} \end{pmatrix} \ , \tag{62}$$

Both the, Einstein frame, metric $g_{\mu\nu}$ and the four-form $A^{(4)}$ are inert under this $SL(2, \mathbb{R})$ transformation.

A discrete subgroup $SL(2, \mathbb{Z})$ of the full classical $SL(2, \mathbb{R})$ is believed to be an exact symmetry of the full string theory. The exact imbedding of the discrete subgroup in the full $SL(2, \mathbb{R})$ depends on the vacuum expectation value of the RR scalar.

The particular transformation

$$g = \begin{pmatrix} 0 & 1 \\ -1 & 0 \end{pmatrix} , \tag{63}$$

maps ϕ into $-\phi$ (when $l = 0$), and B into $A^{(2)}$. This means that the string coupling

$$g_s \rightarrow \frac{1}{g_s} \tag{64}$$

This is a strong/weak coupling type of duality, similar to the electromagnetic duality in that sense. The standard name for it is an *S-duality* type of transformation, mapping the ordinary string with NS charge, to another string with RR charge (which then must be a D-1-brane, and is correspondingly called a *D-string*), and, from there, is connected to all other D-branes by T-duality.

Using the fact that upon compactification on S^1, IIA at R_A is equivalent to IIB at $R_B \equiv 1/R_A$, and the fact that the effective action carries a factor of $e^{-2\phi}$ we get

$$R_A g_B^2 = R_B g_A^2 , \tag{65}$$

which combined with our previous result, $g_A = R_{11}^{3/2}$ implies that $g_B = \frac{R_{11}^{3/2}}{R_A}$. Now the Kaluza-Klein ansatz implies that from the eleven dimensional viewpoint the compactification radius is measured as

$$R_{10}^2 \equiv R_A^2 e^{-2\phi/3} , \tag{66}$$

yielding

$$g_B = \frac{R_{11}}{R_{10}} . \tag{67}$$

From the effective actions written above it is easy to check that there is a (S-duality type) field transformation mapping the SO(32) Type I open string into the SO(32) Heterotic one namely

$$\begin{aligned} g_{\mu\nu} &\rightarrow e^{-\phi} g_{\mu\nu}^{Het} , \\ \phi &\rightarrow -\phi , \\ B' &\rightarrow B . \end{aligned} \tag{68}$$

This means that physically there is a strong/weak coupling duality, because coupling constants of the compactified theories would be related by

$$\begin{aligned} g_{het} &= 1/g_I , \\ R_{het} &= R_I/g_I^{1/2} . \end{aligned} \tag{69}$$

9 Summary: the State of the Art in Quantum Gravity

In the loop approach one is working with nice candidates for a quantum theory. The theories are interesting, probably related to topological field theories [17] and background independence as well as diffeomorphism invariance are clearly implemented. On the other hand, it is not clear that their low energy limit is related to Einstein gravity.

Strings start from a perturbative approach more familiar to a particle physicist. However, they carry all the burden of supersymmetry and Kaluza-Klein. It has proved to be very difficult to study nontrivial non-supersymmetric dynamics.

Finally, and this applies to all approaches, the holographic ideas seem intriguing; there are many indications of a deep relationship between gravity and gauge theories.

We would like to conclude by insisting on the fact that although there is not much we know for sure on quantum effects on the gravitational field, even the few things we know are a big feat, given the difficulty to do physics without experiments.

Progress could be made if we could derive semiclassical gravity in such a way that corrections to it can be reliably estimated, for example

$$R_{\mu\nu} - \frac{1}{2} R g_{\mu\nu} = 2\kappa^2 \langle\psi|\hat{T}_{\mu\nu}|\psi\rangle + \frac{1}{L^2}\Delta\,, \tag{70}$$

when working at a certain scale of distances, say L. In order to understand those equations, we would had to know something about the operator of which the first member is the expectation value; something about the state on which the expectation value is computed (In particular, if it is the vacuum, how is it to be defined?) and finally, something about the definition of the energy momentum tensor as a composite operator. A question of obvious physical interest is the estimate of the size of the corrections: Is the expected error at a given scale of distance L

$$\Delta \sim \frac{\hbar G}{c^3 L^2} \tag{71}$$

or, does it depend of the characteristic energy of the source?

$$\Delta \sim \frac{G E^2}{\hbar c^5} \tag{72}$$

measured with respect to what?

It is painfully clear that there is still a large margin for improving our understanding of effective quantum field theories. For example, there is still no convincing derivation of Hawking radiation without transplanckian modes appearing at some point (this particular example is related to the existence of the nice slices mentioned above). Besides, we do not understand the cosmological constant, which is clearly related to the estimate of Δ.

The observational prospects are rather poor. In many models, in particular in the loop approach (and also in strings, with some qualifications) deviations from the lorentzian dispersion relations are expected:

$$E^2 = \boldsymbol{p}^2 + m^2 + E^2 \sum_{n=1} c_n \left(\frac{E}{m_P}\right)^n \tag{73}$$

Other contributions will undoubtedly analyze those in much more detail. Let us now simply mention that noncommutative models make similar predictions.

Winding states are stringy phenomena, and its observation would be very interesting. Stringy predictions, however, are in general difficulty to disentangle from predictions of supersymmetry (SUSY). Namely, SUSY has to be broken, and this scale spoils almost all differences between strings and QFT models.

With the great triumph of particle physics at the end of the seventies, namely the *experimental* discovery of the intermediate bosons related to electroweak interactions, the standard model was confirmed in all its essential traits, waiting only for the Higgs to be discovered (at LHC?) and the theoretical effort has concentrated in more and more speculative topics, and experimental guidance has become correspondingly scarce. The net result is that, even more so that in the old days of the hunting for the theory of strong interactions, theoretical physics is divided in almost disconnected clans.

All this is even more true when talking about quantum gravity, a paradise of speculation.

This is the reason why all efforts such as the one in the present workshop, aiming at making contact with experiment and/or observation are welcome, and will eventually redirect physics on a healthier track when we learn to recognize the physically relevant facts that presumably lie in front of our eyes.

Acknowledgements

I am grateful to Giovanni Amelino-Camelia and to Jerzy Kowalski-Glikman for organizing such a stimulating workshop and for allowing me to take part in it. I am also grateful to Carlos Muñoz for lending his picture of a stringer. Finally, I am indebted to Jorge Conde for reading the manuscript.

References

1. R. Arnowitt, S. Deser, and C. W. Misner, "Canonical Variables for General Relativity" Phys. Rev. 117, 1595-1602 (1960)
2. J. Alfaro and G. Palma, "Loop quantum gravity corrections and cosmic rays decays," Phys. Rev. D 65 (2002) 103516 [arXiv:hep-th/0111176]
   "Loop quantum gravity and ultra high energy cosmic rays," Phys. Rev. D 67 (2003) 083003 [arXiv:hep-th/0208193]

3. E. Alvarez, "Quantum Gravity: A Pedagogical Introduction To Some Recent Results," Rev. Mod. Phys. 61 (1989) 561
4. E. Alvarez, "Low-Energy Effects Of Quantum Gravity," FTUAM-89-15 *Lectures given at Mtg. on Recent Developments in Gravitation, Barcelona, Spain, Sep 5-8, 1989*
5. E. Alvarez, "Some general problems in quantum gravity," CERN-TH-6257-91 *Lectures at the 22nd Gift Int. Seminar in Theoretical Physics: Quantum Gravity and Cosmology, S. Feliu De Guixols, Jun 3–6, 1991*
6. E. Alvarez, "Some general problems in quantum gravity. 2. The Three-dimensional case," Int. J. Mod. Phys. D 2 (1993) 1 [arXiv:hep-th/9211050]
7. E. Alvarez, L. Alvarez-Gaume and Y. Lozano, "An introduction to T duality in string theory," Nucl. Phys. Proc. Suppl. 41 (1995) 1 [arXiv:hep-th/9410237]
8. L. Alvarez-Gaume and M. A. Vazquez-Mozo, "Topics in string theory and quantum gravity," arXiv:hep-th/9212006
9. I. Antoniadis, N. Arkani-Hamed, S. Dimopoulos and G. R. Dvali, "New dimensions at a millimeter to a Fermi and superstrings at a TeV," Phys. Lett. B 436 (1998) 257 [arXiv:hep-ph/9804398]
10. A. Ashtekar, "New Hamiltonian Formulation Of General Relativity," Phys. Rev. D 36 (1987) 1587
11. A. Ashtekar, "Quantum Geometry And Black Hole Entropy", gr-qc/0005126 [=Adv.Theor.Math.Phys. 4 (2000) 1]
"Quantum Geometry Of Isolated Horizons And Black Hole Entropy", Phys.Rev. D57 (1998) 1009 [=gr-qc/9705059]
12. J.Baez, "An Introduction To Spin Foam Models Of Quantum Gravity And Bf Theory", gr-qc/0010050
13. J. F. Barbero, "Real Ashtekar variables for Lorentzian signature space times," Phys. Rev. D 51 (1995) 5507 [arXiv:gr-qc/9410014]
14. J. W. Barrett and L. Crane, "A Lorentzian signature model for quantum general relativity," Class. Quant. Grav. 17 (2000) 3101 [arXiv:gr-qc/9904025]
15. J. Bekenstein, *Black Holes And Entropy* Phys.Rev. D7 (1973) 2333
(with V. Mukhanov), "Spectroscopy Of The Quantum Black Hole", Commun.Math.Phys. 125 (1989) 417
16. Z. Bern, "Perturbative quantum gravity and its relation to gauge theory," Living Rev. Rel. **5** (2002) 5 [arXiv:gr-qc/0206071]
17. M. Blau, "Topological Gauge Theories Of Antisymmetric Tensor Fields", hep-th/9901069 [=Adv.Theor.Math.Phys. 3 (1999) 1289]
18. C. H. Brans, "Exotic smoothness structures in physics," *Prepared for 1st Mexican School on Gravitation and Mathematical Physics, Guanajuato, Mexico, 12–16 Dec 1994*
19. S. R. Coleman and S. L. Glashow, "High-energy tests of Lorentz invariance," Phys. Rev. D 59 (1999) 116008 [arXiv:hep-ph/9812418]
20. B. S. De Witt, "Quantum Theory Of Gravity. 1. The Canonical Theory," Phys. Rev. **160** (1967) 1113
21. J. F. Donoghue, "General Relativity As An Effective Field Theory: The Leading Quantum Corrections," Phys. Rev. D **50** (1994) 3874 [arXiv:gr-qc/9405057]
22. M. J. Duff, "Inconsistency Of Quantum Field Theory In Curved Space-Time," ICTP/79-80/38 *Talk presented at 2nd Oxford Conf. on Quantum Gravity, Oxford, Eng., Apr 1980*

23. L. Freidel, K. Krasnov and R. Puzio, "BF description of higher-dimensional gravity theories," Adv. Theor. Math. Phys. 3 (1999) 1289 [arXiv:hep-th/9901069]
24. L. Freidel and L. Smolin, "The linearization of the Kodama state," arXiv:hep-th/0310224
25. D. Friedan, "Nonlinear Models In Two Epsilon Dimensions," Phys. Rev. Lett. 45 (1980) 1057
26. F. Gliozzi, J. Scherk and D. I. Olive, "Supersymmetry, Supergravity Theories And The Dual Spinor Model," Nucl. Phys. B 122 (1977) 253
27. M. H. Goroff and A. Sagnotti, "The Ultraviolet Behavior Of Einstein Gravity," Nucl. Phys. B **266**, 709 (1986)
28. M. B. Green and J. H. Schwarz, "Anomaly Cancellation In Supersymmetric D=10 Gauge Theory And Superstring Theory," Phys. Lett. B 149 (1984) 117
29. M. B. Green, J. H. Schwarz and E. Witten, "Superstring Theory. Vol. 1: Introduction,"
"Superstring Theory. Vol. 2: Loop Amplitudes, Anomalies And Phenomenology,"
30. J. B. Hartle, "Space-time quantum mechanics and the quantum mechanics of space-time," arXiv:gr-qc/9304006
31. S. Hawking, *Particle Creation By Black Holes* Commun.Math.Phys. 43 (1975) 199
32. G. W. Gibbons and S. W. Hawking, "Euclidean Quantum Gravity,"
33. G. 't Hooft and M. J. G. Veltman, "One Loop Divergencies In The Theory Of Gravitation," Annales Poincare Phys. Theor. A **20** (1974) 69
34. G. T. Horowitz, "Quantum gravity at the turn of the millennium," arXiv:gr-qc/0011089
35. G. Horowitz, "Exactly Soluble Diffeomorphism Invariant Theories", Annals Phys. 205 (1991) 130
36. C. M. Hull and P. K. Townsend, "Unity of superstring dualities," Nucl. Phys. B 438 (1995) 109 [arXiv:hep-th/9410167]
37. G. Immirzi, "Quantum gravity and Regge calculus," Nucl. Phys. Proc. Suppl. 57 (1997) 65 [arXiv:gr-qc/9701052]
38. C. J. Isham, "Structural issues in quantum gravity," arXiv:gr-qc/9510063
39. J. Julve and M. Tonin, "Quantum Gravity With Higher Derivative Terms," Nuovo Cim. B **46** (1978) 137
40. H. Kawai, D. C. Lewellen and S. H. H. Tye, "A Relation Between Tree Amplitudes Of Closed And Open Strings," Nucl. Phys. B **269** (1986) 1
41. H. Kodama, "Holomorphic Wave Function Of The Universe," Phys. Rev. D 42 (1990) 2548
42. M. Luscher, R. Narayanan, P. Weisz and U. Wolff, "The Schrodinger functional: A Renormalizable probe for nonAbelian gauge theories," Nucl. Phys. B 384 (1992) 168 [arXiv:hep-lat/9207009]
43. Y. M. Makeenko and A. A. Migdal, "Exact Equation For The Loop Average In Multicolor QCD," Phys. Lett. B 88 (1979) 135 [Erratum-ibid. B 89 (1980) 437]
44. J. Maldacena, *The large N limit of superconformal field theories and supergravity*, Adv.Theor.Math.Phys.2:231-252,1998, hep-th/9711200.
45. C. Montonen and D. I. Olive, "Magnetic Monopoles As Gauge Particles?," Phys. Lett. B **72** (1977) 117

46. R. C. Myers and M. Pospelov, "Experimental challenges for quantum gravity," arXiv:hep-ph/0301124
47. H. Osborn, "Topological Charges For N=4 Supersymmetric Gauge Theories And Monopoles Of Spin 1," Phys. Lett. B **83** (1979) 321
48. R. Penrose, *Angular momentum: an approach to combinatorial spacetime*,in *Quantum theory and beyond*, T. Bastin ed.(Cambridge University Press, 1971)
49. G. Perelman, "The entropy formula for the Ricci flow and its geometric applications"
50. H. Pfeiffer, arXiv:gr-qc/0404088
51. J. Polchinski, "Dirichlet-Branes and Ramond-Ramond Charges," Phys. Rev. Lett. 75 (1995) 4724 [arXiv:hep-th/9510017]
52. J. Polchinski, "String Theory. Vol. 1: An Introduction To The Bosonic String," "String Theory. Vol. 2: Superstring Theory And Beyond,"
53. A. M. Polyakov, "Quantum Geometry Of Bosonic Strings," Phys. Lett. B 103 (1981) 207
54. C. Rovelli, "Loop quantum gravity," Living Rev. Rel. 1 (1998) 1 [arXiv:gr-qc/9710008]
55. C. Rovelli *Loop Quantum Gravity* gr-qc/9606088 [=Phys.Lett. B380 (1996) 257] "The Immirzi Parameter In Quantum General Relativity", gr-qc/9505012 [=Phys.Lett. B360 (1995) 7]
56. C. Rovelli and L. Smolin, "Spin networks and quantum gravity," Phys. Rev. D 52 (1995) 5743 [arXiv:gr-qc/9505006]
57. C. Rovelli and L. Smolin, "Discreteness of area and volume in quantum gravity," Nucl. Phys. B 442 (1995) 593 [Erratum-ibid. B 456 (1995) 753] [arXiv:gr-qc/9411005]
58. J. Scherk and J. H. Schwarz, "Dual Models For Nonhadrons," Nucl. Phys. B 81 (1974) 118
59. L. Smolin, "How far are we from the quantum theory of gravity?," arXiv:hep-th/0303185
60. L. Smolin, "Quantum gravity with a positive cosmological constant," arXiv:hep-th/0209079
61. K. S. Stelle, "Renormalization Of Higher Derivative Quantum Gravity," Phys. Rev. D **16** (1977) 953
62. A. Strominger and C. Vafa, "Microscopic Origin of the Bekenstein-Hawking Entropy," Phys. Lett. B 379 (1996) 99 [arXiv:hep-th/9601029]
63. L. Susskind, "The World as a hologram," J. Math. Phys. 36 (1995) 6377 [arXiv:hep-th/9409089]
64. G. 't Hooft, "Dimensional Reduction In Quantum Gravity," arXiv:gr-qc/9310026
65. G. 't Hooft and M. J. Veltman, "One Loop Divergencies In The Theory Of Gravitation," Annales Poincare Phys. Theor. A 20 (1974) 69
66. T. Thiemann, "Introduction to modern canonical quantum general relativity," arXiv:gr-qc/0110034
67. T. Thiemann, "Anomaly-free formulation of non-perturbative, four-dimensional Lorentzian quantum gravity," Phys. Lett. B 380 (1996) 257 [arXiv:gr-qc/9606088].
"Gauge field theory coherent states (GCS). I: General properties," Class. Quant. Grav. 18 (2001) 2025 [arXiv:hep-th/0005233]

68. V. G. Turaev and O. Y. Viro, "State Sum Invariants Of 3 Manifolds And Quantum 6j Symbols," Topology 31 (1992) 865
"Quantum Theory of Gravity. II. The Manifestly Covariant Theory", Phys. Rev. 162, 1195-1239 (1967)
69. E. Witten, *Anti de Sitter Space and Holography*, `hep-th/9802150`;
Anti-de Sitter space, thermal phase transition and confinement in gauge theories `hep-th/9803131`
70. E. Witten, "(2+1)-Dimensional Gravity As An Exactly Soluble System," Nucl. Phys. B 311 (1988) 46
71. E. Witten, "A note on the Chern-Simons and Kodama wavefunctions," arXiv:gr-qc/0306083

Introduction to Quantum-Gravity Phenomenology

G. Amelino-Camelia

Dipartimento di Fisica, Universitá di Roma "La Sapienza" and Sez. Roma1 INFN, P.le Moro 2, 00185 Roma, Italy
amelino@roma1.infn.it

After a brief review of the first phase of development of Quantum-Gravity Phenomenology, I argue that this research line is now ready to enter a more advanced phase: while at first it was legitimate to resort to heuristic order-of-magnitude estimates, which were sufficient to establish that sensitivity to Planck-scale effects can be achieved, we should now rely on detailed analyzes of some reference test theories. I illustrate this point in the specific example of studies of Planck-scale modifications of the energy/momentum dispersion relation, for which I consider two test theories. Both the photon-stability analyzes and the Crab-nebula synchrotron-radiation analyzes, which had raised high hopes of "beyond-Plankian" experimental bounds, turn out to be rather ineffective in constraining the two test theories. Examples of analyzes which can provide constraints of rather wide applicability are the so-called "time-of-flight analyzes", in the context of observations of gamma-ray bursts, and the analyzes of the cosmic-ray spectrum near the GZK scale.

1 From "Quantum Gravity Beauty Contests" to Quantum Gravity Phenomenology

The "quantum-gravity problem" has been studied for more than 70 years [1] assuming that no guidance could be obtained from experiments. This in turn led to the assumption that the most promising path toward the solution of the problem would be the construction and analysis of very ambitious theories (some would call them "theories of everything"), capable of solving at once all of the issues raised by the coexistence of gravity and quantum mechanics. In other research areas the availability of experimental data challenging the current theories encourages theorists to propose phenomenological models which solve the experimental puzzles, even when some aspects of the models are not fully satisfactory from a conceptual perspective. Often those apparently unsatisfactory models turn out to provide an important starting point

G. Amelino-Camelia: *Introduction to Quantum-Gravity Phenomenology*, Lect. Notes Phys. **669**, 59–100 (2005)
www.springerlink.com

for the identification of the correct (and conceptually satisfactory) theoretical description of the new phenomena. But in this quantum-gravity research area, since there was no experimental guidance, it was inevitable for theorists to be tempted into trying to identify the correct theoretical framework relying exclusively on some criteria of conceptual compellingness. Of course, tempting as it may seem, this strategy would not be acceptable for a scientific endeavor. Even the most compelling and conceptually satisfying theory could not be adopted without experimental confirmation.

The mirage (occasionally mentioned at relevant seminars) that one day within an ambitious quantum-gravity theory one might derive from first principles a falsifiable prediction for the mundane realm of doable experiments could give some "scientific legitimacy" to these research programmes, but this possibility never materialized. It may indeed be just a mirage. There are several occasions when a debate between advocates of different ambitious quantum-gravity theories shapes up in a way similar to the discussion between advocates of different religions. And often in the media the different approaches are compared on the basis of the "support" they have in the community: one says "the most popular approach to the quantum gravity problem" rather than "the approach that has had better success reproducing experimental results". So, it would seem, the Quantum Gravity problem is to be solved by an election, by a beauty contest, by a leap of faith.

Over the last few years a growing number of research groups have attempted to tackle the quantum-gravity problem with an approach which is more consistent with the traditional strategy of scientific work. Simple (in some cases even simple-minded) non-classical pictures of spacetime are being analyzed with strong emphasis on their observable predictions. Certain classes of experiments have been shown to have extremely high sensitivity to some non-classical features of spacetime. We now even have (see later) some first examples of experimental puzzles whose solution is being sought also within simple ideas involving non-classical pictures of spacetime. The hope is that by trial and error, both on the theory side and on the experiment side, we might eventually stumble upon the first few definite (experimental!) hints on the quantum-gravity problem.

Quantum gravity phenomenology requires of course a combination of theory and experiments. It does not adopt any particular prejudice concerning the structure of spacetime at short distances (in particular, "string theory" [2, 3], "loop quantum gravity" [4, 5, 6, 7] and "noncommutative geometry" [8, 9] are seen as equally deserving mathematical-physics programmes), but of course must follow as closely as possible the few indications that these ambitious quantum-gravity theories provide. One here is guided by the expectation that quantum-gravity research should proceed just in the old-fashioned way of scientific work: through small incremental steps starting from what we know and combining mathematical-physics studies with experimental studies to reach deeper and deeper layers of understanding of the problem at hand (in this case the short-distance structure of spacetime and the laws that govern it).

The most popular quantum-gravity approaches, such as string theory and loop quantum gravity, could be described as "top-to-bottom approaches" since they start off with some key assumption about the structure of spacetime at scales that are some 17 orders of magnitude beyond the scales presently accessible experimentally, and then they should work their way back to the realm of doable experiments. With "quantum gravity phenomenology" I would like to refer to all studies that are intended to contribute to a "bottom-to-top approach" to the quantum-gravity problem. Since the problem at hand is really difficult (arguably the most challenging problem ever faced by the physics community) it appears likely that the two complementary approaches might combine in a useful way: for the "bottom-to-top approach" it is important to get some guidance from the (however tentative) indications emerging from the "top-to-bottom approaches", while for "top-to-bottom approaches" it might be useful to be alerted by quantum-gravity phenomenologists with respect to the type of new effects that could be most stringently tested experimentally (it is hard for "top-to-bottom approaches" to obtain a complete description of "real" physics, but perhaps it would be possible to dig out predictions on some specific spacetime features that appear to deserve special attention in light of the corresponding experimental sensitivities).

In these lectures I give a "selected-topics" introduction to this "Quantum Gravity Phenomenology". I will in particular stress that, while the first few years of work in this area, the "dawn" of quantum-gravity phenomenology [10], were necessarily based on rather preliminary analyzes, with the only objective of establishing the point that Planck-scale sensitivity could be achieved in some cases, we should now gear up for a more "mature" phase of work on quantum-gravity phenomenology, in which the development and analysis of some carefully crafted test theories takes center stage.

2 Quantum Gravity Phenomenology

In this section I describe the key objectives of quantum-gravity phenomenology and sketch out its strategy in the search of the first manifestation of a quantum property of spacetime. I also start introducing my argument that we should now move from the "dawn" of quantum-gravity phenomenology to a more "mature" quantum-gravity phenomenology, in which a key role is played by the development and analysis of some carefully crafted test theories.

2.1 Planck-Scale Quantum Properties of Spacetime

The first step for the identification of experiments relevant for quantum gravity is of course the identification of the characteristic scale of this new physics. This is a point on which we have relatively robust guidance from theories and theoretical arguments: the characteristic scale at which non-classical properties of spacetime physics become large (as large as the classical properties they compete with) should be the Planck length $L_p \sim 10^{-35}$ m (or equivalently its inverse, the Planck scale $E_p \sim 10^{28}$ eV). The key challenge for quantum-gravity phenomenology must be the one of establishing ways to provide sensitivity to Planck-scale non-classical properties of spacetime.

I will call "quantum" properties of spacetime all effects which represent departures from a classical picture of spacetime. This is after all what is commonly done in the literature, where authors often use the name "quantum properties of spacetime" because of the expectation that some of the familiar features of quantization, which showed up everywhere else in physics, should eventually also play a role in the description of spacetime. There is no guarantee that the non-classical properties of spacetime will take the shape of some sort of proper spacetime quantization. But, as long as this is understood, the use of the spacetime-quantization terminology does no arm.

Of course, the search of a solution of the quantum-gravity problem can benefit also from other types of experimental insight, and therefore the scopes of quantum-gravity phenomenology must go even beyond its key quantum-spacetime challenge. In particular, quantum gravity should also provide a consistent description of the quantum properties of particles in presence of strong (or anyway non-negligible) classical gravity fields. This type of context at the "Interface of Quantum and [classical] Gravitational Realms" [11] has been the subject of a rather sizeable literature for several decades. When quantum properties of spacetime are not relevant for the analysis the insight one can gain for the quantum-gravity problem is of more limited impact, but it is of course still valuable. Indeed a valuable debate on the fate of the Equivalence Principle in quantum gravity was ignited already in the mid 1970s with the renowned experiment performed by Colella, Overhauser and Werner [12]. That experiment has been followed by several modifications and refinements (often labelled "COW experiments" from the initials of the scientists involved in the first experiment) all probing the same basic physics, i.e. the validity of the Schrödinger equation

$$\left[-\left(\frac{\hbar^2}{2\,M_I}\right)\boldsymbol{\nabla}^2 + M_G\,\phi(\boldsymbol{r})\right]\psi(t,\boldsymbol{r}) = i\,\hbar\,\frac{\partial\,\psi(t,\boldsymbol{r})}{\partial t} \tag{1}$$

for the description of the dynamics of matter (with wave function $\psi(t,\boldsymbol{r})$) in presence of the Earth's gravitational potential $\phi(\boldsymbol{r})$. [In (1) M_I and M_G denote the inertial and gravitational mass respectively.]

The COW experiments exploit the fact that the Earth's gravitational potential puts together the contributions of a very large number of particles and as a result, in spite of its per-particle weakness, the overall gravitational field is large enough[1] to introduce observable effects. The relevance of these experiments for the debate on the Equivalence Principle will not be discussed here,

[1] Actually the effect turns out to be observably large because of a double "amplification": the first, and most significant, amplification is the mentioned coherent addition of gravitational fields generated by the particles that compose the Earth, the second amplification [13] involves the ratio between the wavelength of the

but has been discussed in detail by several authors (see, e.g., [14, 15, 16]). I here just bring to the reader's attention a recent experiment which appears to indicate a violation of the Equivalence Principle [17] (but the reliability of this experimental result is still being debated), and some ideas for intriguing new experiments [13, 18] of the COW type. I should also mention for completeness the related work on the interplay between classical general relativity and quantum mechanics of non-gravitational degrees of freedom reported in [19, 20].

Another possibility that, even though it is in contrast with the idea of Planck-scale quantum properties of spacetime, deserves some exploratory effort by those working in quantum-gravity phenomenology is the one of scenarios in which the standard estimate of the quantum-gravity scale as the Planck scale turns out to be too pessimistic. There is (at present) no compelling argument in support of the idea that the quantum-gravity scale should be effectively lowered, but this possibility cannot be excluded. In particular, some recent studies [21] found a mechanism that would allow to lower significantly the quantum-gravity energy scale, several orders of magnitude below the Planck energy scale. This mechanism relies of the hypothesis of "large extra dimensions" which is not in any way "natural" (not even in the eyes of the scientists who proposed it), but it can be used to provide an example of a workable scenario for a low scale of quantum-gravity effects.

For the rest of these lectures I will however focus on what I described as the key challenge: the search of Planck-scale quantum properties of spacetime.

2.2 Identification of Experiments

Unfortunately, in spite of more than 70 years of theory work on the quantum-gravity problem, and a certain proliferation of theoretical frameworks being considered, there is only a small number of physical effects that have been considered in the quantum-gravity literature. Moreover, most of these effects concern strong-gravity contexts, such as black-hole physics and big-bang physics, which are exciting at the level of conceptual analysis and development of formalism, but are not very promising for the actual (experimental) discovery of manifestations of non-classical properties of spacetime.

While it is likely that the largest quantum-gravity effects should be present in large-curvature situations, it only takes a little reasoning to realize that we should give priority to quantum-gravity effects that modify our description of (quasi-)Minkowski spacetime. The effects will perhaps be smaller than, say, in black hole physics (in some aspects of black hole physics quantum-gravity effects might be as large as classical physics effects), but we are likely to be better off considering quasi-Minkowski spacetimes, for which the quality of the data we can obtain is extremely high.

---

particles used in the COW experiments and some larger length scales involved in the experimental setup.

In the analysis of flat-spacetime processes, involving particles with energies that are inevitably much lower than the Planck energy scale, we will have to deal with a large suppression of quantum-gravity effects, a suppression which is likely to take the form of some power of the ratio between the Planck length and the wavelength of the particles involved. The presence of these suppression factors on the one hand reduces sharply our chances of finding quantum-gravity effects, but on the other hand simplifies the problem of identifying promising experimental contexts, since these experimental contexts must enjoy very special properties which would not go easily unnoticed. For laboratory experiments even an optimistic estimate of these suppression factors leads to a suppression of order 10^{-16}, which one obtains by assuming (probably already using some optimism) that at least some quantum-gravity effects are only linearly suppressed by the Planck length, and taking as particle wavelength the shorter wavelengths we are able to produce ($\sim 10^{-19}$ m). In astrophysics (which however limits one to "observations" rather than "experiments") particles of shorter wavelength are being studied, but even for the highest energy cosmic rays, with energy of $\sim 10^{20}$ eV and therefore wavelengths of $\sim 10^{-27}$ m, a suppression of the type L_p/λ would take values of order 10^{-8}. It is mostly as a result of this type of considerations that traditional quantum-gravity reviews considered the possibility of experimental studies of Planck-scale effects with unmitigated pessimism [22].

However, the presence of large suppression factors surely cannot suffice for drawing any conclusions. Even just looking within the subject of particle physics we know that certain types of small effects can be studied, as illustrated by the example of the remarkable limits obtained on proton instability. Outside of fundamental physics more success stories of this type are easily found. Think for example of brownian motion, where some unobservably small micro-processes lead to an effect which is observable on macroscopic scales.

It is hard but clearly not impossible to find experimental contexts in which there is effectively an amplification of the small effect one intends to study. The prediction of proton decay within certain grandunified theories of particle physics is really a small effect, suppressed by the fourth power of the ratio between the mass of the proton and grandunification scale, which is only three orders of magnitude smaller than the Planck scale. In spite of this horrifying suppression, of order $[m_{proton}/E_{gut}]^4 \sim 10^{-64}$, with a simple idea we have managed to acquire full sensitivity to the new effect: the proton lifetime predicted by grandunified theories is of order 10^{39} s and "quite a few" generations of physicists should invest their entire lifetimes staring at a single proton before its decay, but by managing to keep under observation a large number of protons (think for example of a situation in which 10^{33} protons are monitored) our sensitivity to proton decay is dramatically increased. In that context the number of protons is the dimensionless quantity that works as "amplifier" of the new-physics effect. Similar considerations explain the success of brownian-motion studies already a century ago.

We should therefore focus our attention [10] on experiments which have something to do with spacetime structure and that host an ordinary-physics dimensionless quantity large enough that (if we are "lucky") it could amplify the extremely small effects we are hoping to discover. So there is clearly a first level of analysis in which one identifies experiments with this rare quality, and a second level of analysis in which one tries to establish whether indeed the candidate "amplifier" could possibly amplify effects connected with spacetime structure.

2.3 Prehistory of Quantum Gravity Phenomenology

Clearly a good phenomenological programme must be able to falsify theories. Although it is already noteworthy that some candidate quantum-gravity effects could at all be looked for in the data, this would not be so significant if we were not able to use these data to constrain the work of theorists, to falsify some theoretical pictures. The fact that, toward the end of the 1990s, it was convincingly argued that this could be done brought the idea of "Planck-scale tests" to center stage in quantum-gravity research. The fact that we could plausibly gain insight on Planck-scale physics is now widely acknowledged in the quantum-gravity community. Up to 1997 or 1998 there had already been some works on the possibility to find experimental evidence of some Planck-scale effects, but the relevant data analyzes did not in reverse have the capability to falsify any quantum-gravity picture and the relevant research remained at the margins of the mainstream quantum-gravity literature.

A first example of these works of the "prehistory of quantum-gravity phenomenology" is provided by a certain type of investigation of Planck-scale departures from CPT symmetry using the neutral-kaon and the neutral-B systems [23, 24, 25, 26]. These pioneering works were based on the realization that in the relevant neutral-meson systems a Planck-scale departure from CPT symmetry could in principle be amplified; in particular, the neutral-kaon system hosts the peculiarly small mass difference between long-lived and the short-lived kaons $|M_L - M_S|/M_{L,S} \sim 7{\cdot}10^{-15}$. The quantum-gravity picture usually advocated in these studies is the one of a variant of the string-theory picture, which relies on noncritical strings, in the so-called "Liouville" approach [25, 27]. This is an ambitious attempt for a theory of everything, which, while based on an appealing view of the quantum-gravity problem, is for the most part untreatable, at least with current techniques. The departures from CPT symmetry cannot be derived from the theory, but one can provide tentative evidence that the structure of the theory should accommodate such departures. As a result one is forced to set up a multi-parameter phenomenology which looks for the new effects, but a negative outcome of the experiments could not be used to falsify the framework which is at the root of the analysis.

Similar remarks apply to the other pioneering studies reported in [28], which find their original motivation in some aspects of "string field theory" [28].

Also the string-field-theory formalism is very ambitious and too hard to handle. A multi-parameter phenomenology is necessarily set up [28], and a negative outcome of the experiments could not be used to falsify the framework which is at the root of the analysis. Besides the falsifiability issue, this phenomenology may not appeal to many quantum-gravity researchers because it mainly focuses on the "Standard Model Extension", whose key assumption [29] is the renormalizability of the underlying field theory. The assumption of renormalizability limits one to effects that area described in terms of operators of dimension 4 and lower, whereas most quantum-gravity researchers expect Planck-scale-suppressed effects described in terms of operators of dimension 5 and higher.

A third equally-deserving entry in my list of pioneers of the "prehistory of quantum-gravity phenomenology" is the work reported in [30, 31] which explored the general issue of how certain effectively stochastic properties of spacetime would affect the evolution of quantum-mechanical states. The guiding idea was that stochastic processes could provide an effective description of quantum spacetime processes. The implications of these stochastic properties for the evolution of quantum-mechanical states were modelled in [30, 31] via the formalism of "primary state diffusion", but only rather crude models turned out to be treatable. As also emphasized by the authors, the crudeness of the models is such that all conclusions are to be considered as tentative at best, and this is one more instance in which a negative outcome of the experiments could not be used to falsify the framework which is at the root of the analysis.

The three research lines I discussed in this subsection as examples of "prehistory of quantum-gravity phenomenology" showed convincingly that the possibility of stumbling upon an experimental manifestation of Planck-scale effects could not be excluded. On the other hand they proved to be insufficient for the birth a genuine, fully articulated, phenomenological programme. In that regard their key common limitation was the mentioned fact that it appeared that the relevant experiments could not falsify the relevant theories. Moreover, it often appeared that these studies were establishing that some not-much-studied quantum-gravity approaches could lead to observable effects, as a way to distinguish them from the most popular quantum-gravity ideas which would remain untestable. Of course, the interest of the community grew when it became apparent that a rather large variety of quantum-gravity ideas could lead to observable effects (and could be falsified). A sizeable community now works under the assumption that the presence of observably-large quantum-gravity effects is not a peculiar feature of some out-of-mainstream quantum-gravity approaches: it is a property of most quantum-gravity approaches, including some of the most popular ones.

2.4 The Dawn of Quantum Gravity Phenomenology

The research lines discussed in the previous subsections had been establishing that it was not inconceivable to use data within our reach (inevitably involving particles with energies much lower than the Planck energy scale) to find evidence of a Planck-scale effect. However, while these research did ignite a lively interest by some experimentalists (see, e.g., [32, 33]), they went largely unnoticed by mainstream quantum-gravity research. As stressed above, this was likely due to the fact that they were incomplete proposals from the viewpoint of phenomenology, because the test theories could not be really falsified, and because the relevant test theories were all outside mainstream quantum-gravity research, so that the fact that Planck scale effects could be seen appeared to be a peculiar property of out-of-mainstream theories. On the other hand, clearly those research lines were starting set the stage for a wider and more developed phenomenological effort, which indeed came to existence toward the end of the 1990s. When, indeed starting toward the end of the 1990s, the case for falsifiability of some Planck-scale models started to be built, and first evidence of testability of mainstream quantum-gravity proposals emerged, a corresponding quick growth of interest emerged in the community. This is perhaps best illustrated by comparing authoritative quantum-gravity reviews published up to the mid 1990s (see, e.g., [22]) and the corresponding reviews published over the last couple of years [34, 35, 36, 37].

This "dawn" of quantum-gravity phenomenology has revolved around a growing number of experimental contexts in which Planck-scale effects are being sought. Among the most popular such proposals let me mention, as a few noteworthy examples, the studies of in-vacuo dispersion using gamma-ray astrophysics [38, 39], studies of laser-interferometric limits on quantum-gravity effects [40, 41, 42, 43, 44], studies of the role of quantum-gravity effects in the determination of the energy-momentum-conservation threshold conditions for certain particle-physics processes [45, 46, 47, 48, 49], and studies of the role of quantum gravity in the determination of particle-decay amplitudes [50, 51].

The idea of looking for Planck-scale departures from CPT symmetry continues to be pursued, but in that context we are still lacking an analysis showing how a quantum-gravity model could be falsified on the basis of such CPT studies. This is essentially due to some technical challenges in establishing what exactly happens to CPT symmetry within a given Planck-scale picture. It is often easy to see that CPT is affected, but one is then unable to establish how it is affected.

As I shall stress again later, among all these research lines a special role in the development of quantum-gravity phenomenology is being played by studies of the role of quantum-gravity effects in the determination of the energy-momentum-conservation threshold conditions for certain particle-physics processes. In fact, in these studies we have stumbled upon a first example of experimental puzzle whose solution could plausibly be sought within quantum-gravity phenomenology. This of course marked an important milestone for quantum-gravity phenomenology. The relevant context is the one of the process of photopion production, $p + \gamma \rightarrow p + \pi^0$, which, as discussed later in these lectures, plays a crucial role in the analysis of the cosmic-ray spectrum. An apparent "anomaly" in the observed cosmic-ray spectrum could be naturally described in terms of Planck-scale effects. Of course, it is not unlikely that this "anomaly" might fade away, as better data on cosmic rays become available, but it is nonetheless an important sign of maturity for quantum-gravity phenomenology that some data invite interpretation as a possible manifestation of Planck-scale physics. Chances are the first few such "candidate anomalies" will turn out to be incorrect, but eventually one lucky instance could be encountered.

2.5 The Maturity of Quantum Gravity Phenomenology: Test Theories

The fact that quantum-gravity phenomenology is already being considered in attempts to solve present experimental puzzles is indeed a clear indication of progress toward the maturity of the field, but in many respects the field is still rather immature. The first challenge for quantum-gravity phenomenologists was to establish convincingly that there is a chance to test Planck-scale effects, and this type of argument can legitimately be based on intuitive order-of-magnitude analyzes. However, at this point a rather large community acknowledges that quantum-gravity phenomenology has a chance, so the first challenge was successfully overcome, and we must now shift gear. There is very little more to be gained through rudimental back-of-the-envelope analyzes. The standards of quantum-gravity phenomenology must be raised to the ones adopted in other branches of phenomenology, such as particle-physics phenomenology.

In these lectures I shall in particular emphasize the importance of adopting some reference test theories. If an effect is described only vaguely, without the support of an associated test theory, then the experimental limits that can be claimed are of correspondingly uncertain significance. As it has happened in the recent quantum-gravity-phenomenology literature, different authors may end up claiming different limits on "the same effect" simply because they are actually adopting different test theories and therefore they are truly analyzing different effects. This type of phenomenology clearly would not help us gain any insight on Planck-scale physics. The main task of phenomenology is to provide to the theorists working at the development of the theories information on what is and what is not consistent with experimental data. Phenomenology essentially provides some boundaries within which formal theorists are then forced to work. A theory which would predict effects inconsistent with some data is abandoned. But if this boundaries are not clearly drawn, if the experimental limits are placed on "effects" which are not rigorously defined within the context of a test theory, then they are correspondingly useless for the development of theories.

The discussion here reported in Sect. 4 will illustrate this point in a specific context.

3 Some Candidate Quantum-Gravity Effects

Before focusing, in the next section, on an example of "quantum-gravity-phenomenology exercise", it seems appropriate to list at least a few of the candidate quantum-gravity effects that find motivation in the literature.

Testing these effects will be the main task of quantum-gravity phenomenology. While here I will discuss these effects at a rather rudimentary and intuitive level, so that my remarks would apply to a variety of approaches to the quantum-gravity problem, clearly in each theory these effects may take a different form, and in setting up a phenomenology for these effects it will be crucial to develop some corresponding test theories.

In providing motivation for the study of these effects I could use a large variety of arguments; however, I find preferable to show that these effects can be motivated already on the basis of the most plausible of all hypotheses concerning the quantum-gravity problem: the hypothesis that some of the incarnations of the "quantum" idea (such as discretization and noncommutativity of observables) should find place also in the description of spacetime.

3.1 Planck-Scale Departures from Lorentz Symmetry

Perhaps the most debated possibility for a quantum spacetime, possibly intended as Planck-scale discrete or Planck-scale noncommutative spacetime, is the one of Planck-scale departures from Lorentz symmetry.

The continuous symmetries of a spacetime reflect of course the structure of that spacetime. Ordinary Lorentz symmetry is governed by the single scale that sets the structure of classical Minkowski spacetime, the speed-of-light scale c. If one introduces additional structure in a flat spacetime its symmetries will be accordingly affected. This is particularly clear for some simple ideas concerning a Planck-scale discretization of spacetime [52]. Continuous symmetry transformations are clearly at odd with a discrete network of points.

For different reasons, Lorentz symmetry is also often at odds with spacetime noncommutativity. In particular, it appears that in certain cases the noncommutativity length scale [53] (possibly the Planck scale), in addition to c, affects the laws of transformation between inertial observers, and infinitesimal symmetry transformations are actually described in terms of the new language of Hopf algebras [54, 55], rather than by the Poincaré Lie algebra. The type of spacetime quantization provided by noncommutativity may therefore lead to a corresponding "symmetry quantization": the concept of Lie-algebra symmetry is in fact replaced by the one of Hopf-algebra symmetry.

In a large number of recent studies of noncommutative spacetimes it has indeed been found that the Lie-algebra Poincaré symmetries are either broken to a smaller symmetry Lie algebra or deformed into Hopf-algebra symmetries.

For what concerns the idea of spacetime discretization the most developed quantum-gravity picture is the one of Loop Quantum Gravity, which does not predict a rigid discrete network of spacetime points, but introduces discretization in a more sophisticated way: the spectra of areas and volumes are discretized, while spacetime points loose all possible forms of identity. It appears that even this more advanced form of discretization is incompatible with classical Lorentz symmetry; in fact, a growing number of loop-quantum-gravity studies has been reporting [56, 57] evidence of Planck-scale departures from Lorentz symmetry (although the issue remains subject to further scrutiny).

3.2 Planck-Scale Departures from CPT Symmetry

The fact that our low-energy[2] observations are consistent with CPT symmetry is not a miracle: as codified by the CPT theorem, a Lorentz-invariant local quantum field theory is inevitably CPT invariant. The fact that quantum gravity, the "unification" of gravity and quantum theory, invites us to consider Planck-scale departures from Lorentz symmetry (as stressed above) and Planck-scale departures from locality (as natural in a discrete-spacetime theory) opens the door for Planck-scale departures from CPT symmetry.

While this general argument is rather robust, it is not always easy to establish what is the fate of CPT symmetry in a given quantum-gravity approach. For example in Loop Quantum Gravity the analysis of (the various alternative ideas) on coupling ordinary particles to gravity has not yet advanced to the point of allowing a robust description of C transformations. On the other hand there are examples in which some progress in the analysis of CPT transformations has been achieved and evidence of departures from CPT symmetry is found. This is for example the case of κ-Minkowski noncommutative spacetime, where one can clearly see [58] a modification of P transformations.

Since I am not considering CPT symmetry in the remainder of these lectures let me mention here that, besides the neutral-kaon and neutral-B systems, already briefly discussed in the previous section, also neutrinos are being considered [59, 60, 61] as a possible laboratory for tests of Planck-scale departures from CPT symmetry.

---

[2] Since it is often obvious from the context, I will sometimes avoid specifying "energies that are low with respect to the Planck energy scale" and simply write "low-energy". With "high-energy particles" instead I will not mean "particles with energy higher than the Planck energy scale" (a situation which we never encounter), but rather the case of particles with energy rather close to (but still lower than) the Planck scale.

3.3 Distance Fuzziness

As one last example of effect that one could plausibly expect from quantum gravity, I consider here "distance fuzzyness". Once again one is exploring the possibility that some ideas from quantum theory would apply to spacetime physics. A key characteristic of quantum theory is the emergence of uncertainties, and one might expect that the "distance observable" would also be affected by uncertainties. Actually various heuristic arguments suggest that for such a "distance observable" the uncertainties might be more pervasive: in ordinary quantum theory one is still able to measure sharply any given observable, though at the cost of renouncing all information on a conjugate observable, but it appears plausible that a quantum-gravity "distance observable" would be affected by irreducible uncertainties. Most authors would consider a $\delta D \geq L_p$ relation, meaning that the uncertainty in the measurement of distances could not be reduced below the Planck-length level, but measurability bounds of other forms, generically of the type $\delta D \geq f(D, L_p)$ (with f some function such that $f(D,0) = 0$) are also being considered.

The presence of such an irreducible measurement uncertainty could be significant in various contexts. For example, these ideas would suggest that the noise levels in the readout of a laser interferometer would receive an irreducible (fundamental) contribution from quantum-gravity effects. Interferometric noise can in principle be reduced to zero in classical physics, but already the inclusion in the analysis of the ordinary quantum properties of matter introduces an extra noise contribution with respect to classical physics. A fundamental Planck-scale-induced uncertainty in the length of the arms of the interferometer would introduce another source of noise, and the possibility of testing this idea is presently under investigation (see, e.g., [40, 41, 42, 43]).

3.4 Aside on the Differences Between Systematic and Nonsystematic Effects

It is perhaps useful to stress the differences between systematic and nonsystematic Planck-scale effects, which I can illustrate using the the type of effects discussed in the previous parts of this section.

An example of systematic effect is given by the departures from Lorentz symmetry encountered in certain noncommutative spacetimes (on which I shall return later in these lectures). There the Planck-scale structure of spacetime can introduce a systematic dependence of the speed of photons on their wavelength. After a journey of duration T the difference between the expected position of the photon and the Planck-scale-corrected position could take the form $\Delta x \sim T\delta v \sim cTL_p/\lambda$, where λ is the photon wavelength.

If we instead focus on how "distance fuzziness" could affect the propagation of photons it is natural to expect that a group of photons would all travel the same average distance in a given time T (and this average distance is still given by cT), but for each individual photon the distance travelled might be slightly different from the average, as a result of distance fuzziness. This is an example of nonsystematic effect. Just to be more specific let us imagine that distance fuzziness effectively introduces a Planck-length uncertainty in position per each Planck time of travel. Then the final position uncertainty would be of the type $\Delta x \sim \sqrt{cTL_p}$. The square root here (assuming a random-walk-type description) is the result of the fact that nonsystematic effects do not add linearly, but rather according to rules familiar in the analysis of stochastic processes.

4 A Prototype Exercise: Modified Dispersion Relations

In the previous sections I tried to give a general, but rough, description of how one works in quantum-gravity phenomenology. I will now discuss a specific example of quantum-gravity-phenomenology study, with the objective of illustrating in more detail the type of challenges that one must face and some strategies that can be used. The example I am focusing on is the one of Planck-scale modifications of the energy-momentum dispersion relation, which has been extensively studied from the quantum-gravity-phenomenology perspective. I will start with a brief description of how modified dispersion relations arise[3] in the study of noncommutative spacetimes and in the study of loop quantum gravity. I will then discuss some test theories which might play a special role in the development of the relevant phenomenology. And finally I will discuss some observations in astrophysics which can be used to set limits on the test theories.

4.1 Modified Dispersion Relations in Canonical Noncommutative Spacetime

The noncommutative spacetimes in which modifications of the dispersion relation are being most actively considered all fall within the following rather general parametrization of noncommutativity of the spacetime coordinates:

$$[x_\mu, x_\nu] = i\theta_{\mu\nu} + i\rho^\beta_{\mu\nu} x_\beta \ . \tag{2}$$

It is convenient to first focus on the special case $\rho = 0$, the "canonical noncommutative spacetimes"

$$[x_\mu, x_\nu] = i\theta_{\mu\nu} \ . \tag{3}$$

[3] I discuss noncommutative spacetimes and the Loop Quantum Gravity approach, which are the best understood Planck-scale frameworks in which it appears that the dispersion relation is Planck-scale modified. But other types of intuitions about the quantum-gravity problem may lead to modified dispersion relations, including some realizations of the idea of "spacetime foam" [38, 62, 27], which allow an analogy with the laws of particle propagation in a thermal environment [38, 62, 63].

Of course, the natural first guess for introducing dynamics in these spacetimes is a quantum-field-theory formalism. And indeed, for the special case $\rho = 0$, an approach to the construction of a quantum field theory has been developed rather extensively [64, 65]. While most aspects of these field theories closely resemble their commutative-spacetime counterparts, a surprising feature that emerges is the so-called "IR/UV mixing" [64, 65, 66]: the high-energy sector of the theory does not decouple from the low-energy sector. Connected with this IR/UV mixing is the type of modified dispersion relations that one encounters in field theory on canonical noncommutative spacetime, which in general take the form

$$m^2 \simeq E^2 - \boldsymbol{p}^2 + \frac{\alpha_1}{p^\mu \theta_{\mu\nu} \theta^{\nu\sigma} p_\sigma} + \alpha_2 m^2 \ln \left(p^\mu \theta_{\mu\nu} \theta^{\nu\sigma} p_\sigma\right) + \dots \,, \tag{4}$$

where the α_i are parameters, possibly taking different values for different particles (the dispersion relation is not "universal"), that depend on various aspects of the field theory, including its field content and the nature of its interactions. The fact that this dispersion relation can be singular in the infrared is a result of the IR/UV mixing. A part of the infrared singularity could be removed by introducing (exact) supersymmetry, which typically leads to $\alpha_1 = 0$.

The implications of this IR/UV mixing for dynamics are still not fully understood, and there is still justifiable skepticism [67] toward the correctness of the type of field-theory construction adopted so far. I think it is legitimate to even wonder whether a field-theoretic formulation of the dynamics is at all truly compatible with the canonical spacetime noncommutativity. The Wilson decoupling between IR and UV degrees of freedom is a crucial ingredient of most applications of field theory in physics, and it is probably incompatible with canonical noncommutativity: the associated uncertainty principle of the type $\Delta x_\mu \Delta x_\nu \geq \theta_{\mu\nu}$ implies that it is not possible to probe short distances (small, say, Δx_1) without probing simultaneously the large-distance regime ($\Delta x_2 \geq \theta_{2,1}/\Delta x_1$).

In any case, the presence of modified dispersion relations in canonical noncommutative spacetime should be expected, since Lorentz symmetry is "broken" by the tensor $\theta_{\mu\nu}$. An intuitive characterization of this Lorentz-symmetry breaking can be obtained by looking at wave exponentials. The Fourier theory in canonical noncommutative spacetime is based [68] on simple wave exponentials $e^{ip^\mu x_\mu}$ and from the $[x_\mu, x_\nu] = i\theta_{\mu\nu}$ noncommutativity relations one finds that

$$e^{ip^\mu x_\mu} e^{ik^\nu x_\nu} = e^{-\frac{i}{2} p^\mu \theta_{\mu\nu} k^\nu} e^{i(p+k)^\mu x_\mu} \,, \tag{5}$$

i.e. the Fourier parameters p_μ and k_μ combine just as usual, but there is the new ingredient of the overall θ-dependent phase factor. The fact that momenta combine in the usual way reflects the fact that the transformation rules for energy-momentum from one (inertial) observer to another are still the familiar, undeformed, Lorentz transformation rules. However, the product of wave exponentials depends on $p^\mu \theta_{\mu\nu} k^\nu$; it depends on the "orientation" of the energy-momentum vectors p^μ and k^ν with respect to the $\theta_{\mu\nu}$ tensor. The $\theta_{\mu\nu}$ tensor plays the role of a background that identifies a preferred class of inertial observers[4]. Different particles can be affected by the presence of this background in different ways, leading to the emergence of different dispersion relations. All this is consistent with indications of the mentioned popular field theories in canonical noncommutative spacetimes.

4.2 Modified Dispersion Relations in κ-Minkowski Noncommutative Spacetime

In canonical noncommutative spacetimes Lorentz symmetry is "broken" and there is growing evidence that Lorentz symmetry breaking occurs for most choices of the tensors θ and ρ. It is at this point clear, in light of several recent results, that the only way to preserve Lorentz symmetry is the choice $\theta = 0 = \rho$, i.e. the case in which there is no noncommutativity and one is back to the familiar classical commutative Minkowski spacetime. When noncommutativity is present Lorentz symmetry is usually broken, but recent results suggest that for some special choices of the tensors θ and ρ Lorentz symmetry might be deformed, in the sense of the recently proposed "doubly-special relativity" scenario [53], rather than broken. In particular, this appears to be the case for the Lie-algebra κ-Minkowski [54, 55, 58, 71, 72, 73] noncommutative spacetime $(l, m = 1, 2, 3)$

$$[x_m, t] = \frac{i}{\kappa} x_m \ , \qquad [x_m, x_l] = 0 \ . \tag{6}$$

κ-Minkowski is a Lie-algebra spacetime that clearly enjoys classical space-rotation symmetry; moreover, at least in a Hopf-algebra sense (see, e.g., [72]), κ-Minkowski is invariant under "noncommutative translations". Since I am focusing here on Lorentz symmetry, it is particularly noteworthy that in κ-Minkowski boost transformations are necessarily modified [72]. A first hint of this comes from the necessity of a deformed law of composition of momenta, encoded in the so-called coproduct (a standard structure for a Hopf algebra). One can see this clearly by considering the Fourier tranform. It turns out [58, 71] that in the κ-Minkowski case the correct formulation of the Fourier theory requires a suitable ordering prescription for wave exponentials. From

[4] Note that these remarks apply to canonical noncommutative spacetimes as studied in the most recent (often String-Theory inspired) literature, in which $\theta_{\mu\nu}$ is indeed simply a tensor (for a given observer, an antisymmetric matrix of numbers). I should stress however that the earliest studies of canonical noncommutative spacetimes (see [69] and follow-up work) considered a $\theta_{\mu\nu}$ with richer mathematical properties, notably with nontrivial algebra relations with the spacetime coordinates. In that earlier, and more ambitious, setup it is not obvious that Lorentz symmetry would be broken: the fate of Lorentz symmetry may depend on the properties (dynamics?) attributed to $\theta_{\mu\nu}$.

$$: e^{ik^\mu x_\mu} :\equiv e^{ik^m x_m} e^{ik^0 x_0} \,, \tag{7}$$

as a result of $[x_m, t] = ix_m/\kappa$ (and $[x_m, x_l] = 0$), it follows that the wave exponentials combine in a nontrivial way:

$$(: e^{ip^\mu x_\mu} :)(: e^{ik^\nu x_\nu} :) =: e^{i(p\dot{+}k)^\mu x_\mu} : \quad . \tag{8}$$

The notation "$\dot{+}$" here introduced reflects the behavior of the mentioned "coproduct" composition of momenta:

$$p_\mu \dot{+} k_\mu \equiv \delta_{\mu,0}(p_0 + k_0) + (1 - \delta_{\mu,0})(p_\mu + e^{\lambda p_0} k_\mu) \,. \tag{9}$$

As argued in [53] the nonlinearity of the law of composition of momenta might require an absolute (observer-independent) momentum scale, just like upon introducing a nonlinear law of composition of velocities one must introduce the absolute observer-independent scale of velocity c. The inverse of the noncommutativity scale λ should play the role of this absolute momentum scale. This invites one to consider the possibility [53] that the transformation laws for energy-momentum between different observers would have two invariants, c and λ, as required in "doubly-special relativity" [53].

On the basis of (9) one is led [54, 55, 58] to the following result for the form of the energy/momentum dispersion relation

$$\left(\frac{2}{\lambda}\sinh\frac{\lambda m}{2}\right)^2 = \left(\frac{2}{\lambda}\sinh\frac{\lambda E}{2}\right)^2 - e^{\lambda E}\boldsymbol{p}^2 \,, \tag{10}$$

which for low momenta takes the approximate form

$$m^2 \simeq E^2 - \boldsymbol{p}^2 - \lambda E \boldsymbol{p}^2 \,. \tag{11}$$

Actually, the precise form of the dispersion relation may depend on the choice of ordering prescription for wave exponentials [72] ((10) follows form (7)), and this point deserves further scrutiny, but even setting aside this annoying ordering ambiguity, there appear to be severe obstructions [71, 72] for a satisfactory formulation of a quantum field theory in κ-Minkowski. The techniques that were rather straightforwardly applied for the construction of field theory in canonical noncommutative spacetime do not appear to be applicable in the κ-Minkowski case. It is not implausible that the "virulent" κ-Minkowski noncommutativity may require some departures from a standard field-theoretic setup.

4.3 Modified Dispersion Relation in Loop Quantum Gravity

Loop Quantum Gravity is one of the most ambitious approaches to the quantum-gravity problem, and its understanding is still in a relatively early stage. As presently understood, Loop Quantum Gravity predicts an inherently discretized spacetime [4, 5, 6], and this occurs in a rather compelling way: it is not that one introduces by hand an a priori discrete background spacetime; it is rather a case in which a background-independent analysis ultimately leads, by a sort of self-consistency, to the emergence of discretization. There has been much discussion recently, prompted by the studies [38, 56, 57], of the possibility that this discretization might lead to broken Lorentz symmetry and a modified dispersion relation. Although there are cases in which a discretization is compatible with the presence of continuous classical symmetries [74, 75], it is of course natural, when adopting a discretized spacetime, to put Lorentz symmetry under careful scrutiny. Arguments presented in [56, 57] suggest that Lorentz symmetry might indeed be broken in Loop Quantum Gravity.

Moreover, very recently Smolin, Starodubtsev and I proposed [76] (also see the follow-up study in [77]) a mechanism such that Loop Quantum Gravity would be described at the most fundamental level as a theory that in the flat-spacetime limit admits deformed Lorentz symmetry, in the sense of the "doubly-special relativity" scenario [53]. Our argument originates from the role that certain quantum symmetry groups ("q-deformed algebras") have in the Loop-Quantum-Gravity description of spacetime with a cosmological constant, and observing that in the flat-spacetime limit (the limit of vanishing cosmological constant) these quantum groups might not contract to a classical Lie algebra, but rather contract to a quantum (Hopf) algebra.

All these studies point to the presence of a modified dispersion relation, although different arguments lead to different intuition for the form of the dispersion relation. A definite result might have to wait for the solution of the well-known "classical-limit problem" of Loop Quantum Gravity. We are presently unable to recover from this full quantum-gravity theory the limiting case in which the familiar quantum-field-theory description of particle-physics processes in a classical background spacetime applies. Some recent studies appear to suggest [78] that in the same contexts in which departures from Lorentz symmetry may be revealed one should adopt a density-matrix formalism, and then the whole picture would collapse to the familiar Lorentz-invariant quantum-field-theory description in contexts involving both relatively low energies and relatively low boosts with respect to the center-of-mass frame (e.g. the particle-physics collisions studied at several particle accelerators).

4.4 Some Issues Relevant for the Proposal of Test Theories

In these lectures I am attempting to stress in particular the need for quantum-gravity phenomenology to establish that some Planck-scale pictures of spacetime are falsifiable and the need to rely on some reference test theories in the analysis of the progress of experimental limits as better data become available.

The results I briefly summarized in the previous three subsections provide a good indication of the fact that falsifiability is within reach. Both in the analysis of noncommutative spacetimes and in the analysis of Loop Quantum

Gravity there are a few open issues which do not at present allow us to describe in detail a falsifiable prediction, but, in light of the progress achieved over the last few years, the nature of these open issue encourages us to think that we should soon achieve falsifiability.

In the meantime quantum-gravity phenomenology will have to push the limits on the type of effects that are emerging, and this effort should be guided by the objective of falsifiability. The analyzes should avoid relying on assumptions which are likely to prove incorrect for the relevant formalisms. And when the open issues confront us with some alternative scenarios, the phenomenology work should attempt to "cover all possibilities", i.e. push the limits in all directions that are still compatible with the present understanding of the formalism (so that when the ambiguity is resolved there will be a class of data ready for comparison with theory).

In this situation it will be crucial for the development of the phenomenological programme to adopt some suitably structured test theories, which should also be useful for bridging the gap between the experimental data and the, still incomplete, falsifiability analysis. These test theories should be our common language in assessing the progresses made in improving the sensitivity of experiments, a language that must also be suitable for access from the side of those working at the development of the quantum-gravity/quantum-spacetime theories.

As we contemplate the challenge of developing such carefully-balanced test theories it is important to observe that the most robust part of the results I summarized in the previous three subsections is clearly the emergence of modified dispersion relations. Therefore if one could set up experiments testing directly the dispersion relation the resulting limits would have wide applicability. In principle one could investigate the form of the dispersion relation directly by making simultaneous measurements of energy and space-momentum; however, it is easy to see that achieving Planck-scale sensitivity in such a direct test is well beyond our capabilities.

Useful test theories on which to base the relevant phenomenology must therefore combine the ingredient of the dispersion relation with other ingredients. As I shall discuss in greater detail later in this section, there are three key issues for this test-theory development:

(i) in presence of the modified dispersion relation should we still assume the validity of of the relation $v = dE/dp$ between the speed of a particle and its dispersion relation? (here dE/dp is the derivative of the function $E(p)$ which of course is implicitly introduced through the dispersion relation)
(ii) in presence of the modified dispersion relation should we still assume the validity of the standard law of energy-momentum conservation?
(iii) in presence of the modified dispersion relation which formalism should be adopted for the description of dynamics?

The fact that these are key issues is also a consequence of the type of data that we expect to have access to, as I shall discuss later in this section.

Unfortunately on these three key points the quantum-spacetime pictures which are providing motivation for the study of Planck-scale modifications of the dispersion relation, reviewed in the previous three subsections, are not providing much guidance yet.

For example, in Loop Quantum Gravity, while we do have evidence that the dispersion relation should be modified, we do not yet have a clear indication concerning whether the law of energy-momentum conservation should also be modified and we also cannot yet robustly establish whether the relation $v = dE/dp$ should be preserved. Moreover, perhaps most importantly, some recent studies [78] invite us to consider the possibility that in the same contexts in which Loop-Quantum-Gravity departures from Lorentz symmetry may be revealed one should also adopt a density-matrix formalism, and then the whole picture might reduce to the familiar Lorentz-invariant quantum-field-theory description in contexts involving both relatively low energies and relatively low boosts with respect to the center-of-mass frame. We should therefore be prepared for surprises in the description of dynamics.

Similarly in the analysis of noncommutative spacetimes we are close to establishing in rather general terms that some modification of the dispersion relation is inevitable, but other aspects of the framework have not yet been clarified. While most of the literature for canonical noncommutative spacetimes assumes [64, 65] that the law of energy-momentum conservation should not be modified, most of the literature for κ-Minkowski spacetime argues in favor of a modification (perhaps consistent with the corresponding doubly-special-relativity criteria [53]) of the law of energy-momentum conservation. There is also still no consensus on the relation between speed and dispersion relation, and particularly in the κ-Minkowski literature some departures from the $v = dE/dp$ relation are actively considered [79, 80, 81, 82]. And concerning the formalism to be used for the description of dynamics in a noncommutative spacetime, while everybody's first guess is the field-theoretic formalism, the fact that attempts at a field theory formulation encounter so many difficulties (the IR/UV mixing for the canonical-noncommutative spacetime case and the even more pervasive shortcomings of the proposals for a field theory in κ-Minkowski) must invite one to consider possible alternative formulations of dynamics.

Clearly the situation on the theory side invites us to be prudent: if a given phenomenological picture relies on too many assumptions on Planck-scale physics it is likely that it might not reproduce any of the mentioned quantum-gravity and/or quantum-spacetime models (when these models are eventually fully understood they will give us their own mix of Planck-scale features, which is difficult to guess at the present time). On the other hand it is necessary for the robust development of a phenomenology to adopt well-defined test theories. Without reference to a well-balanced set of test theories it is impossible to compare the limits obtained in different experimental contexts, since each experimental context may require different "ingredients" of Planck-scale physics. And it is of course meaningless to compare limits obtained on the basis of different conjectures for the Planck-scale regime, especially since our very limited understanding of the Planck scale regime should encourage us to be prudent when formulating any assumption (virtually any assumption about the Planck-scale regime could turn out to be incorrect, once theories are better understood).

4.5 A Test Theory for Pure Kinematics

The majority (see, e.g., [39, 45, 46, 47, 48]) of studies concerning Planck-scale modifications of the dispersion relation adopt the phenomenological formula

$$m^2 \simeq E^2 - \mathbf{p}^2 + \eta \mathbf{p}^2 \left(\frac{E^n}{E_p^n} \right) + O\left(\frac{E^{n+3}}{E_{QG}^{n+1}} \right) , \qquad (12)$$

with real η of order 1 and integer n. This formula is compatible with some of the results obtained in the Loop-Quantum-Gravity approach and reflects the results obtained in κ-Minkowski and other noncommutative spacetimes (but, as mentioned, in the special case of canonical noncommutative spacetimes one encounters a different, infrared singular, dispersion relation).

As stressed above, on the basis of the status on the theory side, a prudent approach in combining the dispersion relation with other ingredients is to be favored. Since basically all experimental situations will involve some aspects of kinematics that go beyond the dispersion relation (while there are some cases in which the dynamics, the interactions among particles, does not play a role), and taking into account the mentioned difficulties in establishing what is the correct formalism for the description of dynamics[5] at the Planck scale, most authors prefer to prudently combine the dispersion relation with other "purely kinematical" aspects of Planck-scale physics.

Already in the first studies [38, 83] that proposed a phenomenology based on (12) it was assumed that even at the Planck scale the familiar description of "group velocity", obtained from the dispersion relation according to $v = dE/dp$, should hold[6].

[5] I am here using the expression "dynamics at the Planck scale" with some license. Of course, in our phenomenology we will not be sensitive directly to the dynamics at the Planck scale. However, as I discuss in greater detail in the next subsection, if the arguments that encourage the use of new descriptions of dynamics at the Planck scale are correct, then a sort of "order of limits problem" clearly arises. Our experiments will involve energies much lower than the Planck scale, and we know that in the infrared limit the familiar formalism with field-theoretic description of dynamics and Lorentz invariance will hold. So we would need to establish whether experiments that are sensitive to Planck-scale departures from Lorentz symmetry could also be sensitive to Planck-scale departures from the field-theoretic description of dynamics. Since we still know very little about this alternative descriptions of dynamics a prudent approach, avoiding any assumption about the description of dynamics is certainly preferable.

[6] As mentioned, this assumption is not guaranteed to apply to the formalisms of interest, and indeed several authors have considered alternatives [79, 80, 81, 82].

In other works motivated by the analysis reported in [38] another key kinematical feature was introduced: starting with the studies reported in [45, 46, 47, 48] the dispersion relation (12) and the velocity relation $v = dE/dp$ were combined with the assumption that the law of energy-momentum conservation should not be modified at the Planck scale, so that, for example, in a $a + b \to c + d$ particle-physics process one would have

$$E_a + E_b = E_c + E_d \ , \tag{13}$$

$$\boldsymbol{p}_a + \boldsymbol{p}_b = \boldsymbol{p}_c + \boldsymbol{p}_d \ . \tag{14}$$

Most authors work within this kinematic framework assuming "universality" of the dispersion relation (on which I shall return in the next subsection), but some have allowed [49, 85] for a particle-dependence and possibly an helicity-polarization dependence of the coefficients η, n of the dispersion relation.

The elements I described in this subsection compose a quantum-gravity phenomenology test theory that has already been widely considered in the literature, even though it was never previously characterized in detail. In the following I will refer to this test theory as the "AEMNS test theory"[7], and I will assume that experimental bounds on this test theory should be placed by using only the following assumptions:

(AEMNS.1) the dispersion relation is of the form

$$m^2 \simeq E^2 - \boldsymbol{p}^2 + \eta_a \boldsymbol{p}^2 \left(\frac{E^{n_a}}{E_p^{n_a}}\right) + O\left(\frac{E^{n+3}}{E_{QG}^{n+1}}\right) \ , \tag{15}$$

where η_a and n_a can in general take different values for different particles and for different helicities-polarizations of the same particle (the index spans over particles and helicities/polarizations);

(AEMNS.2) the velocity of a particle can be obtained from the dispersion relation using $v = dE/dp$;

While the studies advocating alternatives to $v = dE/dp$ rely of a large variety of arguments (some more justifiable some less), in my own perception [84] a key issue here is whether quantum gravity leads to a modified Heisenberg uncertainty principle, $[x, p] = 1 + F(p)$. Assuming a Hamiltonian description is still available, $v = dx/dt \sim [x, H(p)]$, the relation $v = dE/dp$ essentially follows from $[x, p] = 1$. But if $[x, p] \neq 1$ then $v = dx/dt \sim [x, H(p)]$ would not lead to $v = dE/dp$. And there is much discussion in the quantum-gravity community of the possibility of modifications of the Heisenberg uncertainty principle at the Planck scale.

[7] I am using "AEMNS" on the basis of the initials of the names of the authors in [38], which first proposed a phenomenology based on the dispersion relation (12). But as mentioned the full test theory, as presently used in most studies, only emerged gradually in follow-up work. In particular, there was no discussion of energy-momentum conservation in [38]. Unmodified energy-momentum conservation was introduced in [45, 46, 47, 48].

(AEMNS.3) the law of energy-momentum conservation is not modified;

(AEMNS.4) nothing is assumed about dynamics (i.e. the analysis of this test theory will be limited to classes of experimental data that involve pure kinematics, without any role for dynamics).

4.6 The Minimal AEMNS Test Theory

On the basis of the results we presently have, at least within Loop Quantum Gravity and the study of certain noncommutative spacetimes, the formulation of the "AEMNS test theory" discussed in the previous subsection is general enough that we should expect it to be relevant for most quantum-spacetime pictures in which Lorentz symmetry is broken. Since, as mentioned, the analysis of these models is still in progress we might eventually be forced to consider further generalizations, including a possible modifications of the energy-momentum conservation law[8] and/or of the law $v = dE/dp$.

Rather then prematurely considering this possible even wider parameter space, at present it is more reasonable to focus on a "minimal version" of the AEMNS test theory, in which the universality of the dispersion relation is assumed. It is in fact natural to expect that universality will be preserved in most of the relevant quantum-spacetime pictures. Moreover, as long as this minimal AEMNS test theory is not ruled out, clearly its more general nonuniversal version discussed in the previous subsection cannot be ruled. And it will be very useful to have a simple two-parameter space to use as reference in keeping track of the gradual improvement of the experimental sensitivities.

In order to be self-contained let me list here the characteristics of this "minimal AEMNS test theory":

(minAEMNS.1) the dispersion relation is of the form

$$m^2 \simeq E^2 - \boldsymbol{p}^2 + \eta \boldsymbol{p}^2 \left(\frac{E^n}{E_p^n} \right) + O\left(\frac{E^{n+3}}{E_{QG}^{n+1}} \right) , \tag{16}$$

where η and n are universal (same value for every particle and for both helicities/polarizations of a given particle);

(minAEMNS.2) the velocity of a particle can be obtained from the dispersion relation using $v = dE/dp$;

(minAEMNS.3) the law of energy-momentum conservation is not modified;

[8] In a doubly-special-relativity framework with modified dispersion relation the law of energy-momentum conservation must be *correspondingly modified* in order to preserve the equivalence of inertial observers [53]. Instead in a framework in which Lorentz symmetry is actually broken, with the associated loss of equivalence among inertial observers, modifications of the dispersion relation are in principle compatible with an unmodified law of energy-momentum conservation. Still, even in the broken-Lorentz-symmetry case, a modification of the law of energy-momentum conservation is possible.

(minAEMNS.4) nothing is assumed about dynamics (i.e. the analysis of this test theory will be limited to classes of experimental data that involve pure kinematics, without any role for dynamics).

4.7 A Test Theory Based on Low-Energy Effective Field Theory

The AEMNS test theory has the merit of relying only on a relatively small network of assumptions on kinematics, without assuming anything about the role of the Planck scale in dynamics. However, of course, this justifiable prudence turns into a severe limitation on the class of experimental contexts which can be used to constrain the parameters of the test theory. It is in fact rather rare that a phenomenological analysis can be completed without using (more or less explicitly) any aspects of the interactions among the particles involved in the relevant processes. The desire to be able to analyze a wider class of experimental contexts is therefore providing motivation for the development of test theories more ambitious than the AEMNS test theory, with at least some elements of dynamics. This is understandable but, in light of the situation on the theory side, work with one of these more ambitious test theories should proceed with the awareness that there is a high risk that it may turn out that none of the quantum-gravity approaches which are being pursued is reflected in the test theory.

One reasonable possibility to consider, when the urge to analyze data that involve some contamination from dynamics cannot be resisted, is the one of describing dynamics within the framework of low-energy effective field theory. In this subsection I want to discuss a test theory which is indeed based on low-energy effective field theory, and has emerged from the work recently reported in [86] (which is rooted in part in the earlier [56]).

Before a full characterization of this test theory I should first warn the reader that there might be some severe limitations for the applicability of low-energy effective field theory to the investigation of Planck-scale physics, especially when departures from Lorentz symmetry are present.

A significant portion of the quantum-gravity community is in general, justifiably, skeptical about the results obtained using low-energy effective field theory in analyzes relevant for the quantum-gravity problem. After all the first natural prediction of low-energy effective field theory in the gravitational realm is a value of the energy density which is some 120 orders of magnitude greater than allowed by observations[9]. Somewhat related to this "cosmological constant problem" is the fact that a description of possible Planck-scale departures from Lorentz symmetry within effective field theory can only be

[9] And the outlook of low-energy effective field theory in the gravitational realm does not improve much through the observation that exact supersymmetry could protect from the emergence of any energy density. In fact, Nature clearly does not have supersymmetry at least up to the TeV scale, and this would still lead to a natural prediction of the cosmological constant which is some 60 orders of magnitude too high.

developed with a rather strongly pragmatic attitude; in fact, while one can introduce Planck-scale suppressed effects at tree level, one expects that loop corrections would typically lead to inadmissibly large departures from ordinary Lorentz symmetry. Indeed some studies, notably [87, 88], have shown mechanisms such that, within an effective-field-theory formulation, loop effects would lead to inadmissibly large departures from ordinary Lorentz symmetry, which could be avoided only by introducing a large level of fine tuning.

It is rather amusing that alongside with numerous researchers who are skeptical about any results obtained using low-energy effective field theory in analyzes relevant for the quantum-gravity problem, there are also quite a few researchers interested in the quantum-gravity problem who are completely serene in assuming that all quantum-gravity effects should be describable in terms of effective field theory in low-energy situations. The (quasi-)rationale behind this assumption is that field theory works well at low energies without gravity, and quantum gravity of course must reproduce field theory in an appropriate limit, so one might expect that at least at low energies the quantum-gravity effects could be described in the language of field theory as correction terms to be added to standard lagrangians.

I feel that, while of course an effective-field-theory description may well turn out to be correct in the end, the a priori assumption that a description in terms of effective low-energy field-theory should work is rather naive. If the arguments that encourage the use of new descriptions of dynamics at the Planck scale are correct, then a sort of "order of limits problem" clearly arises. Our experiments will involve energies much lower than the Planck scale, and we know that in some limit (a limit that characterizes our most familiar observations) the field-theoretic description and Lorentz invariance will hold. So we would need to establish whether experiments that are sensitive to Planck-scale departures from Lorentz symmetry could also be sensitive to Planck-scale departures from the field-theoretic description of dynamics. As an example, let me mention the possibility (not unlikely in a context which is questioning the fate of Lorentz symmetry) that quantum gravity would admit a field-theory-type description only in reference frames in which the process of interest is essentially occurring in its center of mass (no "Planck-large boost" [89] with respect to center-of-mass frame). The field theoretic description could emerge in a sort of "low-boost limit", rather than the expected low-energy limit. The regime of low boosts with respect the center-of-mass frame is often indistinguishable with respect to the low-energy limit. For example, from a Planck-scale perspective, our laboratory experiments (even the ones conducted at, e.g. CERN, DESY, SLAC...) are both low-boost (with respect to the center of mass frame) and low-energy. However, the "UHE cosmic-ray paradox", for which a quantum-gravity origin has been conjectured (see later), occurs in a situation where all the energies of the particles are still tiny with respect to the Planck energy scale, but the boost with respect to the center-of-mass frame (as measured by the ratio E/m_{proton} between the proton energy and the proton mass) could be considered to be "large" from a Planck-scale perspective ($E/m_{proton} \gg E_p/E$).

These concerns are strengthen by looking at the literature available on the quantum pictures of spacetime that provide motivation for the study of modified dispersion relations, which usually involve either noncommutative geometry or Loop Quantum Gravity, where, as mentioned, the outlook of a low-energy effective-field-theory description is not encouraging. The construction of field theories in noncommutative spacetimes requires the introduction of several new technical tools, which in turn lead to the emergence of several new physical features, even at low energies. I guess that these difficulties arise from the fact that a spacetime characterized by an uncertainty relation of the type $\delta x\, \delta y \geq \theta(x, y)$ never really behaves has a classical spacetime, not even at very low energies. In fact, some low-energy processes will involve soft momentum exchange in the x direction (large δx) which however is connected with the exchange of a hard momentum in the y direction ($\delta y \geq \theta/\delta x$), and this feature cannot be faithfully captured by our ordinary field-theory formalisms. In the case of canonical noncommutative spacetimes one does obtain a plausible-looking field theory [65], but the results actually show that it is not possible to rely on an ordinary effective low-energy quantum-field-theory description. In fact, the "IR/UV mixing" [64, 65, 66] is such that the high-energy sector of the theory does not decouple from the low-energy sector, and this in turn affects very severely [66] the outlook of analyzes based on an ordinary effective low-energy quantum-field-theory description. For other (non-canonical) noncommutative spacetimes we are still struggling in the search of a satisfactory formulation of a quantum field theory [71, 72], and it is at this point legitimate to suspect that such a formulation of dynamics in those spacetimes does not exist.

Incidentally let me observe that the issues encountered in dealing with the IR/UV mixing may be related to my concerns about the large-boost limit of quantum gravity. In a theory with IR/UV mixing nothing peculiar might be expected for, say, a collision between two photons both of MeV energy, but the boosted version of this collision, where one photon has, say, energy of 100 TeV and the other photon has energy of 10^{-2} eV, could be subject to the IR/UV mixing effects, and be essentially untreatable from a low-energy effective-field-theory perspective.

And noncommutative spacetimes are not the only cases where an ordinary field-theory description may be inadequate. As mentioned, the assumption of availability of an ordinary effective low-energy quantum-field-theory description finds also no support in Loop Quantum Gravity. Indeed, so far, in Loop Quantum Gravity all attempts to find a suitable limit of the theory which can be described in terms of a quantum-field-theory in background spacetime have failed. And on the basis of the recent results of [78] it appears plausible that in several contexts in which one would naively expect a low-energy field theory description Loop Quantum Gravity might instead require a density-matrix description.

Of course, in phenomenology this type of concerns can be set aside, since one is primarily looking for confrontation with experimental data, rather than theoretical prejudice. It is clearly legitimate to set up a test theory exploring the possibility of Planck-scale departures from Lorentz symmetry within the formalism of low-energy effective field theory. But one must then keep in mind that the implications for most quantum-gravity research lines of the experimental bounds obtained in this way might be very limited. This will indeed be the case if we discover that, as some mentioned preliminary results suggest, the limit in which the full quantum-gravity theory reproduces a description in terms of effective field theory in classical spacetime is also the limit in which the departures from Lorentz symmetry must be neglected.

Having provided this long warning, let me now proceed to a characterization of the test theory which I see emerging from the works reported in [56, 86]. These studies explore the possibility of a linear-in-L_p modification of the dispersion relation

$$m^2 \simeq E^2 - \boldsymbol{p}^2 + \eta \boldsymbol{p}^2 L_p E \ , \tag{17}$$

i.e. the case $n = 1$ of (12). The key assumption in [86, 56] is that such modifications of the dispersion relation should be introduced consistently with an effective low-energy field-theory description of dynamics. The implications of this assumption were explored in particular for fermions and photons. It became quickly clear that in such a setup universality cannot be assumed, since one must at least accommodate a polarization dependence for photons: in the field-theoretic setup it turns out that when right-circular polarized photons satisfy the dispersion relation $E^2 \simeq p^2 + \eta_\gamma p^3$ then necessarily left-circular polarized photons satisfy the "opposite sign" dispersion relation $E^2 \simeq p^2 - \eta_\gamma p^3$. For spin-1/2 particles the analysis reported in [86] does not necessarily suggest a similar helicity dependence, but of course in a context in which photons experience such a complete correlation of the sign of the effect with polarization it would be awkward to assume that instead for electrons the effect is completely helicity independent. One therefore introduces two independent parameters η_+ and η_- to characterize the modification of the dispersion relation for electrons.

In the following I will refer to this test theory as the "GPMP test theory" (from the initials of the authors of [86, 56]), and I will assume that experimental bounds on this test theory should be placed by using only the following assumptions:

(GPMP.1) for right-circular polarized photons are governed by the dispersion relation

$$m^2 \simeq E^2 - \boldsymbol{p}^2 + \eta_\gamma \boldsymbol{p}^2 \left(\frac{E}{E_p}\right) \ , \tag{18}$$

while left-circular polarized photons are governed by the dispersion relation

$$m^2 \simeq E^2 - \boldsymbol{p}^2 - \eta_\gamma \boldsymbol{p}^2 \left(\frac{E}{E_p}\right) \ ; \tag{19}$$

(GPMP.2) for fermions the dispersion relation takes the form

$$m^2 \simeq E^2 - \boldsymbol{p}^2 + \eta_R^a \boldsymbol{p}^2 \left(\frac{E}{E_p}\right) , \tag{20}$$

in the positive-helicity case, while for negative-helicity fermions

$$m^2 \simeq E^2 - \boldsymbol{p}^2 + \eta_L^a \boldsymbol{p}^2 \left(\frac{E}{E_p}\right) ; \tag{21}$$

the index a here reflecting a possible particle dependence;

(GPMP.3) dynamics is described in terms of effective low-energy field theory.

4.8 The Minimal GPMP Test Theory

As in the case of the AEMNS test theory, while a large parameter space should be considered in order to achieve full generality, it appears wise to first focus the phenomenology on a reduced version of the test theory, reflecting some natural physical assumptions. As in the case of the AEMNS test theory, a reduced two-parameter space would be ideal for the first-level description of the gradual improvement of the experimental sensitivities. As usual, once the reduced version of the test theory is falsified one can contemplate its possible generalization (if the developments on the pure-theory side still justify such an effort from the perspective of falsification of the theories).

In introducing a reduced GPMP test theory I believe that a key point of naturalness comes from the observation that the effective-field-theory setup imposes for photons a modification of the dispersion relation which has the same magnitude for both polarizations but opposite sign: it is then natural to give priority to the hypothesis that for fermions a similar mechanism would apply, i.e. the modification of the dispersion relation should have the same magnitude for both signs of the helicity, but have a correlation between the sign of the helicity and the sign of the dispersion-relation modification. This would correspond to the natural-looking assumption that the Planck-scale effects are such that in a beam composed of randomly selected particles the average speed in the beam is still governed by ordinary special relativity (the Planck scale effects average out summing over polarization/helicity).

A further "natural" reduction of the parameter space is achieved by assuming that all fermions are affected by the same modification of the dispersion relation.

The reduced GPMP test theory that emerges from this requirements is perhaps the most natural among the possible two-parameter reduction of the GPMP test theory. In the following I refer to this reduced GPMP test theory as the "minimal GPMP test theory"[10], characterized by the following ingredients:

[10] Whereas for the AEMNS test theory there is clearly only one obvious way to set up the reduction to a two-dimensional parameter space, within the GPMP test

(minGPMP.1) right-circular polarized photons are governed by the dispersion relation

$$m^2 \simeq E^2 - \boldsymbol{p}^2 + \eta_\gamma \boldsymbol{p}^2 \left(\frac{E}{E_p}\right) , \qquad (22)$$

while left-circular polarized photons are governed by the dispersion relation

$$m^2 \simeq E^2 - \boldsymbol{p}^2 - \eta_\gamma \boldsymbol{p}^2 \left(\frac{E}{E_p}\right) ; \qquad (23)$$

(minGPMP.2) for fermions the dispersion relation takes the form

$$m^2 \simeq E^2 - \boldsymbol{p}^2 + \eta_f \boldsymbol{p}^2 \left(\frac{E}{E_p}\right) , \qquad (24)$$

in the positive-helicity case, while for negative-helicity fermions

$$m^2 \simeq E^2 - \boldsymbol{p}^2 - \eta_f \boldsymbol{p}^2 \left(\frac{E}{E_p}\right) , \qquad (25)$$

with the same value of η_f for all fermions;

(minGPMP.3) dynamics is described in terms of effective low-energy field theory.

4.9 Derivation of Limits from Analysis of Gamma-Ray Bursts

Both in the AEMNS test theory and in the GPMP test theory one expects a wavelength dependence of the speed of photons, by combining the modified dispersion relation and the relation $v = dE/dp$. At "intermediate energies" ($m < E \ll E_p$) this velocity law will take the form

$$v \simeq 1 - \frac{m^2}{2E^2} + \eta \frac{n+1}{2} \frac{E^n}{E_p^n} . \qquad (26)$$

Whereas in ordinary special relativity two photons ($m = 0$) emitted simultaneously would always reach simultaneously a far-away detector, according to (26) two simultaneously-emitted photons should reach the detector at different times if they carry different energy. Moreover, in the case of the GPMP test theory even photons with the same energy would arrive at different times if they carry different polarization. In fact, while the minimal AEMNS test theory assumes universality, and therefore a formula of this type would apply to photons of any polarization, in the GPMP test theory, as mentioned,

theory, with its automatic polarization dependence of the effects for photons, one could probably envision more than one way to set up the reduction to a two-dimensional parameter space. In a certain sense the two-dimensional parameter space on which I propose to focus for the AEMNS test theory is *the* minimal AEMNS test theory, whereas here I am proposing *a* minimal GPMP test theory.

the sign of the effect is correlated with polarization. As a result, while the AEMNS test theory is best tested by comparing the arrival times of particles of different energies, the GPMP test theory is best tested by considering the highest-energy photons available in the data and looking for a sizeable spread in times of arrivals (which one would then attribute to the different speeds of the two polarizations).

This time-of-arrival-difference effect can be significant [38, 39] in the analysis of short-duration gamma-ray bursts that reach us from cosmological distances. For a gamma-ray burst it is not uncommon that the time travelled before reaching our Earth detectors be of order $T \sim 10^{17}$ s. Microbursts within a burst can have very short duration, as short as 10^{-3} s (or even 10^{-4} s), and this means that the photons that compose such a microburst are all emitted at the same time, up to an uncertainty of 10^{-3} s. Some of the photons in these bursts have energies that extend at least up to the GeV range. For two photons with energy difference of order $\Delta E \sim 1$ GeV a $\eta \Delta E/E_p$ speed difference over a time of travel of 10^{17} s would lead to a difference in times of arrival of order

$$\Delta t \sim \eta T \Delta \frac{E}{E_p} \sim 10^{-2} s \;, \tag{27}$$

which is significant (the time-of-arrival differences would be larger than the time-of-emission differences within a single microburst).

For the AEMNS test theory the Planck-scale-induced time-of-arrival difference could be revealed [38, 39] upon comparison of the "average arrival time" of the gamma-ray-burst signal (or better a microburst within the burst) in different energy channels. The GPMP test theory would be most effectively tested by looking for a dependence of the time-spread of the bursts that grows with energy (at low energies the effect is anyway small, so the polarization dependence is ineffective, whereas at high energies the effect may be nonnegligible and an overall time-spread of the burst could result from the dependence of speed on polarization).

The sensitivities achievable [90] with the next generation of gamma-ray telescopes, such as GLAST [90], could allow to test very significantly (26) in the case $n = 1$, by possibly pushing the limit on η far below 1 (whereas the effects found in the case $n = 2$, $|\eta| \sim 1$ are too small for GLAST). Whether or not these levels of sensitivity to the Planck-scale effects are actually achieved may depend on progress in understanding other aspects of gamma-ray-burst physics. In fact, the Planck-scale-effect analysis would be severely affected if there were poorly understood at-the-source correlations between energy of the photons and time of emission. In the recent [91] it was emphasized that it appears that one can infer such an energy/time-of-emission correlation from available gamma-ray-burst data. The studies of Planck-scale effects will be therefore confronted with a severe challenge of "background/noise removal". At present it is difficult to guess whether this problem can be handled successfully. We do have a good card to play in this analysis: the Planck-scale picture predicts that the times of arrival should depend on energy in a way that grows in exactly linear way with the distance of the source. One may therefore hope that, once a large enough sample of gamma-ray bursts (with known source distances) becomes available, one might be able disentangle the Planck-scale propagation effect from the at-the-source background.

An even higher sensitivity to possible Planck-scale modifications of the velocity law could be achieved by exploiting the fact that, according to current models [92], gamma-ray bursters should also emit a substantial amount of high-energy neutrinos. Some neutrino observatories should soon observe neutrinos with energies between 10^{14} and 10^{19} eV, and one could, for example, compare the times of arrival of these neutrinos emitted by gamma-ray bursters to the corresponding times of arrival of low-energy photons. One could use this strategy to test rather stringently[11] the case of (26) with $n = 1$, an even perhaps gain some access to the investigation of the case $n = 2$.

In order to achieve these sensitivities with neutrino studies once again some technical and conceptual challenges should be overcome. Also this type of analysis requires an understanding of gamma-ray bursters good enough to establish whether there are typical at-the-source time delays. The analysis would loose much of its potential if one cannot exclude some systematic tendency of gamma-ray bursters to emit high-energy neutrinos with, say, a certain delay with respect to microbursts of photons. But also in this case one could hope to combine several observations from gamma-ray bursters at different distances in order to disentangle the possible at-the-source effect.

4.10 Derivation of Limits from Analysis of UHE Cosmic Rays

With a given dispersion relation and a given rule for energy-momentum conservation one has a complete "kinematic scheme" for the analysis of the kinematical requirements for particle production in collisions or decay processes. Both the AEMNS test theory and the GPMP test theory involve modified dispersion relations and unmodified laws of energy-momentum conservation (the fact that the law of energy-momentum conservation is not modified is explicitly among the ingredients of the AEMNS test theory, while in the GPMP test theory it follows from the adoption of low-energy effective field theory).

In these lectures I am not discussing in detail the case of modified dispersion relations introduced within a "doubly-special relativity" scenario [53, 70]. For clarity of the presentation, I thought it would be best to limit to two the number of test theories I consider. Test theories for doubly-special relativity scenarios with modified dispersion relations are under consideration (see, e.g., [93]), but I will not make room for them here. It is appropriate however to

[11] Note however that in an analysis mixing the properties of different particles the sensitivity that can be achieved will depend strongly on whether universality of the modification of the dispersion relation is assumed. For example, for the GPMP test theory a comparison of times of arrival of neutrinos and photons could only introduce a bound on some combination of the dispersion-relation-modification parameters for the photon and for the neutrino sectors.

stress here that the assumption of modified dispersion relations and unmodified laws of energy-momentum conservation is inconsistent with the doubly-special relativity principles, since it inevitably [53] gives rise to a preferred class of inertial observers. A doubly-special relativity scenario with modified dispersion relations must necessarily have a modified law of energy-momentum conservation.

Going back to the AEMNS and GPMP test theories which I am considering, in this subsection I want to stress that combining a modified dispersion relation with unmodified laws of energy-momentum conservation one naturally finds a modification of the threshold requirements for certain particle-producing processes. Let us for example consider, from the AEMNS perspective, the dispersion relation (12), with $n = 1$, in the analysis of a collision between a soft photon of energy ϵ and a high-energy photon of energy E that creates an electron-positron pair: $\gamma\gamma \rightarrow e^+e^-$. For given soft-photon energy ϵ, the process is allowed only if E is greater than a certain threshold energy E_{th} which depends on ϵ and m_e^2. For $n = 1$, combining (12) with unmodified energy-momentum conservation, this threshold energy (assuming $\epsilon \ll m_e \ll E_{th} \ll E_p$) is estimated as

$$E_{th}\epsilon + \eta \frac{E_{th}^3}{8E_p} = m_e^2 \ . \tag{28}$$

The special-relativistic result $E_{th} = m_e^2/\epsilon$ corresponds of course to the $\eta \rightarrow 0$ limit of (28). For $|\eta| \sim 1$ the Planck-scale correction can be safely neglected as long as $\epsilon > (m_e^4/E_p)^{1/3}$. But eventually, for sufficiently small values of ϵ and correspondingly large values of E_{th}, the Planck-scale correction cannot be ignored [46, 47, 48, 49, 50]

And the process $\gamma\gamma \rightarrow e^+e^-$ is not the only case in which this type of Planck-scale modification can be important. There has been strong interest [45, 46, 47, 48, 49, 50, 51, 94] in "photopion production", $p\gamma \rightarrow p\pi$, where again the combination of (12) with unmodified energy-momentum conservation leads to a modification of the minimum proton energy required by the process (for fixed photon energy). In the case in which the photon energy is the one typical of CMBR photons one finds that the threshold proton energy can be significantly shifted upward (for negative η), and this in turn should affect at an observably large level the expected "GZK cutoff" for the observed cosmic-ray spectrum. Observations reported by the AGASA [95] cosmic-ray observatory provide some encouragement for the idea of such an upward shift of the GZK cutoff, but the issue must be further explored. Forthcoming cosmic-ray observatories, such as Auger [96], should be able [45, 48] to fully investigate this possibility.

In this context the comparison of the AEMNS test theory and the GPMP test theory is rather straightforward. We are in fact considering a purely kinematical effect: the shift of a threshold requirement. For the minimal AEMNS test theory there is a clear prediction that for negative η there should be an upward shift of the GZK threshold. For the GPMP test theory one would predict an increase of the GZK threshold if any one (or both) of the two helicities of the proton has dispersion relation of "negative η" type. If both helicities have dispersion relation of negative-η type then the effect looks rather similar to the corresponding effect in the AEMNS test theory. For the situation which I proposed as the "minimal GPMP test theory", where for one of the helicities the dispersion relation is of negative-η type and for the other helicity the dispersion relation is of positive-η type, one would expect roughly one half of the UHE protons to evade the GZK cutoff, so the cutoff would still be violated but in a softer way than in the case of the AEMNS test theory with negative η.

It appears likely that, if the Auger data should actually show evidence of the expected GZK cutoff, we would then be in a position to rule out the case of negative η for the minimal AEMNS test theory, and to rule out both the positive-η_f and negative-η_f case for the minimal GPMP test theory. In fact, in the minimal AEMNS test theory violations of the GZK cutoff are predicted for negative η (while they are not present in the positive-η case), while in the minimal GPMP test theory violations of the GZK cutoff (although less numerous than expected in the minimal AEMNS test theory with negative η) are always expected, independently of the sign of η_f (depending on the sign of η_f the protons that violate the GZK cutoff would have a corresponding helicity).

I should stress that these studies of the cosmic-ray GZK threshold provide an example in which the fact that we do not really identify some of the particles in the relevant particle-physics processes, an analysis which could in principle be involving pure kinematics, ends up being exposed to the risk of contamination from some aspects of dynamics. If the only background radiation available for photopion production was the CMBR, then the prediction of an upward shift of the GZK cosmic-ray cutoff within the AEMNS test theory, for negative η, would be completely robust. But background radiation has many components and one could contemplate the possibility to combine AEMNS kinematics with an unspecified description of dynamics such that interactions of cosmic rays with other components of the background radiation would lead to a net result that does not change the numerical value of the GZK threshold. While this possibility must be contemplated, I also want to stress that, at least for $n = 1$ and negative η of order 1, this "conspiracy scenario" is so unbelievable that it should be dismissed. In fact, for $n = 1$ and negative η of order 1 the AEMNS kinematics allows the interaction of cosmic rays only with photons of energy higher than the TeV scale (see [48]), and the density of such high-energy background photons is so low that, even in a prudent phenomenology, this "conspiracy scenario" can indeed be dismissed.

For the GPMP test theory there is of course no issue of possible conspiracies, since the field-theoretic setup allows to evaluate cross sections.

4.11 Derivation of Limits from Analysis of Photon Stability

As in the case of the GZK cutoff for UHE cosmic rays there are several examples in which a given process is allowed in presence of exact Lorentz symmetry but can be kinematically forbidden in presence of certain departures from Lorentz symmetry. The opposite is also possible: some processes that are kinematically forbidden in presence of exact Lorentz symmetry become kinematically allowed in presence of certain departures from Lorentz symmetry. The fact that a process is kinematically allowed of course does not guarantee that it occurs at an observable rate: it depends on the laws of dynamics and the amplitudes they predict.

Certain observations in astrophysics, which allow us to establish that photons of energies up to $\sim 10^{14}$ eV are not unstable, can be particularly useful [49, 50, 51, 97] in setting limits on some schemes for departures from Lorentz symmetry. Let us for example analyze the process $\gamma \to e^+e^-$ from the AEMNS perspective, using the dispersion relation (12), with $n = 1$, and unmodified energy-momentum conservation. One easily finds a relation between the energy E_γ of the incoming photon, the opening angle θ between the outgoing electron-positron pair, and the energy E_+ of the outgoing positron (of course the energy of the outgoing electron is simply given by $E_\gamma - E_+$). For the region of phase space with $m_e \ll E_\gamma \ll E_p$ this relation takes the form

$$\cos(\theta) \simeq \frac{E_+(E_\gamma - E_+) + m_e^2 - \eta E_\gamma E_+(E_\gamma - E_+)/E_p}{E_+(E_\gamma - E_+)} , \qquad (29)$$

where m_e is the electron mass.

The fact that for $\eta = 0$ (29) would require $cos(\theta) > 1$ reflects the fact that, if Lorentz symmetry is preserved, the process $\gamma \to e^+e^-$ is kinematically forbidden. For $\eta < 0$ the process is still forbidden, but for positive η high-energy photons can decay into an electron-positron pair. In fact, for $E_\gamma \gg (m_e^2 E_p/|\eta|)^{1/3}$ one finds that there is a region of phase space where $\cos(\theta) < 1$, i.e. there is a physical phase space available for the decay.

The energy scale $(m_e^2 E_p)^{1/3} \sim 10^{13}$ eV is not too high for testing, since, as mentioned, in astrophysics we see photons of energies up to $\sim 10^{14}$ eV that are not unstable (they clearly travel safely some large astrophysical distances).

Within AEMNS kinematics, for $n = 1$ and positive η of order 1, it would have been natural to expect that such photons with $\sim 10^{14}$ eV energy would not be stable. Once again, before claiming that $n = 1$ and positive η of order 1 is ruled out, one should be concerned about possible conspiracies. The fact that the decay of 10^{14} eV photons is allowed by AEMNS kinematics (for $n = 1$ and positive η of order 1) of course does not guarantee that these photons should rapidly decay. It depends on the relevant probability amplitude, whose evaluation goes beyond the reach of kinematics. I am unable to provide an intuition for how big of a conspiracy would be needed to render 10^{14} eV photons stable compatibly with AEMNS kinematics with $n = 1$ and $\eta = 1$.

My tentative conclusion is that $n = 1$ with positive η of order 1 is ruled out "up to conspiracies", but unlike the case of the GZK-threshold analysis I am unprepared to argue that the needed conspiracy is truly unbelievable.

For the GPMP test theory the photon stability analysis is weakened because of other reasons. There one does have the support of the effective-field-theory description of dynamics, and within that framework one can exclude huge suppression by Planck scale effects of the interaction vertex needed for $\gamma \to e^+e^-$ around $\sim 10^{13}$ eV, $\sim 10^{14}$ eV. So the limit-setting effort is not weakened by the absence of an interaction vertex. However, as mentioned, consistency with the effective-field-theory setup requires that the two polarizations of the photon acquire opposite-sign modifications of the dispersion relation. We observe in astrophysics some photons of energies up to $\sim 10^{14}$ eV that are stable over large distances, but as far as we know those photons could be all, say, right-circular polarized (or all left-circular polarized). I postpone a detailed analysis to future work, but let me note here that there is a region of minimal-GPMP parameter space where both polarizations of a $\sim 10^{14}$ eV photon are unstable (a subset of the region with $|\eta_f| > |\eta_\gamma|$). That region of the minimal-GPMP parameter parameter space is of course excluded by the photon-stability data.

4.12 Derivation of Limits from Analysis of Synchrotron Radiation

A recent series of papers [85, 98, 99, 100, 101, 102, 103] has focused on the possibility to set limits on Planck-scale modified dispersion relations focusing on their implications for synchrotron radiation. By comparing the content of the first estimates[12] produced in this research line [98] with the understanding that emerged from follow-up studies [85, 99, 100, 101, 102, 103] one can gain valuable insight on the risks involved in analyzes based on simplistic order-of-magnitude estimates, rather than careful comparison with meaningful test theories. In [98] the starting point is the observation that in the conventional (Lorentz-invariant) description of synchrotron radiation one can estimate the characteristic energy E_c of the radiation through a heuristic analysis [104] leading to the formula

$$E_c \simeq \frac{1}{R{\cdot}\delta{\cdot}[v_\gamma - v_e]} , \tag{30}$$

where v_e is the speed of the electron, v_γ is the speed of the photon, δ is an angle obtained from the opening angle between the direction of the electron

[12] Reference [98] is at this point obsolete, since the relevant manuscript has been revised for the published version [100] and the recent [103] provides an even more detailed analysis. It is nevertheless useful to consider this series of manuscripts [98, 100, 103] as an illustration of how much the outlook of a phenomenological analysis may change in going from the level of simplistic order-of-magnitude estimates to the level of careful comparison with meaningful test theories.

and the direction of the emitted photon, and R is the radius of curvature of the trajectory of the electron.

Assuming that the only Planck-scale modification in this formula should come from the velocity law (described using $v = dE/dp$ in terms of the modified dispersion relation), one finds that in some instances the characteristic energy of synchrotron radiation may be significantly modified by the presence of Planck-scale departures from Lorentz symmetry. As an opportunity to test such a modification of the value of the synchrotron-radiation characteristic energy one can hope to use some relevant data [98, 100] on photons detected from the Crab nebula. This must be done with caution since the observational information on synchrotron radiation being emitted by the Crab nebula is rather indirect: some of the photons we observe from the Crab nebula are attributed to synchrotron processes on the basis of a promising conjecture, and the value of the relevant magnetic fields is also conjectured (not directly measured).

Assuming that indeed the observational situation has been properly interpreted, and relying on the mentioned assumption that the only modification to be taken into account is the one of the velocity law, one could basically rule out [98] the case $n = 1$ with negative η for a modified dispersion relation of the type (12).

This observation led at first to some excitement, but more recent papers are starting to adopt a more prudent viewpoint. The lack of comparison with a meaningful test theory represents a severe limitation of the original analysis. In particular, synchrotron radiation is due to the acceleration of the relevant electrons and therefore implicit in the derivation of the formula (30) is a subtle role for dynamics [99]. From a field-theory perspective the process of synchrotron-radiation emission can be described in terms of Compton scattering of the electrons with the virtual photons of the magnetic field. One would therefore be looking deep into the dynamical features of the theory.

The minimal AEMNS test theory does assume a modified dispersion relation of the type (12) universally applied to all particles, but it is a pure-kinematics framework and, since the analysis crucially involves some aspects of dynamics, it cannot be tested using a Crab-nebula synchrotron-radiation analysis.

The GPMP test theory relies on a description of dynamics within the framework of effective low-energy theory, but, as mentioned, this in turn ends up implying that it is not possible to assume that a dispersion relation of the type (12) universally applies to all particles. Actually the two polarizations of photons must, within this framework, satisfy different (opposite-sign Planck-scale corrections) dispersion relations. And for the description of electrons one naturally encounters at least two more free parameters. The only constraint that one could conceivably obtain for the GPMP test theory from the Crab-nebula synchrotron-radiation analysis would simply exclude[13] that both the electron-dispersion-relation parameters be negative (i.e. exclude that both helicities of the electron would be characterized by a dispersion relation of the type (12) with negative η and $n = 1$).

In particular, the case which I characterized as the "minimal GPMP test theory", where the two helicities of the electrons carry opposite-sign modifications of the dispersion relation, would automatically evade this type of constraint from the Crab-nebula synchrotron-radiation analysis (since the two helicities are affected by opposite-sign modifications of the dispersion relation, at least one of them must be a positive-sign-type modification).

5 Summary and Outlook

Quantum-Gravity Phenomenology has already reached its first goal: a sizeable community now works on the quantum-gravity problem with the awareness that there is a chance to test (at least some) Planck-scale effects. In reaching this first goal it was sufficient (and even, in a certain sense, necessary) to proceed with simple intuitive arguments, but the further development of quantum-gravity phenomenology requires us to adopt the standards of other branches of phenomenology, such as particle-physics phenomenology. In particular, the progress of experimental limits must be charted in terms of commonly-adopted, and carefully crafted, test theories of the new Planck-scale effects.

The fact that some Planck-scale pictures of spacetime physics are falsifiable is more and more robustly established, but in many cases we only see a path toward falsifiability rather having achieved already the results needed for a "critical test of a theory" (a test that could be used, in case of contrary experimental results, to discard the relevant Planck-scale picture of spacetime physics). This point of the falsifiability of some relevant theories is crucial for establishing quantum-gravity research as a truly scientific endeavor. The proposal of test theories must of course reflect the status of our analysis of the falsifiability of quantum-spacetime/quantum-gravity theories. In an appropriate sense the test theories must bridge the gap between quantum-gravity theories and experiments. They must be such that the experimental limits on the parameters of the test theories will naturally translate into direct limits on some relevant quantum-gravity theories, as soon as some falsifiable features of the quantum-gravity theory are fully established.

It is of course meaningless to compare limits obtained within different test theories. And there is no scientific content in an experimental limit claimed

[13] Even the possibility to derive any sort of constraint on the electron-dispersion-relation parameters is not guaranteed. In fact, as observed in the latest version of [103], one might be unable to exclude the possibility that the Crab-nebula synchrotron radiation be due to positron (rather than electron) acceleration.

on a vaguely defined test theory. For example, in the recent literature there has been a proliferation of papers claiming to improve limits on Planck-scale modifications of the dispersion relation, but the different studies were simply considering the same type of dispersion relation within significantly different test theories. These results, which were presented as a gradual improvement in the experimental limits on Planck-scale modifications of the dispersion relation, were actually only a series of papers proposing more and more (some better, some worse) different examples of test theories in which a Planck-scale modification of the dispersion relation can be accommodated. Each paper was proposing a *different* test theory and deriving limits on that specific test theory.

In order to illustrate these issues in the context of a specific example of quantum-gravity-phenomenology work, in the second part of these lectures I focused on the example of the phenomenology of Planck-scale modifications of the dispersion relation. I considered two examples of test theories, the AEMNS test theory and the GPMP test theory. These two test theories, although usually not explicitly fully characterized in the relevant papers, are among the most studied in the case of Planck-scale modifications of the dispersion relation.

I also stressed that a phenomenology should build its strength gradually. Within a given set of hypothesis one first sets up a reduced parameter space, and only once that reduced parameter space is ruled out by data one considers the possibility of wider parameter spaces. In the context here of interest the minimal AEMNS test theory, described in Subsect. 4.6, and the minimal GPMP test theory, described in Subsect. 4.8, appear to provide valuable starting points.

In particular, these two test theories can be representative of two types of attitudes that are emerging in the quantum-gravity-phenomenology community concerning the possibility of describing dynamical effects within the framework of effective low-energy field theory. The fact that both in the study of noncommutative spacetimes and in the study of Loop Quantum Gravity, the two quantum pictures of spacetime that provide the key sources of motivation for research on Planck-scale modifications of the dispersion relation, we are really only starting to understand some aspects of kinematics, but we are still missing any robust result on dynamics, encourages an approach to phenomenology which is correspondingly prudent with respect to the description of dynamics. The phenomenologist is therefore confronted with two options: For those who are most concerned about the status of the description of dynamics, the pure-kinematics minimal AEMNS test theory provides a rather reasonable starting point for phenomenology work. For those who are willing to set aside these concerns, and go ahead with the effective-field-theory description, the minimal GPMP test theory could provide a valuable starting point. It is interesting that, while the phenomenology based on pure kinematics is allowed to start with the assumption of full universality of the modification of the dispersion relation, the choice of describing dynamics in terms of an effective low-energy field theory forces upon us from the very beginning a nonuniversality of the effects, with the correlation between polarization and sign of the modification for photons (and, with the additional natural assumption of no net effect on randomly composed beams, one then can introduce for fermions ana analogous correlation between helicity and sign of the modification). This plays a key role in the phenomenology.

In the Subsects. 4.9, 4.10, 4.11, 4.12 I have considered a few examples of phenomenological analyzes which exposed very clearly the type of differences that one can encounter comparing the indications of preliminary sensitivity estimates and the outcome of more robust analyzes supported by test theories. The time-of-travel analyzes described in Subsect. 4.9 can be used to constrain the photon dispersion relation both in the AEMNS and in the GPMP test theory, but the strategy may be somewhat different: while in the AEMNS test theory one can only exploit the energy dependence of the new effects, in the GPMP test theory the additional polarization dependence can also be exploited. The type of analysis of the cosmic-ray spectrum described in Subsect. 4.10 is also applicable to both test theories, but also in that case some differences must be taken into account. In particular, by obtaining good-quality data on the cosmic-ray spectrum around the GZK scale we might be in a position to completely rule out the minimal GPMP test theory, and to rule out the negative-η case for the minimal AEMNS test theory. The photon-stability analysis described in Subsect. 4.11, which received much attention in the literature, actually turned out to be affected by severe limitations in constraining the parameter spaces of the minimal AEMNS and the minimal GPMP test theories: photon-stability analyzes must be treated prudently from a AEMNS perspective because in principle kinematics is insufficient for establishing the probability of particle decay (whereas kinematics is enough for establishing stability), and photon-stability analyzes only lead to rather weak limits on the minimal-GPMP parameter space because of the polarization dependence expected in that test theory (one would need an ideally polarized beam of ultra-high-energy photons in order to be able to infer some constraint on the GPMP test theory). The Crab-nebula synchrotron-radiation analysis, whose preliminary analysis had also raised high hopes, when set up within the test theories here of interest also proves to be largely ineffective: it is not applicable to the AEMNS test theory (once again because of the role that some aspects of dynamics play in the analysis) and it also leads to no constraint on the minimal GPMP test theory.

While, consistently with the objectives of these lectures, it was for me sufficient here to discuss this comparison of test theories to data at a semi-quantitative level, the striking results of this comparison, showing that the analysis at the test-theory level can have very different outcome with respect to the usual preliminary sensitivity estimates, should provide motivation for future publications with detailed quantitative analyzes of the emerging experimental bounds.

References

1. J. Stachel: Early History of Quantum Gravity. In *Black Holes, Gravitational Radiation and the Universe*, ed by B.R. Iyer, B. Bhawal eds. (Kluwer Academic Publisher, Netherlands, 1999).
2. M.B. Green, J.H. Schwarz and E. Witten: *Superstring theory* (Cambridge Univ. Press, Cambridge, 1987).
3. J. Polchinski: *Superstring Theory and Beyond*, (Cambridge University Press, Cambridge, 1998).
4. C. Rovelli: gr-qc/9710008, Living Rev. Rel. **1**, 1 (1998).
5. A. Ashtekar: gr-qc/0112038.
6. L. Smolin: hep-th/0303185.
7. T. Thiemann: gr-qc/0210094, Lect. Notes Phys. **631**, 41 (2003).
8. A. Connes: *Noncommutative Geometry*, (Academic Press, 1995).
9. S. Majid: *Foundations of Quantum Group Theory*, (Cambridge Univ Pr, 1995).
10. G. Amelino-Camelia: gr-qc/9910089, Lect. Notes Phys. **541**, 1 (2000).
11. D.V. Ahluwalia: gr-qc/0202098.
12. R. Colella, A.W. Overhauser and S.A. Werner: Phys. Rev. Lett. **34**, 1472 (1975).
13. D.V. Ahluwalia: gr-qc/9903074, Nature **398**, 199 (1999).
14. J.J. Sakurai: *Modern Quantum Mechanics* (Addison Wesley, New York 1993).
15. M. Gasperini: Phys. Rev. D **38**, 2635 (1988).
16. G.Z. Adunas, E. Rodriguez-Milla and D.V. Ahluwalia: Gen. Rel. Grav. **33**, 183 (2001).
17. K.C. Littrel, B.E. Allman and S.A. Werner: Phys. Rev. A **56**, 1767 (1997).
18. D.V. Ahluwalia: gr-qc/0009033.
19. J. Anandan: Phys. Lett. A **105**, 280 (1984); Class. Quant. Grav. **1**, 151 (1984).
20. A.K. Jain et al: Phys. Rev. Lett. **58**, 1165 (1987).
21. N. Arkani-Hamed, S. Dimopoulos and G. Dvali: Phys. Lett. B **429**, 263 (1998).
22. C.J. Isham: *Structural issues in quantum gravity*. In *Proceedings of General relativity and gravitation 1995* (World Scientic, Singapore 1997).
23. J. Ellis, J.S. Hagelin, D.V. Nanopoulos and M. Srednicki: Nucl. Phys. B **241**, 381 (1984).
24. P. Huet and M.E. Peskin: Nucl. Phys. B **434**, 3 (1995).
25. J. Ellis, J. Lopez, N.E. Mavromatos and D.V. Nanopoulos: Phys. Rev. D **53**, 3846 (1996).
26. F. Benatti and R. Floreanini: Nucl. Phys. B **488**, 335 (1997).
27. J. Ellis, N. Mavromatos and D.V. Nanopoulos: Phys. Lett. B **293**, 37 (1992); E. Gravanis and N.E. Mavromatos: hep-th/0108008.
28. V.A. Kostelecky and R. Potting: Phys. Rev. D **51**, 3923 (1995); O. Bertolami, D. Colladay, V.A. Kostelecky and R. Potting: Phys.Lett. B **395**, 178 (1997).
29. D. Colladay and V.A. Kostelecky, hep-ph/9809521, Phys. Rev. D **58**, 116002 (1998).
30. I.C. Percival: Physics World **10** (No 3), 43 (1997).
31. I.C. Percival and W.T. Strunz: Proc. Roy. Soc. A **453**, 431 (1997).
32. A. Apostolakis et al: hep-ex/9903005, Phys. Lett. B **452**, 425 (1999).
33. J.M. Link et al: hep-ex/0208034, Phys. Lett. B **556**, 7 (2003).
34. A. Ashtekar: gr-qc/9901023.
35. C. Rovelli: gr-qc/0006061.

36. L. Smolin: Physics World **12**, 79 (1999).
37. S. Carlip: gr-qc/0108040, Rept. Prog. Phys. **64**, 885 (2001).
38. G. Amelino-Camelia, J. Ellis, N.E. Mavromatos, D.V. Nanopoulos and S. Sarkar: astro-ph/9712103, Nature **393**, 763 (1998).
39. S.D. Biller et al: Phys. Rev. Lett. **83**, 2108 (1999).
40. G. Amelino-Camelia: gr-qc/9808029, Nature **398**, 216 (1999); gr-qc/0104086, Nature **410**, 1065 (2001).
41. G. Amelino-Camelia: gr-qc/9903080, Phys. Rev. D **62**, 024015 (2000).
42. Y.J. Ng and H. van Dam: gr-qc/9906003, Found. Phys. **30**, 795 (2000).
43. G. Amelino-Camelia: gr-qc/0104005.
44. G. Amelino-Camelia and C. Lammerzahl: gr-qc/0306019, Class. Quant. Grav. **21**, 899 (2004).
45. T. Kifune: Astrophys. J. Lett. **518**, L21 (1999).
46. R. Aloisio, P. Blasi, P.L. Ghia and A.F. Grillo: Phys. Rev. D **62**, 053010 (2000).
47. R.J. Protheroe and H. Meyer: Phys. Lett. B **493**, 1 (2000).
48. G. Amelino-Camelia and T. Piran: Phys. Rev. D **64**, 036005 (2001).
49. T. Jacobson, S. Liberati and D. Mattingly: hep-ph/0112207, Phys. Rev. D **66**, 081302 (2002).
50. G. Amelino-Camelia: gr-qc/0107086, Phys. Lett. B **528**, 181 (2002).
51. O. Bertolami: hep-ph/0301191.
52. G. 't Hooft: Class. Quant. Grav. **13**, 1023 (1996).
53. G. Amelino-Camelia: gr-qc/0012051, Int. J. Mod. Phys. D **11**, 35 (2002); hep-th/0012238, Phys. Lett. B **510**, 255 (2001).
54. S. Majid and H. Ruegg: Phys. Lett. B **334**, 348 (1994).
55. J. Lukierski, H. Ruegg and W.J. Zakrzewski: Ann. Phys. **243**, 90 (1995).
56. R. Gambini and J. Pullin: Phys. Rev. D **59**, 124021 (1999).
57. J. Alfaro, H.A. Morales-Tecotl and L.F. Urrutia: Phys. Rev. Lett. **84**, 2318 (2000).
58. G. Amelino-Camelia and S. Majid: Int. J. Mod. Phys. A **15**, 4301 (2000).
59. D.V. Ahluwalia: Mod. Phys. Lett. A **13**, 2249 (1998).
60. H. Murayama and T. Yanagida: Phys. Lett. B **520**, 263 (2001).
61. N.E. Mavromatos: hep-ph/0402005.
62. L.J. Garay: Phys. Rev. Lett. **80**, 2508 (1998).
63. G. Amelino-Camelia and S.-Y. Pi: hep-ph/9211211, Phys. Rev. D **47**, 2356 (1993).
64. A. Matusis, L. Susskind and N. Toumbas: JHEP **0012**, 002 (2000).
65. N.R. Douglas and N.A. Nekrasov: Rev. Mod. Phys. **73**, 977 (2001).
66. G. Amelino-Camelia, G. Mandanici and K. Yoshida: hep-th/0209254, JHEP **0401**, 037 (2004).
67. P. Fischer and V. Putz: hep-th/0306099, Eur. Phys. J. C **32**, 269 (2004)
68. J. Madore, S. Schraml, P. Schupp and J. Wess: hep-th/0001203, Eur. Phys. J. C **16**, 161 (2000).
69. S. Doplicher, K. Fredenhagen and J.E. Roberts: Phys. Lett. B **331**, 39 (1994).
70. J. Kowalski-Glikman: hep-th/0102098, Phys. Lett. A **286**, 391 (2001); S. Alexander and J. Magueijo: hep-th/0104093; R. Bruno, G. Amelino-Camelia and J. Kowalski-Glikman: hep-th/0107039, Phys. Lett. B **522**, 133 (2001); J. Magueijo and L. Smolin: gr-qc/0207085, Phys. Rev. D **67**, 044017 (2003); J. Kowalski-Glikman and S. Nowak: hep-th/0304101, Class. Quant. Grav. **20**, 4799 (2003).

71. P. Kosinski, J. Lukierski and P. Maslanka: Czech. J. Phys. **50**, 1283 (2000).
72. G. Amelino-Camelia and M. Arzano: hep-th/0105120, Phys. Rev. D **65**, 084044 (2002); A. Agostini, G. Amelino-Camelia, F. D'Andrea: hep-th/0306013.
73. M. Dimitrijevic, L. Jonke, L. Moller, E. Tsouchnika , J. Wess and M. Wohlgenannt: hep-th/0307149, Eur. Phys. J. C **31**, 129 (2003).
74. C. Rovelli and S. Speziale: gr-qc/0205108.
75. G. Amelino-Camelia: gr-qc/0205125.
76. G. Amelino-Camelia, L. Smolin and A. Starodubtsev: hep-th/0306134, Class. Quant. Grav. **21**, 3095 (2004).
77. L. Freidel, J. Kowalski-Glikman and L. Smolin: hep-th/0307085.
78. R. Gambini, R. Porto and J. Pullin: gr-qc/0305098.
79. P. Kosinski and P. Maslanka: hep-th/0211057, Phys. Rev. D **68**, 067702 (2003).
80. S. Mignemi: hep-th/0302065, Phys. Lett. A **316**, 173 (2003).
81. M. Daszkiewicz, K. Imilkowska and J. Kowalski-Glikman: hep-th/0304027.
82. J. Kowalski-Glikman: hep-th/0312140.
83. G. Amelino-Camelia, J. Ellis, N.E. Mavromatos and D.V. Nanopoulos: hep-th/9605211, Int. J. Mod. Phys. A **12**, 607 (1997).
84. G. Amelino-Camelia, F. D'Andrea and G. Mandanici: hep-th/0211022, JCAP **0309**, 006 (2003).
85. J. Ellis, N.E. Mavromatos and A.S. Sakharov: astro-ph/0308403.
86. R.C. Myers and M. Pospelov: hep-ph/0301124, Phys. Rev. Lett. **90**, 211601 (2003).
87. A. Perez and D. Sudarsky: gr-qc/0306113.
88. J. Collins, A. Perez, D. Sudarsky, L. Urrutia and H. Vucetich: gr-qc/0403053.
89. G. Amelino-Camelia: gr-qc/0402009.
90. J.P. Norris, J.T. Bonnell, G.F. Marani and J.D. Scargle: astro-ph/9912136; A. de Angelis: astro-ph/0009271.
91. T. Piran, astro-ph/0407462.
92. P. Meszaros, S. Kobayashi, S. Razzaque, B. Zhang: astro-ph/0305066.
93. G. Amelino-Camelia, J. Kowalski-Glikman, G. Mandanici and A. Procaccini: gr-qc/0312124.
94. Y.J. Ng, D.S. Lee, M.C. Oh, and H. van Dam: Phys. Lett. B **507**, 236 (2001); G. Amelino-Camelia, Y.J. Ng, and H. van Dam: gr-qc/0204077, Astropart. Phys. **19**, 729 (2003).
95. M. Takeda et al: Phys. Rev. Lett. **81**, 1163 (1998).
96. J. Blumer: J. Phys. G **29**, 867 (2003).
97. T.J. Konopka and S.A. Major: New J. Phys. **4**, 57 (2002); D. Heyman, F. Hinteleitner, and S. Major: gr-qc/0312089.
98. T. Jacobson, S. Liberati and D. Mattingly: arXiv.org/abs/astro-ph/0212190v1
99. G. Amelino-Camelia: gr-qc/0212002.
100. T. Jacobson, S. Liberati and D. Mattingly: arXiv.org/abs/astro-ph/0212190v2, Nature **424**, 1019 (2003).
101. T. Jacobson, S. Liberati and D. Mattingly: gr-qc/0303001.
102. S. Carroll: Nature **424**, 1007 (2003).
103. T.A. Jacobson, S. Liberati, D. Mattingly and F.W. Stecker: astro-ph/0309681.
104. J.D. Jackson, *Classical Electrodynamics*, 3rd edn (J. Wiley & Sons, New York 1999).

Astrophysical Bounds on Planck Suppressed Lorentz Violation

T. Jacobson[1], S. Liberati[2] and D. Mattingly[3]

[1] Institut d'Astrophysique de Paris, France, and Department of Physics, University of Maryland, USA
jacobson@umd.edu
[2] SISSA and INFN, Trieste, Italy
liberati@sissa.it
[3] Department of Physics, University of California at Davis, USA
mattingly@physics.ucdavis.edu

This article reviews many of the observational constraints on Lorentz symmetry violation (LV). We first describe the GZK cutoff and other phenomena that are sensitive to LV. After a brief historical sketch of research on LV, we discuss the effective field theory description of LV and related questions of principle, technical results, and observational constraints. We focus on constraints from high energy astrophysics on mass dimension five operators that contribute to LV electron and photon dispersion relations at order E/M_{Planck}. We also briefly discuss constraints on renormalizable operators, and review the current and future constraints on LV at order $(E/M_{\text{Planck}})^2$.

1 Windows on Quantum Gravity?

In most fields of physics it goes without saying that observation and prediction play a central role, but unfortunately quantum gravity (QG) has so far not fit that mold. Many intriguing and ingenious ideas have been explored, but it seems safe to say that without both observing phenomena that depend on QG, and extracting reliable predictions from candidate theories that can be compared with observations, the goal of a theory capable of incorporating quantum mechanics and general relativity will remain unattainable.

Besides the classical limit, there is one observed phenomenon for which quantum gravity makes a prediction that has received encouraging support: the spectrum of primordial cosmological perturbations. The quantized longitudinal linearized gravitational mode, albeit slave to the inflaton and not a dynamically independent degree of freedom, plays an essential role in this story [1].

T. Jacobson, S. Liberati and D. Mattingly: *Astrophysical Bounds on Planck Suppressed Lorentz Violation*, Lect. Notes Phys. **669**, 101–130 (2005)
www.springerlink.com

What other types of phenomena might be characteristic of a quantum gravity theory? Motivated by tentative theories, partial calculations, intimations of symmetry violation, hunches, philosophy, etc, some of the proposed ideas are: loss of quantum coherence or state collapse, QG imprint on initial cosmological perturbations, scalar moduli or other new fields, extra dimensions and low-scale QG, deviations from Newton's law, black holes produced in colliders, violation of global internal symmetries, and violation of spacetime symmetries. It is this last item, more specifically the possibility of Lorentz violation (LV), that is the focus of these lecture notes.

From the observational point of view, developments are encouraging a new look at the possibility of LV. Increased detector size, space-borne instruments, technological improvement, and technique refinement are permitting observations to probe higher energies, weaker interactions, lower fluxes, lower temperatures, shorter time resolution, and longer distances. It comes as a welcome surprise that the day of true quantum gravity observations may not be so far off [2].

2 Lorentz Violation

Lorentz symmetry is linked to a scale-free nature of spacetime: unbounded boosts expose ultra-short distances, and yet nothing changes. However, suggestions for Lorentz violation have come from: the need to cut off UV divergences of quantum field theory and of black hole entropy, tentative calculations in various QG scenarios (e.g. semiclassical spin-network calculations in Loop QG, string theory tensor VEVs, non-commutative geometry, some brane-world backgrounds), and the possibly missing GZK cutoff [3] on ultra-high energy (UHE) cosmic rays.

The GZK question has generated a lot of interest, and is currently the only observational phenomenon thought to indicate a possible violation of Lorentz symmetry. As an invitation to the subject, we discuss it in this section, before embarking on the rest of the lectures. We also give a list of possible LV phenomena, and a brief historical overview of the subject.

2.1 The GZK Cut-Off

In collisions of ultra high energy protons with cosmic microwave background (CMB) photons there can be sufficient energy in the center of mass frame to create a pion, leading to the the reaction

$$p + \gamma_{\mathrm{CMB}} \rightarrow p + \pi^0 \ . \tag{1}$$

The threshold occurs when the invariant magnitude of the total four momentum is the sum of the proton and pion mass, since at threshold these particles are both at rest in the zero-momentum frame. That is, it occurs when $(p_p + p_\gamma)^2 = (m_p + m_\pi)^2$, or $p_p \cdot p_\gamma = m_p m_\pi + \frac{1}{2} m_\pi^2$, where $p_{p,\gamma}$ are the proton and photon 4-momenta, and $m_{p,\pi}$ are the proton and pion mass. Since $E_p \gg m_p$, and $m_\pi \ll 2m_p$, this yields the proton energy threshold

$$E_{GZK} \simeq \frac{m_p m_\pi}{2E_\gamma} \simeq 3 \times 10^{20} \text{eV} \times \left(\frac{2.7\,\text{K}}{E_\gamma} \right) \tag{2}$$

To get a definite number we have put E_γ equal to the energy of a photon at the CMB temperature, 2.7 K, but of course the CMB contains photons of higher energy,

This process degrades the initial proton energy with an attenuation length of about 50 Mpc. Since plausible astrophysical sources for UHE particles are located at distances larger than 50 Mpc, one expects a cutoff in the cosmic ray proton energy spectrum, which occurs at around 5×10^{19} eV, depending on the distribution of sources [4].

One of the experiments measuring the UHE cosmic ray spectrum, the AGASA experiment, has not seen the cutoff. An analysis [6] from January 2003 concluded that the cutoff was absent at the 2.5 sigma level, while another experiment, HiRes, is consistent with the cutoff but at a lower confidence level. (For a brief review of the data see [4].) The question should be answered in the near future by the AUGER observatory, a combined array of 1600 water Čerenkov detectors and 24 telescopic air fluorescence detectors under construction on the Argentine pampas [7]. The new observatory will see an event rate one hundred times higher, with better systematics.

Many ideas have been put forward to explain the possible absence of the GZK cutoff [4]. For example the cosmic rays might originate closer, in some unexpected way, by astrophysical acceleration or by decay of ultra-heavy exotic particles, or they may be produced by collisions with ultra high energy cosmic neutrinos. Virtually all of these explanations have problems.

In this context, it is intriguing to consider that with even a tiny amount of Lorentz violation the energy threshold for the GZK reaction could be affected. According to (2) the Lorentz invariant threshold is proportional to the proton mass. Thus any LV term added to the proton dispersion relation $E^2 = \mathbf{p}^2 + m^2$ will modify the threshold if it is comparable to or greater than m_p^2 at around the energy E_{GZK}. Modifying the proton and pion dispersion relations, the threshold can be lowered, raised, or removed entirely, or even an upper threshold where the reaction cuts off could be introduced (see e.g. [5] and references therein).

For example, the LV term considered by Coleman and Glashow [8] was of the form $\eta \mathbf{p}^2$, assumed given in a reference frame close to that of the earth, which is natural since we are close to being at rest in the universal rest frame. This would affect the GZK threshold as long as $\eta > (m_p/E_{GZK})^2 \sim 10^{-22}$. Even LV suppressed by two powers of the Planck mass M would affect the threshold: a term of the form $\mathbf{p}^4/M^2$ is comparable to m_p^2 when the proton energy is $(m_p M)^{1/2} \simeq 3 \times 10^{18}$ eV, which is two orders of magnitude below the highest energy cosmic rays. Thus a missing GZK cutoff could be explained by Planck double-suppressed LV. Conversely, observational confirmation of the GZK cutoff can severely constrain such LV.

2.2 Possible LV Phenomena

Trans-GZK cosmic rays are not the only window of opportunity we have to detect or constrain Lorentz violation induced by QG effects. In fact, many phenomena accessible to current observations/experiments are sensitive to possible violations of Lorentz invariance. A partial list is

- sidereal variation of LV couplings as the lab moves with respect to a preferred frame or directions, or cosmological variation
- long baseline dispersion and vacuum birefringence (e.g. of signals from gamma ray bursts, active galactic nuclei, pulsars, galaxies)
- new reaction thresholds (e.g. photon decay, vacuum Čerenkov effect)
- shifted thresholds (e.g. photon annihilation from blazars, GZK reaction)
- maximum velocity (e.g. synchrotron peak from supernova remnants)
- dynamical effects of LV background fields (e.g. gravitational coupling and additional wave modes)

2.3 A Brief History of Some LV Research

We conclude this section with a brief historical overview mentioning some of the more influential papers, but by no means complete.

Suggestions of possible LV in particle physics go back at least to the 1960's, when a number of authors wrote on that idea [10][4]. The possibility of LV in a metric theory of gravity was explored beginning at least as early as the 1970's [12]. Such theoretical ideas were pursued in the '70's and '80's notably by Nielsen and several other authors on the particle theory side [13], and by Gasperini [14] on the gravity side. A number of observational limits were obtained during this period [16].

Towards the end of the 80's Kostelecky and Samuel [17] presented evidence for possible spontaneous LV in string theory, and motivated by this explored LV effects in gravitation. The role of Lorentz invariance in the "trans-Planckian puzzle" of black hole redshifts and the Hawking effect was emphasized in the early 90's [18]. This led to study of the Hawking effect for quantum fields with LV dispersion relations commenced by Unruh [19] and followed up by others. Early in the third millennium this line of research led to work on the related question of the possible imprint of trans-Planckian frequencies on the primordial fluctuation spectrum [20]. Meanwhile the consequences of LV

[4] Remarkably, already in 1972 Kirzhnits and Chechin [10] explored the possibility that an apparent missing cutoff in the UHE cosmic ray spectrum could be explained by something that looks very similar to the recently proposed "doubly special relativity" [11].

for particle physics were being explored using LV dispersion relations e.g. by Gonzalez-Mestres [21].

Four developments in the late nineties seem to have stimulated a surge of interest in LV research. One was a systematic extension of the standard model of particle physics incorporating all possible LV in the renormalizable sector, developed by Colladay and Kostelecký [22]. That provided a framework for computing the observable consequences for any experiment and led to much experimental work setting limits on the LV parameters in the lagrangian [23]. On the observational side, the AGASA experiment reported events beyond the GZK cutoff [24]. Coleman and Glashow then suggested the possibility that LV was the culprit in the possibly missing GZK cutoff [8], and explored many other high energy consequences of renormalizable, isotropic LV leading to different limiting speeds for different particles [25]. In the fourth development, it was pointed out by Amelino-Camelia et al. [26] that the sharp high energy signals of gamma ray bursts could reveal LV photon dispersion suppressed by one power of energy over the mass $M \sim 10^{-3} M_{\mathrm{P}}$, tantalizingly close to the Planck mass.

Together with the improvements in observational reach mentioned earlier, these developments attracted the attention of a large number of researchers to the subject. Shortly after [26] appeared, Gambini and Pullin [27] argued that semiclassical loop quantum gravity suggests just such LV. Some later work supported this notion, but the issue continues to be debated [28, 29]. In any case, the dynamical aspect of the theory is not under enough control at this time to make any definitive statements concerning LV.

A very strong constraint on photon birefringence was obtained by Gleiser and Kozameh [30] using UV light from distant galaxies. If the recent report[31] of polarized gamma rays from a GRB turns out to be correct despite the concerns of [32], this constraint will be further strengthened dramatically [33, 34]. Further stimulus came from the suggestion [35] that an LV threshold shift might explain the apparent under-absorption on the cosmic IR background of TeV gamma rays from the blazar Mkn501, however it is now believed by many that this anomaly goes away when a corrected IR background is used [36].

The extension of the effective field theory framework to include LV dimension 5 operators was introduced by Myers and Pospelov [37], and used to strengthen prior constraints. Also this framework was used to deduce a very strong constraint [38] on the possibility of a maximum electron speed less than the speed of light from observations of synchrotron radiation from the Crab Nebula.

3 Theoretical Framework for LV

Various different theoretical approaches to LV have been taken to pursue the ideas summarized above. Some researchers restrict attention to LV described in the framework of effective field theory (EFT), while others allow for effects not describable in this way, such as those that might be due to stochastic fluctuations of a "space-time foam". Some restrict to rotationally invariant LV, while others consider also rotational symmetry breaking. Both true LV as well as "deformed" Lorentz symmetry (in the context of so-called "doubly special relativity"[11]) have been pursued. Another difference in approaches is whether one allows for distinct LV parameters for different particle types, or proposes a more universal form of LV.

The rest of this article will focus on just one of these approaches, namely LV describable by standard EFT, assuming rotational invariance, and allowing distinct LV parameters for different particles. In exploring the possible phenomenology of new physics, it seems useful to retain enough standard physics so that clear predictions can be made, and so that the possibilities are narrow enough to be meaningfully constrained.

This approach is not universally favored. For example a sharp critique appears in [39]. Therefore we think it is important to spell out the motivation for the choices we have made. First, while of course it may be that EFT is not adequate for describing the leading quantum gravity phenomenology effects, it has proven itself very effective and flexible in the past. It produces local energy and momentum conservation laws, and seems to require for its validity just locality and local spacetime translation invariance above some length scale. It describes the standard model and general relativity (which are presumably not fundamental theories), a myriad of condensed matter systems at appropriate length and energy scales, and even string theory (as perhaps most impressively verified in the calculations of black hole entropy and Hawking radiation rates). It is true that, e.g., non-commutative geometry (NCG) seems to lead to EFT with problematic IR/UV mixing, however this more likely indicates a physically unacceptable feature of such NCG rather than a physical limitation of EFT.

The assumption of rotational invariance is motivated by the idea that LV may arise in QG from the presence of a short distance cutoff. This suggests a breaking of boost invariance, with a preferred rest frame, but not necessarily rotational invariance. Since a constraint on pure boost violation is, barring a conspiracy, also a constraint on boost plus rotation violation, it is sensible to simplify with the assumption of rotation invariance at this stage. The preferred frame is assumed to coincide with the rest frame of the CMB.

Finally why do we choose to complicate matters by allowing for different LV parameters for different particles? First, EFT for first order Planck suppressed LV (see Sect. 3.2) requires this for different polarizations or spin states, so it is unavoidable in that sense. Second, we see no reason *a priori* to expect these parameters to coincide. The term "equivalence principle" has been used to motivate the equality of the parameters. However, in the presence of LV dispersion relations, free particles with different masses travel on different trajectories even if they have the same LV parameters [5, 40]. Moreover, different particles would presumably interact differently with the spacetime microstructure since they interact differently with themselves and with each other. An example of this occurs in the braneworld model discussed in [41], and an extreme version occurs in the proposal of [42] in which only certain particles feel the spacetime foam effects. (Note however that in this proposal the LV parameters fluctuate even for a given kind of particle, so EFT would not be a valid description.)

3.1 Deformed Dispersion Relations

A simple approach to a phenomenological description of LV is via deformed dispersion relations. If rotation invariance and integer powers of momentum are assumed in the expansion of $E^2(\mathbf{p})$, the dispersion relation for a given particle type can be written as

$$E^2 = p^2 + m^2 + \Delta(p), \tag{3}$$

where p is hereafter the magnitude of the three-momentum, and

$$\Delta(p) = \tilde{\eta}_1 p^1 + \tilde{\eta}_2 p^2 + \tilde{\eta}_3 p^3 + \tilde{\eta}_4 p^4 + \cdots \tag{4}$$

Since they are not Lorentz invariant, it is necessary to specify the frame in which these relations are given, namely the CMB frame.

Let us introduce two mass scales, $M = 10^{19}\,\mathrm{GeV} \approx M_{\mathrm{Planck}}$, the putative scale of quantum gravity, and μ, a particle physics mass scale. To keep mass dimensions explicit we factor out possibly appropriate powers of these scales, defining the dimensionful $\tilde{\eta}$'s in terms of corresponding dimensionless parameters. It might seem natural that the p^n term with $n \geq 3$ be suppressed by $1/M^{n-2}$, and indeed this has been assumed in most work. But following this pattern one would expect the $n = 2$ term to be unsuppressed and the $n = 1$ term to be even more important. Since any LV at low energies must be small, such a pattern is untenable. Thus either there is a symmetry or some other mechanism protecting the lower dimension operators from large LV, or the suppression of the higher dimension operators is greater than $1/M^{n-2}$. This is an important issue to which we return in Subsect. 3.3.

For the moment we simply follow the observational lead and insert at least one inverse power of M in each term, viz.

$$\tilde{\eta}_1 = \eta_1 \frac{\mu^2}{M}, \qquad \tilde{\eta}_2 = \eta_2 \frac{\mu}{M}, \qquad \tilde{\eta}_3 = \eta_3 \frac{1}{M}, \qquad \tilde{\eta}_4 = \eta_4 \frac{1}{M^2} \,. \tag{5}$$

In characterizing the strength of a constraint we refer to the η_n without the tilde, so we are comparing to what might be expected from Planck-suppressed LV. We allow the LV parameters η_i to depend on the particle type, and indeed it turns out that they *must* sometimes be different but related in certain ways for photon polarization states, and for particle and antiparticle states, if the framework of effective field theory is adopted. In an even more general setting, Lehnert [43] studied theoretical constraints on this type of LV and deduced the necessity of some of these parameter relations.

The deformed dispersion relations are introduced for elementary particles only; those for macroscopic objects are then inferred by addition. For example, if N particles with momentum $\mathbf{p}$ and mass m are combined, the total energy, momentum and mass are $E_{\rm tot} = NE(p)$, $\mathbf{p}_{\rm tot} = N\mathbf{p}$, and $m_{\rm tot} = Nm$, so that $E^2_{\rm tot} = p^2_{\rm tot} + m^2_{\rm tot} + N^2\Delta(p)$. Although the Lorentz violating term can be large in some fixed units, its ratio with the mass and momentum squared terms in the dispersion relation is the same as for the individual particles. Hence, there is no observational conflict with standard dispersion relations for macroscopic objects.

This general framework allows for superluminal propagation, and spacelike 4-momentum relative to a fixed background metric. It has been argued [44] that this leads to problems with causality and stability. In the setting of a LV theory with a single preferred frame, however, we do not share this opinion. As long as the physics is guaranteed to be causal and the states all have positive energy in the preferred frame, we cannot see any room for such problems to arise.

3.2 Effective Field Theory and LV

The standard model extension (SME) of Colladay and Kostelecký [22] consists of the standard model of particle physics plus all Lorentz violating renormalizable operators (i.e. of mass dimension ≤ 4) that can be written without changing the field content or violating the gauge symmetry. For illustration, the leading order terms in the QED sector are the dimension three terms

$$-\, b_a \bar{\psi}\gamma_5\gamma^a\psi - \frac{1}{2}H_{ab}\bar{\psi}\sigma^{ab}\psi \tag{6}$$

and the dimension four terms

$$-\frac{1}{4}k^{abcd}F_{ab}F_{cd} + \frac{i}{2}\bar{\psi}(c_{ab} + d_{ab}\gamma_5)\gamma^a \overleftrightarrow{D}^b\psi\ , \tag{7}$$

where the dimension one coefficients b_a, H_{ab} and dimensionless k^{abcd}, c_{ab}, and d_{ab} are constant tensors characterizing the LV. If we assume rotational invariance then these must all be constructed from a given unit timelike vector u^a and the Minkowski metric η_{ab}, hence $b_a \propto u_a$, $H_{ab} = 0$, $k^{abcd} \propto u^{[a}\eta^{b][c}u^{d]}$, $c_{ab} \propto u_a u_b$, and $d_{ab} \propto u_a u_b$. Such LV is thus characterized by just four numbers.

The study of Lorentz violating EFT in the higher mass dimension sector was initiated by Myers and Pospelov [37]. They classified all LV dimension five operators that can be added to the QED Lagrangian and are quadratic in the same fields, rotation invariant, gauge invariant, not reducible to a combination of lower and/or higher dimension operators using the field equations, and contribute p^3 terms to the dispersion relation. Just three operators arise:

$$-\frac{\xi}{2M}u^m F_{ma}(u\cdot\partial)(u_n\tilde{F}^{na}) + \frac{1}{M}u^m\bar{\psi}\gamma_m(\zeta_1 + \zeta_2\gamma_5)(u\cdot\partial)^2\psi \tag{8}$$

where $\tilde{F}$ denotes the dual of F, and $\xi, \zeta_{1,2}$ are dimensionless parameters. The sign of the ξ term in (8) is opposite to that in [37], and is chosen so that positive helicity photons have $+\xi$ for a dispersion coefficient (see below). All of these terms violate CPT symmetry as well as Lorentz invariance. Thus if CPT were preserved, these LV operators would be forbidden.

In the limit of high energy $E \gg m$, the photon and electron dispersion relations following from QED with the above terms are [37]

$$\omega_{\pm}^2 = k^2 \pm \frac{\xi}{M} k^3 \tag{9}$$

$$E_{\pm}^2 = p^2 + m^2 + \frac{2(\zeta_1 \pm \zeta_2)}{M} p^3 \,. \tag{10}$$

The photon subscripts $\pm$ refer to helicity, i.e. right and left circular polarization, which it turns out necessarily have opposite LV parameters. The electron subscripts $\pm$ refer to the helicity, which can be shown to be a good quantum number in the presence of these LV terms [33]. Moreover, if we write $\eta_{\pm} = 2(\zeta_1 \pm \zeta_2)$ for the LV parameters of the two electron helicities, those for positrons are given [33] by

$$\eta_{\pm}^{\text{positron}} = -\eta_{\mp}^{\text{electron}} \,. \tag{11}$$

If $\eta_1 = 0$, then the two helicities have opposite LV parameters, $\eta_+ = -\eta_-$, so electron and positron have the *same* LV parameters. If instead $\eta_2 = 0$, then the $\eta_+ = \eta_-$, so electron and positron have *opposite* LV parameters.

3.3 Naturalness of Small LV at Low Energy?

As discussed above in Subsect. 3.1, if LV operators of dimension $n > 4$ are suppressed, as we have imagined, by $1/M^{n-2}$, LV would feed down to the lower dimension operators and be strong at low energies [25, 37, 46, 47], unless there is a symmetry or some other mechanism that protects the lower dimension operators from strong LV. What symmetry (other than Lorentz invariance, of course!) could that possibly be?

In the Euclidean context, a discrete subgroup of the Euclidean rotation group suffices to protect the operators of dimension four and less from violation of rotation symmetry. For example [48], consider the "kinetic" term in the EFT for a scalar field with hypercubic symmetry, $M^{\mu\nu}\partial_\mu\phi\partial_\nu\phi$. The only tensor $M^{\mu\nu}$ with hypercubic symmetry is proportional to the Kronecker delta $\delta^{\mu\nu}$, so full rotational invariance is an "accidental" symmetry of the kinetic operator.

If one tries to mimic this construction on a Minkowski lattice admitting a discrete subgroup of the Lorentz group, one faces the problem that each point has an infinite number of neighbors related by the Lorentz boosts. For the action to share the discrete symmetry each point would have to appear in infinitely many terms of the discrete action, presumably rendering the equations of motion meaningless.

Another symmetry that could do the trick is three dimensional rotational symmetry together with a symmetry between different particle types. For example, rotational symmetry would imply that the kinetic term for a scalar field takes the form $(\partial_t\phi)^2 - c^2(\nabla\phi)^2$, for some constant c. Then, for multiple scalar fields, a symmetry relating the fields would imply that the constant c is the same for all, hence the kinetic term would be Lorentz invariant with c playing the role of the speed of light. Unfortunately this mechanism does not work in nature, since there is no symmetry relating all the physical fields.

Perhaps under some conditions a partial symmetry could be adequate, e.g. grand unified gauge and/or super symmetry. In fact, a recent analysis of Nibbelink and Pospelov [49] presents evidence that supersymmetry (SUSY) together with gauge symmetry might indeed play this role. SUSY here refers to the symmetry algebra that is a kind of square root of the spacetime translation group. The nature of this square root depends upon the Minkowski metric, so is tied to the Lorentz group, but it does not require Lorentz symmetry. It is shown in [49], using the superfield formalism, that the SUSY preserving LV operators that can be added to the SUSY Standard Model first appear at dimension five. Moreover, these operators do not contribute $O(p^3)$ terms to the particle dispersion relations. Of course SUSY is broken in the real world, but the suppression in the SUSY theory may mean that the low dimension LV terms allowed in the presence of soft SUSY breaking are suppressed enough to be compatible with observation. On the other hand, it might also mean that they are so suppressed as to lie beyond the scope of observation.

At this stage we assume the existence of some realization of the Lorentz symmetry breaking scheme upon which constraints are being imposed. If none exists, then our parametrization (5) is misleading, since there should be more powers of $1/M$ suppressing the higher dimension terms. In that case, current observational limits on those terms do not significantly constrain the fundamental theory.

4 Reaction Thresholds and LV

Lorentz violation can have significant effects on energy thresholds for particle reactions. Such effects could be signatures of LV, and can be used to put constraints on LV. In the presence of LV, standard properties of LI threshold configurations (e.g. angles and momentum distributions) may not be preserved. Hence a careful study of properties of LV threshold configurations is needed before signatures and constraints can be considered. In this section we review some basic results concerning LV thresholds.

Threshold configurations and new phenomena in the presence of LV dispersion relations were systematically investigated in [25, 50] (see also references therein). We give here a brief summary of the results. We shall consider reactions with two initial and two final particles (results for reactions with only one incoming or outgoing particle can be obtained as special cases). Following our previous choice of EFT we allow each particle to have an independent dispersion relation of the form (3) with $E(p)$ a monotonically increasing non-negative function of the magnitude p of the 3-momentum $\mathbf{p}$. While the assumption of monotonicity could perhaps be violated at the Planck scale, it is satisfied for any reasonable low energy expansion of a LV dispersion relation. EFT further implies that energy and momentum are additive for multiple particles, and conserved.

Consider a four-particle interaction where a target particle of 3-momentum $\mathbf{p}_2$ is hit by a particle of 3-momentum $\mathbf{p}_1$, with an angle α between the two momenta, producing two particles of momenta $\mathbf{p}_3$ and $\mathbf{p}_4$. We call β the angle between $\mathbf{p}_3$ and the total incoming 3-momentum $\mathbf{p}_{in} = \mathbf{p}_1 + \mathbf{p}_2$. We define the notion of a threshold relative to a fixed value of the magnitude of the target momentum p_2. A lower or upper threshold corresponds to a value of p_1 (or equivalently the energy E_1) above which the reaction starts or stops being allowed by energy and momentum conservation.

We now introduce a graphical interpretation of the energy-momentum conservation equation that allows the properties of thresholds to be easily understood. For given values of $(p_1, p_2, \alpha, \beta, p_3)$, momentum conservation determines p_4. Since p_3 and p_4 determine the final energies E_3 and E_4, we can thus define the final energy function $E_f^{\alpha,\beta,p_3}(p_1)$. (Since p_2 is fixed we drop it from the labelling.) Energy conservation requires that E_f be equal to $E_i(p_1)$, the initial energy (again, we do not indicate the dependence on the fixed momentum p_2).

Now consider the region $\mathcal{R}$ in the (E, p_1) plane covered by plotting $E_f^{\alpha,\beta,p_3}(p_1)$ for all possible configurations (α, β, p_3). An example is shown in Fig. 1. The region $\mathcal{R}$ is bounded below by $E = 0$ since the particle energies are assumed non-negative, hence it has some bounding curve $E_B(p_1)$. Similarly one can plot $E_i(p_1)$. The reaction is allowed (i.e. there is a solution to the energy and momentum conservation equations) when this latter curve lies inside the region $\mathcal{R}$. A lower or upper threshold occurs when $E_i(p_1)$ enters or leaves $\mathcal{R}$.

This graphical representation demonstrates that in any threshold configuration (lower or upper) occurring at some p_1, the parameters (α, β, p_3) are such that the final energy function $E_f^{\alpha,\beta,p_3}(p_1)$ is minimized. That is, the configuration always yields the minimum final particle energy configuration conserving momentum at fixed p_1 and p_2. From this fact, it is easy to deduce two general properties of these configurations:

1. All thresholds for processes with two outgoing particles occur at parallel final momenta ($\beta = 0$).
2. For a two-particle initial state the momenta are antiparallel at threshold ($\alpha = \pi$).

These properties are in agreement with the well known case of Lorentz invariant kinematics. Nevertheless, LV thresholds can exhibit new features not

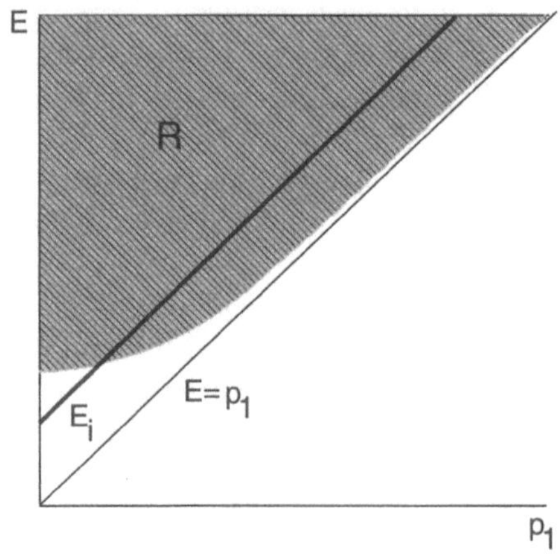

Fig. 1. $\mathcal{R}$ is the region covered by all final energy curves $E_f^{\alpha,\beta,p_3}(p_1)$ for some fixed p_2, with p_4 determined by momentum conservation. The curve $E_i(p_1)$ is the initial energy for the same fixed p_2. Where the latter curve lies inside $\mathcal{R}$ there is a solution to the energy and momentum conservation equations

present in the Lorentz invariant theory, in particular upper thresholds and asymmetric pair creation.

Figure 1 clearly shows that LV allows for a reaction to not only to start at some lower threshold but also to end at some upper threshold where the curve E_i exits the region $\mathcal{R}$. It can even happen that E_i enters and exits $\mathcal{R}$ more than once, in which case there are what one might call "local" lower and and upper thresholds.

Another interesting novelty is the possibility to have a (lower or upper) threshold for pair creation with an unequal partition of the initial momentum p_{in} into the two outgoing particles (i.e. $p_3 \neq p_4 \neq p_{in}/2$). Equal partition of momentum is a familiar result of Lorentz invariant physics, which follows from the fact that the final particles are all at rest in the zero-momentum frame at threshold. This has often been (erroneously) presumed to hold as well in the presence of LV dispersion relations.

A reason for the occurrence of asymmetric LV thresholds can be seen graphically, as shown in Fig. 2. Suppose the dispersion relation for a massive outgoing particle $E_{out}(\mathbf{p})$ has negative curvature at $p = p_{in}/2$, as might be the case for negative LV coefficients. Then a small momentum-conserving displacement from a symmetric configuration can lead to a net decrease in the final state energy. According to the result established above, the symmetric configuration cannot be the threshold one in such a case. A lower p_1 could satisfy both energy and momentum conservation with an asymmetric final configuration. A *sufficient* condition for the pair-creation threshold configuration to be asymmetric is that the final particle dispersion relation has negative curvature at $p = p_{in}/2$. This condition is not necessary however, since it could

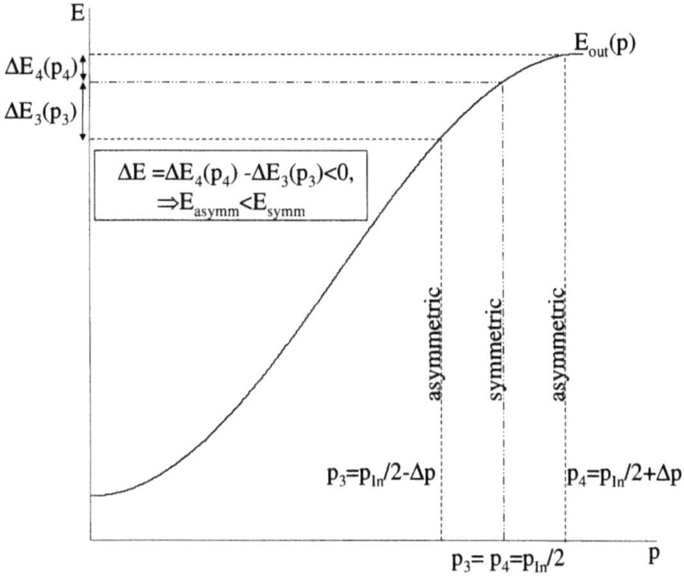

Fig. 2. Asymmetric pair production. The negative curvature of the outgoing particle dispersion relation allows the energy of the outgoing pair to be reduced by distributing the initial momentum p_{in} un-equally between the two particles

happen that the energy is locally but not globally minimized by the symmetric configuration.

5 Constraints

Observable effects of LV arise, among other things, from (1) sidereal variation of LV couplings due to motion of the laboratory relative to the preferred frame, (2) dispersion and birefringence of signals over long travel times, (3) anomalous reaction thresholds. We will often express the constraints in terms of the dimensionless parameters η_n introduced in (5). An order unity value might be considered to be expected in Planck suppressed LV.

The possibility of interesting constraints in spite of Planck suppression arises in different ways for the different types of observations. In the laboratory experiments looking for sidereal variations, the enormous number of atoms allow variations of a resonance frequency to be measured extremely accurately. In the case of dispersion or birefringence, the enormous propagation distances would allow a tiny effect to accumulate. In the anomalous threshold case, the creation of a particle with mass m would be strongly affected by a LV term when the momentum becomes large enough for this term to be comparable to the mass term in the dispersion relation.

We briefly mention first constraints on the renormalizable Standard Model Extension, then focus on LV suppressed by one or two powers of the ratio E/M.

5.1 Constraints on Renormalizable Terms

For the $n = 2$ term in (4,5), the absence of a strong threshold effect yields a constraint $\eta_2 \lesssim (m/p)^2(M/\mu)$. If we consider protons and put $\mu = m = m_p \sim$ 1 GeV, this gives an order unity constraint when $p \sim \sqrt{mM} \sim 10^{19}$ eV. Thus the GZK threshold, if confirmed, can give an order unity constraint, but multi-TeV astrophysics yields much weaker constraints. The strongest laboratory constraints on dimension three and four operators come from clock comparison experiments using noble gas masers [51]. The constraints limit a combination of the coefficients for dimension three and four operators for the neutron to be below 10^{-31} GeV (the dimension four coefficients are weighted by the neutron mass, yielding a constraint in units of energy). For more on such constraints see e.g. [23, 52]. Astrophysical limits on photon vacuum birefringence give a bound on the coefficients of dimension four operators of 10^{-32} [53].

5.2 Summary of Constraints on LV in QED at $O(E/M)$

Since we do not assume universal LV coefficients, different constraints cannot be combined unless they involve just the same particle types. To achieve the strongest combined constraints it is thus preferable to focus on processes involving a small number of particle types. It also helps if the particles are very common and easy to observe. This selects electron-photon physics, i.e. QED, as a useful arena.

The current constraints on the three LV parameters at order E/M – one in the photon dispersion relation and two in the electron dispersion relation – will now be summarized. These are equivalent to the parameters in the dimension five operators (8) written down by Myers and Pospelov.

For $n = 3$, a strong effect on energy thresholds involving only electrons and photons can occur when the LV term $\eta p^3/M$ in the electron or photon dispersion relation is comparable to or greater than the electron mass term m^2. This happens when

$$p \simeq 14\,\mathrm{TeV}\,\eta_3^{-1/3}\,. \tag{12}$$

We can thus obtain order unity and even much stronger constraints from high energy astrophysics, where such energies are reached and exceeded.

In Fig. 3 (from [33]) constraints on the photon (ξ) and electron (η) LV parameters are plotted on a logarithmic scale to allow the vastly differing strengths to be simultaneously displayed. For negative parameters, the negative of the logarithm of the absolute value is plotted, and a region of width 10^{-18} is excised around each axis. The synchrotron and Čerenkov constraints must both be satisfied by at least one of the four quantities $\pm\eta_\pm$. The IC

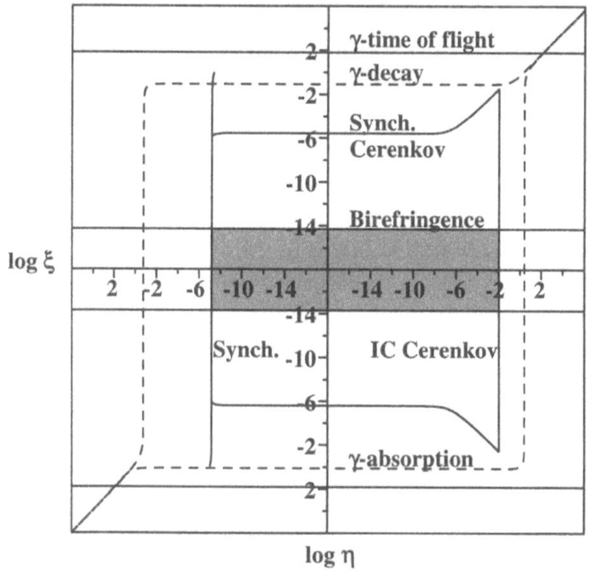

Fig. 3. Constraints on LV in QED at $O(E/M)$ (figure from [33])

and synchrotron Čerenkov lines are truncated where they cross. Prior photon decay and absorption constraints are shown in dashed lines since they do not account for the EFT relations between the LV parameters.

We now briefly review the physics and observations behind these and other constraints.

Electron Helicity Dependence and "Helicity Decay"

The constraint $|\eta_+ - \eta_-| < 4$ on the difference between the positive and negative electron helicity parameters was deduced by Myers and Pospelov [37] using a previous spin-polarized torsion pendulum experiment [54] that looked for diurnal changes in resonance frequency. They also determined a numerically stronger constraint using nuclear spins, however this involves four different LV parameters, one for the photon, one for the up-down quark doublet, and one each for the right handed up and down quark singlets. It also requires a model of nuclear structure.

It is possible that an interesting constraint could be obtained from the process of "helicity decay"[33]. If η_+ and η_- are unequal, say $\eta_+ > \eta_-$, then a positive helicity electron can decay into a negative helicity electron and a photon, even when the LV parameters do not permit the vacuum Čerenkov effect. In this process, the large R or small $(O(m/E))$ L component of a positive helicity electron is coupled to the small R or large L component of a negative helicity electron respectively. Such "helicity decay" has no threshold energy, so whether this process can be used to set a constraint is solely a matter of the decay rate. It can be shown (assuming $|\xi| \lesssim 10^{-3}$) that for electrons of energy less than the transition energy $(m^2M/(\eta_+ - \eta_-))^{1/3}$, the lifetime of an electron susceptible to helicity decay is greater than $4\pi M/(\eta_+ - \eta_-)e^2m^2$. At the limit of the best current bound $|\eta_+ - \eta_-| < 4$, the transition energy is approximately 10 TeV and the lifetime for electrons below this energy is greater than 10^4 seconds. This is long enough to preclude any terrestrial experiments from seeing the effect. The lifetime above the transition energy is instead bounded below by E/e^2m^2, which is 10^{-11} seconds for energies just above 10 TeV. The lifetime might therefore be short enough to provide new constraints. Such a constraint might come from the Crab Nebula, as explained below.

Vacuum Birefringence

The birefringence constraint arises from the fact that the LV parameters for left and right circular polarized photons are opposite (10). The phase velocity thus depends on both the wavevector and the helicity. Linear polarization is therefore rotated through an energy dependent angle as a signal propagates, which depolarizes any initially linearly polarized signal. Hence the observation of linearly polarized radiation coming from far away can constrain the magnitude of the LV parameter.

In more detail, with the dispersion relation (10) the direction of linear polarization is rotated through the angle

$$\theta(t) = [\omega_+(k) - \omega_-(k)]\, t/2 = \xi k^2 t/2M \tag{13}$$

for a plane wave with wave-vector k over a propagation time t. The difference in rotation angles for wave-vectors k_1 and k_2 is thus

$$\Delta\theta = \xi(k_2^2 - k_1^2)d/2M \ , \tag{14}$$

where we have replaced the time t by the distance d from the source to the detector (divided by the speed of light). Note that the effect is *quadratic* in the photon energy, and proportional to the distance travelled.

This effect has been used to constrain LV in the dimension three (Chern-Simons) [55], four [53] and five [30, 33, 34] terms. The constraint shown in the figure derives from the recent report [31] of a high degree of polarization of MeV photons from GRB021206. The data analysis has been questioned [32], so we shall have to wait and see if it is confirmed. The next best constraint on the dimension five term was deduced by Gleiser and Kozameh [30] using UV light from distant galaxies. While ten orders of magnitude weaker, it is still very strong, $|\xi| \lesssim 2 \times 10^{-4}$.

Photon Time of Flight

Photon time of flight constraints [57] limit differences in the arrival time at Earth for photons originating in a distant event [26, 56]. Time of flight can vary with energy since the LV term in the group velocity is $\xi k/M$. The arrival time difference for wave-vectors k_1 and k_2 is thus

$$\Delta t = \xi(k_2 - k_1)d/M \; , \tag{15}$$

which is proportional to the energy difference and the distance travelled. Using the EFT result (10), the velocity difference of the two polarizations at a given energy is $2|\xi|k/M$, at least twice as large as the one arising from energy differences. However, the time of flight constraint remains many orders of magnitude weaker than the birefringence one from polarization rotation. In Fig. 3 we use the EFT improvement of the constraint of Biller et al. [57] (this is the best constraint to date for which a reliable distance is known), which yields $|\xi| < 63$.

Vacuum Čerenkov Effect, Inverse Compton Electrons

In the presence of LV the process of vacuum Čerenkov radiation $e \to e\gamma$ can occur. For example, if the photon dispersion is unmodified and the electron parameter η (for one helicity) is positive, then the electron group velocity $v_g = 1 - (m^2/2p^2) + (\eta p/M) + \cdots$ exceeds the speed of light when

$$p_{\rm th} = (m^2 M/2\eta)^{1/3} \simeq 11\,{\rm TeV}\,\eta^{-1/3} \; . \tag{16}$$

This turns out to be the threshold energy for the vacuum Čerenkov process with emission of a zero energy photon, which we call the soft Čerenkov threshold. There is also the possibility of a hard Čerenkov threshold [5, 58]. For example, if the electron dispersion is unmodified and the photon parameter ξ is negative then at sufficiently high electron energy the emission of an energetic positive helicity photon is possible. This hard Čerenkov threshold occurs at $p_{\rm th} = (-4m^2M/\xi)^{1/3}$, and the emitted photon carries away half the incoming electron momentum. It turns out that the threshold is soft when both $\eta > 0$ and $\xi \geq -3\eta$, while it is hard when both $\xi < -3\eta$ and $\xi < \eta$. The hard threshold in the general case is given by $p_{\rm th} = (-4m^2M(\xi+\eta)/(\xi-\eta)^2)^{1/3}$, and the photon carries away a fraction $(\xi - \eta)/2(\xi + \eta)$ of the incoming momentum. In the general case at threshold, neither the incoming nor outgoing electron group velocity is equal to the photon group velocity, so the hard Čerenkov effect cannot simply be interpreted as being due to faster than light motion of a charged particle.

The inverse Compton (IC) Čerenkov constraint uses the electrons of energy up to 50 TeV inferred via the observation of 50 TeV gamma rays from the Crab nebula which are explained by IC scattering. (The implications of a possible high energy population of positrons is discussed below.) Since the vacuum Čerenkov rate is orders of magnitude higher than the IC scattering rate, that process must not occur for these electrons [5, 25]. (For a study of the vacuum Čerenkov process in the Maxwell-Chern-Simons limit of the standard model extension see [59].) The absence of the soft Čerenkov threshold up to 50 TeV produces the vertical IC Čerenkov line in Fig. 3. One can see from (16) that this yields a constraint on η of order $(11\ \text{TeV}/50\ \text{TeV})^3 \sim 10^{-2}$. It could be that only one electron helicity produces the IC photons and the other loses energy by vacuum Čerenkov radiation. Hence we can infer only that at least one of η_+ and η_- satisfies the bound.

We do not indicate the hard IC Čerenkov threshold constraint in Fig. 3 since it is superseded by the hard synchrotron Čerenkov constraint discussed below.

Crab Synchrotron Emission

A constraint complementary to the Čerenkov one was derived in [38] by making use of the very high energy electrons that produce the highest frequency synchrotron radiation in the Crab nebula. For negative values of η the electron has a maximal group velocity less than the speed of light, hence there is a maximal synchrotron frequency that can be produced regardless of the electron energy [38]. In the Lorentz invariant case these electrons must have an energy of at least 1500 TeV, which suggests that we should be able to obtain a constraint many orders of magnitude stronger than the IC Čerenkov one. We now explain how this indeed comes about.

Cycling electrons in a magnetic field B emit synchrotron radiation with a spectrum that sharply cuts off at a frequency ω_c given by the formula

$$\omega_c = \frac{3}{2} eB \frac{\gamma^3(E)}{E} , \tag{17}$$

where $\gamma(E) = (1 - v^2(E)/c^2)^{-1/2}$. Here $v(E)$ is the electron group velocity, and c is the usual low energy speed of light. (As shown in [38] the photon energy is low enough to neglect any possible LV correction as long as $|\xi| \lesssim 10^{11}(-\eta)^{4/3}$.) The formula (17) is based on the electron trajectory for a given energy in a given magnetic field, the radiation produced by a given current, and the relativistic relation between energy and velocity. As explained in [38], and also in some more detail in [60] (which was written in response to the criticism of [39]), only the last of these ingredients is significantly affected by LV in the EFT framework we are considering. (See also [61] for another demonstration that the electron trajectory is essentially unchanged.) Hence (17) holds in that framework.

In standard relativistic physics, $E = \gamma m$, so the energy dependence in (17) is entirely through the factor γ^2, which grows without bound as the energy grows. In the LV case, the maximum synchrotron frequency $\omega_c^{\max}$ is obtained by maximizing $\o_c$ (17) with respect to the electron energy, which amounts to maximizing $\gamma^3(E)/E$. Using the difference of group velocities

$$c - v(E) \simeq \frac{m^2}{2E^2} - \eta \frac{E}{M} , \tag{18}$$

we find that this maximization yields

$$\omega_c^{\max} = 0.34\,\frac{eB}{m}(-\eta m/M)^{-2/3}\,. \tag{19}$$

This maximum frequency is attained at the energy $E_{\max} = (-2m^2M/5\eta)^{1/3} = 10\,(-\eta)^{-1/3}$ TeV. This is higher than the energy that produces the same cutoff frequency in the Lorentz invariant case, but only by a factor of order unity.

The rapid decay of synchrotron emission at frequencies larger than ω_c implies that most of the flux at a given frequency in a synchrotron spectrum is due to electrons for which ω_c is above that frequency. Thus $\omega_c^{\max}$ must be greater than the maximum observed synchrotron emission frequency $\o_{\rm obs}$. This yields the constraint

$$\eta > -\frac{M}{m}\left(\frac{0.34\,eB}{m\o_{\rm obs}}\right)^{3/2}\,. \tag{20}$$

The Crab synchrotron emission has been observed to extend at least up to energies of about 100 MeV [62], just before the inverse Compton hump begins to contribute to the spectrum. The magnetic field in the emission region has been estimated by several methods which agree on a value between 0.15–0.6 mG (see e.g. [63] and references therein.) Two of these methods, radio synchrotron emission and equipartition of energy, are insensitive to Planck suppressed Lorentz violation, hence we are justified in adopting a value of this order for the purpose of constraining Lorentz violation. We use the largest value 0.6 mG for B, since it yields the weakest constraint.

Our prior work assumed the high energy Crab radiation was produced purely by electrons, not positrons. We consider here first this case. Then we infer that at least one of the two parameters $\eta_\pm$ must be greater than -7×10^{-8}. We cannot constrain both η parameters in this way since it could be that all the Crab synchrotron radiation is produced by electrons of one helicity.

Combined Synchrotron and IC Čerenkov Constraint

The η satisfying the synchrotron constraint *must be the same* η *as satisfies the IC Čerenkov constraint discussed above.* If the synchrotron η did not satisfy the IC Čerenkov constraint, the energy of these synchrotron electrons would necessarily be under 50 TeV, rather than over the Lorentz invariant value of 1500 TeV. The Crab spectrum is well accounted for with a single population of electrons responsible for both the synchrotron radiation and the IC γ-rays. If there were enough extra electrons to produce the observed synchrotron flux with thirty times less energy per electron, then those of the other helicity would produce too many IC γ-rays [33], unless they were far fewer in number in just the right proportion to agree with the self-consistent single population model. While possible, such a conspiracy seems highly unlikely. It is important that the *same* η, i.e. either η_+ or η_-, satisfies both the synchrotron and the IC Čerenkov constraints. Otherwise, both constraints could have been satisfied by having one of these two parameters arbitrarily large and negative, and the other arbitrarily large and positive.[5]

Possible Helicity Dependence Constraint

As alluded to above, a constraint on helicity dependence of the electron parameter η might be possible using the Crab Nebula. Suppose that η_- is below the synchrotron constraint (i.e. $\eta_- < -7\times10^{-8}$), so that η_+ must satisfy both the synchrotron and Čerenkov constraints as explained above. Then positive helicity electrons must have an energy of at least 50 TeV to produce the observed synchrotron radiation. These must not decay to negative helicity electrons (since those would be unable to produce the synchrotron emission). This would require that the transition energy (discussed in the helicity dependence section above) be greater than 50 TeV if the decay rate is fast enough. This would yield the constraint $\eta_+ - \eta_- < 10^{-2}$.

Possible Role of Positrons

If the population of high energy charges includes positrons as well as electrons, as in some models [64], then the above constraint analysis must be modified. The reasoning discussed so far implies only that at least one of the four parameters $\pm\eta_\pm$ satisfies both the synchrotron and IC Čerenkov constraints, since the emitting charge could be either an electron or a positron. In effect, this reduces to the statement that one of $|\eta_\pm|$ satisfies the IC Cerenkov constraint. We are currently investigating what constraints can be inferred if the amount of radiation produced by each of the four populations of charges is accounted for more quantitatively.

Vacuum Čerenkov Effect, Synchrotron Electrons

The existence of the synchrotron producing electrons can be exploited to extend the vacuum Čerenkov constraint. For a given η satisfying the synchrotron bound, some definite electron energy $E_{\rm synch}(\eta)$ must be present to produce the observed synchrotron radiation. (This is higher for negative η and lower for positive η than the Lorentz invariant value [38].) Values of $|\xi|$ for which the vacuum Čerenkov threshold is lower than $E_{\rm synch}(\eta)$ for either photon helicity can therefore be excluded [33]. This is always a hard photon threshold, since the soft photon threshold occurs when the electron group velocity reaches the low energy speed of light, whereas the velocity required to produce any finite synchrotron frequency is smaller than this.

[5] We thank G. Amelino-Camelia for focusing our attention on this point

Photon Decay

In the presence of LV the process of photon decay $\gamma \to e^+e^-$ can occur. For example, if the electron dispersion is unmodified and the photon parameter ξ is positive, the positive helicity photon decays above the threshold energy $k_{\text{th}} = (4m^2M/\xi)^{1/3}$. If instead the photon dispersion is unmodified and if electron and positron have the same dispersion with $\eta < 0$, then the threshold occurs at $k_{\text{th}} = (-8m^2M/\eta)^{1/3}$. The threshold for general ξ and η is found in [5, 58].

Contrary to relativistic intuition, it turns out that when $\eta < \xi < 0$ the electron and positron are not created with the same momentum. The reason (cf. Sect. 4) is the electron and positron energy functions $E(p)$ have negative curvature at the threshold value of p. If the two momenta were equal, the energy of the final state at fixed momentum could be lowered by making the momentum of one particle smaller and one larger by an equal amount.

Previous work on observational constraints using photon decay and photon absorption (to be discussed below) were carried out before it was known how the dispersion depends on helicity and particle vs. anti-particle. Since these constraints are in any case not competitive now with others, we have not attempted to fully account for these relations. Here we just make a few remarks.

The strongest limit on photon decay came from the highest energy photons known to propagate, which at the moment are the 50 TeV photons observed from the Crab nebula [5, 58]. These photons must not decay before reaching the earth, so we can rule out those LV parameters that lead to a threshold below 50 TeV, provided the decay rate is fast enough.

Since we do not know the polarization of the observed photons however, we can only exclude regions where *both* photon polarizations decay. Recall that according to (10) positive and negative helicity photons have opposite parameters $\pm\xi$. A positive helicity photon carries a spin angular momentum of one along the direction of motion. At threshold, where all momenta are aligned, the electron and positron must therefore both have positive helicity. Likewise a left-handed photon decays at threshold into a negative helicity pair. Consider first the case $\eta_- = -\eta_+$ so that, according to (11), the electron and positron have the same dispersion parameter. Then the outgoing pair both have parameter η_+ for a positive helicity incoming photon and $-\eta_+$ for a negative helicity one. We can then exclude those parameters for which *both* (ξ, η_+) and $(-\xi, -\eta_+)$ lead to photon decay thresholds below 50 TeV. The allowed region is the intersection of that from the old photon decay constraint [5, 58] with its reflection about the ξ and η axes. It is a pair of wedges in the upper-right and bottom left quadrants. Numerical work shows that this wedge pattern is maintained for different choices of η_- relative to η_+, however the exact orientation and shape of the wedges varies. A complete analysis of constraints would also require examination of above threshold processes when the outgoing particles have orbital angular momentum and hence helicities that are not determined solely by the incoming photon.

Photon Absorption

A process related to photon decay is photon absorption, $\gamma\gamma \to e^+e^-$. Unlike photon decay, this is allowed in Lorentz invariant QED. If one of the photons has energy ω_0, the threshold for the reaction occurs in a head-on collision with the second photon having the momentum (equivalently energy) $k_{\rm LI} = m^2/\omega_0$. For $k_{\rm LI} = 10$ TeV (which will be relevant for the observational constraints) the soft photon threshold ω_0 is approximately 25 meV, corresponding to a wavelength of 50 microns.

In the presence of Lorentz violating dispersion relations the threshold for this process is in general altered, and the process can even be forbidden. Moreover, as noticed by Kluźniak [65], in some cases there is an upper threshold beyond which the process does not occur.[6] The lower and upper thresholds for photon annihilation as a function of the two parameters ξ and η were obtained in [5], before the helicity dependence required by EFT was appreciated. As the soft photon energy is low enough that its LV can be ignored, this corresponds to the case where electrons and positrons have the same LV terms. The analysis is rather complicated. In particular it is necessary to sort out whether the thresholds are lower or upper ones, and whether they occur with the same or different pair momenta.

The photon absorption constraint, neglecting helicity dependent effects, came from the fact that LV can shift the standard QED threshold for annihilation of multi-TeV γ-rays from nearby blazars, such as Mkn 501, with the ambient infrared extragalactic photons [5, 58, 60, 65, 66, 67, 68]. LV depresses the rate of absorption of one photon helicity, and increases it for the other. Although the polarization of the γ-rays is not measured, the possibility that one of the polarizations is essentially unabsorbed appears to be ruled out by the observations which show the predicted attenuation [68]. The electron and positron spin angular momenta add to at most one. At threshold, where the collision is head-on, the photons must therefore have opposite helicity, and hence the electron and positron have opposite helicity. According to (11), they therefore have opposite LV parameters. The threshold analysis has not been redone to account for this.

Vacuum Photon Splitting

Another forbidden QED process that is allowed in the presence of LV is vacuum photon splitting into N photons, $\gamma \to N\gamma$. Unlike the other processes considered here, this would be a loop effect. The lowest order Feynman diagram contributing would be a fermion loop with various photon lines attached.

[6] Our results agree with those of [65] only in certain limiting cases.

The process has no threshold, so whether or not it can be used to set constraints depends on the rate.

Aspects of vacuum photon splitting have been examined in [5, 69]. An estimate of the rate, independent of the particular form of the Lorentz violating theory, was given in [5]. It was argued that a lower bound on the lifetime is $\delta^{-4}E^{-1}$, where δ is a Lorentz violating factor. For a photon of energy 50 TeV, this is $10^{-29}\delta^{-4}$ seconds. Such 50 TeV photons arrive from the Crab nebula, about 10^{13} seconds away, so the best constraint (i.e. if there is is no further small parameter such as α^N or $1/16\pi^2$ in the decay rate) we could possibly get on δ from photon splitting is $\delta \lesssim 10^{-10}$.

For a p^n LV term with $n = 2$ in the dispersion relation, this is not competitive with the other constraints already obtained. For higher n, each contribution arising from an operator of dimension greater than four will be suppressed by at least one inverse power of the scale M. For example, contributions from $n = 3$ would yield $\delta \sim \xi E/M$. In this case the strongest conceivable constraint on ξ would be of order $\xi \lesssim 10^4$, and even this is not competitive with the other constraints.

5.3 Constraints at $O(E/M)$ from UHE Cosmic Rays

If ultra-high energy cosmic rays (UHECR) are (as commonly assumed) protons, then we can derive strong constraints on $n = 3$ type dispersion by a) the absence of a vacuum Čerenkov effect at GZK energies and b) the position of the GZK cutoff. For a soft emitted photon with a long wavelength, the partonic structure of a UHECR proton is presumably irrelevant. In this case we can treat the proton as a point particle as in the QED analysis. With a GZK proton of energy 5×10^{19} eV the constraint from the absence of a vacuum Čerenkov effect is $\eta < O(10^{-14})$. For a hard emitted photon, the partonic nature of the proton is important and the relevant mass scale will involve the quark mass. The exact calculation considering the partonic structure for $n = 3$ has not been performed, however the threshold region will be similar to that in [5]. The allowed region in the $\eta - \xi$ plane will be bounded on the right by the ξ axis (within a few orders of magnitude of 10^{-14}) and below by the line $\xi = \eta$ [5]. These constraints apply to only one helicity of proton and photon, since the UHECR could consist all of a single helicity. Also the different quarks could have different dispersion parameters. See however Sect. 5.4 for remarks on the approach of [9] which can be applied to deduce combined constraints in this case.

If the GZK cutoff is observed in its predicted place, this will place limits on the parameters η_p and η_π. For example, if the GZK cutoff is eventually observed to be somewhere between 2 and 7 times 10^{19} eV then there are strong constraints of $O(10^{-11})$ on η_p and η_π [5]. As a final comment, an interesting possible consequence of LV is that with upper thresholds, one could possibly reconcile the AGASA and Hi-Res/Fly's Eye experiments. Namely, one can place an upper threshold below 10^{21} eV while keeping the GZK threshold near 5×10^{19} eV. Then the cutoff would be "seen" at lower energies but extra flux would still be present at energies above 10^{20} eV, potentially explaining the AGASA results [5]. The region of parameter space for this scenario is terribly small, however, again of $O(10^{-11})$.

5.4 Constraints at $O(E^2/M^2)$?

As previously mentioned, CPT symmetry alone could exclude the dimension five LV operators that give $O(E/M)$ modifications to particle dispersion relation, and in any case the constraints on those have become nearly definitive. Hence it is of interest to ask about the $O(E^2/M^2)$ modifications. We close with a brief discussion of the constraints that might be possible on those, i.e. constraints at $O(E^2/M^2)$.

As discussed above, the strength of constraints can be estimated by the requirement $\eta_4 p^4/M^2 \lesssim m^2$, which yields

$$\eta_4 \lesssim \left(\sqrt{\frac{m}{1\,\mathrm{eV}}}\frac{100\,\mathrm{TeV}}{p}\right)^4 . \tag{21}$$

Thus, for electrons, an energy around 10^{17} eV is needed for an order unity constraint on η_4, and we are probably not going to see any effects directly from such electrons.

For protons an energy $\sim 10^{18}$ eV is needed. This is well below the UHE cosmic ray energy cutoff, hence if and when Auger [7] confirms the identity of UHE cosmic rays as protons at the GZK cutoff, we will obtain an impressive constraint of order $\eta_4 \lesssim 10^{-5}$ from the absence of vacuum Čerenkov radiation for 10^{20} eV protons. From the fact that the GZK threshold is not shifted, we will obtain a constraint of order $\eta_4 \gtrsim -10^{-2}$, assuming equal η_4 values for proton and pion.

In fact, if one assumes the cosmic rays near but below the GZK cutoff are hadrons, one already obtains a strong bound [9]. Depending on the species dependence of the LV coefficients, bounds of order 10^{-2} or better can be placed on η_4. The bounds claimed in [9] are actually two sided, and it is worthwhile to explain how such bounds come about for a single source particle. Up to this point it has been necessary to use at least *two* reactions with different source particles to derive a two sided bounds. For example, the Crab constraints rely on the existence of both 50 TeV electrons and photons, treating each as a fundamental particle with its own LV coefficient. In contrast, the two sided bounds in [9] are derived by using a parton model for particles where the LV coefficients apply to the constituent partons. By considering many different outgoing particle spectra from an incoming hadron in combination with the parton approach the authors of [9] are able to find sets of reactions that yield two sided bounds. Hence, the parton approach is extremely useful as it dramatically increases the number of constraints that can be derived from a single incoming particle. However, it also requires more assumptions about the behavior of the parton distributions at cosmic ray energies.

Impressive constraints might also be obtained from the absence of neutrino vacuum Čerenkov radiation: putting in 1 eV for the mass in (21) yields an order unity constraint from 100 TeV neutrinos, but only if the Čerenkov *rate* is high enough. The rate will be low, since it proceeds only via the non-local charge structure of the neutrino. Recent calculations [70] have shown that the rate is not high enough at that energy. However, for 10^{20} eV UHE neutrinos, which may be observed by the proposed EUSO and/or OWL satellite observatories, the rate will be high enough to derive a strong constraint. The value of the constraint would depend on the emission rate, which has not yet been computed. For a *gravitational* Čerenkov reaction, the rate (which is lower but easier to compute than the electromagnetic rate) would be high enough for a neutrino from a distant source to radiate provided $\eta_4 \gtrsim 10^{-2}$. Hence in this case one might obtain a constraint of order $\eta_4 \lesssim 10^{-2}$, or stronger in the electromagnetic case.

A time of flight constraint at order $(E/M)^2$ might be possible [71] if gamma ray bursts produce UHE ($\sim 10^{19}$ eV) neutrinos, as some models predict, via limits on time of arrival differences of such UHE neutrinos vs. soft photons (or gravitational waves). Another possibility is to obtain a vacuum birefringence constraint with higher energy photons [34], although such a constraint would be less powerful since EFT does not imply that the parameters for opposite polarizations are opposite at order $(E/M)^2$. If future GRB's are found to be polarized at ~ 100 MeV, that could provide a birefringence constraint $|\xi_{4+} - \xi_{4-}| \lesssim 1$.

6 Conclusion

At present there are only hints, but no compelling evidence for Lorentz violation from quantum gravity. Moreover, even if LV is present, the use of EFT for its low energy parametrization is not necessarily valid. Nevertheless, we believe that the constraints derived from the simple ideas discussed here are still important. They allow tremendous advances in observational reach to be applied in a straightforward manner to limit reasonable possibilities that might arise from fundamental Planck scale physics. Such guidance is especially welcome for the field of quantum gravity, which until the past few years has had little connection with observed phenomena.

References

1. V. F. Mukhanov, H. A. Feldman and R. H. Brandenberger, "Theory Of Cosmological Perturbations. Part 1. Classical Perturbations. Part 2. Quantum Theory Of Perturbations. Part 3. Extensions," Phys. Rept. **215**, 203 (1992).

2. G. Amelino-Camelia, "Are we at the dawn of quantum-gravity phenomenology?," Lect. Notes Phys. **541**, 1 (2000) [arXiv:gr-qc/9910089].
3. K. Greisen, "End To The Cosmic Ray Spectrum?," Phys. Rev. Lett. **16**, 748 (1966); G. T. Zatsepin and V. A. Kuzmin, "Upper Limit Of The Spectrum Of Cosmic Rays," JETP Lett. **4**, 78 (1966) [Pisma Zh. Eksp. Teor. Fiz. **4**, 114 (1966)].
4. F. W. Stecker, "Cosmic physics: The high energy frontier," J. Phys. G **29**, R47 (2003) [arXiv:astro-ph/0309027].
5. T. Jacobson, S. Liberati and D. Mattingly, "Threshold effects and Planck scale Lorentz violation: Combined constraints from high energy astrophysics," Phys. Rev. D **67**, 124011 (2003) [arXiv:hep-ph/0209264].
6. D. De Marco, P. Blasi and A. V. Olinto, "On the statistical significance of the GZK feature in the spectrum of ultra high energy cosmic rays," Astropart. Phys. **20**, 53 (2003) [arXiv:astro-ph/0301497].
7. http://www.auger.org/
8. S. R. Coleman and S. L. Glashow, "Evading the GZK cosmic-ray cutoff," [arXiv:hep-ph/9808446]. For previous suggestions relating LV to modification of the GZK cutoff see Kirzhnits and Chechin[10] and Gonzalez-Mestres[21].
9. O. Gagnon and G. D. Moore, "Limits on Lorentz violation from the highest energy cosmic rays," arXiv:hep-ph/0404196.
10. See, e.g. P. A. M. Dirac, "Is there an aether?," Nature **168**, 906–907 (1951); J. D. Bjorken, "A dynamical origin for the electromagnetic field," Ann. Phys. **24**, 174 (1963); P. Phillips, "Is the graviton a Goldstone boson?," Physical Review **146**, 966 (1966); D.I. Blokhintsev, Usp. Fiz. Nauk. **89**, 185 (1966) [Sov. Phys. Usp. **9**, 405 (1966)]; L.B. Rédei, "Validity of special relativity at small distances and the velocity dependence of the muon lifetime," Phys. Rev. 162, 1299-1301 (1967); T. G. Pavlopoulos, "Breakdown Of Lorentz invariance," Phys. Rev. **159**, 1106 (1967); D.A. Kirzhnits and V.A. Chechin, "Ultra-High-Energy Cosmic Rays and a Possible Generaliation of Relativistic Theory," Yad. Fiz. **15**, 1051 (1972) [Sov. J. Nucl. Phys. **15**, 585 (1972)].
11. G. Amelino-Camelia, "Relativity in space-times with short-distance structure governed by an observer-independent (Planckian) length scale," Int. J. Mod. Phys. D **11**, 1643 (2002) [arXiv:gr-qc/0210063].
12. C.M. Will and K. Nordvedt, Jr., "Conservation Laws and Preferred Frames in Relativistic Gravity. I. Preferred-Frame Theories and an Extended PPN Formalism," Astrophys. J. **177**, 757 (1972); K. Nordvedt, Jr. and C.M. Will, "Conservation Laws and Preferred Frames in Relativistic Gravity. II. Experimental Evidence to Rule Out Preferred-Frame Theories of Gravity," Astrophys. J. **177**, 775 (1972); R.W. Hellings and K. Nordvedt, Jr., "Vector-metric theory of gravity," Phys. Rev. **D7**, 3593 (1973).
13. H. B. Nielsen and M. Ninomiya, "Beta Function In A Noncovariant Yang-Mills Theory," Nucl. Phys. B **141**, 153 (1978); S. Chadha and H. B. Nielsen, "Lorentz Invariance As A Low-Energy Phenomenon," Nucl. Phys. B **217**, 125 (1983); H. B. Nielsen and I. Picek, "Lorentz Noninvariance," Nucl. Phys. B **211**, 269 (1983) [Addendum-ibid. B **242**, 542 (1984)]; J. R. Ellis, M. K. Gaillard, D. V. Nanopoulos and S. Rudaz, "Uncertainties In The Proton Lifetime," Nucl. Phys. B **176**, 61 (1980). A. Zee, "Perhaps Proton Decay Violates Lorentz Invariance," Phys. Rev. D **25**, 1864 (1982).

14. See, for example, M. Gasperini, "Singularity prevention and broken Lorentz symmetry", Class. Quantum Grav. **4**, 485 (1987); "Repulsive gravity in the very early Universe", Gen. Rel. Grav. **30**, 1703 (1998); and references therein.
15. T. Jacobson and D. Mattingly, "Generally covariant model of a scalar field with high frequency dispersion and the cosmological horizon problem," Phys. Rev. D **63**, 041502 (2001). [hep-th/0009052];
"Gravity with a dynamical preferred frame," Phys. Rev. D **64**, 024028 (2001).
16. See M. Haugan and C. Will, Physics Today, May 1987; C.M. Will, *Theory and Experiment in Gravitational Physics* (Cambridge Univ. Press, 1993), and references therein.
17. V. A. Kostelecky and S. Samuel, "Spontaneous Breaking Of Lorentz Symmetry In String Theory," Phys. Rev. D **39**, 683 (1989).
18. T. Jacobson, "Black hole evaporation and ultrashort distances," Phys. Rev. D **44**, 1731 (1991).
19. W. G. Unruh, "Dumb holes and the effects of high frequencies on black hole evaporation," Phys. Rev. D **51**, 2827-2838 (1995), [arXiv:gr-qc/9409008].
20. See J. Martin, this volume; or, e.g., J. Martin and R. Brandenberger, "On the dependence of the spectra of fluctuations in inflationary cosmology on trans-Planckian physics," Phys. Rev. D **68**, 063513 (2003) [arXiv:hep-th/0305161] and references therein.
21. L. Gonzalez-Mestres, "Lorentz symmetry violation and high-energy cosmic rays,", [arXiv:physics/9712005].
22. D. Colladay and V. A. Kostelecky, "Lorentz-violating extension of the standard model," Phys. Rev. D **58**, 116002 (1998), [arXiv:hep-ph/9809521].
23. V. A. Kostelecky, *Proceedings of the Second Meeting on CPT and Lorentz Symmetry*, Bloomington, USA, 15-18 August 2001. Singapore, World Scientific (2002).
24. M. Takeda et al., "Extension of the cosmic-ray energy spectrum beyond the predicted Greisen-Zatsepin-Kuzmin cutoff," Phys. Rev. Lett. **81**, 1163 (1998) [arXiv:astro-ph/9807193].
25. S. R. Coleman and S. L. Glashow, "High-energy tests of Lorentz invariance," Phys. Rev. D **59**, 116008 (1999) [arXiv:hep-ph/9812418].
26. G. Amelino-Camelia, J. R. Ellis, N. E. Mavromatos, D. V. Nanopoulos and S. Sarkar, "Potential Sensitivity of Gamma-Ray Burster Observations to Wave Dispersion in Vacuo," Nature **393**, 763 (1998) [arXiv:astro-ph/9712103].
27. R. Gambini and J. Pullin, "Nonstandard optics from quantum spacetime," Phys. Rev. D **59**, 124021 (1999) [arXiv:gr-qc/9809038].
28. C. N. Kozameh and M. F. Parisi, "Lorentz invariance and the semiclassical approximation of loop quantum gravity," Class. Quant. Grav. **21**, 2617 (2004) [arXiv:gr-qc/0310014].
29. J. Alfaro, M. Reyes, H. A. Morales-Tecotl and L. F. Urrutia, "On alternative approaches to Lorentz violation in loop quantum gravity inspired models," arXiv:gr-qc/0404113.
30. R. J. Gleiser and C. N. Kozameh, "Astrophysical limits on quantum gravity motivated birefringence," Phys. Rev. D **64**, 083007 (2001) [arXiv:gr-qc/0102093].
31. W. Coburn and S. E. Boggs, "Polarization of the prompt gamma-ray emission from the gamma-ray burst of 6 December 2002," Nature **423**, 415 (2003) [arXiv:astro-ph/0305377].

32. R. E. Rutledge and D. B. Fox, "Re-Analysis of Polarization in the Gamma-ray flux of GRB 021206," [arXiv:astro-ph/0310385]; S. E. Boggs and W. Coburn, "Statistical Uncertainty in the Re-Analysis of Polarization in GRB021206," [arXiv:astro-ph/0310515]. C. Wigger et al., "Gamma-Ray Burst Polarization: Limits from RHESSI Measurements," [arXiv:astro-ph/0405525].
33. T. A. Jacobson, S. Liberati, D. Mattingly and F. W. Stecker, "New limits on Planck scale Lorentz violation in QED," Phys. Rev. Lett. **93**, 021101 (2004) [arXiv:astro-ph/0309681].
34. I. G. Mitrofanov, "A constraint on canonical quantum gravity", Nature **426**, 139 (2003).
35. R. J. Protheroe and H. Meyer, "An infrared background TeV gamma ray crisis?," Phys. Lett. B **493**, 1 (2000) [arXiv:astro-ph/0005349].
36. A. K. Konopelko, A. Mastichiadis, J. G. Kirk, O. C. de Jager and F. W. Stecker, "Modelling the TeV gamma-ray spectra of two low redshift AGNs: Mkn 501 and Mkn 421," Astrophys. J. **597**, 851 (2003) [arXiv:astro-ph/0302049].
37. R. C. Myers and M. Pospelov, "Experimental challenges for quantum gravity," Phys. Rev. Lett. **90**, 211601 (2003) [arXiv:hep-ph/0301124].
38. T. Jacobson, S. Liberati and D. Mattingly, "Lorentz violation and Crab synchrotron emission: A new constraint far beyond the Planck scale," Nature **424**, 1019 (2003) [arXiv:astro-ph/0212190].
39. G. Amelino-Camelia, "Improved limit on quantum-spacetime modifications of Lorentz symmetry from observations of gamma-ray blazars," [arXiv:gr-qc/0212002]; "A perspective on quantum gravity phenomenology," [arXiv:gr-qc/0402009].
40. E. Fischbach, M. P. Haugan, D. Tadic and H. Y. Cheng, "Lorentz Noninvariance And The Eotvos Experiments," Phys. Rev. D **32**, 154 (1985).
41. C. P. Burgess, J. Cline, E. Filotas, J. Matias and G. D. Moore, "Loop-generated bounds on changes to the graviton dispersion relation," JHEP **0203**, 043 (2002) [arXiv:hep-ph/0201082].
42. J. R. Ellis, N. E. Mavromatos, D. V. Nanopoulos and A. S. Sakharov, "Space-time foam may violate the principle of equivalence," arXiv:gr-qc/0312044.
43. R. Lehnert, "Threshold analyses and Lorentz violation," Phys. Rev. D **68**, 085003 (2003) [arXiv:gr-qc/0304013].
44. V. A. Kostelecky and R. Lehnert, "Stability, causality, and Lorentz and CPT violation," Phys. Rev. D **63**, 065008 (2001) [arXiv:hep-th/0012060].
45. V. A. Kostelecky and S. Samuel, "Gravitational Phenomenology In Higher Dimensional Theories And Strings," Phys. Rev. D **40**, 1886 (1989).
46. A. Perez and D. Sudarsky, "Comments on challenges for quantum gravity," [arXiv:gr-qc/0306113].
47. J. Collins, A. Perez, D. Sudarsky, L. Urrutia and H. Vucetich, "Lorentz invariance: An additional fine-tuning problem," [arXiv:gr-qc/0403053].
48. See e.g. M. Creutz, *Quarks, gluons and lattices* (Cambridge Univ. Press, 1985); G. Moore, "Informal Lectures on Lattice Gauge Theory," http://www.physics.mcgill.ca/~guymoore/latt_lectures.pdf.
49. S. G. Nibbelink and M. Pospelov, "Lorentz violation in supersymmetric field theories," arXiv:hep-ph/0404271.
50. D. Mattingly, T. Jacobson and S. Liberati, "Threshold configurations in the presence of Lorentz violating dispersion relations," Phys. Rev. D **67**, 124012 (2003) [arXiv:hep-ph/0211466].

51. D. Bear, R. E. Stoner, R. L. Walsworth, V. A. Kostelecky and C. D. Lane, “Limit on Lorentz and CPT violation of the neutron using a two-species noble-gas maser,” Phys. Rev. Lett. **85**, 5038 (2000) [Erratum-ibid. **89**, 209902 (2002)] [arXiv:physics/0007049].
52. R. Bluhm, “Lorentz and CPT tests in matter and antimatter,” Nucl. Instrum. Meth. B **221**, 6 (2004) [arXiv:hep-ph/0308281].
53. V. A. Kostelecky and M. Mewes, “Cosmological constraints on Lorentz violation in electrodynamics,” Phys. Rev. Lett. **87**, 251304 (2001) [arXiv:hep-ph/0111026]; Phys. Rev. D **66**, 056005 (2002) [arXiv:hep-ph/0205211].
54. B. R. Heckel et al., “Torsion balance test of spin coupled forces,” Proceedings of the International Conference on Orbis Scientiae, 1999, Coral Gables, (Kluwer, 2000); B. R. Heckel, http://www.npl.washington.edu/eotwash/publications/cpt01.pdf.
55. S. M. Carroll, G. B. Field and R. Jackiw, “Limits On A Lorentz And Parity Violating Modification Of Electrodynamics,” Phys. Rev. D **41**, 1231 (1990).
56. See Pavlopoulos in [10].
57. B. E. Schaefer, “Severe Limits on Variations of the Speed of Light with Frequency,” Phys. Rev. Lett. **82**, 4964 (1999) [astro-ph/9810479]; S. D. Biller et al., “Limits to quantum gravity effects from observations of TeV flares in active galaxies,” Phys. Rev. Lett. **83**, 2108 (1999) [arXiv:gr-qc/9810044]; P. Kaaret, “Pulsar radiation and quantum gravity,” Astronomy and Astrophysics, **345**, L32-L34 (1999) [arXiv:astro-ph/9903464]; J. R. Ellis, N. E. Mavromatos, D. V. Nanopoulos and A. S. Sakharov, “Quantum-gravity analysis of gamma-ray bursts using wavelets,” Astron. Astrophys. **402**, 409 (2003) [arXiv:astro-ph/0210124]; S. E. Boggs, C. B. Wunderer, K. Hurley and W. Coburn, “Testing Lorentz Non-Invariance with GRB021206,” arXiv:astro-ph/0310307, Astrophysical J. **611**, L77 (2004).
58. T. J. Konopka and S. A. Major, “Observational limits on quantum geometry effects,” New J. Phys. **4**, 57 (2002) [arXiv:hep-ph/0201184].
59. R. Lehnert and R. Potting, “Vacuum Cerenkov radiation,” arXiv:hep-ph/0406128; “The Cerenkov effect in Lorentz-violating vacua,” arXiv:hep-ph/0408285.
60. T. Jacobson, S. Liberati and D. Mattingly, “Comments on ‘Improved limit on quantum-spacetime modifications of Lorentz symmetry from observations of gamma-ray blazars’,” [arXiv:gr-qc/0303001].
61. J. R. Ellis, N. E. Mavromatos and A. S. Sakharov, “Synchrotron radiation from the Crab Nebula discriminates between models of space-time foam,” Astropart. Phys. **20**, 669 (2004) [arXiv:astro-ph/0308403].
62. F.A. Aharonian, A.M. Atoyan, “On the mechanisms of gamma radiation in the Crab Nebula”, Mon. Not. R. Astron. Soc. 278, 525 (1996).
63. A. M. Hillas et al., The Spectrum of TeV gamma rays from the crab nebula, Astrophysical J., **503**, 744 (1998).
64. A. Harding, “Gamma-ray pulsars: Models and predictions” in *High energy gamma-ray astronomy: international symposium* Heidelberg, Germany, 26-30 June 2000. Melville, NY, American Institute of Physics (2001) [arXiv:astro-ph/0012268].
65. W. Kluzniak, “Transparency Of The Universe To Tev Photons In Some Models Of Quantum Gravity,” Astropart. Phys. **11**, 117 (1999).

66. G. Amelino-Camelia and T. Piran, "Planck-scale deformation of Lorentz symmetry as a solution to the UHECR and the TeV-gamma paradoxes," Phys. Rev. D **64**, 036005 (2001) [arXiv:astro-ph/0008107].
67. F. W. Stecker and S. L. Glashow, "New tests of Lorentz invariance following from observations of the highest energy cosmic gamma rays," Astropart. Phys. **16**, 97 (2001) [arXiv:astro-ph/0102226].
68. F. W. Stecker, "Constraints on Lorentz invariance violating quantum gravity and large extra dimensions models using high energy gamma ray observations," Astropart. Phys. **20**, 85 (2003) [arXiv:astro-ph/0308214].
69. C. Adam and F. R. Klinkhamer, "Photon decay in a CPT-violating extension of quantum electrodynamics," Nucl. Phys. B **657**, 214 (2003) [arXiv:hep-th/0212028]; V. A. Kostelecky and A. G. M. Pickering, "Vacuum photon splitting in Lorentz-violating quantum electrodynamics," Phys. Rev. Lett. **91**, 031801 (2003) [arXiv:hep-ph/0212382]; C. Adam and F. R. Klinkhamer, "Comment on 'Vacuum photon splitting in Lorentz-violating quantum electrodynamics'," arXiv:hep-ph/0312153.
70. D. Mattingly and B. McElrath, To be published.
71. G. Amelino-Camelia, "Proposal of a second generation of quantum-gravity-motivated Lorentz-symmetry tests: Sensitivity to effects suppressed quadratically by the Planck scale," Int. J. Mod. Phys. D **12**, 1633 (2003) [arXiv:gr-qc/0305057].
72. T. Jacobson, S. Liberati and D. Mattingly, "Quantum gravity phenomenology and Lorentz violation," arXiv:gr-qc/0404067.

Introduction to Doubly Special Relativity

J. Kowalski-Glikman

Institute for Theoretical Physics, University of Wrocław
jurekk@ift.uni.wroc.pl

1 Introduction

What is the fate of Lorentz symmetry at Planck scale? This question was the main theme of the Winter School and, as the reader could see from the proceedings, there are many possible answers. Here I would like to describe one possibility, whose central postulate is that in spite of the fact that departures from Special Relativity are introduced at scales close to Planck scale, one keeps unchanged the central physical message of the theory of relativity, namely the equivalence of all (inertial) observers. This justifies the term *Relativity* in the title.

To be more specific, let us start with the set of postulates of Doubly Special Relativity[1] (I will use the acronym DSR in what follows) or Special Relativity with Two Observer Independent Scales, as proposed in [1, 2] (see also [3, 4].) These postulates can be formulate as follows.

- One assumes that the relativity principle holds, i.e., equivalence of all inertial observers in the sense of Galilean Relativity and Special Relativity is postulated.
- There are *two* observer independent scales: one of velocity c, identified with the speed of light[2], and second of dimension of mass κ (or length $\lambda = \kappa^{-1}$), identified with the Planck mass. Of course, it is assumed that in the limit $\kappa \to \infty$ DSR becomes the standard Special Relativity. This postulate is the reason for the term "Doubly". Since it turns out that the action

[1] Some authors prefer to use the name Deformed Special Relativity, fortunately leading to the same acronym

[2] Some readers may be confused already at this point since it is often claimed that DSR predicts dependence of the speed of massless particles on energy they carry, so that the speed of light is energy (and wavelength) dependent. Then the question arises to which speed this postulate refers to. As I will show below there are, arguably, good reasons to believe that in DSR the speed of light equals 1, independently of the energy.

J. Kowalski-Glikman: *Introduction to Doubly Special Relativity*, Lect. Notes Phys. **669**, 131–159 (2005)
www.springerlink.com

of symmetry generators must be deformed in DSR, one may talk about "Deformed Special Relativity".

It is a quite nontrivial problem, though, how these postulates can be realized in practice, given the fact that at the Planck scale we are to have to do with two scales of length and/or mass. Indeed, we know both from the theory and numerous experiments that in Special Relativity different observers do attribute different lengths and masses to the same measurements: as it is well known, we have to do with Lorentz-FitzGerald contraction and relativistic corrections to mass. How is it then possible to have at the same time relativity principle and the observer-independent scale of length or mass? It turns out that it is possible, but the price to pay is quite high, as one presumably must describe space-time in terms of non-commutative geometry, and to talk about space-time symmetries, one should use the language of quantum groups.

It should be noted also that as an immediate consequence of the postulates DSR theory should possess (like Galilean and Special Relativity theories) a ten dimensional group of symmetries, corresponding to rotations, boosts, and translations, which however, as a result of the presence of the second scale, cannot be just the linear Poincaré group . This immediately poses a problem. Namely, if we have a theory with observer independent scale of mass, then it follows that it should be expected that the standard Special Relativistic Casimir $E^2 - p^2 = m^2$ is to be replaced by some nonlinear mass-shell relation, between energy and three-momentum (which would involve the parameter κ[3].) Thus the second scale κ must be encoded into the mass-shell condition so that it is kept invariant by symmetry transformations. But then it follows that the speed of massless particles defined as $\partial E/\partial p$ would be dependent on the energy they carry, which makes it hard to understand what would be the operational meaning of the observer-independent speed of light. Below I will suggest ways out of this dilemma.

I should warn the reader that the construction of the theory of Doubly Special relativity is not completed yet; in fact we do not even have a single DSR candidate, which would satisfy all the requirements of internal and conceptual self-consistency. Nevertheless during the last three years many results have been obtained, and for example we now control pretty well the one particle sector of the theory, both technically and conceptually. However, many problems remain, for example, we still do not understand the multi-particle sector of DSR theory.

The structure of this notes corresponds to the structure of the lectures I gave at the Winter School. The next section corresponding to the first lecture is devoted to the questions whether and how DSR could emerge as an appropriate limit of quantum gravity. The complete answer to these questions is still unknown but we have some number of evidences suggesting that indeed DSR may be rooted in quantum gravity. The third section of these notes

[3] Note however that there exists a class of models of DSR, in which the dispersion relation between energy and momentum is not deformed (see below.)

is devoted to describing techniques used in a particular, best developed approach to DSR, based on the so-called κ-Poincaré algebra and κ-Minkowski space-time. In Sect. 4 I would like to describe main results obtained in the DSR framework, as well as bunch of open problems, mainly related to the multi-particle processes.

2 DSR from Quantum Gravity?

If the DSR idea is correct, it is quite natural to expect that Doubly Special Relativity emerges somehow as a limit of quantum gravity. It is rather clear why it must be so. In the standard Special Relativity we have only one scale, and there is no natural way in which another scale of mass and/or length could be introduced purely in Special Relativistic setting. On the other hand, in quantum gravity we have, in addition to the velocity scale c, three additional dimensionful constants, G, $\hbar$ (which I often set equal 1 in what follows), and (sometimes) the cosmological constant Λ. The immediate idea is that in the limiting procedure, in which the gravitational interactions as well as quantum effects become negligible, and the space-time becomes effectively flat (at least locally), some trace of these constants remains, giving rise to new observer-independent scale κ. In this section I will try to convince the reader that such scenario may indeed result from quantum theory of gravity.

Usually we take for granted that the $G \to 0$, (and possibly $\Lambda \to 0$ if we start with non-zero cosmological constant) limit of (quantum) gravity is just the Minkowski space-time. But perhaps this is not correct, and we are forced to take the limit (especially in the case in which point particles are present) such that either

1. $\lim_{G,\Lambda\to 0} \sqrt{\frac{G}{\Lambda}} = \kappa^{-1} \neq 0$, or alternatively,
2. $\lim_{G,\hbar\to 0} \sqrt{\frac{\hbar}{G}} = \kappa \neq 0$.

It is not clear which of these scenarios (if any) is realized in Nature, but there are some indirect evidences in favor of the claim that indeed it might be so.

Let us try to investigate the first scenario following the ideas presented in [5]. To this end let us consider first the three-dimensional quantum gravity with positive cosmological constant Λ. Then it is well known [6] that the excitations of $3d$ quantum gravity with cosmological constant transform under representations of the quantum deformed deSitter algebra $SO_q(3,1)$, with $z = \ln q$ behaving in the limit of small[4] $\Lambda\hbar^2/\kappa^2$ as $z \approx \sqrt{\Lambda}\hbar/\kappa$.

I will not discuss at this point the notion of quantum deformed algebras (Hopf algebras) in much details (the book [7] would be a good references for the reader who wants to study this exciting branch of mathematics.) It will

[4] Since in $3d$, the dimension of the gravitational constant is $1/kg$, we write $G = \kappa^{-1}$.

suffice to say that quantum algebras consist of several structures, the most important for our current purposes would be the universal enveloping algebra, which could be understand as an algebra of brackets among generators, which are equal to some analytic functions of them. Thus the quantum algebra is a generalization of a Lie algebra, and it is worth observing that the former reduces to the latter in an appropriate limit. Quantum algebras start playing an important role in various branches of theoretical physics; in particular, in some cases, they can play a role of relativistic symmetries in some field theoretical models (see an excellent, pedagogical exposition in [8].) In the case of quantum algebra $SO_q(3,1)$ the algebraic part looks as follows (the parameter z used below is related to q by $z = \ln q$)

$$\begin{aligned}
[M_{2,3}, M_{1,3}] &= \frac{1}{z}\sinh(zM_{1,2})\cosh(zM_{0,3}) \\
[M_{2,3}, M_{1,2}] &= M_{1,3} \\
[M_{2,3}, M_{0,3}] &= M_{0,2} \\
[M_{2,3}, M_{0,2}] &= \frac{1}{z}\sinh(zM_{0,3})\cosh(zM_{1,2}) \\
[M_{1,3}, M_{1,2}] &= -M_{2,3} \\
[M_{1,3}, M_{0,3}] &= M_{0,1} \\
[M_{1,3}, M_{0,1}] &= \frac{1}{z}\sinh(zM_{0,3})\cosh(zM_{1,2}) \\
[M_{1,2}, M_{0,2}] &= -M_{0,1} \\
[M_{1,2}, M_{0,1}] &= M_{0,2} \\
[M_{0,3}, M_{0,2}] &= M_{2,3} \\
[M_{0,3}, M_{0,1}] &= M_{1,3} \\
[M_{0,2}, M_{0,1}] &= \frac{1}{z}\sinh(zM_{1,2})\cosh(zM_{0,3})
\end{aligned} \tag{1}$$

Since this is our first encounter with quantum algebra let us pause for a moment to discuss its main features. First of all, we observe that on the right hand sides we do not have linear functions generators, as in the Lie algebra case, but some (analytic) functions of them. However we still assume that the brackets are antisymmetric and that Jacobi identity holds.

Exercise 1. Convince yourself by direct inspection that for the algebra (1) Jacobi identities indeed hold.

It follows from this observation that contrary to the Lie algebras case, we are now entitled to use any analytic functions of the initial set of generators as a basis of the quantum algebra (in the Lie algebra case we can only take linear combinations of them.) It should be stressed already at this point that quantum algebras possess more structures than just the enveloping algebra structure (for more details see [7]); some of them will be relevant in what follows. Note that in the limit $z \to 0$ the algebra (1) becomes the standard algebra $SO(3,1)$, and this is the reason for using the term $SO_q(3,1)$.

Exercise 2. Denote by $M^z_{\mu\nu}$ the generators of the algebra (1) and by $M_{\mu\nu}$ the generators of the standard $SO(3,1)$ algebra (obviously the equation $\lim_{z\to 0} M^z_{\mu\nu} = M_{\mu\nu}$ should hold.) Find explicit expressions for $M^z_{\mu\nu}$ as functions of $M_{\mu\nu}$ and z. (If this exercise happens to be too hard do that only up to the next-to-leading order in z.)

The $SO(3,1)$ Lie algebra is the three dimensional de Sitter algebra and it is well known how to obtain the three dimensional Poincaré algebra from it. First of all one has to single out the energy and momentum generators of right physical dimension (note that the generators $M_{\mu\nu}$ of (1) are dimensionless): one identifies three-momenta $P_\mu \equiv (E, P_i)$ ($\mu = 1,2,3,\ i = 1,2$) as appropriately rescaled generators $M_{0,\mu}$ and then one takes the Inömü–Wigner contraction limit (see, for example, [9].)

Let us try therefore to proceed in an analogous way and contract the algebra (1). To this aim we must first rescale some of the generators by an appropriate scale, provided by combination of dimensionful constants present in definition of the parameter z

$$\begin{aligned} E &= \sqrt{\Lambda}\hbar\, M_{0,3} \\ P_i &= \sqrt{\Lambda}\hbar\, M_{0,i} \\ M &= M_{1,2} \\ N_i &= M_{i,3} \end{aligned} \tag{2}$$

Taking now into account the relation $z \approx \sqrt{\Lambda}\hbar/\kappa$, which holds for small Λ, from

$$[M_{2,3}, M_{1,3}] = \frac{1}{z}\sinh(zM_{1,2})\cosh(zM_{0,3})$$

we find

$$[N_2, N_1] = \frac{\kappa}{\hbar\sqrt{\Lambda}}\sinh(\hbar\sqrt{\Lambda}/\kappa M)\cosh(E/\kappa) \tag{3}$$

Similarly from

$$[M_{0,2}, M_{0,1}] = \frac{1}{z}\sinh(zM_{1,2})\cosh(zM_{0,3})$$

we get

$$[P_2, P_1] = \sqrt{\Lambda}\hbar\kappa\,\sinh(\sqrt{\Lambda}\hbar/\kappa\, M)\cosh(E/\kappa) \tag{4}$$

Similar substitutions can be made in other commutators of (1). Now going to the contraction limit $\Lambda \to 0$, while keeping κ constant we obtain the following algebra

$$
\begin{aligned}
&[N_i, N_j] = -M\epsilon_{ij} \cosh(E/\kappa) \\
&[M, N_i] = \epsilon_{ij} N^j \\
&[N_i, E] = P_i \\
&[N_i, P_j] = \delta_{ij}\, \kappa \sinh(E/\kappa) \\
&[M, P_i] = \epsilon_{ij} P^j \\
&[E, P_i] = 0 \\
&[P_2, P_1] = 0
\end{aligned} \tag{5}
$$

This algebra is called the three dimensional κ-Poincaré algebra (in the standard basis.)

It turns out that this contracted algebra is again a quantum algebra, i.e., after the contraction all the additional structures of $SO_q(3,1)$ became the analogous structures of the new algebra (which is not obvious a priori because, in principle, it may happen that during the contraction procedure additional structures of the quantum algebra may become not well defined). This really nontrivial and remarkable result has been obtained in early nineties in [10, 11].

Let us pause for a moment here to make couple of comments. First of all, one easily sees that in the limit $\kappa \to \infty$ from the κ-Poincaré algebra algebra (5) one gets the standard Poincaré algebra. Second, we see that in this algebra both the Lorentz and translation sectors are deformed. However, as I have been stressing already, in the case of quantum algebras one is free to change the basis of generators in arbitrary, analytic way. It turns out that there exists such a change of basis that the Lorentz part of the algebra becomes classical (i.e., undeformed.) Such a basis, derived in [12], is called bicrossproduct (because of the remarkable bicrossproduct structure of the full quantum algebra, see [7]), and the Doubly Special Relativity model (both in 3 and 4 dimensions) based on such an algebra is called DSR1. In this basis the algebra looks as follows

$$
\begin{aligned}
&[N_i, N_j] = -\epsilon_{ij}\, M \\
&[M, N_i] = \epsilon_{ij} N^j \\
&[N_i, E] = P_i \\
&[N_i, P_j] = \delta_{ij}\, \frac{\kappa}{2}\left(1 - e^{-2E/\kappa} + \frac{\boldsymbol{P}^2}{\kappa^2}\right) - \frac{1}{\kappa} P_i P_j \\
&[M, P_i] = \epsilon_{ij} P^j \\
&[E, P_i] = 0 \\
&[P_1, P_2] = 0\ .
\end{aligned} \tag{6}
$$

Exercise 3. Derive explicit transformations from variables in (5) to variables in (6) (solution can be found in [12].)

Note now that the algebra (6) is exactly of the form needed in Doubly Special Relativity. By construction this is the algebra of symmetries of flat space, being an appropriate limit of the algebra of symmetries of states of quantum gravity. Moreover it manifestly contains the observer-independent scale of dimension of mass κ.

Exercise 4. Check that κ is the observer-independent scale in the sense that if $|\boldsymbol{P}| = \kappa$, then $\delta|\boldsymbol{P}| = 0$, where δ denotes the change under infinitesimal action of boosts (solution can be found in [3].)

This shows that, at least in principle, one can try to construct a theory, which satisfies principles of DSR, and that such a theory may be neither inconsistent, nor trivial. Of course to construct a theory of particle kinematics, with symmetries defined by (5), (6) much more is needed; for example we must know how to compose momenta for multiparticle systems, what is the form of conservation laws, etc. I will discuss these issues in the following sections below.

The algebras (5), (6) has been derived from the limit of the algebra of symmetries of three dimensional gravity, which, as it is well known, has some remarkable features, namely it is a topological field theory with no dynamical degrees of freedom. The question arises as to if it is possible to repeat this analysis in the most interesting, four dimensional case. One can expect that this latter case would be much more complex: to go to the appropriate limit reminding the Special Relativistic setting one should first switch off the dynamical degrees of freedom of gravity. The good news is that in the limit, in which the gravitational constant goes to zero, four dimensional gravity becomes a topological field theory again, reminding the three-dimensional situation. However, I must admit that it is not known if there exists a limit of four dimensional quantum gravity, which results in DSR theory. There are, however, some circumstantial evidences in favor of such a claim.

In the four-dimensional case the excitations of ground state[5] of a quantum gravity theory are conjectured to transform under representations of the quantum deformed de Sitter algebra $SO_q(3,2)$, with $z = \ln q$ behaving in the limit of small $\Lambda\kappa^{-2}$ as, $z \approx \Lambda\kappa^{-2}$ [13, 14, 15, 16][6]. Then (see [5] for more details) one again takes the limit, which this time is much more involved, since one must not only rescale variables, as it was done above, but also to renormalize them (see also [10]), in order to get finite result. It turns out that now we have to do with one parameter family of contractions, labelled by real, positive parameter r: for $0 < r < 1$ as a result of contraction one obtains the standard Poincaré algebra, for $r > 1$ the contraction does not exists and only for a single value $r = 1$ the contraction gives the desired four dimensional

[5] We restrict our attention to the ground state, because we are interested only in the limit in which all local degrees of freedom of quantum gravity are switched off. After all our goal is to formulate a theory which is to replace Special Relativity!

[6] From now on I put $\hbar$ equal 1.

κ-Poincaré algebra. It remains therefore an open problem whether and how quantum gravity singles out the value for r and is this value 1?

We see therefore that it is possible to obtain the DSR1 algebra by contracting the algebras of symmetries of quantum gravity, in dimensions 3 and 4. This strongly suggests that indeed this algebra would be an algebra of symmetries of particle kinematics taking part in the flat space. It is interesting therefore that, in some cases at least, there are traces of quantum gravity in this algebra. I must stress, however that it remains to prove rigorously that the algebra $SO_q(3,2)$ indeed plays the conjectured role in quantum gravity.

And now something completely different. In Special Relativity the Poincaré algebra plays dual role: it is an algebra of symmetries of space-time and at the same time it labels momenta and spin of a particle. Deformed Poincaré algebra should also play such a dual role, so now let us investigate the algebras of charges carried by point particles coupled to quantum gravity. As I will show in Sect. 3 below, in the DSR framework it turns out that the four momentum of a particle is not a point in the flat Minkowski space, as in Special Relativity, but instead, the manifold of momenta is a curved manifold of constant curvature, κ^{-2} [18, 20]. But then, by the same token, positions, which are identified with "translations" of momenta, cannot commute, so that the space-time of DSR should necessarily be a non-commutative manifold, called κ-Minkowski space-time [12, 21]. Let us see therefore, how this picture emerges from quantum gravity, this time coupled to point particles, and without cosmological constant.

In what follows I will review the results obtained in [22]. Let us start with the case of three-dimensional quantum gravity now coupled to a point particle. Then it is well known (see the detailed and clear exposition in [23] and references therein) that since in $3d$ gravity does not have any dynamical degrees of freedom, the theory is fully characterized by Poincaré charges carried by the particle. In other words the theory reduces to a theory of the phase space of the particle, which is different from the phase space of free particles, as a result of the modifications induced by topological degrees of freedom of gravity. This phase space is characterized by the following properties [23]

- The coordinates of the particle (understood as variables on the phase space, which are canonically conjugated to momenta) do not commute and instead
$$[x_0, x_i] = -\frac{1}{\kappa}\, x_i, \quad [x_i, x_j] = 0\,. \tag{7}$$
(The bracket above is either the Poisson bracket or the commutator.) Such a non-commutative space-time is called κ-Minkowski.
- The space of (three-) momenta is not the flat R^3 manifold, but the maximally symmetric space of constant curvature $-\kappa$ (anti de Sitter space of momenta).
- Last but not least it has been shown in numerous works on $3d$ quantum gravity that the full Hopf κ-Poincaré algebra with all the quantum group structures plays the role (see e.g., [17] and references therein.)

But as I will show in Chap. 3, these are exactly the properties of phase space of a particle in DSR (in the case of both 3 and 4 dimensional space-time.) Note in passing an interesting duality between curvature and non-commutativity[7]

$$\begin{gathered}\text{Curvature of momentum space}\\ \Updownarrow\\ \text{Non-commutativity of position space}\end{gathered}$$

As I will show below this duality can be understood as a consequence of the co-product structure of quantum Poincaré algebra.

Thus we see again that kinematics of particles in three dimensions is described by the DSR-like structure with observer independent scale. The question arises as to if something similar can happen in four space-time dimensions. I have only circumstantial evidences in favor of such claim, and the argument goes as follows [22].

The main idea is to construct an experimental situation that forces a dimensional reduction from the four dimensional to the 2+1 dimensional theory. It is interesting that this can be done in quantum theory, using the uncertainty principle as an essential element of the argument. Let us consider a free elementary particle in $3+1$ dimensions, whose mass is less than $G^{-1} = \kappa$. The motion of the particle will be linear, at least in some classes of coordinates systems, not accelerating with respect to the natural inertial coordinates at infinity. Let us consider the particle as described by an inertial observer who travels perpendicular to the plane of its motion, which I will call the z direction. From the point of view of that observer, the particle is in an eigenstate of longitudinal momentum, $\hat{P}_z^{total}$, with some eigenvalue P_z. Since the particle is in an eigenstate of $\hat{P}_z^{total}$ its wavefunction will be uniform in z, with wavelength L where (note that I assume here that L is so large that I can trust the standard uncertainty relation; besides this uncertainty relation is not being modified in some formulations of DSR)

$$L = \frac{1}{P_z^{total}} \tag{8}$$

At the same time, we assume that the uncertainties in the transverse positions are bounded a scale r, such that $r \ll 2L$. Then the wavefunction for the the particle has support on a narrow cylinder of radius r which extend uniformly in the z direction. Finally, we assume that the state of the gravitational field is semiclassical, so that to a good approximation, within $\mathcal{C}$ the semiclassical Einstein equations hold.

$$G_{ab} = 8\pi G\langle\hat{T}_{ab}\rangle \tag{9}$$

[7] See the insightful discussion in [7], in which Shahn Majid argues that this duality indicates a deep relation between non-commutativity and quantization of gravity.

Note that we do not have to assume that the semiclassical approximation holds for all states. We assume something much weaker, which is that there are subspaces of states in which it holds. This assumption is, in a sense, analogous to the assumption above that we are interested only in the analysis of ground state of quantum gravity.

Since the wavefunction is uniform in z, this implies that the gravitational field seen by our observer will have a spacelike Killing field $k^a = (\partial/\partial z)^a$.

Thus, if there are no forces other than the gravitational field, the particle described semiclassically by (9) must be described by an equivalent $2+1$ dimensional problem in which the gravitational field is dimensionally reduced along the z direction so that the particle, which is the source of the gravitational field, is replaced by a punctures.

The dimensional reduction is governed by a length d, which is the extent in z that the system extends. We cannot take $d < L$ without violating the uncertainty principle. It is then convenient to take $d = L$. Further, since the system consists of the particle, with no intrinsic extent, there is no other scale associated with their extent in the z direction. We can then identify $z = 0$ and $z = L$ to make an equivalent toroidal system, and then dimensionally reduce along z. The relationship between the four dimensional Newton's constant G^4 and the three dimensional Newton's constant $G^3 = G$ is given by

$$G^3 = \frac{G^4}{L} = \frac{G^4 P_z^{tot}}{\hbar} \tag{10}$$

Thus, in the analogous 3 dimensional system, which is equivalent to the original system as seen from the point of view of the boosted observer, the Newton's constant depends on the longitudinal momentum.

Of course, in general there will be an additional scalar field, corresponding to the dynamical degrees of freedom of the gravitational field. We will for the moment assume that these are unexcited, but exciting them will not affect the analysis so long as the gravitational excitations are invariant also under the Killing field and are of compact support.

Now we note that, if there are no other particles or excited degrees of freedom, the energy of the system can to a good approximation be described by the hamiltonian H of the two dimensional dimensionally reduced system. This is described by a boundary integral, which may be taken over any circle that encloses the particle. But it is well known that in $3d$ gravity H is bounded from above. This may seem strange, but it is easy to see that it has a natural four dimensional interpretation.

The bound is given by

$$M < \frac{1}{4G^3} = \frac{L}{4G^4} \tag{11}$$

where M is the value of the ADM hamiltonian, H. But this just implies that

$$L > 4G^4 M = 2R_{Sch} \tag{12}$$

i.e. this has to be true, otherwise the dynamics of the gravitational field in 3+1 dimensions would have collapsed the system to a black hole! Thus, we see that the total bound from above of the energy in 2+1 dimensions is necessary so that one cannot violate the condition in 3+1 dimensions that a system be larger than its Schwarzschild radius.

Note that we also must have

$$M > P_z^{tot} = \frac{\hbar}{L} \tag{13}$$

Together with (12) this implies $L > l_{Planck}$, which is of course necessary if the semiclassical argument we are giving is to hold.

Now, we have put no restriction on any components of momentum or position in the transverse directions. So the system still has symmetries in the transverse directions. Furthermore, the argument extends to any number of particles, so long as their relative momenta are coplanar. Thus, we learn the following.

Let $\mathcal{H}^{QG}$ be the full Hilbert space of the quantum theory of gravity, coupled to some appropriate matter fields, with $\Lambda = 0$. Let us consider a subspace of states $\mathcal{H}^{weak}$ which are relevant in the low energy limit in which all energies are small in Planck units. We expect that this will have a symmetry algebra which is related to the Poincaré algebra $\mathcal{P}^4$ in 4 dimensions, by some possible small deformations parameterized by G^4 and $\hbar$. Let us call this low energy symmetry group $\mathcal{P}^4_G$.

Let us now consider the subspace of $\mathcal{H}^{weak}$ which is described by the system we have just constructed . It contains the particle, and is an eigenstate of $\hat{P}_z^{tot}$ with large P_z^{tot} and vanishing longitudinal momentum. Let us call this subspace of Hilbert space $\mathcal{H}_{P_z}$.

The conditions that define this subspace break the generators of the (possibly modified) Poincaré algebra that involve the z direction. But they leave unbroken the symmetry in the 2 + 1 dimensional transverse space. Thus, a subgroup of $\mathcal{P}_G^{3+1}$ acts on this space, which we will call $\mathcal{P}_G^{2+1} \subset \mathcal{P}_G^{3+1}$.

We have argued that the physics in $\mathcal{H}_{P_z}$ is to good approximation described by an analogue system in of a particle in 2 + 1 gravity. However, we know from the results mentioned above that the symmetry algebra acting there is not the ordinary 3 dimensional Poincaré algebra, but the κ-Poincaré algebra in 3 dimensions, with

$$\kappa^{-1} = \frac{4G^4 P_z^{tot}}{\hbar} \tag{14}$$

Now we can note the following. Whatever $\mathcal{P}^4_G$ is, it must have the following properties:

- It depends on G^4 and $\hbar$, so that it's action on *each* subspace $\mathcal{H}_{P_z}$, for each choice of P_z, is the κ deformed 3*d* Poincaré algebra, with κ as above.
- It does not satisfy the rule that momenta and energy add, on all states in $\mathcal{H}$, since they are not satisfied in these subspaces.
- Therefore, whatever $\mathcal{P}^4_G$ is, it is not the classical Poincaré group.

Thus the theory of particle kinematics at ultra high energies is not Special Relativity, and the arguments presented above suggest that it might be Doubly Special Relativity. So it is good time now to start discussing the structures of this theory.

3 Doubly Special Relativity and the κ-Poincaré Algebra

Soon after pioneering papers of Amelino-Camelia [1, 2] it was realized in [3] and [4] that the κ-Poincaré algebra [10, 11, 12] is a perfect mathematical setting to describe one particle kinematics in DSR. Let us recall from the preceding section that in particular, in the bicrossproduct basis the brackets of rotations M_i, boosts N_i, and the components of momenta P_μ read[8]

$$[M_i, M_j] = i\,\epsilon_{ijk}M_k, \quad [M_i, N_j] = i\,\epsilon_{ijk}N_k\ ,$$

$$[N_i, N_j] = -i\,\epsilon_{ijk}M_k\ , \tag{15}$$

$$[M_i, P_j] = i\,\epsilon_{ijk}P_k, \quad [M_i, P_0] = 0\ , \tag{16}$$

$$[N_i, P_j] = i\,\delta_{ij}\left(\frac{\kappa}{2}\left(1 - e^{-2P_0/\kappa}\right) + \frac{1}{2\kappa}\boldsymbol{P}^2\right) - i\,\frac{1}{\kappa}P_iP_j \tag{17}$$

$$[N_i, P_0] = i\,P_i \tag{18}$$

It is important to note that the algebra of M_i N_i is just the standard Lorentz algebra, so one of the first conclusions is that the Lorentz sector of κ-Poincaré algebra is not deformed. Therefore in DSR theories, in accordance with the first postulate above, the Lorentz symmetry *is not broken* but merely nonlinearly realized in its action on momenta. This simple fact has lead some authors (see e.g., [24, 25]) to the claim that DSR is nothing but the standard Special Relativity in non-linear disguise. As we will see this view is clearly wrong, simply because the algebra (15)–(18) describes only half of the phase space of the particle, and the full phase space algebra cannot be reduced to the one of Special Relativity.

[8] From now on I will be discussing the four-dimensional case only. However, the reader can easy convince her(him)self that what will be said here applies with minor and obvious modifications in other dimensions as well. Notice that now I use the QFT convention of adding the "i" on the right-hand-side of the algebra.

As one can easily check, the Casimir of the κ-Poincaré algebra reads

$$\kappa^2 \cosh \frac{P_0}{\kappa} - \frac{\boldsymbol{P}^2}{2} e^{P_0/\kappa} = M^2 . \tag{19}$$

Exercise 5. Check that (19) is indeed the Casimir of the algebra (15)–(18) i.e., its commutators with all the generators of κ-Poincaré algebra vanish. Is it the only possible Casimir of this algebra? Compute the velocity $v = \partial P_0/\partial|\boldsymbol{P}|$. How the behavior of this velocity depends on the sign of κ?

It follows from (19) that the value of three-momentum $|\boldsymbol{P}| = \kappa$ corresponds to infinite energy $P_0 = \infty$. One can check easily (see Exercise 4 above) that in this particular realization of DSR κ is indeed observer independent [3, 4] (i.e., if a particle has momentum $|\boldsymbol{P}| = \kappa$ for some observer, it has the same momentum for all, Lorentz related, observers.) One also sees that the speed of massless particles, naively defined as derivative of energy over momentum, increases monotonically with momentum and diverges for the maximal momentum $|\boldsymbol{P}| = \kappa$, if κ is positive. As I mentioned already in the DSR terminology, the theory based on the algebra (15)–(18) with Casimir (19) is sometimes called DSR1.

One should note at this point that the bicrossproduct algebra above is not the only possible realization of DSR. For example, in [26, 27] Magueijo and Smolin proposed and carefully analyzed another DSR proposal, called sometimes DSR2. In DSR2 the Lorentz algebra is still not deformed and there are no deformations in the brackets of rotations and momenta. The boosts–momenta generators have now the form

$$[N_i, p_j] = i\left(\delta_{ij}p_0 - \frac{1}{\kappa}p_i p_j\right) , \tag{20}$$

and

$$[N_i, p_0] = i\left(1 - \frac{p_0}{\kappa}\right) p_i . \tag{21}$$

It is easy to check that the Casimir for this algebra has the form

$$M^2 = \frac{p_0^2 - \boldsymbol{p}^2}{\left(1 - \frac{p_0}{\kappa}\right)^2} . \tag{22}$$

Exercise 6. Check that (22) is indeed the Casimir of the DSR2 algebra (20), (21). Compute the velocity $v = \partial P_0/\partial|\boldsymbol{P}|$. Find relations between DSR1 and DSR2 momentum variables (the answer can be found in [19, 21].)

Moreover there is a basis of DSR, closely related to the famous Snyder theory [28], in which the energy-momentum space algebra is purely classical (it was first found in [29] and further analyzed in [19, 21].)

Exercise 7. Find explicit transformation from DSR1 to the classical basis, in which all the brackets are identical to those of the standard Poincaré algebra. (See [19, 21], where the relation of the DSR algebra in classical basis and Snyder's theory is analyzed in details.)

3.1 Space-Time of DSR

The formulation of DSR in the energy-momentum space is clearly incomplete, as it lacks any description of the structure of space-time. DSR has been formulated in a somehow unusual way: one started with the energy–momentum space and only then the problem of construction of space-time had been considered. Usually we do the opposite, for example in the standard formulation of Special Relativity one starts with clear operational definition of space-time notions (distance, time interval) and only then the energy-momentum space and phase space is being constructed.

Exercise 7. (Difficult[9].) Formulate Special Relativity in the operational way, taking as a starting point the space of energy and momenta.

There are in principle many ways how the phase space can be constructed. For example in [30] one constructs the position space along the same lines as the energy-momentum space has been constructed in [26, 27]. Here, following [21], I take another route. As I have been stressing in the preceding section, one of the distinctive features of the κ-Poincaré algebra is that it possesses additional structures that make it a Hopf algebra. Namely one can construct the so called co-products for the rotation, boosts, and momentum generators, which, in turn, can be used to provide a procedure to construct the phase space in a unique way.

The co-product is the mapping from the algebra $\mathcal{A}$ to the tensor product $\mathcal{A} \otimes \mathcal{A}$ satisfying some requirements that make it in a sense dual to algebra multiplication (see [7] for details), which essentially provides a rule how the algebra acts on products (of functions, and, in physical applications, on multiparticle states.) For the bicrossproduct κ-Poincaré algebra (15)–(18) the co-products read

$$\Delta(P_0) = \mathbb{1} \otimes P_0 + P_0 \otimes \mathbb{1} \tag{23}$$

$$\Delta(P_k) = P_k \otimes \mathrm{e}^{-P_0/\kappa} + \mathbb{1} \otimes P_k \tag{24}$$

$$\Delta(M_i) = M_i \otimes \mathbb{1} + \mathbb{1} \otimes M_i \tag{25}$$

$$\Delta(N_i) = \mathbb{1} \otimes N_i + N_i \otimes \mathrm{e}^{-P_0/\kappa} - \frac{1}{\kappa}\epsilon_{ijk}\, M_j \otimes P_k \tag{26}$$

In order to construct the one-particle phase space we must first introduce objects that are dual to M_i, N_i, and P_μ. These are the matrix $\Lambda^{\mu\nu}$ and the vector X^μ. Let us briefly interpret their physical meaning. X^μ are to be dual to momenta P_μ, which clearly indicates that they should be interpreted as translation of momenta, in other words the positions. The duality between $\Lambda^{\mu\nu}$ and $M_{\mu\nu} = (M_i, N_i)$ is a bit more tricky. However if one interprets $M_{\mu\nu}$ in analogy to the interpretation of momenta, i.e., as Lorentz charge carried

[9] By this I mean that I do not quite know how to solve it (as a matter of fact I believe nobody does)!

by the particle, that is its angular momentum, then the dual object $\Lambda^{\mu\nu}$ has clear interpretation of Lorentz transformation. Thus we have the structure of the form $G \times \mathcal{MP}$, where G is the Poincaré group acting on the space of Poincaré charges of the particle $\mathcal{MP}$. We see therefore that we can make use of the powerful mathematical theory of Lie-Poisson groups and co-adjoint orbits (see, for example, [31, 32]) and their quantum deformations.

Following [33] and [34] we assume the following form of the co-product on the group

$$\Delta(X^\mu) = \Lambda^\mu{}_\nu \otimes X^\nu + X^\mu \otimes 1\!\!1 \tag{27}$$

and

$$\Delta(\Lambda^\mu{}_\nu) = \Lambda^\mu{}_\rho \otimes \Lambda^\rho{}_\nu \tag{28}$$

The next step is to define the pairing between elements of the algebra and of the group in a canonical way that establish the duality between these two structures.

$$\langle P_\mu, X^\nu \rangle = i\delta^\nu_\mu \tag{29}$$

$$\langle M^{\alpha\beta}, \Lambda^\mu{}_\nu \rangle = i\left(g^{\alpha\mu}\delta^\beta_\nu - g^{\beta\mu}\delta^\alpha_\nu\right) \tag{30}$$

$$\langle \Lambda^\mu{}_\nu, 1 \rangle = \delta^\mu_\nu \tag{31}$$

In (30) $g^{\alpha\mu}$ is the Minkowski space-time metric. This pairing must be consistent with the co-product structure in the following sense

$$\langle A, XY \rangle = \langle A_{(1)}, X \rangle \langle A_{(2)}, Y \rangle \,, \tag{32}$$

$$\langle AB, X \rangle = \langle A, X_{(1)} \rangle \langle B, X_{(2)} \rangle \,, \tag{33}$$

The rules (29)–(33) make it possible to construct the commutator algebra of the phase space. To this end one makes use of the Heisenberg double procedure [32, 34], that defines the brackets in terms of the pairings as follows (no summation over repeated indices here!)

$$[X^\mu, P_\nu] = P_{\nu(1)} \left\langle X^\mu_{(1)}, P_{\nu(2)} \right\rangle X^\mu_{(2)} - P_\nu X^\mu \,, \tag{34}$$

$$[X^\mu, M^\rho{}_\sigma] = M_{(1)}{}^\rho{}_\sigma \left\langle X_{(1)}{}^\mu, M_{(2)}{}^\rho{}_\sigma \right\rangle X_{(2)}{}^\mu - M^\rho{}_\sigma X^\mu, \tag{35}$$

and analogously for $\Lambda^\mu{}_\nu$ commutators, where on the right hand side we make use of the standard ("Sweedler") notation for co-product

$$\Delta\mathcal{T} = \sum \mathcal{T}_{(1)} \otimes \mathcal{T}_{(2)} \,.$$

As an example let us perform these steps in the case of the bicrossproduct κ-Poincaré algebra of DSR1. It follows from (24), and (29), and (33) that

$$\langle P_i, X_0 X_j \rangle = -\frac{1}{\kappa}\,\delta_{ij}, \quad \langle P_i, X_j X_0 \rangle = 0 \,,$$

from which one gets

$$[X_0, X_i] \equiv X_0 X_i - X_i X_0 = -\frac{i}{\kappa} X_i \,. \tag{36}$$

Similarly, using (34) we get the standard relations

$$[P_0, X_0] = -i \,, \quad [P_i, X_j] = i \, \delta_{ij} \,. \tag{37}$$

It turns out that the phase space algebra contains one more non-vanishing commutator (which can be, of course, also obtained from Jacobi identity), namely

$$[P_i, X_0] = -\frac{i}{\kappa} P_i \,. \tag{38}$$

Thus we have constructed the phase space of the bicrossproduct κ-Poincaré algebra of DSR1. Let us stress that this construction relies heavily on the form of co-product. However, as it will turn out below, some of the commutators are sensitive to the particular form of the DSR, while the others are not. In particular we will see that the non-commutativity of positions (36) is to large extend universal for a whole class of DSR theories. The non-commutative space-time with such Lie-like type of non-commutativity is called κ-Minkowski space-time.

Exercise 8. Using Jacobi identity derive the brackets of boosts and positions, assuming that they form a Lie algebra. Which algebra is it? (The answer can be found below.)

3.2 From DSR Theory to DSR Theories

The introduction of invariant momentum (or mass) scale κ has immediate consequences. The most important is that there is nothing sacred about the bicrossproduct DSR presented above, as one can simply use κ to define new energy and momentum (new basis of DSR) as analytic functions of the old ones, to wit

$$\mathcal{P}_i = \mathcal{F}_i(P_i, P_0; \kappa), \quad \mathcal{P}_0 = \mathcal{F}_0(P_i, P_0; \kappa) \,, \tag{39}$$

the only restrictions being that the equations in (39) transform covariantly under rotations and that in the $\kappa \to \infty$ limit $\mathcal{P}_\mu = P_\mu$, because we insist on the right low energy limit in all the bases. Observe that such a change of energy and momentum is not possible in a theory without any mass scale, like special relativity and Newtonian mechanics, in which the energy momentum spaces are linear, and the mass shell conditions are expressed by quadratic form.

Then a natural question arises: which momenta are the "right" ones? The hope is that the theory of quantum gravity or some other fundamental theory, from which DSR is descending will tell what is the correct physical choice. One can also contemplate the possibility that in the final, complete formulation of

DSR one will have to do with some kind of "energy-momentum general covariance", i.e., that physical observables do not depend on a particular realization of (39), like observables in general relativity do not depend on coordinate system. Then a natural question arises: is it possible to understand transformations (39) as coordinate transformations on some (energy-momentum) space?

Surprisingly enough the answer to this question is in the positive: indeed the transformations between DSR theories, described by (39) are nothing but coordinate transformation of the constant curvature manifold, on which momenta live. To reach this conclusion one observes first [19, 21] that it follows from the Heisenberg double construction that both the κ-Minkowski commutator (36) and the commutators between Lorentz charges $M_{\mu\nu}$ and positions X_μ are left invariant by the transformations (39). This follows from the fact that the transformations (39) a severely constrained by assumed rotational invariance and the fact that in the $\kappa \to \infty$ limit the new energies and momenta must must be the same as in the standard Special Relativity. Since the bicrossproduct DSR variables satisfy this requirement it follows that the new variables cannot differ from the DSR1 ones in the κ^0 order. Therefore, in the leading order, they must be of the form

$$\mathcal{P}_i \approx P_i + \alpha \frac{1}{\kappa} P_i P_0 + O\left(\frac{1}{\kappa^2}\right), \quad \mathcal{P}_0 = P_0 + \beta \frac{1}{\kappa} P_0^2 + O\left(\frac{1}{\kappa^2}\right) \tag{40}$$

where α and β are numerical parameters. It turns out that in computing the brackets of positions X and the ones of positions with boosts Heisenberg double procedure picks up only the first terms in this expansion, and thus the form of the commutators remains unchanged. Of course, the position-momenta commutators are changed by the transformations (39), (40).

Exercise 9. Using expansion (40) derive the brackets of positions and four-momenta $\mathcal{P}_\mu$. It would help to notice that co-product is a homomorphism and thus $\Delta(ab) = \Delta(a)\Delta(b)$.

Next it was realized in [18, 20] that the algebra of positions and Lorentz charges is nothing but de Sitter $SO(4,1)$ algebra. The positions and Lorentz transformations are, in turn, nothing but the transformations of the manifold, whose points are energy and momenta (energy-momentum manifold.) On this manifold positions are generators of translational symmetry, while boosts and rotations generate Lorentz transformations. Thus the energy–momentum manifold is a four-dimensional manifold with ten-parameter group of symmetries and thus it must be a maximally symmetric space of constant curvature. It follows from the well known theorem of differential geometry that such a manifold must be locally diffeomorphic to one of the three spaces of constant curvature, and since the group of symmetries is $SO(4,1)$, this manifold must be de Sitter space[10]. Then it follows that the algebra of positions and Lorentz transformations is just an algebra of symmetries of de Sitter space, and therefore it is, of course, independent of a coordinate system we use to describe this space.

De Sitter space of momenta can be constructed as a four dimensional surface of constant curvature κ in the five dimensional Minkowski space with coordinates η_A, $A = 0, \ldots, 4$, to wit

$$-\eta_0^2 + \eta_1^2 + \cdots + \eta_4^2 = \kappa^2 . \tag{41}$$

The $SO(4,1)$ generators can be decomposed into positions X_μ and Lorentz charges $M_{\mu\nu}$, which act on η_A variables as follows

$$[X_0, \eta_4] = \frac{i}{\kappa}\,\eta_0 , \quad [X_0, \eta_0] = \frac{i}{\kappa}\,\eta_4 , \quad [X_0, \eta_i] = 0 , \tag{42}$$

$$[X_i, \eta_4] = [X_i, \eta_0] = \frac{i}{\kappa}\,\eta_i , \quad [X_i, \eta_j] = \frac{i}{\kappa}\,\delta_{ij}(\eta_0 - \eta_4) , \tag{43}$$

and

$$[M_i, \eta_j] = i\epsilon_{ijk}\eta_k , \quad [N_i, \eta_j] = i\,\delta_{ij}\,\eta_0 , \quad [N_i, \eta_0] = i\,\eta_i , \tag{44}$$

It should be noted that there is another decomposition of $SO(4,1)$ generators [18, 20], in which the resulting algebra is exactly the one considered by Snyder [28].

On the space (41) one can built various co-ordinate systems, each related to some DSR theory. In particular, one recovers the bicrossproduct DSR1 with the following coordinates (which are, accidentally, the standard "cosmological" coordinates on de Sitter space)

$$\begin{aligned}
\eta_0 &= -\kappa \sinh \frac{P_0}{\kappa} - \frac{\boldsymbol{P}^2}{2\kappa}\, e^{\frac{P_0}{\kappa}} \\
\eta_i &= -P_i\, e^{\frac{P_0}{\kappa}} \\
\eta_4 &= \kappa \cosh \frac{P_0}{\kappa} - \frac{\boldsymbol{P}^2}{2\kappa}\, e^{\frac{P_0}{\kappa}} .
\end{aligned} \tag{45}$$

Using (45), (43), and the Leibnitz rule, one easily recovers the commutators (15)–(18).

Exercise 10. Check this explicitly.

Other coordinates systems, are possible, of course.

---

[10] It turns out that all other spaces of constant curvature are also possible, if one generalizes somehow the definition of κ-Poincaré algebra, i.e., the phase space associated with κ-Poincaré algebra can have positive, zero, and negative curvature (see [35] for details.)

Exercise 11. Find the coordinates on de Sitter space of momenta, corresponding to DSR2.

In particular one can choose the "standard basis" in which

$$\mathcal{P}_\mu = \eta_\mu/\eta_4 \,. \tag{46}$$

Note that in this basis (or classical DSR) the commutators of all Poincaré charges, $\mathcal{P}_\mu$ and $M_{\mu\nu}$ are purely classical. However, the positions brackets, as well as the momenta/positions cross-relations are still non-trivial.

Exercise 12. Compute the bracket of positions with energy and momenta in the classical basis.

This means that in the classical bases of DSR the (observer-independent) scale κ disappears completely from the Lorentz sector, but is still present in the translational one. Thus such a theory fully deserves the name DSR.

De Sitter space setting reveals the geometrical structure of DSR theories. As we saw the energy momentum space of DSR is a four dimensional manifold of positive constant curvature, and the curvature radius equals the scale κ. The Lorentz charges and positions are identified with the set of ten tangent vectors to the de Sitter energy-momentum space, and as an immediate consequence of this their algebra is independent of any particular coordinate system on this space. However the latter seems to be, at least naively, physically relevant. Each such coordinate system defines for us (up to the redundancy discussed in [20]) the physical energy and momentum. In one-particle sector the particular choice may not be relevant, but it seems that it would be of central importance for the proper understanding of many particles phase spaces, in particular in analysis of the phenomenologically important issue of particles scattering and conservation laws.

Having obtained the one-particle phase space of DSR, it is natural to proceed with construction of the field theory. Here two approaches are possible. One can try to construct field theory on the non-commutative κ-Minkowski space-time. Attempts to construct such a theory has been reported, for example, in [36] and references therein, as well as in [37, 38]. This line of research is, however, far from being able to give any definite results, though some partial results, like an interesting, nontrivial vertex structure reported in [37, 38] may shed some light on physics of the scattering processes. The major obstacle seems to be lack of the understanding of functional analysis on the spaces with Lie-type of non-commutativity, which is most likely a deep and hard mathematical problem (already the definition of appropriate differential and integral calculi is a mater of discussion.) Therefore it seems simpler (and in fact more along the line of the DSR proposal, where the energy momentum space is more fundamental than the space-time structures) to try to built (quantum) field theory in energy-momentum space directly. This would amount to understand how to define (quantum) fields on the curved energy-momentum space, but, in principle, for spaces of constant curvature at least functional analysis is well understood. It should be noted that such an idea has been contemplated for a long time, and in fact it was one of the main motivations of [28]. Field theories with curved energy-momentum manifold has been intensively investigated by Kadyshevsky and others [39], without any conclusive results, though.

4 Physics with Doubly Special Relativity

Till now I have been discussing formal aspects of Doubly Special Relativity in a particular formulation, in which quantum algebras and non-commutative space-time played the fundamental role. Now it is time to try to turn to more physical questions, related with possible experimental signatures of quantum gravity. In other contributions to this volume, the reader can find much more detailed discussion of the "quantum gravity phenomenology", here I would like to concentrate on those physical aspects and problems that are directly related to a particular formulation of DSR in terms of κ-Poincaré algebras.

4.1 Time-of-Flight Experiments and the Issue of Velocity in DSR

One of the simplest experimental tests of quantum gravity phenomenology is the time-of-flight experiment. In this experiment which is to be performed in a near future with good accuracy by GLAST satellite (see e.g., [40] and references therein) one measures the energy-dependence of velocity of light coming from a distant source. Naively, most DSR models predicts positive signal in such an experiment (for details see [41].) Indeed, in DSR $\partial E(p)/\partial p$ does, with an exception of the classical bases, depend on energy, which suggest that velocity of massless particles may depend on the energy they carry. This is the case, for example, both in the bicrossproduct DSR1 and in the Magueijo-Smolin DSR2 model.

Of course, the velocity formula should be derived from the first principles. In the careful analysis reported in [44] (based on the calculations presented some time ago in [42]) the authors construct the wave packet from plane waves moving on the κ-Minkowski space-time, and then calculate the group velocity of such a packet, which, they claim, turns-out to be exactly $v^{(g)} = \partial E(p)/\partial p$[11]. This result is puzzling in view of the phase space calculation of velocity, which I will present below. Therefore let us analyze this calculation in more details.

The authors of [44] consider the wave packet built of waves moving in non-commutative κ-Minkowski space-time, centered at $(\omega_0, \boldsymbol{k}_0)$, to wit

$$\Psi_{(\omega_0,\boldsymbol{k}_0)}(\boldsymbol{x}, t) = \int e^{i\boldsymbol{k}\cdot\boldsymbol{x}} e^{-i\omega t} d\mu \tag{47}$$

[11] Notice however that similar analysis presented in [43] resulted in different conclusion. I will discuss below the reason for this discrepancy.

Here the plane waves have been ordered so that the time variable appear on the right, and $d\mu$ is an appropriate measure on the space of three-momenta, whose detailed form will be irrelevant to what follows. We assume that the plane waves in the integral satisfy appropriate field equations so that ω is a given function of $\boldsymbol{k}$ such that for the pair $(\boldsymbol{k}, \omega(\boldsymbol{k}))$ the Casimir vanishes identically. Let us assume that the integral in (47) has support on small neighborhood

$$\omega_0 - \Delta\omega \leq \omega \leq \omega_0 + \Delta\omega$$
$$\boldsymbol{k}_0 - \Delta\boldsymbol{k} \leq \boldsymbol{k} \leq \boldsymbol{k}_0 + \Delta\boldsymbol{k}$$

Factoring out the phases e^{ik_0x} to the left and $e^{-i\omega_0 t}$ to the right one gets

$$\Psi^m_{(\omega_0,\boldsymbol{k}_0)}(\boldsymbol{x},t) = e^{i\boldsymbol{k}_0\cdot\boldsymbol{x}}\left[\int e^{i\Delta\boldsymbol{k}\cdot\boldsymbol{x}}e^{-i\Delta\omega t}d\mu\right]e^{-i\omega_0 t} \tag{48}$$

Now the integral in the middle carries the information about the group velocity of the wave packet. Indeed it follows that the group velocity equals (in deriving the expression above one should make use of the fact that in the limit $\Delta\omega, \Delta\boldsymbol{k} \to 0$, the commutator $[e^{-i\Delta\boldsymbol{k}\cdot\boldsymbol{x}}, e^{i\Delta\omega t}] = 0$)

$$v^{(g)} = \lim_{\Delta\boldsymbol{k}\to 0}\frac{\Delta\omega}{|\Delta\boldsymbol{k}|} = \frac{d\omega}{d|\boldsymbol{k}|} = \frac{dE}{d|\boldsymbol{P}|}\,. \tag{49}$$

The expression (48) is, however, ambiguous because the middle, amplitude term does not commute with the exponents on the left and on the right as a result of the identity

$$e^{i\boldsymbol{k}\cdot\boldsymbol{x}}e^{-i\omega t} = e^{-i\omega t}e^{i\,e^{-\omega/\kappa}\,\boldsymbol{k}\cdot\boldsymbol{x}}$$

Thus instead of (48) we can use

$$\Psi^r_{(\omega_0,\boldsymbol{k}_0)}(\boldsymbol{x},t) = e^{i\boldsymbol{k}_0\cdot\boldsymbol{x}}\,e^{-i\omega_0 t}\left[\int e^{i\,e^{-\omega/\kappa}\,\Delta\boldsymbol{k}\cdot\boldsymbol{x}}e^{-i\Delta\omega t}d\mu\right] \tag{50}$$

or

$$\Psi^l_{(\omega_0,\boldsymbol{k}_0)}(\boldsymbol{x},t) = \left[\int e^{i\Delta\boldsymbol{k}\cdot\boldsymbol{x}}e^{-i\Delta\omega t}d\mu\right]\,e^{i\boldsymbol{k}_0\cdot\boldsymbol{x}}\,e^{-i\omega_0 t} \tag{51}$$

where in the last expression, we neglected the $e^{-\Delta\omega/\kappa}$ term in the exponent (it goes to zero in the relevant limit.)

We see therefore that the group velocity depends on the ordering of the wave packet (48), (50), (51) and equals

$$v^{(g)} = \begin{cases} \frac{d\omega}{d|\boldsymbol{k}|} & \text{in the cases m, l} \\ \frac{d\omega}{d|\boldsymbol{k}|}\,e^{-\omega/\kappa} & \text{in the case r} \end{cases} \tag{52}$$

Using the fact that for massless particles ω and $\boldsymbol{k}$ are related by (see (19))

$$\kappa^2 \cosh \frac{\omega}{\kappa} - \frac{\boldsymbol{k}^2}{2}\, e^{\omega/\kappa} = \kappa^2 \,. \tag{53}$$

we find easily

$$v^{(g)} = \begin{cases} \frac{\kappa}{\kappa - |\boldsymbol{k}|} & \text{in the cases m, l} \\ 1 & \text{in the case r} \end{cases} \tag{54}$$

Thus we see that the ordering ambiguity in the derivation leads to the ambiguity in the prediction of DSR1 concerning one of the few effects that might be in principle observed. In particular, for one ordering we have velocity of massless particles growing with the energy, while for other we have constant speed of light, as in Special Relativity. The only way out, therefore, is to compute the velocity in a different, though physically equally appealing framework.

To this aim let us try to compute the velocity starting from the phase space of DSR theories. This computation has been presented in [45] (see also [46] and [47].)

The idea is to start with the commutators (42)–(44). Note first that since the for the variable η_4, $[M_i, \eta_4] = [N_i, \eta_4] = 0$, $\kappa\, \eta_4$ is a Casimir (cf. (45)) and can be therefore naturally identified with the relativistic Hamiltonian $\mathcal{H}$ for free particle in any DSR basis as it is by construction Lorentz-invariant, and reduces to the standard relativistic particle hamiltonian in the large κ limit. Indeed, using the fact that for P_μ small compared to κ, in any DSR theory $\eta_\mu \sim P_\mu + O(1/\kappa)$ we have

$$\kappa\eta_4 = \kappa^2 \sqrt{1 + \frac{P_0^2 - \boldsymbol{P}^2}{\kappa^2}} \sim \kappa^2 + \frac{1}{2}\left(P_0^2 - \boldsymbol{P}^2\right) + O\left(\frac{1}{\kappa^2}\right) \tag{55}$$

Then it follows from (43) that

$$\eta_\mu = [x_\mu, \kappa\eta_4] = [x_\mu, \mathcal{H}] \equiv \dot{x}_\mu \tag{56}$$

can be identified with four velocities u_μ. The Lorentz transformations of four velocities are then given by (44) and are identical with those of Special Relativity. Moreover, since

$$u_0^2 - \boldsymbol{u}^2 \equiv \mathcal{C} = M^2 \tag{57}$$

by the standard argument the three velocity equals $v_i = u_i/u_0$ and the speed of massless particle equals 1. Let me stress here once again that this result is DSR model independent, though, of course, the relation between three velocity of massive particles and energy they carry depends on a particular DSR model one uses.

Exercise 12. Compute the velocity of massless particles for DSR1 directly. Use $\kappa\eta_4$ as the hamiltonian and explicit expressions for η_μ as functions of energy and momenta (45). (The answer can be found in [45].)

Thus this calculation indicates that GLAST should not see any signal of energy dependent speed of light, at least if it is correct to think of photons as of point massless classical particles, as I have implicitly assumed in the derivation above.

It should be stressed that the issue of velocity of *physical* particles is not completely settled on the theoretical ground, and thus any experimental input would be extremely valuable.

4.2 Remarks on Multi-Particle Systems

Having obtained the one-particle phase space of DSR it is natural to try to generalize this result to find the two- and multi-particles phase spaces. It turns out however that such a generalization is very difficult, and in spite of many attempts not much about multi-particles kinematics is known. On the other hand the control over particle scattering processes is of utmost relevance in the analysis of seemingly one of the most important windows to quantum gravity phenomenology, provided by Ultra High Energy Cosmic Rays and possible violations of predictions of Special Relativity in UHECR physics (see e.g., [40, 41] for more detailed discussion and the list of relevant references.)

Ironically, we have in our disposal the mathematical structure that seem to provide a tool to solve multi-particle the problem directly. This structure is co-product. Recall that the co-product is a mapping from the algebra to the tensor product

$$\Delta : \mathcal{A} \to \mathcal{A} \otimes \mathcal{A} \tag{58}$$

and thus it provides the rule how the algebra acts on tensor products of its representations. We know that in ordinary quantum mechanics two-particles states are described as a tensor product of single-particle ones[12]. Note that this is a very strong physical assumption: in making it we claim that any two-particle system is nothing but two particles in a black box, i.e., that the particles preserve their identities even in multi-particle states. But it is well possible that multi-particle states differ qualitatively from the single-particle ones, for example as a result of non-local interactions. Let us, however, assume that in also DSR to obtain the multi-particle states one should only tensor the single-particle ones, and let us try to proceed.

In the case of classical algebras the co-product is trivial: $\Delta G = G \otimes 1 + 1 \otimes G$ which means that the group action on two particle states just respects Leibnitz rule. For example the total momentum of two particles in Special Relativity is just the sum of their momenta:

[12] In fact there is more to the description of multi-particles states than just the tensor product, namely one should impose somehow the statistics by symmetrizing or anti-symmetrizing the product. It is well known that in 4 dimensions these are the only possibilities, but the proof relies heavily on the assumption of Poincaré invariance. It is not known if relaxing this assumption by replacing the Poincaré with κ-Poincaré invariance can result in some other, braided statistics.

$$\begin{aligned}\Delta(P_\mu)\,|1+2\rangle &= \Delta(P_\mu)\,|P^{(1)}\rangle \otimes |P^{(2)}\rangle \\ &= (P_\mu \otimes 1\!\!1 + 1\!\!1 \otimes P_\mu)\,|P^{(1)}\rangle \otimes |P^{(2)}\rangle \\ &= \left(P^{(1)}_\mu + P^{(2)}_\mu\right)\,|P^{(1)}\rangle \otimes |P^{(2)}\rangle \end{aligned} \tag{59}$$

In the case of quantum algebras the co-product is non-trivial and non-symmetric by definition (if the co-product was symmetric we would have to do instead with just a classical Lie algebra in nonlinear disguise). This immediately leads to the problem, as I will argue below.

Before turning to this problem let us point out yet another one, relevant for DSR1 as well as for DSR2. Namely the co-product has been constructed so that two-particle states transform as the single-particle ones (for example in Special Relativity total momentum is Lorentz vector.) Indeed if we calculate the total energy and momentum of two-particles system using the co-product addition rule of DSR1 from

$$\Delta(P_0) = 1\!\!1 \otimes P_0 + P_0 \otimes 1\!\!1 \tag{60}$$

$$\Delta(P_k) = P_k \otimes \mathrm{e}^{-P_0/\kappa} + 1\!\!1 \otimes P_k \tag{61}$$

we find

$$P_0^{1+2} = P_0^{(1)} + P_0^{(2)}, \quad P_k^{1+2} = P_k^{(1)} e^{-P_0^{(2)}/\kappa} + P_k^{(2)}\,. \tag{62}$$

But then it follows that total momentum must satisfy the same mass shell relation as the single particle does.

Exercise 13. Check that P_0^{1+2} and P_k^{1+2} satisfy the dispersion relation of DSR1 if $P_0^{(1/2)}$, $P_k^{(1/2)}$ do.

We know however that in the case of the DSR1 we have to do with maximal momentum for particles, of order of Planck mass. While acceptable for Planck scale elementary particles, this is certainly violated for macroscopic bodies. To prove this, the reader can perform a nice quantum gravity phenomenology experiment just by kicking a soccer ball! So we know that there is an experimental proof that either our procedure of attributing momentum to composite system by tensoring and applying co-product, or the bicrossproduct DSR, or both are wrong.

To investigate things further let us turn to the DSR theory, which does not suffer from the "soccer ball problem" namely to the classical basis DSR with standard dispersion relation $\mathcal{P}_0^2 - \mathcal{P}_i^2 = m^2$, for which de Sitter coordinates are given by (46). The co-product for this basis has been calculated in [21] and up to the leading terms in $1/\kappa$ expansion read

$$\Delta(\mathcal{P}_0) = 1\!\!1 \otimes \mathcal{P}_0 + \mathcal{P}_0 \otimes 1\!\!1 + \frac{1}{\kappa}\,\mathcal{P}_i \otimes \mathcal{P}_i + \dots \tag{63}$$

$$\Delta(\mathcal{P}_i) = 1\!\!1 \otimes \mathcal{P}_i + \mathcal{P}_i \otimes 1\!\!1 + \frac{1}{\kappa}\,\mathcal{P}_0 \otimes \mathcal{P}_i + \dots \tag{64}$$

Using this we see that according to the co-product addition rule the total momentum of two-particles system is

$$\mathcal{P}_0^{(1+2)} = \mathcal{P}_0^{(1)} + \mathcal{P}_0^{(2)} + \frac{1}{\kappa}\, \mathcal{P}_i^{(1)} \mathcal{P}_i^{(2)} \tag{65}$$

$$\mathcal{P}_i^{(1+2)} = \mathcal{P}_i^{(1)} + \mathcal{P}_i^{(2)} + \frac{1}{\kappa}\, \mathcal{P}_0^{(1)} \mathcal{P}_i^{(2)} \tag{66}$$

As it stands, the formulas (65, 66) suffer from two problems: first of all, recalling that $\mathcal{P}_\mu$ transforms as a Lorentz vector for single particle, these expressions look terribly non-covariant. Second, even though (65) is symmetric in exchanging particles labels $1 \leftrightarrow 2$, (66) is not. How do we know which particle is first and which is second? Let us try to resolve these puzzles in turn.

That the first puzzle is just an apparent paradox follows immediately from the consistency of the quantum algebra. As I said above the action of boosts on two-particle state is such that total momentum transforms exactly as the single-particle momentum does. This is in fact the very reason of the "soccer ball problem" in the DSR1. In fact the boosts do not only act on $\mathcal{P}_\mu^{(1)}$ and $\mathcal{P}_\mu^{(2)}$ independently; they also mix them in a special way. This feature was to be expected, since the co-product addition rule mixes single-particle states in a non-trivial way. More specifically, note that boosts must act on two-particle states by co-product as well, therefore in order to find out how a two-particle state changes when we boost it we must compute the commutator $[\Delta(N), \Delta(\mathcal{P})]$. Recall now that the co-product of boosts reads (again up to the leading terms in $1/\kappa$ expansion)

$$\Delta(N_i) = \mathbb{1} \otimes N_i + N_i \otimes \mathbb{1} - \frac{1}{\kappa}\, N_i \otimes \mathcal{P}_0 - \frac{1}{\kappa}\, \epsilon_{ijk}\, M_j \otimes \mathcal{P}_k \tag{67}$$

Using this one easily checks explicitly that

$$[\Delta(N_i), \Delta(\mathcal{P}_j)] = \delta_{ij} \Delta(\mathcal{P}_0), \quad [\Delta(N_i), \Delta(\mathcal{P}_0)] - \Delta(\mathcal{P}_i) \tag{68}$$

from which it follows that $\mathcal{P}_0^{(1+2)}$ and $\mathcal{P}_i^{(1+2)}$ do transform covariantly, as they should[13]. Of course equation (68) holds to all orders, as it just reflects the defining property of the co-product.

Let us now turn to the second puzzle, the apparent dependence of the total energy/momentum on physically arbitrary labelling of particles. Here I have much less to say, as this paradox has not been yet solved. One should however mention an interesting result obtained in the case of the analogous problem in deformed, non-relativistic model. In the paper [48] the authors find that even though there is an apparent asymmetry in particle labels due to the asymmetry of the co-product, the representations with flipped labels are related to the original ones by unitary transformation, and are therefore

[13] This holds, of course, for a DSR theory in any basis, not just in the classical one.

physically completely equivalent. In the similar spirit in [49] one uses the fact of such an equivalence in 1+1 dimensions to demand that the action of generators on two particles (bosonic) states is through symmetrized co-product.

During this Winter School Aurelio Grillo and Fernando Mendez produced another interesting puzzle concerning the validity of co-product based momenta addition rule. This puzzle reminds somehow the entanglement problem in quantum mechanics and it can be described as follows.

Suppose we use (62) to formulate conservation rule for two-to-two particles scattering, which would therefore take the following form

$$P_0^{(1)} + P_0^{(2)} = P_0^{(3)} + P_0^{(4)} \tag{69}$$

$$P_k^{(1)} e^{-P_0^{(2)}/\kappa} + P_k^{(2)} = P_k^{(1)} e^{-P_0^{(3)}/\kappa} + P_k^{(4)} . \tag{70}$$

But what about all other particles in the Universe (spectators)? In principle, their presence would contribute non-trivially to the conservation laws (69), (70), to wit

$$P_0^{(1)} + P_0^{(2)} + P_0^{(univ)} = P_0^{(3)} + P_0^{(4)} + P_0'^{(univ)} \tag{71}$$

$$\begin{aligned}\left(P_k^{(1)} e^{-P_0^{(2)}/\kappa} + P_k^{(2)}\right) e^{-P_0^{(univ)}/\kappa} + P_k^{(univ)} \\ = \left(P_k^{(1)} e^{-P_0^{(3)}/\kappa} + P_k^{(4)}\right) e^{-P_0'^{(univ)}/\kappa} + P_k'^{(univ)} .\end{aligned} \tag{72}$$

In the standard, Special Relativistic case we neglect the influence of the rest of the Universe, because we believe that the processes are (at least approximately) *local*, but here we have non-local influence of one particle on another all the time, independently of their separation (in the formulas (71, 72) there is no information concerning separation of particles in space and time.) Thus, the final construction of DSR theory must necessarily solve this spectator problem as well!

5 Conclusion

There is a growing hope that some form of DSR theory indeed describes Nature in the kinematical regime, where the energies of the particles became close to the Planck energy scale and at the same time one could neglect local degrees of gravity, described by (still to be constructed) Quantum Theory of Gravity. This hope is based on the analogy between the ground state of 4d quantum gravity and 3d quantum gravity, both being described by topological quantum field theory.

As we saw, we seem to know some of the ingredients of the DSR theory, and we can even predict some (testable, in principle) DSR phenomenology. It seems however that there would be very hard to derive the complete form of

DSR just from the first principles, the hope being that soon we will be able to derive DSR as an appropriate limit of (Loop) Quantum Gravity.

Four years ago during the Winter School entitled "Towards Quantum Gravity" Giovanni Amelino-Camelia asked the insightful question "Are we at the dawn of Quantum Gravity Phenomenology?". This year we devoted the whole Winter School to discuss possible observable signals of Quantum Gravity. I hope that in four years we will meet to discuss numbers coming from Quantum Gravity experiments that would be already running and producing data. I also hope that it will turn out that these data would agree with the final form of Doubly Special Relativity.

References

1. G. Amelino-Camelia, "Testable scenario for relativity with minimum-length," Phys. Lett. B **510**, 255 (2001) [arXiv:hep-th/0012238].
2. G. Amelino-Camelia, "Relativity in space-times with short-distance structure governed by an observer-independent (Planckian) length scale," Int. J. Mod. Phys. D **11**, 35 (2002) [arXiv:gr-qc/0012051].
3. J. Kowalski-Glikman, "Observer independent quantum of mass," Phys. Lett. A **286**, 391 (2001) [arXiv:hep-th/0102098].
4. N. R. Bruno, G. Amelino-Camelia and J. Kowalski-Glikman, "Deformed boost transformations that saturate at the Planck scale," Phys. Lett. B **522**, 133 (2001) [arXiv:hep-th/0107039].
5. G. Amelino-Camelia, L. Smolin and A. Starodubtsev, "Quantum symmetry, the cosmological constant and Planck scale phenomenology," arXiv:hep-th/0306134.
6. J. E. Nelson, T. Regge and F. Zertuche, "Homotopy Groups And (2+1)-Dimensional Quantum De Sitter Gravity," Nucl. Phys. B **339** (1990) 516.
7. S. Majid, *Introduction to Quantum Groups*, Cambridge University Press, 1995.
8. A. Agostini, G. Amelino-Camelia and F. D'Andrea, "Hopf-algebra description of noncommutative-spacetime symmetries," arXiv:hep-th/0306013.
9. A. O. Barut and R. Raczka, *Theory of Group Representations and Applications*, PWN, Warsaw, 1977.
10. J. Lukierski, H. Ruegg, A. Nowicki and V. N. Tolstoi, "Q deformation of Poincare algebra," Phys. Lett. B **264** (1991) 331.
11. J. Lukierski, A. Nowicki and H. Ruegg, "Real forms of complex quantum anti-De Sitter algebra U_q(Sp(4:C)) and their contraction schemes," Phys. Lett. B **271** (1991) 321 [arXiv:hep-th/9108018].
12. S. Majid and H. Ruegg, "Bicrossproduct structure of kappa Poincare group and noncommutative geometry," Phys. Lett. B **334** (1994) 348 [arXiv:hep-th/9405107]; J. Lukierski, H. Ruegg and W. J. Zakrzewski, "Classical quantum mechanics of free kappa relativistic systems," Annals Phys. **243** (1995) 90 [arXiv:hep-th/9312153].
13. L. Smolin, "Linking Topological Quantum Field Theory and Nonperturbative Quantum Gravity," J. Math. Phys. 36 (1995) 6417 [arXiv:gr-qc/9505028].
14. J. C. Baez, "An Introduction to Spin Foam Models of Quantum Gravity and BF Theory," Lect. Notes Phys. 543 (2000) 25 [arXiv:gr-qc/9905087].

15. S. Major, L. Smolin, "Quantum deformation of quantum gravity," Nucl.Phys. B473 (1996) 267 [arXiv:gr-qc/9512020].
16. A. Starodubtsev, "Topological excitations around the vacuum of quantum gravity I: the symmetries of the vacuum," hep-th/0306135.
17. L. Freidel and D. Louapre, "Ponzano-Regge model revisited. I: Gauge fixing, observables and interacting spinning particles," arXiv:hep-th/0401076.
18. J. Kowalski-Glikman, "De Sitter space as an arena for doubly special relativity," Phys. Lett. B **547** (2002) 291 [arXiv:hep-th/0207279].
19. J. Kowalski-Glikman and S. Nowak, "Doubly special relativity theories as different bases of kappa-Poincare algebra," Phys. Lett. B **539** (2002) 126 [arXiv:hep-th/0203040].
20. J. Kowalski-Glikman and S. Nowak, "Doubly special relativity and de Sitter space," Class. Quant. Grav. **20** (2003) 4799 [arXiv:hep-th/0304101].
21. J. Kowalski-Glikman and S. Nowak, "Non-commutative space-time of doubly special relativity theories," Int. J. Mod. Phys. D **12** (2003) 299 [arXiv:hep-th/0204245].
22. L. Freidel, J. Kowalski-Glikman and L. Smolin, "2+1 gravity and doubly special relativity," Phys. Rev. D **69** (2004) 044001 [arXiv:hep-th/0307085].
23. H. J. Matschull and M. Welling, "Quantum mechanics of a point particle in 2+1 dimensional gravity," Class. Quant. Grav. **15** (1998) 2981 [arXiv:gr-qc/9708054].
24. J. Lukierski and A. Nowicki, "Doubly Special Relativity versus κ-deformation of relativistic kinematics," Int. J. Mod. Phys. A **18** (2003) 7 [arXiv:hep-th/0203065].
25. D. V. Ahluwalia–Khalilova, "Operational indistinguishabilty of doubly special relativities from special relativity," arXiv:gr-qc/0212128.
26. J. Magueijo and L. Smolin, "Lorentz invariance with an invariant energy scale," Phys. Rev. Lett. **88** (2002) 190403 [arXiv:hep-th/0112090].
27. J. Magueijo and L. Smolin, "Generalized Lorentz invariance with an invariant energy scale," Phys. Rev. D **67** (2003) 044017 [arXiv:gr-qc/0207085].
28. H. S. Snyder, "Quantized Space-Time," Phys. Rev. **71** (1947) 38.
29. P. Kosinski, J. Lukierski, P. Maslanka and J. Sobczyk, "The Classical basis for kappa deformed Poincare (super)algebra and the second kappa deformed supersymmetric Casimir," Mod. Phys. Lett. A **10** (1995) 2599 [arXiv:hep-th/9412114].
30. D. Kimberly, J. Magueijo and J. Medeiros, "Non-Linear Relativity in Position Space," arXiv:gr-qc/0303067.
31. A. A. Kirillov, *Elements of the Theory of Representations*, Springer 1976.
32. A. Yu. Alekseev and A. Z. Malkin, "Symplectic structures associated with Lie-Poisson groups", Comm. Math. Phys. **162** (1994) 147.
33. P. Kosinski and P. Maslanka, "The $\kappa-$Weyl group and its algebra", arXiv:q-alg/9512018.
34. J. Lukierski and A. Nowicki, " Heisenberg double description of κ-Poincaré algebra and κ-deformed phase space", Proceedings of Quantum Group Symposium at Group 21, (July 1996, Goslar) Eds. H.-D. Doebner and V.K. Dobrev, Heron Press, Sofia, 1997, p. 186, [arXiv:q-alg/9702003].
35. A. Blaut, M. Daszkiewicz, J. Kowalski-Glikman and S. Nowak, "Phase spaces of doubly special relativity," Phys. Lett. B **582** (2004) 82 [arXiv:hep-th/0312045].
36. P. Kosinski, P. Maslanka, J. Lukierski and A. Sitarz, "Generalized kappa-deformations and deformed relativistic scalar fields on noncommutative Minkowski space," arXiv:hep-th/0307038.

37. G. Amelino-Camelia and M. Arzano, "Coproduct and star product in field theories on Lie-algebra non-commutative space-times," Phys. Rev. D **65** (2002) 084044 [arXiv:hep-th/0105120].
38. G. Amelino-Camelia, M. Arzano and L. Doplicher, "Field theories on canonical and Lie-algebra noncommutative spacetimes," arXiv:hep-th/0205047.
39. V. G. Kadyshevsky et. al., "Quantum field theory and a new universal high-energy scale", Nuovo Cim. **87 A** (1985) 324; Nuovo Cim. **87 A** (1985) 350; Nuovo Cim. **87 A** (1985) 373, and references therein.
40. G. Amelino-Camelia, "Quantum-gravity phenomenology: Status and prospects," Mod. Phys. Lett. A **17** (2002) 899 [arXiv:gr-qc/0204051].
41. G. Amelino-Camelia, J. Kowalski-Glikman, G. Mandanici and A. Procaccini, "Phenomenology of doubly special relativity," arXiv:gr-qc/0312124.
42. G. Amelino-Camelia and S. Majid, "Waves on noncommutative spacetime and gamma-ray bursts," Int. J. Mod. Phys. A **15** (2000) 4301 [arXiv:hep-th/9907110].
43. T. Tamaki, T. Harada, U. Miyamoto and T. Torii, "Particle velocity in noncommutative space-time," Phys. Rev. D **66** (2002) 105003 [arXiv:gr-qc/0208002].
44. G. Amelino-Camelia, F. D'Andrea and G. Mandanici, "Group velocity in noncommutative spacetime," JCAP **0309** (2003) 006 [arXiv:hep-th/0211022].
45. M. Daszkiewicz, K. Imilkowska and J. Kowalski-Glikman, "Velocity of particles in doubly special relativity," Phys. Lett. A **323** (2004) 345 [arXiv:hep-th/0304027].
46. P. Kosinski and P. Maslanka, "On the definition of velocity in doubly special relativity theories," Phys. Rev. D **68** (2003) 067702 [arXiv:hep-th/0211057].
47. S. Mignemi, "On the definition of velocity in theories with two observer-independent scales," Phys. Lett. A **316** (2003) 173 [arXiv:hep-th/0302065].
48. P. Kosinski and P. Maslanka, "Deformed Galilei symmetry," [arXiv:math.QA/9811142].
49. F. Bonechi, E. Celeghini, R. Giachetti, E. Sorace and M. Tarlini, "Inhomogeneous Quantum Groups As Symmetries Of Phonons," Phys. Rev. Lett. **68** (1992) 3718 [arXiv:hep-th/9201002].
50. G. Amelino-Camelia, "Are we at the dawn of quantum-gravity phenomenology?," Lect. Notes Phys. **541** (2000) 1 [arXiv:gr-qc/9910089].

Interferometry as a Universal Tool in Physics

C. Lämmerzahl

ZARM, University of Bremen, Am Fallturm, 28359 Bremen, Germany
laemmerzahl@zarm.uni-bremen.de

The wide range of applications of atomic interferometry and of laser interferometry in the search for quantum gravity induced effects is presented. These effects consists of the exploration of relativistic gravity theories, tests of the Einstein Equivalence principle, of searches for quantum gravity induced deviations of the ordinary dispersion relation and of the search for fundamental fluctuations.

1 Introduction

In the quantum domain, interferometry is a very basic and universal tool. First, it is the very characteristic feature of quantum properties of matter [1]. It can be used for the study of properties of quantum states as well as for the exploration of fields interacting with quantum states. With interferometry the superposition principle underlying the formalism of quantum theory as well as notions like entanglement has been verified. Furthermore, applying interferometry the interaction with external fields like gravitational, inertial and electromagnetic fields can be explored. In many cases the accuracy of these quantum measurements are better than the corresponding classical measurements. Also new types of interactions which are not present in classical physics like the non-local Aharonov–Bohm like effects are accessible with interferometry. Being such a powerful scheme, interferometry also plays an important role in the search for deviations from standard physics given by Special and General Relativity and the Standard Model of particle physics. In fact, the modern precise tests of Special Relativity, namely the tests of the isotropy and constancy of the speed of light, are carried through with interferometric setups, the most precise clocks are atomic clocks based on atomic interferometry, and very precise tests of the Universality of Free Fall (the weak equivalence principle) can also be performed with atomic interferometry. Therefore, with interferometry all interactions and basic principles of physics can be tested and explored.

C. Lämmerzahl: *Interferometry as a Universal Tool in Physics*, Lect. Notes Phys. **669**, 161–198 (2005)
www.springerlink.com

With interferometry also the classically appearing forces can be explored. On the quantum level, forces appear as potentials. If the splitting is in the configuration space, then the result compares the potentials at two different positions. In the limit of infinitesimal splitting this then gives directly the force. Therefore, interferometry gives results which can be obtained by classical devices but is, as alredy noted, more powerful in exploring further properties of interactions.

Here we give a list of all the interferometric tests

1. tests of principles of quantum mechanics
   (a) superposition principle
   (b) 2π–rotation of spinors
   (c) which–path information
   (d) delayed–choice experiments
   (e) decoherence
2. tests exploring interactions
   (a) measurement of gravitational acceleration
   (b) measurement of the gravity gradient
   (c) measurement of inertial acceleration
   (d) measurement of rotation (Sagnac effect)
   (e) measurement of the electromagnetic interaction
   (f) Aharonov–Bohm effect
   (g) Aharonov–Casher effect
   (h) measurement of Berry phase
3. tests exploring the strucure of physics
   (a) Michelson–Morley tests
   (b) Kennedy–Thorndike tests
   (c) search for anomalous spin interaction
   (d) test of the Universality of Free Fall
   (e) tests of the Universality of the Gravitational Redshift

Technical applications are

1. gyroscopes (based either on laser or atomic interferometry – laser gyroscopes are commercially available, atomic gyroscopes are still laboratory devices)
2. accelerometers (based on atomic interferometry – laboratory device)
3. gravity gradiometers (based on atomic interferometry – laboratory device)
4. material sciences

Since interferometry is a very sensitive tool for the exploration of interactions and the structure of quantum mechanics it can be used for the search of quantum gravity effects. Since the typical laboratory energies are of the order of 1 eV and the quantum gravity energy scale is assumed to be of the order of the Planck energy which is about 10^{28} eV, the quantum gravity effects in laboratory experiments are likely to be of the order of 10^{-28} which looks very unlikely to be accessible in laboratory experiments. However, there are several reasons for nevertheless pursuing this way:

- Since there is no finally worked out theory of quantum gravity available, all statements regarding the quantum gravity energy scale of 10^{28} eV have the status of a hypothesis only. Nobody knows about the "true" energy scale of the final quantum gravity theory.
- As constantly emphasized by G. Amelino–Camelia, there are sometimes mechanisms at work which magnify the quantum gravity induced effects through some multipliers. This happens, e.g., for proposals for deviations from Newton's law at small distances where the mechanism which enhances the effect results from the assumption of higher dimensions which introduces additional constants, see e.g. [2]. Another example is the effect of quantum gravity induced fluctuations in interferometers. In very general models of such fluctuations, the magnitude of these fluctuations increase for small frequencies, that is, for long measurement times ($1/f$–noise) [3]. Therefore, searching for fundamental noises in high precision long-term stable devices (like optical resonators) may give new access to this domain of quantum gravity effects [4]. Further examples are the predictions from quantum gravity induced dilaton scenarios [5, 6] that the Universality of Free Fall might be violated already at the 10^{-13} level and that the PPN parameter γ which in ordinary Einsteinian gravity is exactly 1 might be different from unity by up to 10^{-5} – predictions which are of considerably larger order than the first guess of 10^{-28}.
- Using very high precision devices it might be possible even in laboratory experiments to achieve a sensitivity which approaches, at least in principle, the 10^{-28} range. Such devices are gravitational wave interferometers, for example.

Therefore, it is mandatory that for the search of quantum gravity induced effects *all kinds of experimental tests should be considered and constantly tried to be improved.*

2 The Importance of Interferometry

The interference of photons and electrons, neutrons and other particles is a very strange effect which possesses no classical counterpart [1]. It is a statistical effect where the particles are correlated with each other. The main part of the experimental setup consists of a wall with two slits and a photographic plate behind that wall which registeres each particle hiting that plate. Particles like photons, electrons, neutrons, etc. can be sent through the double-slit setup. For one single particle there is no way to predict the position of the particle hiting the screen.

There are two ways to obtain these interference fringes: (i) Particles will be sent one after another through this setup in a way that every particle reaches the screen before the next particle is launched. In an extreme case we may have one particle each day. Then, after many days, an interference pattern builds up, see Fig. 1. One consequence of this is, that all the particles in some sense "know" about each other. They show a correlation which cannot be understood in classical terms because there is always one particle only in the interferometer. In the limit of an infinite number of particles one gets the ideal interfernce pattern. (ii) In a second setup one may have thousands of identical interferometers which may be distributed around the world, on the Moon, on the Mars etc. Through any of these interferometers only *one* particle will be sent. After that, all the positions of the particles on the various screens will be collected. What comes out is an interference pattern as before. As a consequence, in classical terms again all particles must know something from the other particles: The particle on the Moon somehow knows the position of the other particles on their screens. Again, for an infinite number of interferometers the ideal interference pattern comes out.

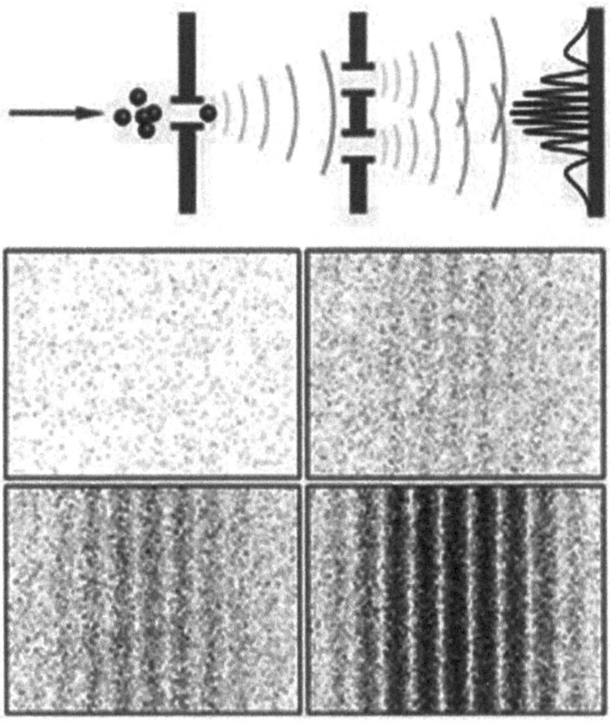

Fig. 1. Evolving interference fringes. Even though particles (photons, electrons, neutrons, etc.) traverse the double slit setup one after another, an interference patters evolves in time

One way to interpret and to describe these results is to accept that only the interference pattern has physical significance and, thus, is accessible to a deterministic description, and that the interference pattern somehow maps out the geometry of the whole setup. If only the interference pattern is assumed to have physical reality, then this can be related to the propagation of waves. The superposition of waves very easily leads to an interference pattern. In fact, assuming states of the form $\psi_\pm(\boldsymbol{x}) = \frac{\sqrt{I_0}}{|\boldsymbol{x}\pm\boldsymbol{d}/2|} e^{-ik|\boldsymbol{x}\pm\boldsymbol{d}/2|}$, where I_0 and $\boldsymbol{k}$ are the intensity and wave vector of this wave and $\boldsymbol{d}$ the separation between the two slits, then we get, for large distances compared to the separation of the slits $x \gg d$ and for small $z \ll x$, the interference pattern

$$I(\boldsymbol{x}) = |\psi_+(\boldsymbol{x}) + \psi_-(\boldsymbol{x})|^2 = I_0\left(1 + \cos\left(k\frac{d}{x}z\right)\right) \tag{1}$$

on the screen. The cosine is characteristic for the interference. The more complicated exact result which is also valid for larger z and d leads to the function plotted in Fig. 1.

The notion of interferometry is not only restricted to the physical setup with the double slit described above where the interference pattern is related to the spatial position. It extends to any parameter the physical state can be characterized with: to spin, energy (frequency), momentum, or any other internal quantum number, see Fig. 2. Accordingly, we start with a quantum state $\psi(a_1, t_0)$ characterized by the parameter a_1. This state evolves in time: $\psi(a_1, t) = U(t, t_0)\psi(a_1, t_0)$ where $U(t, t_0)$ is the evolution operator. At $t = t_1$ the state will be split in a coherent way into two states with respect to the parameter a:

$$\psi(a_1, t_1) \to \psi(t_1^+) = U_a^{(1)}\psi(a_1, t_1) = \alpha^{(1)}\psi(a_1, t_1) + \beta^{(1)}\psi(a_2, t_2)\,, \tag{2}$$

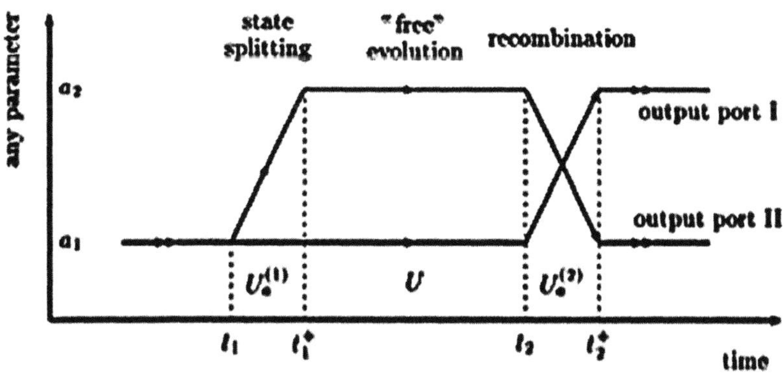

Fig. 2. Interferometry: A splitting of a qantum state with respect to any parameter characterizing this quantum state and subsequent "free" evolution followed by a recombination (superposition) of the splitted states. The measurement of the intensity (probablility) of the outgoing states (port I or port II) shows an interference pattern

where $t_1^+ = t_1 + \delta t$ with δt the time the beam splitting process needs. Probability conservation requires $|\alpha^{(1)}|^2 + |\beta^{(1)}|^2 = 1$. For a 50:50 splitting we have $|\alpha^{(1)}| = |\beta^{(1)}| = 1/\sqrt{2}$. Therefore, up to an overall phase,

$$\psi(t_1^+) = \frac{1}{\sqrt{2}} \left(\psi(a_1, t_1) - e^{-i\varphi_1}\psi(a_2, t_1)\right) , \tag{3}$$

with some phase φ_1. This superposition of states again evolves and gives

$$\begin{aligned} \psi(t) &= U(t, t_1^+)\psi(t_t^+) \\ &= \alpha^{(1)} U_1(t, t_1^+)\psi(a_1, t_1^+) + \beta^{(1)} U_2(t, t_1^+)\psi(a_2, t_1^+) \\ &= \alpha^{(1)}\psi(a_1, t) + \beta^{(1)}\psi(a_2, t) . \end{aligned} \tag{4}$$

Here we assume that the evolution governed by $U(t, t_0)$ (called "free" evolution) does not mix the states with different quantum numbers a but, in general, may depend on the states.

Since only states with the same quantum characteristics can superpose one has to apply again a beam splitter (now called recombiner) which brings $\psi(a_1, t_2)$ into the $\psi(a_2, t_2)$ state and vice versa:

$$\psi(a_1, t_2) \to \psi_1(t_2^+) = U_a^{(2)}\psi(a_1, t_2) = \alpha^{(2)}\psi(a_1, t_2) + \beta^{(2)}\psi(a_2, t_2) \tag{5}$$
$$\psi(a_2, t_2) \to \psi_2(t_2^+) = U_a^{(2)}\psi(a_2, t_2) = \gamma^{(2)}\psi(a_1, t_2) + \delta^{(2)}\psi(a_2, t_2) . \tag{6}$$

Conservation of the overall probability requires $U_a^+ U_a = 1$ which gives $|\alpha^{(2)}|^2 + |\gamma^{(2)}|^2 = 1$, $|\beta^{(2)}|^2 + |\delta^{(2)}|^2 = 1$, $(\alpha^{(2)})^*\beta^{(2)} - (\gamma^{(2)})^*\delta^{(2)} = 0$ and, thus, $|\delta^{(2)}| = |\alpha^{(2)}|$ and $|\beta^{(2)}| = |\gamma^{(2)}|$. Again we assume a 50:50 splitter which, if a state with a definite quantum number a_1 or a_2 comes in, distributes this states with equal probability onto the two states. Then $\alpha^{(2)} = e^{i\varphi_\alpha}/\sqrt{2}$, $\gamma^{(2)} = e^{i\varphi_\beta}/\sqrt{2}$, $\gamma^{(2)} = e^{i\varphi_\gamma}/\sqrt{2}$, and $\delta^{(2)} = e^{i\varphi_\delta}/\sqrt{2}$. This leads, up to overall phase factors, to

$$U_a^{(2)}\psi(a_1, t_2) = \frac{1}{\sqrt{2}} \left(\psi(a_1, t_2) + e^{i\varphi_2}\psi(a_2, t_2)\right) \tag{7}$$
$$U_a^{(2)}\psi(a_2, t_2) = \frac{1}{\sqrt{2}} \left(-e^{-i\varphi_2}\psi(a_1, t_2) + \psi(a_2, t_2)\right) . \tag{8}$$

In matrix form this means

$$\psi(t_2^+) = U_a^{(2)}\psi(t_2) , \tag{9}$$

where the 2×2-matrix

$$U_a^{(2)} = \frac{1}{\sqrt{2}} \begin{pmatrix} 1 & e^{i\varphi_2} \\ -e^{-i\varphi_2} & 1 \end{pmatrix} \tag{10}$$

acts on the two-level state

$$\psi(t_2) = \begin{pmatrix} \psi(a_1, t_2) \\ \psi(a_2, t_2) \end{pmatrix} . \tag{11}$$

Also the first beam splitting can be expressed as $\psi(t_1^+) = U_a^{(1)} \psi(t_1)$ with $U_a^{(1)}$ given by (10) with the φ_2 replaced by φ_1. It is interesting, and also of practical importance, that the subsequent use of two beam splitters described by (10) gives an inversion of the quantum states:

$$U_a^{(2)} U_a^{(2)} = \begin{pmatrix} 0 & e^{i\varphi_2} \\ e^{-i\varphi_2} & 0 \end{pmatrix} . \tag{12}$$

Such a transformation can be interpreted as "mirror".

Putting everything together, we have

$$\psi(t_2^+) = U_a^{(2)} U(t_2, t_1^+) U_a^{(1)} \psi(t_1) , \tag{13}$$

where

$$U(t_2, t_1^+) = \begin{pmatrix} U_1(t_2, t_1^+) & 0 \\ 0 & U_2(t_2, t_1^+) \end{pmatrix} . \tag{14}$$

In our approach we have chosen $\psi(t_1) = \begin{pmatrix} \psi(a_1, t_1) \\ 0 \end{pmatrix}$ which can be generalized to any ingoing state. In interference experiments the intensity of the outgoing state corresponding to a_1 or a_2 is being measured, that is, $|\psi(a_1, t_2)|^2$ or $|\psi(a_2, t_2)|^2$. A calculation gives

$$|\psi(a_1, t_2)|^2 = 2|\psi(a_1, t_1)|^2 \left(1 - \mathrm{Re} \frac{\psi^*(a_1, t_1) e^{i\delta\varphi} U_1^{-1} U_2 \psi(a_1, t_1)}{|\psi(a_1, t_1)|^2}\right) \tag{15}$$

$$|\psi(a_2, t_2)|^2 = 2|\psi(a_1, t_1)|^2 \left(1 + \mathrm{Re} \frac{\psi^*(a_1, t_1) e^{i\delta\varphi} U_1^{-1} U_2 \psi(a_1, t_1)}{|\psi(a_1, t_1)|^2}\right) , \tag{16}$$

with $\delta\varphi = \varphi_2 - \varphi_1$. The real part on the right hand side is always smaller than 1 so that it can be represented as a cosine:

$$|\psi(a_1, t_2)|^2 = 2|\psi(a_1, t_1)|^2 (1 - \cos\vartheta) \tag{17}$$

$$|\psi(a_2, t_2)|^2 = 2|\psi(a_1, t_1)|^2 (1 + \cos\vartheta) \tag{18}$$

with $\cos\vartheta = \mathrm{Re}\left(\psi^*(a_1, t_1) e^{i\delta\varphi} U_1^{-1} U_2 \psi(a_1, t_1)\right) / |\psi(a_1, t_1)|^2$. In the case that the relative evolution $U_1^{-1} U_2$ gives a pure phase shift, $U_1^{-1} U_2 \psi(a_1, t_1) = e^{i\xi(t_2 - t_1^+)} \psi(a_1, t_1)$, then we have $\vartheta = \delta\varphi + \xi(t_2 - t_1^+)$.

This scheme also extends to the case of stationary waves propagating in position space. Then, t has to be replaced by x and the beam splitting can be described as scattering process of waves at periodic potential walls, see [7].

In any case, the beam splitter relies on a certain interaction of the wave with some external field. In modern atomic interferometers, for example, the beam splitting is a consequence of the interaction of a laser beam with the atomic beam. In neutron interferometry, the beam splitting is a consequence of the interaction of the neutron wave with the periodic potential of the silicon crystal. By changing of the interaction time of the atoms with the lasers or by changing the laser beam intensity, one can manipulate the properties of the beam splitters. For a certain intensity or duration of the interaction one may get, for example, not only a 50:50 splitting of the incoming atomic beam but, instead, a complete transformation in the other state. This then plays the role of an atomic mirror.

All kinds of interferometers with 50:50 beam splitters are special cases of the above general treatment. We will specify this general formalism to the case of atom interferometry for exploring gravito–inertial interactions as well as to interferometry with light for tests of Special and General Relativity.

3 Atomic Interferometry

Atomic interferometry has been first realized by Carnal and Mlynek in 1991 using a Young double slit setup [8]. Modern setups use laser cooled atoms and laser beams as beam splitters [9]. With these latter techniques atom interferometers are a very precise tools for exploring the gravito–inertial interactions, for making precise tests of foundations of quantum mechanics. They also enabled the most precise determination of the fine structure constant [10]. These devices find applications as accelerometers [11, 12, 13], gyroscopes [14, 15] and gravity gradiometers [16]. Recent studies show that with atomic interferometry it also should be possible to map the gravitomagnetic field of the Earth in an orbiting satellite [17].

3.1 The Beam Splitter

In atomic interferometry the beam splitting is accomplished by the atom–laser interaction. The corresponding dipole–interaction yields that photons can transmit momenta to the atoms in units of $\hbar\boldsymbol{k}$, where $\boldsymbol{k}$ is the wave vector of the laser beam [18]. That means that for an incoming atomic beam with momentum $\boldsymbol{p}$ the interaction with a laser beam gives atoms with momenta $\boldsymbol{p} + n\hbar\boldsymbol{k}$, where n is an integer. It can be shown that for appropriately chosen laser frequencies the possible atomic states effectively reduce to a two-level system which is characterized by the two momenta $\boldsymbol{p}$ and $\boldsymbol{p} + \hbar\boldsymbol{k}$. The atomic beam splitter therefore acts in momentum space and can be described by means of the matrix [12]

$$(U^{\pi/2}a)(t_1+) = \frac{1}{\sqrt{2}} \begin{pmatrix} 1 & -ie^{i\delta_0} e^{i\delta(\boldsymbol{p})(t_1-t_0)} \\ -ie^{-i\delta_0} e^{-i\delta(\boldsymbol{p})(t_1-t_0)} & 1 \end{pmatrix} a(t_0) . \quad (19)$$

The corresponding laser pulse is called a $\pi/2$-pulse. The laser pulse which has double duration or is twice as intensive acts as mirror

$$(U^\pi a)(t_1+) = \begin{pmatrix} 0 & -ie^{i\delta_0}e^{i\delta(\boldsymbol{p})(t_1-t_0)} \\ -ie^{-i\delta_0}e^{-i\delta(\boldsymbol{p})(t_1-t_0)} & 0 \end{pmatrix} a(t_0) \tag{20}$$

and is called a π-pulse. With these atom-optical elements we can describe atom interferometric experiments.

3.2 Interaction with the Gravitational Field

We discuss the interaction with a homogeneous gravitational field, or, equivalently, with the acceleration of the interferometer. This is described by the Schrödinger equation describing the "free" evolution between the beam splitters

$$H = \frac{1}{2m}\boldsymbol{p}^2 - m\boldsymbol{g}\cdot\boldsymbol{x}\,. \tag{21}$$

Here m is the mass of the atoms and $\boldsymbol{g}$ the gravitational acceleration. This operator does not induce any transitions between the upper and the lower atomic levels. The corresponding evolution operator can be given in closed form [19]

$$U(t,t_0) = e^{-\frac{i}{\hbar}\left(\omega_{\text{atom}}-\frac{\boldsymbol{p}^2}{2m}\right)(t-t_0)}e^{\frac{i}{\hbar}\boldsymbol{\xi}(t)\cdot\boldsymbol{p}}e^{-\frac{i}{\hbar}\dot{\boldsymbol{\xi}}\cdot\boldsymbol{x}}e^{-\frac{i}{\hbar}\int^t\left(\frac{m}{2}\dot{\boldsymbol{\xi}}(t')^2+\boldsymbol{g}\cdot\boldsymbol{\xi}\right)dt'} \tag{22}$$

with $\ddot{\boldsymbol{\xi}} = \boldsymbol{g}$. We calculate the phase shift for the Kasevich–Chu-setup using a sequence of $\pi/2$–, π, and $\pi/2$-laser pulses as beam splitter, mirror and recombiner. For simplicity, we do not take into account any modification of the beam splitting process due to the gravitational field (see [20, 21]). One obtains the phase shift [12, 22]

$$\delta\phi = \boldsymbol{k}\cdot\boldsymbol{g}\,T^2\,. \tag{23}$$

where T is the time between two laser pulses.

This phase shift does not depend on the mass of the atoms. In the case of uniform gravitational acceleration the observable phase shift for *quantum matter* which evolution is governed by the usual Schrödinger equation, *exactly fulfills the weak equivalence principle (EP)* [23]. This is remarkable because the solutions of the Schrödinger equation in a homogeneous gravitational field depend on the mass [24]. Consequently, a mass dependence of the solutions does not imply that observables depend on the mass and therefore does not indicate a break-down of the Weak Equivalence Principle.

Therefore this kind of experiment can be used to test the Weak Equivalence Principle in the quantum domain. If one introduces an inertial mass m_i and a gravitational mass m_g in the kinetic and the gravitational part of the Schrödinger equation then one obtains the modified phase shift

$$\delta\phi = -\frac{m_g}{m_i}\delta\boldsymbol{g}\cdot\boldsymbol{k}\,T^2\,. \tag{24}$$

Thus, with different kinds of atoms one can perform direct tests of the equivalence principle. With the accuracy of the Kasevich–Chu interferometer [13] a null-result will verify the EP in the sense of the Eötvös–ratio $\eta = (m_g/m_i)^{(a)} - (m_g/m_i)^{(b)}$ with the accuracy[1] $|\eta| \leq 10^{-9}$ (here (a) and (b) indicate the two different kinds of atoms). This kind of experiment will be a genuine *quantum mechanical test* of the weak EP. In the derivation of (23) and (24) no approximations have been used.

One also can turn the line reasoning the other way around: If one assumes an interaction of the atom with a potential $V(\boldsymbol{x})$ which structure is undetermined and requires the validity of the Equivalence Principle for the resulting phase shift then one can show [26] that the potential must be of the form $mU(\boldsymbol{x})$ where $U(\boldsymbol{x})$ is a function which does not depend on the mass m of the quantum object (m is the mass from the kinetic term in the Schrödinger equation). This kind of reasoning also applies to vectorial and tensorial interactions. Therefore, in the same way as for point particles, the Equivalence Principle makes it possible to determine the structure of the gravitational interaction with quantum matter and, even more important, to assign the gravitational interaction also in the quantum domain a geometrical structure which is independent of the quantum matter under consideration.

Another striking feature of (23) is the fact that, though it is an exact quantum result, it does not depend on Planck's constant $\hbar$. This is because the wave vector $\boldsymbol{k}$ and the interaction time T are given by the experimentator, $\delta\phi$ is the measured quantity, and $\boldsymbol{g}$ the quantity calculated from the experimental input and the measured quantity. There is no room for having an $\hbar$. Only by introducing *formally* the classical notions "length" $\boldsymbol{l} = v_0 T$ where $v_0 = \langle\widehat{\boldsymbol{v}}\rangle_0$ is the mean value of the velocity of the atomic beam at the first beam splitter, and "height" $\boldsymbol{h} = \hbar\boldsymbol{k}T/m$ we can reformulate the phase shift (23) so that it acquires the well known ordinary COW–form $\delta\phi = mghl/(\hbar v_0)$. However, this is purely formal, because there is neither an operational separation of the atomic beams and, thus, no operational realization of the "height" of the interferometer. The only experimentally given quantities are the wave vector of the laser beams and the time between the laser pulses.

In the same way, for neutron interferometry the gravitationally induced phase shift can be recovered in the form $\delta\phi = \boldsymbol{G}\cdot\boldsymbol{g}\,T^2$ where only the reciprocal lattice vector $\boldsymbol{G}$ and the time of flight T are the relevant experimental quantities [23]. This result also shows that for neutron interferometry in a uniform gravitational field the weak equivalence principle holds, too.

3.3 Rotation – The Sagnac Effect

The Sagnac effect originally describes the influence of the rotation of the interferometer on the interference pattern of light. It was first predicted by Sagnac and first observed by Michelson and Gale [27]. The fact that a rotating

[1] For bulk matter the accuracy is presently 5×10^{-13}, see [25].

interferometer also effects the phase shift of matter waves was first predicted by Heer [28] (see also Page [29, 30]) and has been confirmed by interference experiments for neutrons [31, 32], electrons [33, 34, 35] and atoms [9].

For spinless matter the corresponding Hamiltonian attached to the rotating interferometer is given by

$$H = \frac{\boldsymbol{p}^2}{2m} - \boldsymbol{\Omega} \cdot (\boldsymbol{x} \times \boldsymbol{p}) \tag{25}$$

where $\boldsymbol{\Omega}$ is the angular velocity of the interferometer. The phase shift results from the solution of this Schrödinger equation given by

$$U_{\mathrm{I}}(t, t_0) = \exp\left(-\frac{i}{\hbar}\frac{p^2}{2m}(t - t_0)\right) \exp\left(\frac{i}{\hbar}\boldsymbol{\Omega} \cdot (\boldsymbol{x} \times \boldsymbol{p})(t - t_0)\right) \tag{26}$$

which successively interacts with a sequence of $\pi/2$–, π–, and $\pi/2$-pulses. The corresponding phase shift is called the *Sagnac-effect.* To first order in the rotation $\boldsymbol{\Omega}$ it turns out to be

$$\delta\phi = 2T^2 \boldsymbol{k} \cdot (\langle \boldsymbol{v} \rangle_0 \times \boldsymbol{\Omega}) \tag{27}$$

where $\langle \boldsymbol{v} \rangle_0$ denotes the expectation value of the atomic velocity at the position of the first beam splitter. Because of the large energy due to the rest mass of the atoms this effect is larger than the corresponding effect for light with a scaling factor $mc^2/h\nu$. As already discussed in connection for the acceleration induces phase shift, in (27) there appears neither m nor $\hbar$. Again, with the introduction of classical notions "height" and "length" the above phase shift acquires the well-known form of the Sagnac effect

$$\delta\phi = 2\frac{m}{\hbar}\boldsymbol{\Omega} \cdot \boldsymbol{A}\,. \tag{28}$$

The Sagnac effect is also the basis of the Schiff effect [17, 18].

In [14] the measurement of the Sagnac effect induced by the Earth's rotation with atom beam interferometry has been reported. They arrived a sensitivity to rotations of $2 \cdot 10^{-8}$ rad sec$^{-1}/\sqrt{\mathrm{Hz}}$.

3.4 Coupling to the Gravity Gradient

We can expand the interaction with the Newtonian potential to the second order in the position:

$$U(\boldsymbol{x}) = U(\boldsymbol{x}_0) + \boldsymbol{x} \cdot \boldsymbol{\nabla} U(\boldsymbol{x}_0) + \frac{1}{2} r^i r^j \partial_i \partial_j U(\boldsymbol{x}_0) \tag{29}$$

with $\boldsymbol{r} = \boldsymbol{x} - \boldsymbol{x}_0$ where $\boldsymbol{x}_0$ is the expectation value of the position of the atomic beam at time t_0. The second term is the uniform gravitational acceleration described above and the third term can be interpreted as Newtonian part of the Riemannian space-time curvature tensor. Consequently, the influence of the largest part of the Riemannian space-time curvature on quantum matter is in lowest order given by the interaction term $H_{\rm I} = -\frac{1}{2} r^i r^j \partial_i \partial_j U$. This interaction corresponds to a particle in an three-dimensional anisotropic oscillator potential. This case can again be treated in an exact manner. Generalizing the procedure in [19] we get as solution for the time-evolution operator

$$U(t, t_0) = \exp\left(-\frac{i}{\hbar}\lambda_{ij} r_i r^j\right) \exp\left(-\frac{i}{\hbar}\mu^{ij} p_i p_j\right) \times \exp\left(-\frac{i}{\hbar}\nu^i_j (p_i r^j + r^j p_i)\right) \tag{30}$$

where the functions $\lambda_{ij}(t)$, $\mu^{ij}(t)$, and $\nu^i_j(t)$ are solutions of

$$\frac{d}{dt}\lambda_{ij}(t) = 2\frac{1}{m}\lambda^l_i(t)\lambda_{lj}(t) - \frac{1}{2}\partial_i\partial_j U \tag{31}$$

$$\frac{d}{dt}\mu^{ij}(t) = \frac{\delta^{ij}}{2m} + 2\frac{\hbar}{m^2}\mu^{ik}(t)\lambda^j_k(t) \tag{32}$$

$$\nu^i_j(t) = \int_{t_0}^t \lambda^i_j(t')dt' \tag{33}$$

with the initial conditions $\lambda_{ij}(t_0) = 0$ and $\mu^{ij}(t_0) = 0$ which can be derived from the initial condition $U(t_0, t_0) = 1$. For the geometry of the atom beam interferometer of Kasevich and Chu we get as phase shift to lowest order in the curvature

$$\delta\phi = -k^i T^2 \left(g_i - U_{ij} T\left(\frac{\hbar k^j}{2m} + \langle\widehat{v}^j\rangle_0 - g^j \frac{37}{12} T\right)\right) . \tag{34}$$

This phase shift has also been discussed in [36], where it has been shown that it may be possible to measure this phase shift for a space-time curvature which is created by masses aligned in the laboratory. Since the curvature is proportional to the matter density masses in the laboratory are better suited for that purpose than the Earth.

The phase shift (34) is the *quantum analogue of the geodesic deviation.* In the same manner as in classical mechanics (where one calculates the relative acceleration of two freely falling point particles, see e.g. [37]) it describes the relative motion of two "parts" of a quantum system which are subject to a gravitational field corresponding to a curved space-time.

It is of importance for the formalism of quantum mechanics in curved space-time to realize experimentally this effects. This would be for the first time a test of a *single* quantum system with the (Newtonian parts of the) curvature of space-time. Since the corresponding phase shift is not the result of a transport along a closed path but instead an effect which results from an overall defined quantum field, it is a pure curvature effect. In addition, this would constitute a measurement of the space-time curvature on a very small scale ($\sim$1 cm^{-3}) which is not possible using classical methods.

3.5 Tests of Relativistic Gravity

Relativistic Approximation of the Klein–Gordon Equation

We derive a Hamilton operator for a spinless quantum field within the PPN framework [38] which is a test theory for a very broad class of gravitational theories. Within this formalism the space-time metric depends on the matter in the universe according to the gravitational theory under consideration. The matter in the universe is given by its rest mass, its currents, pressure, et. Therefore, each gravitational theory is characterized by some set of parameters which describe the coupling of the metric to the various components of the matter. Consequently, any equation which depends on the metric as, e.g., the Klein–Gordon or Dirac equation depends on these parameters. Comparison of the outcome of experiments with the theoretical description within this general PPN framework will give information about the PPN parameters. We first perform a relativistic approximation of the minimally coupled Klein–Gordon equation in an arbitrary gravitational field [39], where gravity is described in a PPN approximation which is consistent with the approximation scheme employed.

The Klein–Gordon equation minimally coupled to gravity is ($\mu, \nu = 0, \ldots, 3$)

$$0 = g^{\mu\nu}\hbar^2 D_\mu D_\nu \varphi_{\mathrm{KG}} - m^2 c^2 \varphi_{\mathrm{KG}} \,. \tag{35}$$

We insert the PPN metric [38] into (35) and make the ansatz [40]

$$\varphi_{\mathrm{KG}}(x,t) = \exp\left(c^2 S_0(x,t) + S_1(x,t) + c^{-2} S_2(x,t) + \ldots\right) \tag{36}$$

and compare powers of the expansion parameter c^2. The lowest order term is $\partial_i S_0 = 0$ implying that S_0 is a function of t only. The next order $\left(\frac{\partial S_0}{\partial t}\right)^2 - m^2 = 0$ has the solutions $S_0 = \pm mt$ from which we choose $S_0 = -mt$. The next order gives after the substitution $\varphi_1 := \exp\left(\frac{i}{\hbar} S_1\right)$ the Schrödinger equation for φ_1 coupled to the Newtonian potential

$$i\hbar \frac{\partial \varphi_1}{\partial t} = \frac{\boldsymbol{p}^2}{2m}\varphi_1 - mU(\boldsymbol{x})\varphi_1 \,. \tag{37}$$

The next order is an equation for S_1 and S_2. Replacing S_1 by φ_1 by means of the above substitution and introducing a new wave function $\varphi := \varphi_1 \exp\left(\frac{iS_2}{\hbar c^2}\right)$ gives the relativistic correction of order c^{-2}

$$\begin{aligned} H = & \frac{\boldsymbol{p}^2}{2m} - \frac{\boldsymbol{p}^4}{8m^3c^2} - mU - (2\gamma+1)\frac{U}{c^2}\frac{\boldsymbol{p}^2}{2m} - \left(\frac{1}{2} - \beta\right)\frac{mU^2}{c^2} \\ & - \frac{3i\gamma\hbar\partial_t U}{2c^2} - \frac{i\hbar\delta^{ij}}{2c^2 m}\gamma\partial_i U p_j - \frac{\hbar^2\nabla^2 U}{4c^2 m} + H_{\mathrm{grav-magn}} + H_\Phi + H_\alpha \,, \end{aligned} \tag{38}$$

where

$$H_{\text{grav-magn}} := \frac{7}{4}\frac{\Delta_1}{c^2}\{V^i, p_i\} + \frac{1}{4}\frac{\Delta_2}{c^2}\{W^i, p_i\} \tag{39}$$

$$\begin{aligned} H_\Phi := &-\left(\frac{\zeta_1}{2} - \xi\right)\frac{m}{c^2}\mathcal{A} + \left(1 + \frac{\alpha_3}{2} + \gamma - \xi + \frac{\zeta_1}{2}\right)\frac{m}{c^2}\Phi_1 \\ &+ (1 - 2\beta + 3\gamma + \xi + \zeta_2)\frac{m}{c^2}\Phi_2 + (1 + \zeta_3)\frac{m}{c^2}\Phi_3 \\ &+ (3\gamma - 2\xi + 3\zeta_4)\frac{m}{c^2}\Phi_4 - \xi\frac{m}{c^2}\Phi_W \end{aligned} \tag{40}$$

$$\begin{aligned} H_\alpha := &-(\alpha_2 + \alpha_3 - \alpha_1)\frac{mw^2U}{2c^2} - (2\alpha_3 - \alpha_1)\frac{mw^iV_i}{2c^2} \\ &+\alpha_2\frac{mw^iw^jU_{ij}}{2c^2} + \frac{\alpha_1 - 2\alpha_2}{4c^2}w^i\{U, \widehat{p}_i\} + \frac{\alpha_2}{2c^2}w^i\{U_{ji}, p_j\}\,. \end{aligned} \tag{41}$$

$H_{\text{grav-magn}}$ describes gravitomagnetic effects and H_α are preferred frame contributions. All external fields U, U^{ij}, V^i, W^i, Φ, etc. in general depend on the position $\boldsymbol{x}$ and time t and are fixed by the boundary condition that they vanish at spatial infinity. In the following all statements are valid in our order of approximation only.

The scalar product which is necessary for a quantum interpretation of our dynamical equation (38) can be obtained from the relativistic approximation of the conserved quantity $const = \int j^\mu n_\mu d^3V$ with $j^\mu = g^{\mu\nu}(D_\nu\varphi^*_{\text{KG}}\varphi_{\text{KG}} - \varphi^*_{\text{KG}}D_\nu\varphi_{\text{KG}}$ where $(\cdot)^*$ denotes complex conjugation. Here n_μ is a normalized time-like 1-form: $g^{\mu\nu}n_\mu n_\nu = -1$ and $d^3V = \sqrt{^{(3)}g}d^3x$ is the invariant 3-volume in the $t = const$ hypersurface. If we align $n_\mu \sim \delta^0_\mu$ we have $n_\mu = \sqrt{-g^{00}}\delta^0_\mu$. Inserting the Klein–Gordon wave function $\varphi_{\text{KG}} = e^{-i\frac{mc^2}{\hbar}t}\varphi$ yields

$$j^0 = \varphi^*\varphi + \frac{1}{2mc^2}\left((H\varphi)^*\varphi + \varphi^*(H\varphi)\right) \tag{42}$$

where we normalized the wave function in such a manner that the leading non-relativistic term acquires the form of the Schrödinger probability density. Terms containing $g^{0\hat{\mu}}$ drop out.

Using this conserved quantity we define the scalar product

$$\begin{aligned} \langle\psi \mid \varphi\rangle &= \int\left[\psi^*\varphi\left(1 - \frac{U}{c^2}\right) + \psi^*\frac{\boldsymbol{p}^2}{2m^2c^2}\varphi\right]\frac{\sqrt{^{(3)}g}}{\sqrt{-g_{00}}}d^3x \\ &= \int\left[\psi^*\varphi + \psi^*\mathbf{p}^2 2m^2c^2\varphi\right]\sqrt{^{(3)}g}d^3x\,. \end{aligned} \tag{43}$$

The norm is defined by $\langle\varphi \mid \varphi\rangle$. The scalar product and, thus, the norm is conserved in our order of approximation. Since $\boldsymbol{p}^2$ is a positive operator, the norm $\langle\varphi \mid \varphi\rangle$ is positive definite. The expectation value of an operator O is defined as $\langle\varphi \mid O \mid \varphi\rangle$. As usual, O is hermitian if $\langle O\psi \mid \varphi\rangle - \langle\psi \mid O\varphi\rangle = 0$ for all ψ and φ. The Hamiltonian H (38) is hermitian with respect to (43) if and only if $\partial_t U = 0$ what can also be stated as $\partial_t\sqrt{^{(3)}g} = 0$.

It is convenient [41, 42] to transform the wave functions and the Hamiltonian H to a "flat Schrödinger" scalar product. In a first step we absorb the 3-volume element into the wave functions and operators by defining "flat" wave functions, scalar product, operators and the Hamiltonian by means of

$$\varphi \to \varphi_{\mathrm{f}} = ({}^{(3)}g)^{\frac{1}{4}}\varphi \tag{44}$$

$$O \to O_{\mathrm{f}} = ({}^{(3)}g)^{\frac{1}{4}} O ({}^{(3)}g)^{-\frac{1}{4}} \tag{45}$$

so that

$$\langle \psi \mid \varphi \rangle = \int \left[\psi_{\mathrm{f}}^* \varphi_{\mathrm{f}} + \psi_{\mathrm{f}}^* \frac{\boldsymbol{p}^2}{2m^2c^2} \varphi_{\mathrm{f}} \right] d^3x \tag{46}$$

$$\begin{aligned} H_{\mathrm{f}} &= ({}^{(3)}g)^{\frac{1}{4}} H ({}^{(3)}g)^{-\frac{1}{4}} + i\hbar \frac{\partial}{\partial t} ({}^{(3)}g)^{-\frac{1}{4}} \\ &= \frac{\boldsymbol{p}^2}{2m} - \frac{\boldsymbol{p}^4}{8m^3c^2} - mU - (2\gamma+1)\frac{U}{c^2}\frac{\boldsymbol{p}^2}{2m} - \left(\frac{1}{2} - \beta\right)\frac{mU^2}{c^2} \\ &\quad + \frac{i\hbar}{c^2 m}\gamma \delta^{ij} \nabla_i U p_j - (1-3\gamma)\frac{\hbar^2 \nabla^2 U}{4c^2 m} + H_{\text{L-T}} + H_{\Phi} + H_{\alpha} \end{aligned} \tag{47}$$

The time derivative of the Newtonian potential dropped out. The total probability as well as the expectation values are not affected by this transformation.

In the second step we perform a non-local transformation

$$\psi_{\mathrm{f}} \to \psi_{\mathrm{f,S}} := \left(1 + \frac{\boldsymbol{p}^2}{m^2c^2}\right)^{\frac{1}{4}} \psi_{\mathrm{f}} \tag{48}$$

$$O_{\mathrm{f}} \to O_{\mathrm{f,S}} = \left(1 + \frac{\boldsymbol{p}^2}{m^2c^2}\right)^{\frac{1}{4}} O_{\mathrm{f}} \left(1 + \frac{\boldsymbol{p}^2}{m^2c^2}\right)^{-\frac{1}{4}} \tag{49}$$

so that

$$\langle \psi \mid \varphi \rangle = \int \psi_{\mathrm{f,S}}^* \varphi_{\mathrm{f,S}} d^3x \tag{50}$$

$$\begin{aligned} H_{\mathrm{f,S}} &= \left(1 + \frac{\boldsymbol{p}^2}{m^2c^2}\right)^{\frac{1}{4}} H_{\mathrm{f}} \left(1 + \frac{\boldsymbol{p}^2}{m^2c^2}\right)^{-\frac{1}{4}} \\ &\quad - i\hbar \left(1 + \frac{\boldsymbol{p}^2}{4m^2c^2}\right) \partial_t \left(1 - \frac{\boldsymbol{p}^2}{4m^2c^2}\right) \\ &= \frac{\boldsymbol{p}^2}{2m} - \frac{\boldsymbol{p}^4}{8m^3c^2} - mU + (2\gamma+1)\frac{1}{2mc^2}\left(-U\boldsymbol{p}^2 + i\hbar\delta^{ij}\partial_i U p_j\right) \\ &\quad - \left(\frac{1}{2} - \beta\right)\frac{mU^2}{c^2} + 3\gamma\frac{\hbar^2 \nabla^2 U}{4c^2 m} + H_{\text{grav-magn}} + H_{\Phi} + H_{\alpha} \,. \end{aligned} \tag{51}$$

The Hamiltonian $H_{\mathrm{f,S}}$ is manifest hermitian. This means that by choosing the flat scalar product in the Schrödinger form automatically gives a Hamiltonian which is hermitian. This means in particular that even for the treatment of relativistic corrections the usual interpretation and handling according to the non-relativistic Schrödinger equation is possible. Therefore this representation of the Hamiltonian is most preferable for the following discussions[2]

The general structure of the part of the Hamiltonian (51) describing the interaction with the gravitational field is

$$H_{\text{int}}^{\text{grav}} = \Psi^{ab}(x) p_a p_b + \Psi^a(x) p_a + \Psi(x) \tag{52}$$

where the coefficients $\Psi^{ab}(x)$, $\Psi^a(x)$, and $\Psi(x)$ can be read off from (51). We expand the gravitational field with respect to a certain reference point x_0 which we take to be the position of the beam splitter. Neglecting terms containing the second derivative combined with c^{-2} we have

$$\Psi(x) = \Psi + \Psi_a \hat{x}^a + \Psi_{ab} \hat{x}^a \hat{x}^b \tag{53}$$

$$\Psi^a(x) = \Psi^a + \Psi^a_b \hat{x}^b \tag{54}$$

$$\Psi^{ab}(x) = \Psi^{ab} + \Psi^{ab}_c \hat{x}^c \tag{55}$$

with $\hat{x}^a = x^a - x^a_0$ and $\Psi_a = \partial_a \Psi(x_0)$, $\Psi_{ab} = \partial_a \partial_b \Psi(x_0)$, $\Psi^a = \Psi^a(x_0)$, etc. Then the Hamiltonian is

$$\begin{aligned} H_{\text{c.m.}} &= H_0(p) + \Psi + \Psi_a x^a + \Psi_{ab} x^a x^b + \Psi^a p_a \\ &\quad + \Psi^a_b x^b p_a + \Psi^{ab} p_a p_b + \Psi^{ab}_c x^c p_a p_b \end{aligned} \tag{56}$$

with $H_0(p) = \frac{\boldsymbol{p}^2}{2m} - \frac{\boldsymbol{p}^4}{8m^3c^2}$ and

$$\begin{aligned} \Psi &= -mU - \left(\frac{1}{2} - \beta\right) mU \frac{U}{c^2} - (\alpha_2 + \alpha_3 - \alpha_1) \frac{mw^2 U}{2c^2} \\ &\quad - (2\alpha_3 - \alpha_1) \frac{mw^i V_i}{2c^2} \varphi_2 + \alpha_2 \frac{mw^i w^j U_{ij}}{2c^2} - \frac{1}{2} i\hbar \Psi^a_a + H_\Phi \end{aligned} \tag{57}$$

$$\begin{aligned} \Psi_a &= -m\partial_a U - 2(1-\beta) m \partial_a U \frac{U}{c^2} + \partial_a H_\Phi - (\alpha_2 + \alpha_3 - \alpha_1) \frac{mw^2 \partial_a U}{2c^2} \\ &\quad - (2\alpha_3 - \alpha_1) \frac{mw^i \partial_a V_i}{2c^2} \varphi_2 + \alpha_2 \frac{mw^i w^j \partial_a U_{ij}}{2c^2} \end{aligned} \tag{58}$$

---

[2] In the "flat Schrödinger" representation it is also convenient to take x^i as position operator. Transforming this operator back to the "curved" representation one gets the non-local position operator $\widehat{x}^i = x^i + \frac{\hbar^2}{2m^2c^2} \delta^{ij} \nabla_j$ which is hermitian with respect to the "curved" scalar product (43). This position operator gives us the possibility to define a center-of-mass x_0 of the matter field representing the *coordinate* position of the atom $x_0^i := \langle \varphi \mid x^i \mid \varphi \rangle = \int \varphi^*_{\text{f,S}} x^i \varphi_{\text{f,S}} d^3x$.

$$\Psi_{ab} = -m\partial_a\partial_b U \tag{59}$$

$$\Psi^a = \frac{7}{2}\frac{\Delta_1}{c^2}V^a + \frac{1}{2}\frac{\Delta_2}{c^2}W^a + \frac{\alpha_1 - 2\alpha_2}{2c^2}Uw^a + \alpha_2\frac{U^a{}_b}{c^2}w^b - i\hbar\Psi^{ab}_b \tag{60}$$

$$\Psi^a_b = \frac{7}{2}\frac{\Delta_1}{c^2}\partial_b V^a + \frac{1}{2}\frac{\Delta_2}{c^2}\partial_b W^a + \frac{\alpha_1 - 2\alpha_2}{2c^2}\partial_b U w^a + \alpha_2\frac{\partial_b U^a{}_c}{c^2}w^c \tag{61}$$

$$\Psi^{ab} = -2\gamma\frac{U}{2mc^2}\delta^{ab} \tag{62}$$

$$\Psi^{ab}_c = -(1+2\gamma)\frac{\partial_c U}{2mc^2}\delta^{ab}\,, \tag{63}$$

where all gravitational fields are evaluated at the position of the beam splitter.

The Solution

Now we solve the equation of motion for vanishing laser beam. Then we have for our 2-level system the equations ($r = 1, 2$ and $p^{(1)} = p$, $p^{(2)} = p + \hbar k$)

$$\frac{d}{dt}a_{r,\boldsymbol{p}^{(r)}} = -\frac{i}{\hbar}\left(\Xi(\boldsymbol{p}^{(r)}, t, t_0)a_{r,\boldsymbol{p}^{(r)}} + i\hbar\Xi_a(\boldsymbol{p}^{(r)}, t, t_0)\frac{\partial}{\partial p_a}a_{r,\boldsymbol{p}^{(r)}}\right.$$
$$\left. -\hbar^2\Xi_{ab}(\boldsymbol{p}^{(r)})\frac{\partial^2}{\partial p_a\partial p_b}a_{r,\boldsymbol{p}^{(r)}}\right)\,, \tag{64}$$

with

$$\Xi(\boldsymbol{p}, t, t_0) = \Psi + \Psi^a p_a + \Psi^{ab}p_a p_b + i\hbar\Psi^a_a + 2i\hbar\Psi^{ab}_b p_a$$
$$+(\Psi_a + \Psi^b_a p_b + \Psi^{bc}_a p_b p_c)\frac{\partial H_0(\boldsymbol{p})}{\partial p_a}(t - t_0)$$
$$+\Psi_{ab}\left(\frac{\partial H_0(\boldsymbol{p})}{\partial p_a}\frac{\partial H_0(\boldsymbol{p})}{\partial p_b}(t - t_0) + i\hbar\frac{\partial^2 H_0(\boldsymbol{p})}{\partial p_a\partial p_b}\right)(t - t_0) \tag{65}$$

$$\Xi_a(p, t, t_0) = \Psi_a + \Psi^c_a p_c + \Psi^{bc}_a p_b p_c + 2\Psi_{ab}\frac{\partial H_0(\boldsymbol{p})}{\partial p_b}(t - t_0) \tag{66}$$

$$\Xi_{ab}(p) = \Psi_{ab} + \Psi^c_{ab}p_c + \Psi^{cd}_{ab}p_c p_d\,. \tag{67}$$

We will treat the quantities $\mathrm{Re}\Psi$ and Ψ_a in an exact manner and all other quantities in first order. We also split $H_0(\boldsymbol{p}) = \boldsymbol{p}^2/2m + h(\boldsymbol{p})$ where the function $h(\boldsymbol{p})$ is small.

We can solve that parts of (64) containing Ψ and Ψ_a exactly. With this well known solution we can now solve the full equation (64) and get

$$a_{r,\boldsymbol{p}^{(r)}}(t) = e^{-\frac{i}{\hbar}\phi(\boldsymbol{p}^{(r)}, t, t_0)}\left(1 - \frac{i}{\hbar}A(\boldsymbol{p}^{(r)}, t, t_0)a_{r,\boldsymbol{p}^{(r)}_a + \Psi_a(t-t_0)}(t_0)\right.$$
$$+B_a(\boldsymbol{p}^{(r)}, t, t_0)\frac{\partial}{\partial p_a}a_{r,\boldsymbol{p}^{(r)}_a + \Psi_a(t-t_0)}(t_0)$$
$$\left. +i\hbar C_{ab}(\boldsymbol{p}^{(r)}, t, t_0)\frac{\partial^2}{\partial p_a\partial p_b}a_{r,\boldsymbol{p}^{(r)}_a + \Psi_a(t-t_0)}(t_0)\right)\,, \tag{68}$$

with

$$
\begin{aligned}
A(\boldsymbol{p}^{(r)},t,t_0) \approx & \left(\frac{i\hbar}{2}\Psi^a_a + i\hbar\Psi^{ab}_b p^{(r)}_a + \mathrm{Re}\Psi^a p_a + \Psi^{ab} p_a p_b\right)(t-t_0) \\
& + \frac{1}{2}\left((\Psi^\mu_\rho p_\mu + \Psi^{\mu\nu}_\rho p_\mu p_\nu)\frac{\delta^{\rho\tau} p_\tau}{m} + \Psi_\rho \frac{\partial h(\boldsymbol{p})}{\partial p_\rho} + \Psi_{\rho\sigma} i\hbar \frac{\delta^{\rho\sigma}}{m}\right)(t-t_0)^2 \\
& + \frac{1}{3}\left(\Psi_{\rho\sigma}\frac{\delta^{\rho\nu}p_\nu}{m}\frac{\delta^{\sigma\tau}p_\tau}{m} + (\Psi^c_a p_c + \Psi^{bc}_a p_b p_c)\frac{1}{2m}\delta^{ab}\Psi_b\right)(t-t_0)^3 \\
& + \frac{1}{4m^2}\Psi_{ab}\delta^{bc}p_c\Psi_f\delta^{af}(t-t_0)^4 + \frac{1}{5}\Psi_{ab}\frac{1}{4m^2}\delta^{ac}\Psi_c\delta^{bd}\Psi_d(t-t_0)^5 \qquad (69)
\end{aligned}
$$

$$
\begin{aligned}
B_a(\boldsymbol{p}^{(r)},t,t_0) \approx & (\Psi^c_a p_c + \Psi^{bc}_a p_b p_c)(t-t_0) \\
& + \Psi_{ab}\frac{\delta^{bf}p_f}{m}(t-t_0)^2 + \frac{\Psi_{ab}\Psi_e\delta^{be}}{3m}(t-t_0)^3 \qquad (70)
\end{aligned}
$$

$$
C_{ab}(\boldsymbol{p}^{(r)},t,t_0) = \Psi_{ab}(t-t_0)\,. \qquad (71)
$$

The Phase Shift

Again we assume an interferometer geometry of Kasevich and Chu type with a sequence of $\pi/2-\pi-\pi/2$ pulses with a time T between two pulses. Between the pulses the states evolve according to the Hamiltonian with gravitational interaction only. Therefore we have the following evolution from the initial state $a(t_0) = \begin{pmatrix} a_{1,\boldsymbol{p}}(t_0) \\ a_{2,\boldsymbol{p}+\hbar\boldsymbol{k}}(t_0) \end{pmatrix}$ to the final state after the last $\pi/2$ pulse:

$$
a(t_f) = U^{\pi/2}(t_3)U^{\mathrm{grav}}(t_3,t_2)U^{\pi}(t_2)U^{\mathrm{grav}}(t_2,t_1)U^{\pi/2}(t_1)U^{\mathrm{grav}}(t_1,t_0)a(t_0) \qquad (72)
$$

The probability $I_2 := \int a^*_{2,\boldsymbol{p}+\hbar\boldsymbol{k}}(t_f)a_{2,\boldsymbol{p}+\hbar\boldsymbol{k}}(t_f)d^3p$ to observe this state in terms of the probability $I_1 := \int a^*_{1,\boldsymbol{p}}(t_0)a_{1,\boldsymbol{p}}(t_0)d^3p$. With the initial conditions $a_{2,\boldsymbol{p}+\hbar\boldsymbol{k}}(t_0) = 0$, $a_{1,\boldsymbol{p}}(t_0) = 1$ we get

$$
I_2 = I_1(1-\cos\phi) \qquad (73)
$$

where

$$
\begin{aligned}
\phi = T^2 \Bigg\{ k^a \Bigg[& -\nabla_a U\left(1 + 2(1-\beta)\frac{U}{c^2} - (\alpha_2+\alpha_3-\alpha_1)\frac{w^2}{2c^2}\right) \\
& -2\nabla_a\nabla_b U\delta^{bc}T\left(\frac{\langle p_c\rangle}{m} - \frac{5}{4}\partial_c U\right) + \frac{\nabla_a H_\Phi}{m} + \alpha_2\frac{w^i w^j \nabla_a U_{ij}}{2c^2}\Bigg] \qquad (74) \\
& -T\left(\frac{7}{2}\frac{\Delta_1}{c^2}\partial_c V^a + \frac{1}{2}\frac{\Delta_2}{c^2}\partial_c W^a + \frac{\alpha_1-2\alpha_2}{2c^2}\partial_c U w^a + \alpha_2\frac{\partial_c U^a{}_b}{c^2}w^b\right)4k^c\frac{\langle p_d\rangle}{m}\Bigg\}
\end{aligned}
$$

Here we defined $\langle p_a\rangle = \int \varphi * p_a \varphi d^3x$ and replaced the derivative with respect to the coordinates by a derivative with respect to the proper length: $\frac{\partial}{\partial x^a} = \sqrt{g_{aa}(\boldsymbol{x}_0)}\frac{\partial}{\partial r^a}$ with $r^a = \sqrt{g_{aa}(\boldsymbol{x}_0)}dx^a$. We denote $\nabla_a := \frac{\partial}{\partial r^a}$ and raise indices by δ^{ab}. For simplicity we also neglected terms of the order $\boldsymbol{v}^2/c^2$ where $\boldsymbol{v}$ may be $\hbar\boldsymbol{k}/m$ or $\langle\boldsymbol{p}\rangle/m$.

Equation (74) describes the phase shift in terms of the transferred momentum, the interaction time and the various parts of the gravitational field. This result is exact with respect to quantum mechanics; we performed no quasi-classical limit. This result is also exact with respect to the external field $\nabla_a U$ and to first order with respect to all other parts of the gravitational interaction.

The first line consists of a modification of the non-relativistic Newtonian gravitational acceleration. The first modification tests the PPN parameter β and the second is a preferred frame effect: the acceleration depends on the velocity of the laboratory with respect to the rest frame of the universe. The first term in the second line is a curvature term. It describes the coupling to the second derivative of the Newtonian potential. The next two terms in the second line give further post Riemannian effects. The last line consists in gravitomagnetic effects, and additional velocity dependent effects.

Numerical Estimates

In the following we want to estimate PPN parameters for the phase shift to be expected for the atomic beam interferometer of the Kasevich-Chu type assuming a null-result and using an accuracy of $\Delta\phi/\phi = 10^{-8}$. We assume that the experiment takes place on the surface of the Earth. The atoms used are characterized by their mass $m = 10^{-26}$ kg, their velocity $v \approx 10^{-1}$ m/sec. The laser wave vector is taken to be $k = 10^6 \mathrm{m}^{-1}$.

The various contributions to the phase shift are given in Table 1. The coupling to acceleration has already been tested [11] and the curvature term has been discussed in [36].

Assuming a null-result, the calculated phase shift leads using the gravitational potential of our galaxy to an estimate [43] $1-\beta \leq 10^{-5}$ (presently [38] $\beta - 1 \leq 10^{-3}$). Preferred frame effects may be tested using the rotation of the

Table 1. List of effects due to the interaction of a scalar atom with a general PPN parameterized gravitational field

Effect	Typical Form	Tested Parameter	Phase Shift
acceleration	∇U	–	10^6
curvature	$\nabla\nabla U$	–	10^{-3}
rel. corr. to COW	$\nabla U/c^2$	$\frac{1}{2}+\gamma$	10^{-13}
	$U\nabla U/c^2$	$1-\beta$	10^{-2}
frame dragging		$7\Delta_1 + \Delta_2$	10^{-11}
preferred frame		α_2	1
		$(\alpha_2 - \frac{\alpha_1}{2})$	10^{-2}
		$\alpha_3 + \alpha_2 - \alpha_1$	1

earth around the sun where 1-year-variations may occur. A null result leads to $\alpha_2 \leq 10^{-4}$ (today $4 \cdot 10^{-7}$), $\alpha_1 \leq 10^{-2}$ (today $2 \cdot 10^{-4}$), $\alpha_3 \leq 10^{-2}$ (today $4 \cdot 10^{-10}$). Also for γ we cannot get a better estimate than the current one [44]. Optical tests of PPN parameters have been proposed in [45, 46]. However, one has to bear in mind that these are *quantum tests* of gravitational theories and do therefore test gravity on another scale than bulk matter.

3.6 Tests of the Einstein Equivalence Principle

Atomic interferometry is very well suited for testing hypothetical effects which are due to violations of the Einstein Equivalence Principle on the level of the quantum matter equation, that is, the Dirac equation or the Schrödinger equation. See also [47].

The Generalized Dirac Equation

We start with a generalized Dirac equation (see, for example, [48, 49])

$$i\partial_t\varphi = -ic(\widetilde{\alpha}^i\nabla_i + i\Gamma)\varphi + mc^2\widetilde{\beta}\varphi + e\phi\varphi \tag{75}$$

$(i, j = 1, 2, 3)$ where φ is a complex 4-component field. All matrices are complex 4×4-matrices and $\widetilde{\alpha}^i$ and $\widetilde{\beta}$ obey $(\widetilde{\alpha}^i)^+ = \widetilde{\alpha}^i$, $\widetilde{\beta}^+ = \widetilde{\beta}$ but are not assumed to fulfill a Clifford algebra. Furthermore, $\Gamma^+\Gamma + i\partial_i\widetilde{\alpha}^i$. The Maxwell field comes in through minimal coupling $\nabla_i = \partial_i - \frac{ie}{c}A_i$. ϕ and A_i are the electromagnetic scalar and vector potentials. Here we also take the usual form of the Maxwell field as granted (for the experimental status of the electromagnetic field to obey the Einstein Equivalence Principle compare [50, 51]). It is of course also possible to couple the Maxwell field in an anomalous way to the GDE, analogous to the $TH\epsilon\mu$-formalism. However, since any modification of this kind will result in corrections of the same structure as those which we will derive below, we will not take anomalous couplings to the Maxwell field into account.

We also introduced a c which has the dimension of a velocity. This velocity can be introduced by considering the null cones which the GDE defines: c is the maximum speed of propagation (from the physical point of view it is approximately the velocity of light, because any deviation from SRT, if there is any, is small). The purpose of this velocity is twofold: First, it makes the coefficient in front of the spatial derivative dimensionless (what is necessary in order to connect $\widetilde{\alpha}^i$ with space-time geometry), and second, it will be used later as ordering parameter in a Foldy–Wouthuysen transformation leading to the non-relativistic limit of the GDE.

The splitting between the matrices $c\Gamma$ and $mc^2\widetilde{\beta}$ may be defined by means of a WKB approximation (compare [52]). While $mc^2\widetilde{\beta}$ is the "mass"-tensor which appears in the lowest order of approximation, Γ influences the first order only. Both matrices have the dimension of length^{-1}. In order to extract from the "mass"-tensor a dimensionless matrix possessing a geometrical meaning, we introduced a parameter m (so that mc^2 has the dimension time^{-1}) which can also be defined via the WKB approximation.

Equation (75) is general enough to describe violations of basic principles of GR. However, since due to the properties of the matrices $\widetilde{\alpha}^i$ (75) is a symmetric hyperbolic system very general principles of quantum mechanics are still fulfilled, namely (i) the well-posedness of the Cauchy problem, (ii) the superposition principle, (iii) finite propagation speed, and (iv) a conservation law. Indeed, it has been shown that this generalized Dirac equation can be derived from these fundamental principles, see [48] for a review.

If we introduce the quantities

$$\frac{4}{g^{00}} := \mathrm{tr}\widetilde{\beta}^2\,, \qquad \mathbf{g}^{0i} := \frac{g^{0i}}{g^{00}} := \frac{1}{4}\mathrm{tr}(\widetilde{\alpha}^i) \tag{76}$$

$$\mathbf{g}^{ij} := -\frac{g^{ij}}{g^{00}} := \frac{1}{4}\mathrm{tr}(\widetilde{\alpha}^i\widetilde{\alpha}^j) - 2\mathbf{g}^{0i}\mathbf{g}^{0j} \tag{77}$$

then the matrices $\widetilde{\alpha}^i$ and $\widetilde{\beta}$ fulfill

$$\widetilde{\alpha}^{(i}\widetilde{\alpha}^{j)} - \mathbf{g}^{ij}1 - 2\mathbf{g}^{0(i}\widetilde{\alpha}^{j)} = X^{ij} \tag{78}$$

$$\widetilde{\alpha}^i\widetilde{\beta} + \widetilde{\beta}\widetilde{\alpha}^i - 2\mathbf{g}^{0i}\widetilde{\beta} = 2X^i \tag{79}$$

$$\widetilde{\beta}^2 - \frac{1}{g^{00}} = X \tag{80}$$

where the deviation from the usual Clifford algebra is described by the matrices X, X^i, and X^{ij}. (In the case $X = 0$, $X^i = 0$, and $X^{ij} = 0$ one can represent $\alpha^i = (\gamma^0)^{-1}\gamma^i$ and $\beta = (\gamma^0)^{-1}$ with matrices γ^μ fulfilling $\gamma^{(\mu}\gamma^{\nu)} = g^{\mu\nu}$, $\mu, \nu = 0, \dots 3$. Even in the case that the X-matrices do not vanish it can be shown [52, 53] that the matrices $\widetilde{\alpha}^i$ and $\widetilde{\beta}$ fulfill a generalized Clifford algebra.) If the matrices $\widetilde{\alpha}^i$ and $\widetilde{\beta}$ do not fulfill the usual Clifford algebra then the characteristic surfaces, the null cones, and the mass shells (see Figure) of the generalized Dirac equation split and do not longer coincide with the usual light cones and mass shells. It is obvious that in these cases Lorentz invariance is violated. This has also been discussed in [54] (see also [49] and [55] and references therein).

The Generalized Pauli Equation

By performing a non-relativistic limit we arrive at a generalized Pauli equation [56] (see also [57] for the case without gravity and coupling to the electromagnetic field)

$$H\varphi = -\frac{1}{2m}\left(\delta^{ij} + \frac{\delta m_{\mathrm{I}}^{ij}}{m} + \frac{\delta \bar{m}_{\mathrm{I}k}^{ij}}{m}\sigma^k\right)\nabla_i\nabla_j\varphi - \left(\frac{1}{m}a^i_j + cA^i_j\right)\sigma^j i\nabla_i\varphi$$
$$+\left[e\phi(x) + \frac{e}{2m}H_i(x)(K^i + (\delta^i_k + K^i_k)\sigma^k)\right.$$
$$\left. +(mc^2B_i + cT_i)\sigma^i + (1 + C_i\sigma^i)mU(x) + \delta m_{\mathrm{P}ij}U^{ij}(x)\right]\varphi \qquad (81)$$

where $\phi(x)$ is the electrostatic potential and $H_i = \frac{1}{2}\epsilon_{ijk}H_{jk}$ the magnetic field. We introduced in addition a gravitational potential tensor U^{ij} with $\delta_{ij}U^{ij} = U$ [38].

If space-time is endowed with a hypothetical torsion then the usual Dirac equation minimally coupled to metric and torsion gives rise to the quantities a^i_j and T_j. The latter is the space part of the axial torsion vector, and the first is related to the corresponding time component, see [58].

All terms but the U, U^{ij}, ϕ, H_i and the A_i are constant. The tensors $\delta m_{\mathrm{I}}^{ij}$ and $\delta\bar{m}_{\mathrm{I}k}^{ij}$ give spin dependent anomalous inertial mass tensors, a^i_j and A^i_j are spin-momentum couplings, mc^2B_i may be considered as a spin-dependent "rest mass", the T_i may be interpreted as the space-like part of an axial torsion vector, and $\delta m_{\mathrm{P}ij}$ and C_i are anomalous spin dependent gravitational mass tensors. K^i and K^i_k give anomalous modifications of the coupling of the magnetic field to the spin-$\frac{1}{2}$ particle. Due to our systematic approach (81) contains all possible anomalous interactions on the non-relativistic level. The GPE (81) is a non-trivial generalization of Haugan's [59] test theory for matter with spin.

Note that there is no need and no possibility to introduce any $\hbar$. Indeed, also in the usual Schrödinger theory only the ratio $\hbar/m$ enters the equation of motion (see [60]). All our mass-like parameters are to be understood in the sense of being the ratio of mass and $\hbar$. Our mass-like parameter has the dimension of time/length^2, and our Hamilton operator has the dimension 1/time. It is no problem to introduce artificially an $\hbar$ so that the equations acquire the usual form and all parameters have the usual dimensions. T_i and a^i_j have the dimension 1/length. In the following we will neglect the coupling of the magnetic field to anomalous terms. It is not possible to absorb the parameters $\left(\frac{1}{m}a^i_j + cA^i_j\right)\sigma^j$, B_i, and T_i into the inertial or gravitational spin-dependent anomalous mass tensors.

Since (81) can be inferred from the generalized Dirac equation (75) all anomalous terms in (81) are derived in systematic manner. These are the most general anomalous terms on the non-relativistic level which can be derived from a generalized Dirac equation which is the most general equation obeying the very general basic principles listed above. The anomalous terms are necessarily connected with that parts of the matrices $\widetilde{\alpha}^i$, $\widetilde{\beta}$, and $\overset{0}{\Gamma}$ which are responsible for a possible violation of LLI and LPI.

$\delta m_{\mathrm{I}}^{ij}$, $\delta\bar{m}_{\mathrm{I}k}^{ij}$, a_j^i, A_j^i, and B_i give rise to LLI-violation, while C_i and $\delta m_{\mathrm{P}ij}$ are responsible for LPI-violation. If all these coefficients vanish, we recover the usual Schrödinger equation coupled to the Newtonian potential. It is clear that with the energy mc^2 and the characteristic dimensionless quantity U/c^2 describing a gravitational interaction, the generalized Pauli equation is the most general 2nd order differential equation including spin and the gravitational potential tensor.

Matter Wave Interferometry

We propose two kinds of interference experiments, first a experiment with a spin-flip where both parts of the matter wave propagate with the same momentum, and second an interference experiment which measures the acceleration.

Spin-Flip Experiment

We take an atomic beam with a definite spin value along a certain axis propagating with momentum p_i. We split this atomic state into two states and perform with one of these states two spin-flips, one at time t and the second one reverses the first flip at time $t + \Delta t$. The phase shift after the second splin-flip (for convenience, we introduce $\hbar$) $\phi = \frac{1}{\hbar}(H(p,S) - H(p,-S))\Delta t$ is given by (for $A_i = \phi = 0$)

$$\phi = \frac{2}{\hbar}\left(\frac{\delta\bar{m}_{\mathrm{I}k}^{ij}}{2m^2}p_i p_j - \frac{\hbar}{m}a_k^i p_i - cA_k^i p_i + mc^2 B_k + C_k mU + cT_k\right) S^k \Delta t \,, \quad (82)$$

where S^k is the spin of the atoms. To first order we can replace the momentum by the velocity

$$\phi = \frac{1}{\hbar}\left(\frac{\delta\bar{m}_{\mathrm{I}k}^{ij}}{m}p_i\delta_{jk}l^k - 2(cA_j^i m + a_j^i)\delta_{ik}l^k + 2mc^2 B_j\Delta t + 2C_j mU\Delta t + \hbar c T_j\Delta t\right) S^j \,. \quad (83)$$

The coefficients of the spin-momentum coupling enter the phase shift only via the length of propagation between the two spin flipping processes.

For an atom interferometer of Kasevich and Chu type $l \approx 1$ cm, $m \approx 10^{-26}$ kg, $v \approx 10$ cm/sec, $S = \frac{1}{2}$, and $\Delta t \approx 0.1$ sec. With the accuracy $\Delta\phi/\phi = 10^{-8}$ we can estimate in the case that we get a null result: $|\delta\bar{m}_{\mathrm{I}k}^{ij}/m| \leq 10^{-7}$, $|A_j^i| \leq 10^{-17}$, $|a_j^i| \leq 1\ \mathrm{m}^{-1}$, $|B_i| \leq 5\cdot 10^{-27}$, $|C_i| \leq 10^{-17}$, and $|T_i| \leq 3\cdot 10^{-10}\,\mathrm{m}^{-1}$. For the first coefficient we may get a better estimate if we take a large velocity $v = 10^3$ m/sec and $l = 100$ m. We get $|\delta\bar{m}_{\mathrm{I}k}^{ij}/m| \leq 10^{-15}$. This generalises results in [55] (see also [61]). If one of these quantities turns out to be non-null, then we infer a violation of Lorentz invariance and of the Universality of the Gravitational Redshift. However, all these quantities but the A^i_j and a^i_j can be measured better by Hughes–Drever type experiments.

Measurement of Acceleration

For the atom beam interferometer of Kasevich and Chu [11, 12] we get a phase shift $\phi = -k_i a^i T^2$ where T is the time between the laser pulses and a^i the acceleration

$$a^i = -\left(\delta^{ij} + \frac{\delta m_{\mathrm{I}}^{ij}}{m} + 2\left(\frac{\delta \bar{m}_{\mathrm{I}k}^{ij}}{m} + \delta^{ij} C_k\right) S^k\right) \partial_j U - \delta^{ij} \frac{\delta m_{\mathrm{P}kl}}{m} \partial_j U^{kl} . \quad (84)$$

For a spherically symmetric gravitational field the potential U and U^{ij} are easily computed [38]. The phase phase shift then has the structure

$$\delta\phi = -(1+\alpha)\, k T^2 g \cos(\vartheta + b) , \quad (85)$$

where α and b are quantitised which depend on the various coefficients $\delta m_{\mathrm{I}}^{ij}/m$, $\delta \bar{m}_{\mathrm{I}k}^{ij}/m$, C_k, $\delta m_{\mathrm{P}kl}/m$ as well as on the spin S^k [56]. Using atomic interferometry α and b and, thus, all the anomalous coefficients can be determined with an accuracy of the order 10^{-6} which may improve present bounds on $\delta m_{\mathrm{P}kl}/m$ by two orders or magnitude.

4 Test of Anomalous Dispersion Relation

One of the most prominent predictions of quantum gravity schemes is a deviation from the ordinary dispersion relation for photons as well as for massive particles. The modified dispersion relation reads

$$m^2 = E^2 - \boldsymbol{p}^2 + f(E, \boldsymbol{p}) , \quad (86)$$

where we choose the simple model $f(E, \boldsymbol{p}) = \eta(E^3/E_{\mathrm{Pl}})$. Here η is a dimensionless parameter lying between 10^{-3} and 1 coming from the idea that the quantum gravity scale should lie between the grand unification scale with $\sim 10^{25}$ eV and $E_{\mathrm{Pl}} \sim 10^{28}$ eV. The extra term $f(E, \boldsymbol{p})$ can be probed by laser and atomic interferometery.

4.1 Tests with Laser Interferometry

It is amazing but with the next generation of gravitational wave laser interferometers one will reach, at least in principle, the sensitivity needed for the search for quantum gravity effects: The present day sensitivity of, e.g., LIGO is of the order 10^{-21} for the relative difference of the arm lengths $\Delta L/L$ of the Michelson interferometer of 4 km length. The projected sensitivity of the next

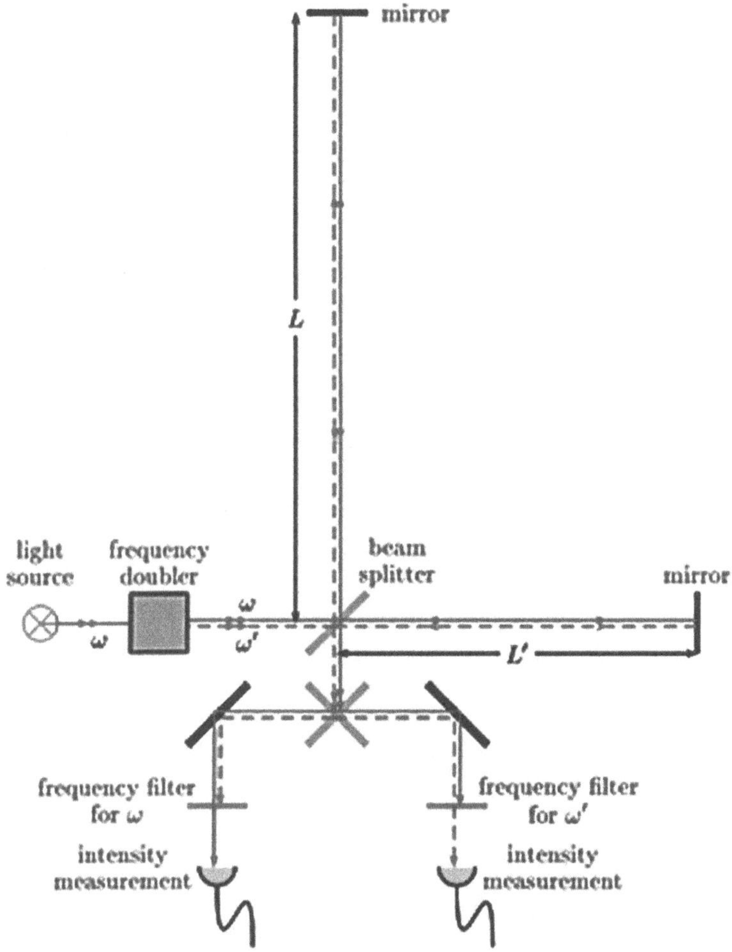

Fig. 3. The unequal-arm interferometer which may be used to search for dispersion effects. The paths for the original ω and the frequency doubled ω' beams photons are displayed by *solid* and *dashed lines*

generation, advanced LIGO, is $\Delta L/L \sim 10^{-24}$ at a frequency of about 10 Hz. If one measured a periodic signal each 0.1 s over one year then, by statistics, the sensitivity is almost $\Delta L/L \sim 10^{-28}$ which coincides with the ratio of laboratory energies and the Planck energy. This is the sensitivity required for a serious laboratory search for quantum gravity effects. This sensitvity to strain is related to the sensitivty to detect intensity differences in the interference pattern or, equivalently, to phase differences.

From the above modified dispersion relation for photons, $m = 0$, we get the modulus of the wave vector

$$k(\omega) = \omega \left(1 + \frac{1}{2}\frac{\omega}{\omega_{QG}}\right) , \quad (87)$$

where $\omega_{QG} = E_{Pl}/\eta$. The main feature of this solution is that the wave vector is no longer a homogeneous function of the frequency. This property of $k(\omega)$ can be explored by performing an interference experiment with two different frequencies [62]. The appropriate interferometer is a Michelson interferometer with unequal arm lengths. For a length difference of $\delta L = L' - L$, the intensity is given by $I(\omega) = \frac{1}{2}(1 + \cos\phi(\omega))$ with $\phi(\omega) = k(\omega)\delta L + \phi_0$ where ϕ_0 is a constant phase. One can now compare the intensity for two waves of different frequencies ω and $\omega' = 2\omega$ travelling simultaneously in the same interferometer for a varying arm length difference (the doubled frequency can be obtained using a second harmonic generator device (SHG)). In the case of ordinary dispersion, the intensity for ω' varies twice as fast for varying δL than for the original frequency ω. For anomalous dipersion, the variation of the intensity for $\omega' = 2\omega$ is slightly different from being twice the variation for ω, see Fig. 4. This difference can be related to a phase difference

$$\phi(2\omega) - \phi(\omega) = \frac{3}{2}\frac{\omega^2}{\omega_{QG}}\delta L . \quad (88)$$

Another way to search for effects induced by an anomalous dispersion relation uses interferometry in frequency space. Due to the properties of the SHG it is possible to set up an interferometer in frequency space: An incoming wave of frequency ω consists of, after passing a SHG, two waves, one with the

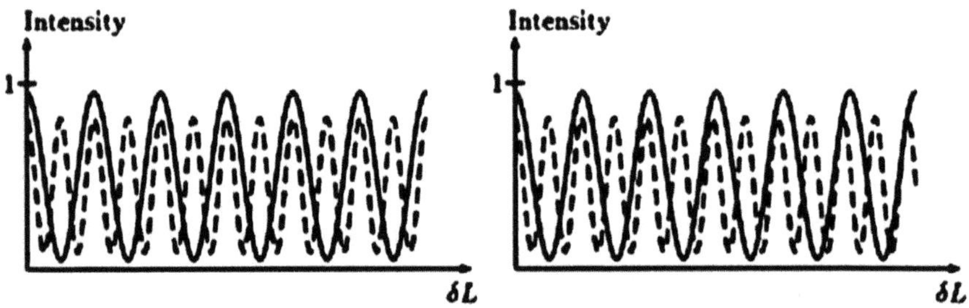

Fig. 4. Qualitative description of the dependence on δL of the intensities $I(\omega)$ and $I(2\omega)$. Different maximum values for $I(\omega)$ and $I(2\omega)$ reflect the fact that the intensities of the two beams that emerge from SHG in general are not identical. *Left:* The case of ordinary dispersion relation showing a particular correlation between the intensities. *Right:* The case of modified dispersion relation. The anomalous term induces a misalignment between the maxima of $I(2\omega)$ and the maxima and minima of $I(\omega)$. This misalignment is proportional to $(\omega^2/\omega_{QG})\delta L$

original frequency ω, the other with the doubled frequency 2ω. If these two waves run through a second SHG[3], then frequency addition and subtraction

[3] We shortly describe the process of frequency addition and subtraction. The basic Maxwell equation in nonlinear dielectric media is given by

$$\nabla^2 E_i - \frac{\epsilon}{c^2}\partial_t^2 E_i = P_i = \chi_{ijk}\mathrm{Re}E_j\mathrm{Re}E_j \ , \tag{89}$$

where χ_{ijk} is the second order electric susceptibility. We now treat the case of medium with no electric field and an incoming wave E^{in}. This wave leads to a polarization inside the dielectric material. This polarization is the source of a new electric field E_i inside the material

$$\nabla^2 E_i - \frac{\epsilon}{c^2}\partial_t^2 E_i = P_i = \chi_{ijk}\mathrm{Re}E_j^{\mathrm{in}}\mathrm{Re}E_j^{\mathrm{in}} \tag{90}$$

We consider the case that incoming wave consists of two frequencies

$$E_i^{\mathrm{in}}(x,t) = E_{0i}^{\mathrm{in},1}e^{i(\omega_1 t - k(\omega_1)x)} + E_{0i}^{\mathrm{in},2}e^{i(\omega_2 t - k(\omega_2)x)} \tag{91}$$

with the wave vector obeying $k^2(\omega) = \epsilon(\omega)\omega^2/c^2$ (it is obvious that anomalous dispersion can be neglected here). The polarization P_i then is

$$\begin{aligned} P_i = 2\chi_{ijk} \Big(& \mathrm{Re}(E_{0i}^{\mathrm{in},1}E_{0j}^{\mathrm{in},1*} + E_{0i}^{\mathrm{in},2}E_{0j}^{\mathrm{in},2*}) \\ & +\mathrm{Re}\left((E_{0i}^{\mathrm{in},1}E_{0j}^{\mathrm{in},2} + E_{0i}^{\mathrm{in},2}E_{0j}^{\mathrm{in},1})e^{i((\omega_1+\omega_2)t-(k(\omega_1)+k(\omega_2))x)}\right) \\ & +\mathrm{Re}\left((E_{0i}^{\mathrm{in},1}E_{0j}^{\mathrm{in},2*} + E_{0i}^{\mathrm{in},2*}E_{0j}^{\mathrm{in},1})e^{i((\omega_1-\omega_2)t-(k(\omega_1)-k(\omega_2))x)}\right) \\ & +\mathrm{Re}\left(E_{0i}^{\mathrm{in},1}E_{0j}^{\mathrm{in},1}e^{i(2\omega_1 t-2k(\omega_1)x)}\right) + \mathrm{Re}\left(E_{0i}^{\mathrm{in},2}E_{0j}^{\mathrm{in},2}e^{i(2\omega_2 t-2k(\omega_2)x)}\right)\Big) \end{aligned} \tag{92}$$

and consists of parts with the frequencies

$$\omega_{\mathrm{shg}} = \{0, \quad 2\omega_1, \quad 2\omega_2, \quad \omega_1+\omega_2, \quad \omega_1-\omega_2\} \tag{93}$$

and the wave vectors

$$k_{\mathrm{shg}} = \{0, \quad 2k(\omega_1), \quad 2k(\omega_2), \quad k(\omega_1)+k(\omega_2), \quad k(\omega_1)-k(\omega_2)\}\,. \tag{94}$$

(Waves with these frequencies and wave vectors not necessarily fulfill the wave equation inside or outside the material. However, we calculated the polarization due to a given electric field, not a propagating wave.)

We choose one of these waves as source. Then the Maxwell equation reads

$$\nabla^2 E_i - \frac{\epsilon}{c^2}\partial_t^2 E_i = P_i = \chi_{ijk}\mathrm{Re}E_j^{\mathrm{in}}\mathrm{Re}E_j^{\mathrm{in}}e^{i(\omega_{\mathrm{shg}}t-k_{\mathrm{shg}}x)} = \widehat{P}_i e^{i(\omega_{\mathrm{shg}}t-k_{\mathrm{shg}}x)} \tag{95}$$

For solving this equation for the electric field E_i we make the ansatz $E_i = E_i(x)e^{i\omega_{\mathrm{shg}}t}$ and get

$$\nabla^2 E_i(x) + \frac{\epsilon}{c^2}\omega_{\mathrm{shg}}^2 E_i(x) = \widehat{P}_i e^{-ik_{\mathrm{shg}}x}\,. \tag{96}$$

We assume a slowly varying amplitude $E_i(x) = \widehat{E}_i(x)e^{-ikx}$. Then can occur. That is, the outgoing waves consists of frequencies 0, ω, 2ω, and 3ω. The latter one is forbidden by the matching condition.

Using a general ansatz for the waves which leave the SHG one can show [62] that the waves leaving the second SHG consists of two frequencies, ω and 2ω. Each of these waves consists of a superposition of two waves which propagated between the two SHGs with different frequencies. Some calculation gives that the intensity of the 2ω wave is given by

---

$$-k^2\widehat{E}_i(x)e^{-ikx} - 2ik\nabla\widehat{E}_i(x)e^{-ikx} + \frac{\epsilon}{c^2}\omega_{\text{shg}}^2\widehat{E}_i(x)e^{-ikx} = \widehat{P}_i e^{-ik_{\text{shg}}x} \tag{97}$$

We identify $k^2 = \frac{\epsilon}{c^2}\omega_{\text{shg}}^2$ and get

$$-2ik\nabla\widehat{E}_i(x) = \widehat{P}_i e^{ikx}e^{-ik_{\text{shg}}x} \tag{98}$$

what can be solved easily with the boundary condition $\widehat{E}_i(x=0)=0$:

$$\widehat{E}_i(x) \sim \frac{1}{k_{\text{shg}}-k}\left(e^{-i(k_{\text{shg}}-k)x}-1\right) \sim \frac{\sin((k_{\text{shg}}-k)x/2)}{k_{\text{shg}}-k}e^{-i(k_{\text{shg}}-k)x/2}\,. \tag{99}$$

Therefore the intensity of the wave generated inside material is

$$I \sim \frac{\sin^2((k_{\text{shg}}-k)x/2)}{(k_{\text{shg}}-k)^2} \tag{100}$$

which sahows maximum growth (linear with x) if the condition

$$k_{\text{shg}} - k = 0 \tag{101}$$

is fulfilled. This condition is called phase matching. Only if this condition is fulfilled, then waves with frequencies ω_{shg} related to the wave vector k_{shg} will be created. (For a given frequency ω phase matching is a problem of the material to possess the appropriate refractive index.)

We can now discuss the phase matching and, thus, the possibility of frequency doubling and frequency subtraction for the two cases we are interested in:

1. Phase matching for $\omega_1 = \omega_2 = \omega$ and frequency doubling, $\omega_{\text{shg}} = 2\omega$:

$$k - k_{\text{shg}} = k(\omega+\omega) - k(\omega) - k(\omega) = k(2\omega) - 2k(\omega) = 0\,. \tag{102}$$

2. Phase matching for $\omega_1 = 2\omega$ and $\omega_2 = \omega$:
   - Frequency addition: $\omega_{\text{shg}} = \omega_1 + \omega_2 = 3\omega$

$$k - k_{\text{shg}} = k(3\omega) - k(2\omega) - k(\omega) = 0\,. \tag{103}$$

   - Frequency subtraction $\omega_{\text{shg}} = 2\omega - \omega = \omega$

$$k - k_{\text{shg}} = k(\omega) - k(2\omega) + k(\omega) = 0\,. \tag{104}$$

Therefore, the phase matching condition for $\omega + \omega \to 2\omega$ is the same as for $2\omega - \omega \to \omega$.

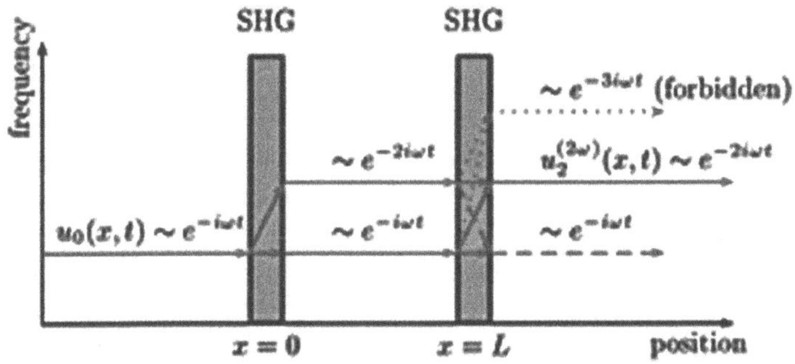

Fig. 5. The waves leaving two SHGs. The two waves leaving the second doubler with frequency 2ω form an interferometer since they propagate between the two doublers at different frequencies. Different paths in the figure only distinguish different frequencies: all beams follow the same path in configuration space

$$I(2\omega) = B_{21}^2 + B_{22}^2 + 2B_{21}B_{22}\cos\phi(2\omega)\,, \tag{105}$$

where B_{21} and B_{22} are the intensity of the waves with frequency ω and 2ω leaving the second SHG if a wave with amplitude 1 and frequency 2ω enters this SHG. The phase is given by

$$\phi(2\omega) = \phi_0 + (k(2\omega) - 2k(\omega))\delta L\,, \tag{106}$$

where δL is the distance between the two SHGs and ϕ_0 is some constant phase. Since the generation of doubled frequencies is not very efficient (may be up to 30%), B_{22} is of the order of B_{21}, so that the visibility of the 2ω-interference pattern may reach unity. Again, with a variation of the distance between the two SHGs one is sensitive to $k(2\omega) - 2k(\omega) = \frac{\omega^2}{\omega_{\text{QG}}}$. The variation of the distance may be accomplished by using a third SHG and by electronically switching on and off in an alternating mode the capability of the second and third SHG to do frequency addition and subtraction. With this setup one avoids mechnical transport of the second SHG which induces a lot of errors.

4.2 Tests with Atomic Interferometry

From the dispersion relation (86) we get to first order in the quantum gravity corrections

$$E = \sqrt{m^2 + p^2}\left(1 + \frac{1}{2}\frac{\sqrt{m^2+p^2}}{E_{\text{QG}}}\right)\,. \tag{107}$$

The group velocity then is

$$v_{\text{gr}} = \frac{p}{\sqrt{m^2+p^2}}\left(1 + \frac{\sqrt{m^2+p^2}}{E_{\text{QG}}}\right)\,. \tag{108}$$

For small velocities v, the qg-correction in the velocity is of the order $\Delta v_{\mathrm{QG}} \sim mc^2/E_{\mathrm{QG}}v$. For an atom we have typically $mc^2 \sim 100$ GeV so that $\Delta v_{\mathrm{QG}} \sim 10^{-17}v$.

In a recent paper [63] an atom interferometer setup has been proposed which should be capable to detect gravitational waves. The gravitational waves are measured through a corresponding change of the velocity of the atoms while flying through the interferometer. This velocity change is given by $\Delta v \sim v_0\dot{h}T$ where v_0 is the initial velocity of the atoms, $\dot{h}$ the time derivative of the amplitude of the gravitational wave, and T the time-of-flight of the atom through the interferometer. Since T should be smaller than a period of the gravitational wave, $\dot{h}T \leq h$, so that $\Delta v \sim v_0 h$. For atomic beams with velocities between 10^3 and 10^4 m/s, we get $\Delta v \sim 10^{-17}$ m/s as a result of the interaction with the gravitational wave.

That means for our quantum gravity induced modification of the velocity, that even for $v = 1\,\mathrm{m/s}$ we have the effect which should be in the range of sensitivity of MIGO. Again, for a search of such an effect one should measure and to compare the phase shift for two different velocities.

5 Test of Space-Time Fluctuations

Beside modifications of well established laws like the Maxwell or the Dirac equation, space-time fluctuations are another prominent prediction of schemes of quantum gravity. space-time fluctuations influence the propagation of, e.g., light with the consequence that the time of flight between two events is not sharp but, instead, fluctuates. A consequence of this is that sharp signals from, e.g., distant stars will wash out [64]. Other consequence are that distance measurements by optical devices will be limited by a fundamental noise and that quantum states may show decoherence or a lowering in the visibility of the interference. These effects will be discussed below.

5.1 Atom and Neutron Interferometry

The quantum mechanical treatment of the Maxwell field implies that there are fluctuations in the electromagnetic field. These fluctuations are present even if there is no "real" Maxwell field; these vacuum fluctuations influence all physical effects and leads, e.g., to shifts in the atomic levels called Lamb shift. If one takes into account that boundary conditions for the electromagnetic field are responsible for the structure of the vacuum fluctuation then it is clear that such vacuum fluctuations can be detected by interferometric methods: one has to split an atomic wave and has to lead one part through a region with some boundary conditions. After leaving this region the two parts of the atomic beam can be brought together leading to an interference pattern which depends on the vacuum fluctuation, or on the boundary conditions. By modifying the boundary conditions the influence can be tested directly[4].

If one carries through a quantization scheme for the gravitational field one expects similar effects. However, there are several ideas of fluctuations of space-time: first there may be a fluctuating metric $\delta g_{\mu\nu}$ leading in some limit to a fluctuating Newtonian potential δU. Second the topology of space-time may be fluctuating leading to some space-time foam. And a third possibility is that the space-time points become unsharp and are fluctuating itself [65].

A reasonable model for the influence of quantum fluctuations of space-time leading to a space-time foam, has been given by Hawking [66] and worked out by Ellis et al. [67]. These fluctuations introduce in test quantum systems a coherence disturbing term. This is described by an equation of motion for the density matrix of the quantum system which has the form of a Markovian master equation

$$i\frac{d}{dt}\rho = [H,\rho] + h\rho \tag{109}$$

where $h\rho \leftrightarrow (h\rho)_{ab} = h_{abcd}\rho_{cd}$. We restrict to finite dimensional systems, e.g. two-level systems, $a, b, \ldots = 1, 2$. For the evolution we require

1. $h\rho$ is hermitian if ρ is,
2. conservation of probability $\mathrm{tr}\rho = 1$,
3. $\mathrm{tr}\rho^2 \leq 1$,
4. conservation of energy.

We assume a Hamilton operator of the form

$$H = \begin{pmatrix} E + \frac{1}{2}\Delta E & 0 \\ 0 & E - \frac{1}{2}\Delta E \end{pmatrix} . \tag{110}$$

The first condition implies $\mathrm{tr}\frac{d}{dt}\rho = 0$ and therefore $h_{aacd} = 0$ (in our two-dimensional model $h_{22cd} = -h_{11cd}$), and the second condition $0 \geq \frac{d}{dt}\mathrm{tr}\rho^2 = 2\mathrm{tr}(\rho h\rho) = 2\rho_{ab}h_{abcd}\rho_{cd}$. The third condition reads for H as the operator corresponding to the observation of the total energy of the quantum system

$$\frac{d}{dt}E = \frac{d}{dt}\mathrm{tr}(H\rho) = \mathrm{tr}\left(H\frac{d}{dt}\rho\right) = -i\mathrm{tr}(Hh\rho) = -i\Delta E h_{11cd}\rho_{cd} \tag{111}$$

so that the diagonal terms of the fluctuation induced term vanish $h_{11cd} = h_{22cd} = 0$. It can further been shown that $h_{abcd} = h_{cdab}$ so that also $h_{cd11} = h_{cd22} = 0$.

One finally arrives at the equations of motion for the components of the density matrix

[4] These quantum fluctuations of the electromagnetic field due to boundary conditions have already been observed using spontaneous decay of atomic states: Since the channels into which a photon can be emitted depends on whether the atom is inside a cavity or outside this leads to a modification of the lifetime of excited atoms.

$$i\frac{d}{dt}\begin{pmatrix}\rho_{11} & \rho_{12}\\ \rho_{21} & \rho_{22}\end{pmatrix} = \begin{pmatrix}0 & -\frac{i}{2}(\alpha+\gamma)+\Delta E\\ -\frac{i}{2}(\alpha+\gamma)-\Delta E=0 & \end{pmatrix}, \tag{112}$$

where $\alpha > 0$ and $\gamma > 0$ are related to the remaining degrees of freedon h_{1212} and h_{2121}. The solution is given by [67]

$$\rho(t) = \frac{1}{2}\begin{pmatrix}1 & e^{-\frac{1}{2}(\alpha+\gamma)t-i\Delta Et}\\ e^{-\frac{1}{2}(\alpha+\gamma)t+i\Delta Et} & 1\end{pmatrix} \tag{113}$$

which clearly describes an evolution where the states become completely mixed for large times.

If one relates the coefficients α and γ to space-time fluctuations then the decay of coherence is an indicator for certain quantum gravity effects. In an interference experiment the above effects decrease the visibility of the interference pattern. Assuming for neutron interferometry a decrease in the visibility of about 20% during a time of flight of $t = 1/3000$ sec, gives an estimate $\alpha + \beta \leq 2 \cdot 10^{-21}$ GeV. For atom interferometry this attenuation of the interference pattern has a competing cause, namely the finite lifetime of used atomic states. However, if one assumes a time of flight which is about four orders of magnitude longer than in neutron interferometry, then this may sharpen the above estimate by four orders.

In this connection we also want to mention the approach of Percival [68, 69] using an alternative quantum theory based on the notion of primary state diffusion. This approach describes open quantum mechanical systems with a similar behaviour than discussed above.

At last we want mention that neutron interferometry in the WKB regime is also capable to detect fluctuations in the Newtonian piotential. The reason is that due to the fact that the gravitational potential appears in a square root, the expansion of the square root gives quadratic effects which do not average to zero but, instead, add up [70].

5.2 Laser Interferometry

Initiated by the research of G. Amelino–Camelia [3] who proposed to use a random-walk ansatz to describe the effect of space-time fluctuations of the propagation of light (see also the lecture of J. Ng in this volume) and to use gravitational wave interferometers to search for these effects, in a recent work [4] optical cavities with a very high long term stability have been used for the search for space-time fluctuations.

The ansatz for the strain noise spectrum is

$$\left(\frac{\Delta L}{L}\right)^2 = \int S_{\rm sf}(\nu)d\nu \qquad \text{with} \qquad S_{\rm sf}(\nu) = \zeta\frac{\Lambda}{c}\left(\frac{L_{\rm Pl}}{\Lambda}\right)^\alpha\left(\frac{\nu}{c/\Lambda}\right)^\gamma, \tag{114}$$

where Λ is a length characteristic of the experimental setup, ν the frequency, and $L_{\rm Pl}$ the Planck length. Following Amelino–Camelia, the above general frame is specified for two random–walk hypotheses:

$$S_{\mathrm{sf}}^{\alpha=1,\gamma=-2} = 5 \cdot 10^{-27} \zeta_{\mathrm{RW1}} \left(\frac{\mathrm{m}}{\Lambda}\right)^2 \left(\frac{\mathrm{Hz}}{\nu}\right)^2 \mathrm{Hz}^{-1} \tag{115}$$

$$S_{\mathrm{sf}}^{\alpha=2,\gamma=-2} = 7 \cdot 10^{-62} \zeta_{\mathrm{RW2}} \left(\frac{\mathrm{m}}{\Lambda}\right)^3 \left(\frac{\mathrm{Hz}}{\nu}\right)^2 \mathrm{Hz}^{-1} , \tag{116}$$

with Λ given in units of m and the frequency in Hz. The overall order of magnitude of these spectral noise densities is determined by the Planck length and the characteristic length of the experimental setup which we took to be 5 cm. The unspecified parameters ζ_{RW1} and ζ_{RW2} are of the order 1. From that expressions it is clear that low frequencies and small devices are preferred setups for the search for these kind of fluctuations. We like to emphasize again that all these models, and so are the assumed order of magnitude for these effecs, are mere hypotheses. However, they show that such kind of effects can signal effects due to quantum gravity and that any improvement of experimental data is important. One may consider the search for fundamental space-time fluctuations in devices like optical cavities or in gravitatinal wave interferometer to be on the same level as searches for an anisotropy of the speed of light or, more generally, as a search for an anisotropy of space.

Amelino–Camelia [3] confronted his RW hypotheses with the possible performance of gravitational wave interferometers. In such interferometers the beam splitter and mirrors are independent and, thus, are subject to seismic noise in the case of Earth–bound interferometers, or to the noises of the drag–free control in the case of LISA. That means that space-time fluctuations can be searchd for in the frequency range larger than 100 Hz for Earth bound devices and for frequencies larger than 1 mHz for LISA. For smaller frequencies these devices do not possess a high long term stability. Such a stability can be provided by optical cavities. With these devices one can access the μHz range.

By using monolithic resonators one has to make the assumption that the space-time fluctuations experienced by light are different, that is, are not compensated, by the space-time fluctuations experienced by the cavity material (spacing material). One way to exclude such compensation is to use different cavity materials.

In the actual search for space-time fluctuations, the data of a frequency comparison between two optical resonators has been analyzed. (One also may use a comparison between a cavity and another clock like an atomic clock.) The comparison of the two frequencies amounts to a comparison of the optical path length of the two resonators $\delta L = \delta(n_1 L_1 - n_2 L_2)$ where $n_{1,2}$ and $L_{1,2}$ are the refraction index and the lengths of the two cavities. Furthermore we have to assume that the fluctuations of the two cavities are not correlated. That means $\delta L = \delta(n_1 L_1) - \delta(n_2 L_2)$. This assumption is not required for cavity–clock comparisons.

The actually measured quantity is the beat frequency $\nu_2 - \nu_1$ for the two cavities where each of these frequencies is of the order 282 THz. The fluctuations in this beat frequency consist of the hypothetical space-time fluctuations as well physical fluctuations coming from temperature fluctuations and quantum uncertainties as well as fluctations in the laser lock:

$$\begin{aligned}\delta\left(\frac{\nu_2-\nu_1}{\nu_1}\right) &= \left(\frac{\delta L_2^{\rm sf}}{L_2}-\frac{\delta L_1^{\rm sf}}{L_1}\right)+\left(\frac{\delta L_2^{\rm phys}}{L_2}-\frac{\delta L_1^{\rm phys}}{L_1}\right)\\ &\quad+\left(\frac{\delta L_2^{\rm lock}}{L_2}-\frac{\delta L_1^{\rm lock}}{L_1}\right)+\ldots\\ &= S_{\rm sf}+S_{\rm phys}+S_{\rm lock}+\cdots \end{aligned} \tag{117}$$

Half of the total noise represents an upper bound to the looked for $S_{\rm sf}$.

Two geometric setups are analyzed: in the first arrangement (setup A) the frequencies of two resonators oriented parallel and located in two different cryostats separated by a distance of 2.5 m have been compared, and in the second setup (setup B) two cavities oriented orthogonally and hosted in the same cryostat separated by 10 cm. The frequency difference of the two cavities has been measured each second. For the second setup the frequency has been averaged over 5 min what does not change the data if one consideres very low frequencies as we do.

From these data sets one gets the spectral noise density shown in Fig. 6. The curves for the spectral noise density can be estimated within the two random walk ansatzes by $S < 7\cdot 10^{-31}/\nu$ and $S < 3\cdot 10^{-34}\ {\rm Hz}/\nu^2$. Comparison with the actual ansatzes (115) and (116), one gets estimates for the undetermined parameters $\zeta_{\rm RW1}$ and $\zeta_{\rm RW2}$,

$$\zeta_{\rm RW1}\leq 2\cdot 10^{-13}\qquad \zeta_{\rm RW2}\leq 4\cdot 10^{20}\,. \tag{118}$$

Since these parameters are of the order 1, this clearly rules out the random walk hypothesis 1.

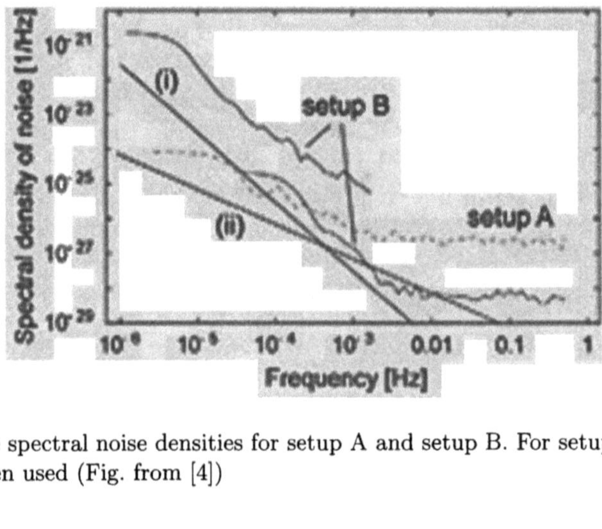

Fig. 6. The spectral noise densities for setup A and setup B. For setup B two data set have been used (Fig. from [4])

Acknowledgement

For continuous fruitful collaboration I like to thank G. Amelino–Camelia, Ch. Bordé, H. Dittus, F. Hehl, A. Macias, H. Müller, A. Peters, St. Schiller, and A. Wicht. Financial support of the German Space Agency DLR is greatfully acknowledged.

References

1. R.P. Feynman, R.B. Leighton, and M.L. Sands. *Lectures on Physics, Vol. 3: Quantum Mechanics.* Addison–Wesley, Reading, 1966.
2. I. Antoniadis. Physics with large extra dimensions and non-newtonian gravity at sub–mm distances.
3. G. Amelino-Camelia. Gravity-wave interferometers as probes of a low-energy effective quantum gravity. *Phys. Rev.*, D 62:0240151, 2000.
4. S. Schiller, C. Lämmerzahl, H. Müller, C. Braxmaier, S. Herrmann, and A. Peters. Experimental limits for low-frequency space-time fluctuations from ultra-stable optical resonators. *Phys. Rev.*, D 69:027504, 2004.
5. T. Damour, F. Piazza, and G. Veneziano. Runaway dilaton and equivalence principle violations. *Phys. Rev. Lett.*, 89:081601, 2002.
6. T. Damour, F. Piazza, and G. Veneziano. Violations of the equivalence principle in a dilaton-runaway scenario. *Phys. Rev.*, D 66:046007, 2002.
7. Ch. Bordé and C. Lämmerzahl. Atomic interferometry as two-level particle scattering by a periodic potential. *Ann. Physik (Leipzig)*, 8:83, 1999.
8. O. Carnal and J. Mlynek. Young's double slit experiment with atoms: a simple atom interferometer. *Phys. Rev. Lett.*, 66:2689, 1991.
9. F. Riehle, Th. Kisters, Witte A., J. Helmcke, and Ch.J. Bordé. Optical ramsey spectroscopy in a rotating frame: Sagnac effect in a matter wave interferometer. *Phys. Rev. Lett.*, 67:177, 1991.
10. A. Wicht, J.M. Hensley, E. Sarajlic, and S. Chu. . In P. Gill, editor, *Proceedings of the 6th Symposium on Frequency Standards and Metrology*, p. 193. World Scientific, Singapore, 2002.
11. M. Kasevich and S. Chu. Atomic interferometry using stimulated Raman transitions. *Phys. Rev. Lett.*, 67:181, 1991.
12. S. Chu and M. Kasevich. Measurement of the gravitational acceleration of an atom with a light-pulse atom interferometer. *Appl. Phys.*, B 54:321, 1992.
13. A. Peters, K.Y. Chung, and S. Chu. Measurement of gravitational acceleration by dropping atoms. *Nature*, 400:849, 1999.
14. T.L. Gustavson, P. Bouyer, , and M.A. Kasevich. Precision rotation measurements with an atom interferometer gyroscope. *Phys. Rev. Lett.*, 78:2046, 1996.
15. T.L. Gustavson, A. Landragin, and M.A. Kasevich. Rotation sensing with a dual atom–interferometer sagnac gyroscope. *Class. Quantum Grav.*, 17:2385, 2000.
16. J.M. McGuirk, G.T. Foster, J.B. Fixler, M.J. Snadden, and M.A. Kasevich. Sensitive absolute-gravity gradiometry using atom interferometry. *Phys. Rev.*, A 65:033608, 2002.
17. C. Jentsch, T. Müller, E.M. Rasel, and W. Ertmer. High precision atom interferometry on ground and in space. *Gen. Rel. Grav.*, 36:to appear, 2004.

18. C. Lämmerzahl. Atom interferometry and gravitomagnetism. In L. Iorio, editor, *The Measurement of Gravitomagnetism: A Challenging Enterprise*, page to be published. Nova Publishers, Hauppauge, N.Y., 2005.
19. Ch.J. Bordé. Propagation of laser beams and of atomic systems. In J. Dalibard, J.M. Raimond, and J. Zinn-Justin, editors, *Fundamental Systems in Quantum Optics*, page North Holland, Amsterdam, 1992.
20. C. Lämmerzahl and Ch. Bordé. Rabi oscillations in gravitational fields: exact solution. *Phys. Lett.*, A 203:59, 1995.
21. Ch. Bordé and C. Lämmerzahl. Atom beam interferometry in gravitational fields: The beam splitting process. *Gen. Rel. Grav.*, 31:635, 1999.
22. Ch.J. Bordé. Atomic interferometry with internal state labeling. *Phys. Lett.*, A 140:10, 1989.
23. C. Lämmerzahl. On the equivalence principle in quantum mechanics. *Gen. Rel. Grav.*, 28:1043, 1996.
24. D. Greenberger and A.W. Overhauser. Coherence effects in neutron diffraction and gravity experiments. *Rev. Mod. Phys.*, 51:43, 1979.
25. S. Baeßler, B.R. Heckel, E.G. Adelberger, J.H. Gundlach, U. Schmidt, and H.E. Swanson. Improved tes of the equivalence principle for gravitational self-energy. *Phys. Rev. Lett.*, 83:3585, 1999.
26. C. Lämmerzahl. Minimal coupling and the equivalence principle in quantum mechanics. *Acta Phys. Pol.*, 29:1057, 1998.
27. A.A. Michelson and H.G. Gale. The effect of the Earth's rotation on the velocity of light. *Nature*, 115:566, 1925.
28. C.V. Heer. Interference of electromagnetic and matter waves in a nonpermanent gravitational field. *Bull. Am. Phys. Soc.*, 6:393, 1961.
29. L.A. Page. Effect of Earth's rotation in neutron interferometry. *Phys. Rev. Lett.*, 35:543, 1975.
30. L.A. Page. Phase considerations in a rotating system. In Bonse U. and Rauch H., editors, *Neutron Interferometry*, p. 327. Clarendon Press, Oxford, 1979.
31. S.A. Werner, J.-L. Staudenmann, and R. Collella. Effect of Earth's rotation on quantum mechanical phase of the neutron. *Phys. Rev. Lett.*, 42:1103, 1979.
32. D.K. Atwood, M.A. Horne, C.G. Shull, and J. Arthur. Neutron phase shift in a rotating two-crystal interferometer. *Phys. Rev. Lett.*, 52:1673, 1984.
33. F. Hasselbach and M. Nicklaus. An electron optical Sagnac effect. *Physica*, B 151:230, 1988.
34. F. Hasselbach and M. Nicklaus. Observation of the rotational phase shift of electron waves (Sagnac effect). In *Conference on Foundations of Quantum Mechanics to celebrate 30 Years of the Aharonov-Bohm effect.* 1989.
35. F. Hasselbach and M. Nicklaus. Sagnac effect with electrons: Observation of the rotational phase shift of electrons in vacuum. *Phys. Rev.*, A 48:152, 1993.
36. J. Audretsch and K.P. Marzlin. Ramsey fringes in atomic interferometry: Measurability of the influence of space-time curvature. *Phys. Rev.*, A 50:2080, 1994.
37. C.W. Misner, K. Thorne, and J.A. Wheeler. *Gravitation.* Freeman, San Francisco, 1973.
38. C.M. Will. *Theory and Experiment in Gravitational Physics (Revised Edition).* Cambridge University Press, Cambridge, 1993.
39. C. Lämmerzahl. A Hamilton operator for quantum optics in gravitational fields. *Phys. Lett.*, A 203:12, 1995.
40. C. Kiefer and T.P. Singh. Quantum gravitational corrections to the functional Schrödinger equation. *Phys. Rev.*, D 44:1067, 1991.

41. J. Audretsch and G. Schäfer. Quantum mechanics of electromagnetically bounded spin-$\frac{1}{2}$ particles in expanding universes: I. influence of the expansion. *Gen. Rel. Grav.*, 9:243, 1978.
42. E. Fischbach, B.S. Freeman, and W.K. Cheng. General relativistic effects in hydrogenic systems. *Phys. Rev.*, D 23:2157, 1981.
43. C. Lämmerzahl. Relativistic treatment of the Raman light pulse atom beam interferometer with applications in gravity theory. *J. Physique II (France)*, 4:2089, 1994.
44. B. Bertotti, L. Iess, and P. Tortora. A test of general relativity using radio links with the Cassini spacecraft. *Nature*, 425:374, 2003.
45. M.O. Scully, M.S. Zubairy, and M.P. Haugan. Proposed optical test of metric gravitation theories. *Phys. Rev.*, A 24:2009, 1981.
46. W. Schleich and O. Scully. General relativity and modern optics. In Grynberg G. and Stora R., editors, *Tendances actuelles en physique atomique, Les Houches Session XXXVIII*, p. 995. Elsevier Science Publishers, Amsterdam, 1984.
47. C. Lämmerzahl. The Einstein Equivalence Principle and the search for new physics. In D. Giulini, C. Kiefer, and C. Lämmerzahl, editors, *Quantum Gravity – From Theory to Experimental Search*, Lect. Notes Phys. **631**, 367 (2003).
48. J. Audretsch and C. Lämmerzahl. A new constructive axiomatic scheme for the geometry of space-time. In Majer U. and Schmidt H.-J., editors, *Semantical Aspects of Space-Time Geometry*, p. 21. BI Verlag, Mannheim, 1993.
49. U. Bleyer and D.-E. Liebscher. Mach's principle and causal structure. In Barbour J. and Pfister H., editors, *Mach's Principle, From Newton's Bucket to Quantum Gravity*, p. 293. Birkhäuser, Boston, 1995.
50. C. Lämmerzahl. On the experimental foundation of the Maxwell equations. In A. Macias, J. Cervantes, and C. Lämmerzahl, editors, *Recent Developments on Exact Solutions and Scalar Fields in Gravity*, p. 295. Kluver Academic / Plenum Publishers, New York, 2001.
51. A. Kostelecky and M. Mewes. Signals for Lorentz violation in electrodynamics. *Phys. Rev.*, D 66:056005, 2002.
52. C. Lämmerzahl. The geometry of matter fields. In deSabbata V. and Audretsch J., editors, *Quantum Mechanics in Curved Space-Time, NATO ASI series, series B: Physics*, volume 230, p. 23. Plenum Press, New York, 1990.
53. C. Lämmerzahl. A new quantum test theory for gravitational fields. In A. Garcias, C. Lämmerzahl, A. Macias, Matos T., and D. Nuñes, editors, *Recent developments in gravitation and mathematical Physics, Proceedings of the 2nd Mexican School*, pages http://kaluza.physik.uni–konstanz.de/SNP/Books.html. Science Network Publishing, Konstanz, 1997.
54. D.-E. Liebscher. The geometry of the Dirac equation. *Ann. Physik (Leipzig)*, 42:35, 1985.
55. J. Audretsch, U. Bleyer, and C. Lämmerzahl. Testing Lorentz invariance with atomic beam interferometry. *Phys. Rev.*, A 47:4632, 1993.
56. C. Lämmerzahl. Quantum tests of foundations of general relativity. *Class. Quantum Grav.*, 14:13, 1998.
57. V.A. Kostelecky and C.D. Lane. Constraints on Lorentz violation from clock-comparison experiments. *Phys. Rev.*, D 60:116010, 1999.
58. C. Lämmerzahl. Constraints on space-time torsion from Hughes–Drever experiments. *Phys. Lett. A*, A 228:223, 1997.

59. M.P. Haugan. Energy conservation and the principle of equivalence. *Ann. Phys.*, 118:156, 1979.
60. D.S. Weiss, B.C. Young, and S Chu. Precision measurement of $\hbar/m_{\rm Cs}$ based on photon recoil using laser–cooled atoms and atomic interferometry. *Appl. Phys.*, B 59:217, 1994.
61. P.R. Phillips. Seach of of spatial isotropy using a cyrogenic torsion pendulum. *Phys. Rev. Lett.*, 59:1784, 1987.
62. G. Amelino-Camelia and C. Lämmerzahl. Quantum-gravity-motivated Lorentz-symmetry tests with laser interferometers. *Class. Quantum Grav.*, 21:to appear, 2004.
63. R.Y. Chiao. Towards MIGO, the Matter–wave Interferometric Gravitational–wave Observatory, and the intersection of quantum mechanics with general relativity. gr-qc/0312096.
64. G. Amelino-Camelia, J. Ellis, N.E. Mavromatos, D.V. Nanopoulos, and S. Sarkar. Tests of quantum gravity from observations of gamma-ray bursts. *Nature*, 393:763, 1998.
65. C.J. Isham. Conceptual and geometrical problems in quantum gravity. In Mitter H. and Gaustere H., editors, *Recent Aspects of Quantum Fields*, p. 123. Springer Verlag, Berlin, 1991.
66. S.W. Hawking. The unpredictability of quantum gravity. *Comm. Math. Phys.*, 87:395, 1982.
67. J. Ellis, S. Hagelin, D.V. Nanopoulos, and M. Srednicki. Search for violations of quantum mechanics. *Nucl. Phys.*, B 241:381, 1984.
68. C. Percival. Quantum space-time fluctuations and primary state diffusion. *preprint*, quant-ph/9508021, 1995.
69. I.P. Percival and W.T. Strunz. Detection of space-time fluctuations by a model matter interferometer. *Proc. Roy. Soc. London*, A 453:431, 1996.
70. J. Audretsch and C. Lämmerzahl. New interial and gravitational effects made measureable by atom beam interferometry. *Appl. Phys.*, B 54:351, 1992.

Inflationary Cosmological Perturbations of Quantum-Mechanical Origin

J. Martin

Institut d'Astrophysique de Paris, GreCO, FRE 2435-CNRS, 98bis boulevard Arago, 75014 Paris, France
jmartin@iap.fr

This review article aims at presenting the theory of inflation. We first describe the background spacetime behavior during the slow-roll phase and analyze how inflation ends and the Universe reheats. Then, we present the theory of cosmological perturbations with special emphasis on their behavior during inflation. In particular, we discuss the quantum-mechanical nature of the fluctuations and show how the uncertainty principle fixes the amplitude of the perturbations. In a next step, we calculate the inflationary power spectra in the slow-roll approximation and compare these theoretical predictions to the recent high accuracy measurements of the Cosmic Microwave Background radiation (CMBR) anisotropy. We show how these data already constrain the underlying inflationary high energy physics. Finally, we conclude with some speculations about the trans-Planckian problem, arguing that this issue could allow us to open a window on physical phenomena which have never been probed so far.

1 Introduction

Inflation is the most promising theory of the early Universe. It was invented by A. Guth [1] at the beginning of the 80's in order to solve the puzzles of the hot Big Bang theory. A very interesting aspect of the inflationary theory is that it allows us to build a bridge between cosmology and high energy physics. This is particularly valuable in view of the fact that it is difficult to probe physics beyond the standard model of particular physics.

However, the details of the underlying particle physics model are encoded into the fine structure of the cosmological observables. This is why, after the invention of the inflationary scenario and during quite a long time, it was in fact only possible to check the consistency of the inflationary predictions. The situation has now changed drastically with the recent releases of very high accuracy cosmological data. One can now take advantage of the full predictive

J. Martin: *Inflationary Cosmological Perturbations of Quantum-Mechanical Origin*, Lect. Notes Phys. **669**, 199–244 (2005)
www.springerlink.com

power of inflation with the hope to learn about physics in a regime which has never been reached before.

The goal of this review article is to give a general presentation of the inflationary scenario. In particular, we will emphasize how the origin of the inhomogeneities present in our Universe is explained in the framework of inflation. We will see that it is based on an elegant interplay between general relativity and quantum theory. Then, we will present the corresponding predictions made by inflation and will study how the currently available data can already put some constraints on the underlying particle physics models.

This article is organized as follows. In the next section, we describe the evolution of the inflationary background, the slow-roll phase and the reheating. Then, we present the theory of cosmological perturbations of quantum-mechanical origin and compare its predictions to the available data. Finally, we conclude this article with some speculations concerning the trans-Planckian problem of inflation, demonstrating that future astrophysical observations will maybe open a new window on high energy physics.

2 The Inflationary Universe

2.1 Basic Equations

The cosmological principle implies that the Universe is, on large scales, homogeneous and isotropic. As a consequence, the metric tensor which describes the geometry of the Universe is of the Friedman-Lemaître-Robertson-Walker (FLRW) form, namely

$$\mathrm{d}s^2 = -c^2\mathrm{d}t^2 + a^2(t)\gamma_{ij}^{(3)}\mathrm{d}x^i\mathrm{d}x^j = a^2(\eta)\left[-\mathrm{d}\eta^2 + \gamma_{ij}^{(3)}\mathrm{d}x^i\mathrm{d}x^j\right] , \quad (1)$$

where $\gamma_{ij}^{(3)}$ is the metric of the three-dimensional spacelike sections. The three-dimensional sections have a constant scalar curvature. The variable t is the cosmic time while η is the conformal time. They are linked by the relation $c\mathrm{d}t = a\mathrm{d}\eta$. In this article, we will work with dimensionless coordinates x^i and, as a consequence, the scale factor $a(\eta)$ will have the dimension of a length.

The matter is assumed to be a collection of N perfect fluids and therefore its stress-energy tensor is given by the following expression

$$T_{\mu\nu} = \sum_{i=1}^{N} T_{\mu\nu}^{(i)} = (\rho_{\mathrm{T}} + p_{\mathrm{T}})u_\mu u_\nu + p_{\mathrm{T}} g_{\mu\nu} , \quad (2)$$

where ρ_{T} is the (total) energy density and p_{T} the (total) pressure. These two quantities are linked by the equation of state, $p_{\mathrm{T}} = \omega(\rho_{\mathrm{T}})$ [in general, there is an equation of state per fluid considered, i.e. $p_i = \omega_i(\rho_i)$]. The vector u_μ is the four velocity and satisfy the relation $u_\mu u^\mu = -1$. This means that one has $u^\mu = (1/a, 0)$ and $u_\mu = (-a, 0)$. The fact that the stress-energy tensor is conserved, $\nabla^\alpha T_{\alpha\mu} = 0$, amounts to

$$\rho_{\rm T}' + 3\frac{a'}{a}(\rho_{\rm T} + p_{\rm T}) = 0\,. \qquad (3)$$

This expression is obtained from the $\mu = 0$ component. The component $\mu = i$ does not lead to an interesting equation for the background. If one assumes that each fluid is separately conserved then the above equation is valid for each species.

We will assume that gravity is correctly described by the theory of General Relativity even in the very early Universe. This means that the equations which link the geometrical part to the matter part are nothing but the Einstein equations

$$R_{\mu\nu} - \frac{1}{2}R g_{\mu\nu} = \kappa T_{\mu\nu}\,, \qquad (4)$$

where $\kappa \equiv 8\pi G/c^4 = 8\pi/m_{\rm Pl}^2$. These equations, in the case of a FLRW Universe, are differential equations determining the time evolution of the scale factor and read

$$\frac{3}{a^2}\left[\left(\frac{a'}{a}\right)^2 + k\right] = \kappa\sum_{i=1}^{N}\rho_i, \quad -\frac{1}{a^2}\left[2\frac{a''}{a} - \left(\frac{a'}{a}\right)^2 + k\right] = \kappa\sum_{i=1}^{N}p_i\,, \qquad (5)$$

where a prime denotes a derivative with respect to conformal time. In the following, we will use the definition $\mathcal{H} \equiv a'/a$. The parameter $k = 0, \pm 1$ represents the curvature of the spacelike sections. If, in addition, the equation of state of the perfect fluids are provided, then we have a closed system of differential equations and, therefore, the evolution of the corresponding model of the Universe is completely specified.

2.2 The Inflationary Hypothesis

By definition, inflation is a phase of accelerated expansion, i.e. the scale factor satisfies [2]

$$\frac{{\rm d}^2 a}{{\rm d}t^2} > 0\,. \qquad (6)$$

It is interesting to postulate that such a phase took place in the very early Universe because, in this case, one can explain many different seemingly paradoxical facts like, for instance, the horizon problem or the flatness problem. Because of the latter, from now on, we will put $k = 0$ in the Einstein equations. More precisely, one can show the inflationary scenario is satisfactory if the number of e-folds, i.e. the logarithm of the scale factor at the end of inflation to the scale factor at the beginning of inflation is greater than 60 [2],

$$N_{\rm T} > 60\,. \qquad (7)$$

More detailed arguments about the advantages of inflation can be found in [3] but, at this point, it is important to notice the following three facts. Firstly, inflation is convincing because, by means of a single concept or hypothesis, one can solve many different problems. In this sense, inflation is an economical assumption. Secondly, as we will show below, inflation is falsifiable since it makes definite predictions that we will describe. Therefore, there is the hope either to confirm or to exclude this hypothesis and, in any case, there is the certainty to learn something about the early Universe. Thirdly, inflation is defined by the condition (6) but this does not prejudge the physical nature of the matter responsible for the acceleration of the Universe. The only thing one can say is obtained by expressing the acceleration of the scale factor in cosmic time. Using the Einstein equations, one gets

$$\frac{\ddot{a}}{a} = -\frac{\kappa}{6}(\rho + 3p) , \tag{8}$$

where a dot denotes a derivative with respect to the cosmic time. Therefore, the fluid responsible for inflation must be such that

$$p < -\frac{\rho}{3} , \tag{9}$$

i.e. must have a negative pressure. As a consequence, this fluid cannot be a standard fluid, like a gas for instance, but must be somehow "exotic". However, this does not come as a surprise since inflation is supposed to take place at very high energies. At those energies, the natural description of matter is (quantum) field theory. As we are now going to demonstrate, it is quite interesting to remark that the most simple example of a field theory can do the job very well and "produce" the negative pressure which is necessary to inflation. We now discuss this point in more details.

2.3 Implementing the Inflationary Hypothesis

The most simple implementation of the inflationary scenario is to assume that matter is described by a scalar field $\varphi(\eta)$ [1, 2]. This case is nothing but a particular example of a perfect fluid. The corresponding action reads

$$S = -\int \mathrm{d}^4x\sqrt{-g}\left[\frac{1}{2}g^{\mu\nu}\partial_\mu\varphi\partial_\nu\varphi + V(\varphi)\right] . \tag{10}$$

Then, the stress-energy tensor, which is defined by

$$T_{\mu\nu} = -\frac{2}{\sqrt{-g}}\frac{\delta S}{\delta g^{\mu\nu}} , \tag{11}$$

can be written as

$$T_{\mu\nu} = \partial_\mu\varphi\partial_\nu\varphi - g_{\mu\nu}\left[\frac{1}{2}g^{\alpha\beta}\partial_\alpha\varphi\partial_\beta\varphi + V(\varphi)\right] . \tag{12}$$

From this expression, it is clear that the scalar field is indeed a perfect fluid. The energy density and the pressure are defined by $T^0{}_0 = -\rho$, $T^i{}_j = p\delta^i{}_j$ and we obtain

$$\rho = \frac{1}{2}\frac{(\varphi')^2}{a^2} + V(\varphi), \quad p = \frac{1}{2}\frac{(\varphi')^2}{a^2} - V(\varphi) . \tag{13}$$

The conservation equation can be obtained either by re-deriving it from the very beginning or just by inserting the previous expressions of the energy density and pressure into (3). Assuming $\varphi' \neq 0$, this reproduces the Klein-Gordon equation written in a FLRW background, namely

$$\varphi'' + 2\frac{a'}{a}\varphi' + a^2\frac{\mathrm{d}V(\varphi)}{\mathrm{d}\varphi} = 0 . \tag{14}$$

The other equation of conservation expresses the fact that the scalar field is homogeneous and, therefore, does not bring any new information. Finally, a comment is in order about the equation of state. In general, there is no simple link between ρ and p except when the kinetic energy dominates the potential energy, where $\omega \equiv p/\rho \simeq 1$, i.e. the case of stiff matter or, on the contrary, when the potential energy dominates the kinetic energy for which one obtains $\omega \simeq -1$. This last case is of course very interesting since this leads to an inflationary solution. We have thus identified the condition under which inflation can occur: the potential energy must dominate the kinetic energy, i.e.

$$V(\varphi) \gg \frac{1}{2}\frac{(\varphi')^2}{a^2} . \tag{15}$$

We now turn to a systematic study of this regime.

2.4 Slow-roll Inflation

Since the kinetic energy to potential energy ratio and the scalar field acceleration to the scalar field velocity ratio are small, this suggests to view these quantities as parameters in which a systematic expansion is performed. The slow roll regime is controlled by the three (at leading order) slow-roll parameters defined by:

$$\epsilon \equiv 3\frac{\dot{\varphi}^2}{2}\left(\frac{\dot{\varphi}^2}{2} + V\right)^{-1} = -\frac{\dot{H}}{H^2} = 1 - \frac{\mathcal{H}'}{\mathcal{H}^2} , \tag{16}$$

$$\delta \equiv -\frac{\ddot{\varphi}}{H\dot{\varphi}} = -\frac{\dot{\epsilon}}{2H\epsilon} + \epsilon , \quad \xi \equiv \frac{\dot{\epsilon} - \dot{\delta}}{H} . \tag{17}$$

The slow-roll conditions are satisfied if ϵ and δ are much smaller than one and if $\xi = \mathcal{O}(\epsilon^2, \delta^2, \epsilon\delta)$. It is also convenient to re-express the slow-roll parameters in terms of the inflaton potential. Using the equations of motion in the slow-roll approximation, one can show that

$$\epsilon \simeq \frac{m_{\rm Pl}^2}{16\pi}\left(\frac{V'}{V}\right)^2, \quad \delta \simeq -\frac{m_{\rm Pl}^2}{16\pi}\left(\frac{V'}{V}\right)^2 + \frac{m_{\rm Pl}^2}{8\pi}\frac{V''}{V}, \tag{18}$$

where, here, a prime denotes a derivative with respect to the scalar field.

The equations of motion, that is to say the Friedman equation and the Klein-Gordon equation can be re-written exactly as

$$H^2 = \frac{\kappa V}{3-\epsilon}, \quad \frac{\mathrm{d}\varphi}{\mathrm{d}t} = -\frac{1}{(3-\delta)H}\frac{\mathrm{d}V}{\mathrm{d}\varphi}, \tag{19}$$

from which one deduces that, if the slow-roll conditions are satisfied

$$H^2 \simeq \frac{\kappa}{3}V(\varphi) + \mathcal{O}(\epsilon), \quad \frac{\mathrm{d}\varphi}{\mathrm{d}t} \simeq -\frac{1}{3H}\frac{\mathrm{d}V}{\mathrm{d}\varphi} + \mathcal{O}(\delta). \tag{20}$$

These equations are of course easier to analyze and solve than the original ones.

Let us now analyze a concrete example. We choose the following class of potentials

$$V(\varphi) = \frac{3\lambda_n}{8\pi}m_{\rm Pl}^4\left(\frac{\varphi}{m_{\rm Pl}}\right)^n, \tag{21}$$

where n is a free parameter and λ_n the coupling constant. The factors that show up into the definition of the potential have been chosen for future convenience. Let us first try to see under which conditions the slow-roll approximation is valid. We adopt the criterion $\epsilon < 1$ (it is in fact $\epsilon \ll 1$ and, strictly speaking, $\epsilon < 1$ only corresponds to the condition necessary in order to have an accelerated expansion). This amounts to

$$\varphi > \varphi_{\rm end} = \frac{n}{4\sqrt{\pi}}m_{\rm Pl}. \tag{22}$$

Of course, this constraint applies in particular to the initial value of the field. We already conclude, that for this class of models, the values of the field must be at least a few Planck mass. Let us be more precise and evaluate the total number of e-folds during slow-roll inflation. It is given by the formula

$$N_{\rm T} = \ln\left(\frac{a_{\rm end}}{a_{\rm ini}}\right) \simeq -\kappa\int_{\varphi_{\rm ini}}^{\varphi_{\rm end}} \mathrm{d}\varphi V(\varphi)\left(\frac{\mathrm{d}V}{\mathrm{d}\varphi}\right)^{-1}, \tag{23}$$

from which one gets

$$N_{\rm T} = \frac{4\pi}{n}\left(\frac{\varphi_{\rm ini}}{m_{\rm Pl}}\right)^2 - \frac{n}{4}. \tag{24}$$

Let $N_{\rm min}$ the minimum number of e-folds required in order to solve the problems of the hot big-bang model (we have seen before that $N_{\rm min} \simeq 60$) then one has

$$\varphi_{\rm ini} > m_{\rm Pl}\sqrt{\frac{n}{4\pi}\left(N_{\rm min} + \frac{n}{4}\right)}. \tag{25}$$

For $n = 2$, this gives $\varphi_{\rm ini} \simeq 3.1 m_{\rm Pl}$ and for $n = 4$, one obtains $\varphi_{\rm ini} \simeq 4.4 m_{\rm Pl}$. However, it is often argued that "natural" initial conditions (see the last article in [2]) are such that $V(\varphi_{\rm ini}) = m_{\rm Pl}^4$ which amounts to

$$\varphi_{\rm ini} = m_{\rm Pl}\left(\frac{8\pi}{3}\right)^{1/n} \lambda_n^{-1/n} \gg m_{\rm Pl}\,, \tag{26}$$

because, as we will discuss later one, the Cosmic Background Explorer (COBE) normalization implies that the coupling constant is small. In this case, the number of e-folds is a large number, much larger that the minimum required. Of course, the fact that the value of the scalar field must be larger or of the order of the Planck mass has led to many discussions about the model building problem. In this review, we do not address this question. Details about this issue can be found in [4, 5].

Let us now solve the equations of motions in the slow-roll approximation. For the scalar field straightforward calculations lead to ($n \neq 4$)

$$\varphi(t) = \varphi_{\rm ini}\left[1 - \frac{n(4-n)}{2}\frac{\sqrt{\lambda_n}}{8\pi}\left(\frac{m_{\rm Pl}}{\varphi_{\rm ini}}\right)^{(4-n)/2} m_{\rm Pl}(t - t_{\rm ini})\right]^{2/(4-n)} . \tag{27}$$

The last expression can also be expressed in terms of $t_{\rm end}$, the time at which slow-roll inflation stops. One obtains

$$\varphi(t) = \varphi_{\rm ini}\left\{1 - \frac{t - t_{\rm ini}}{t_{\rm end} - t_{\rm ini}}\left[1 - \left(\frac{\varphi_{\rm end}}{\varphi_{\rm ini}}\right)^{(4-n)/2}\right]\right\}^{2/(4-n)} . \tag{28}$$

The advantage of the above equation is to show that for $t < t_{\rm end}$ the argument between braces always remains positive and hence the whole expression well-defined. A negative argument would simply signal the break-down of the slow-roll approximation and, in this case, the above formula cannot be used.

Let us now turn to the scale factor. Integrating the first of (20) leads to

$$a(t) = a_{\rm ini}\exp\left\{-\frac{4\pi}{n m_{\rm Pl}^2}\left[\varphi_0^2(t) - \varphi_{\rm ini}^2\right]\right\} . \tag{29}$$

From this expression, one can also calculate the evolution of the scalar field in terms of the number of e-folds N which is the natural time variable during inflation. One gets

$$\varphi(N) = m_{\rm Pl}\sqrt{\left(\frac{\varphi_{\rm ini}}{m_{\rm Pl}}\right)^2 - \frac{n}{4\pi}N}\,, \tag{30}$$

from which one obtains the formula giving the evolution of the Hubble parameter during inflation

$$H(N) = m_{\rm Pl}\sqrt{\lambda_n}\left[\left(\frac{\varphi_{\rm ini}}{m_{\rm Pl}}\right)^2 - \frac{n}{4\pi}N\right]^{n/4} . \tag{31}$$

This equation is valid until $N = N_{\rm T}$ and, in this regime, the above formula is always well-defined. Indeed, (31) becomes meaningless at $N_{\rm max} = 4\pi(\varphi_{\rm ini}/m_{\rm Pl})^2/n$ but $N_{\rm max} > N_{\rm T}$. The above equation has interesting consequences for our understanding of inflation. It shows that the Hubble parameter can evolve and change significantly during the slow-roll phase. Later on, we will see that a quantity which plays an important role is the value of the Hubble parameter when the scales of astrophysical interest today crossed out the horizon during inflation. This happens 60 e-folds before the end of inflation. This scale is constrained by the observations on the Cosmic Microwave Background Radiation (CMBR) anisotropies to be $H/m_{\rm Pl} < 10^{-5}$. However, this does not mean that the Hubble parameter has not been larger before, especially if the total number of total e-folds is large, as it is the case for the initial conditions discussed in (26). For instance, if we have a massive potential, $n = 2$, and $\varphi_{\rm ini} = 100\, m_{\rm Pl}$ then inflation starts with an initial Hubble parameter of $H_{\rm ini} \simeq 10^{-3} m_{\rm Pl}$ but ends at $H_{\rm end} = m_{\rm Pl}\sqrt{\lambda_n}\,[n/(4\sqrt{\pi})]^{n/2} \simeq 0.28 \times 10^{-5} m_{\rm Pl}$ after $N \simeq 62000$ e-folds, where we have used $\lambda_2 \simeq 10^{-10}$ (corresponding to a mass $m \simeq 10^{-5} m_{\rm Pl}$). Therefore, in summary, it will be important to keep in mind that the observations give indications about the scale of inflation when the relevant scales crossed out the horizon during inflation but cannot, a priori, put constrains on the Hubble parameter in the earliest phases of evolution. This situation is summarized in Fig. 1.

When the field reaches the value $\varphi_{\rm end}$, slow-roll inflation stops and the system enters a new regime that we now briefly describe.

2.5 Reheating

When the scalar field reaches the point where the slow-roll parameter $\epsilon \simeq 1$, for which $\varphi = \varphi_{\rm end}$, inflation stops and the field starts oscillating around its minimum [6, 7]. In this regime, the system is governed by two time scales: the Hubble time H^{-1} and the period of the oscillations around the minimum $(V'')^{-1}$ (here, a prime means derivative with respect to the field). The important point is that these two scales are very different. The frequency of the oscillations is much larger than the Hubble rate, $\omega_{\rm osci} \simeq V'' \gg H$. The scalar field obeys the Klein-Gordon equation which can be put under the following form

$$\frac{{\rm d}\rho}{{\rm d}t} = -3H\dot{\varphi}^2 = -6H(\rho - V) , \tag{32}$$

where the relation $\dot{\varphi}^2 = 2(\rho - V)$ has been used. This equation can be time averaged and one gets

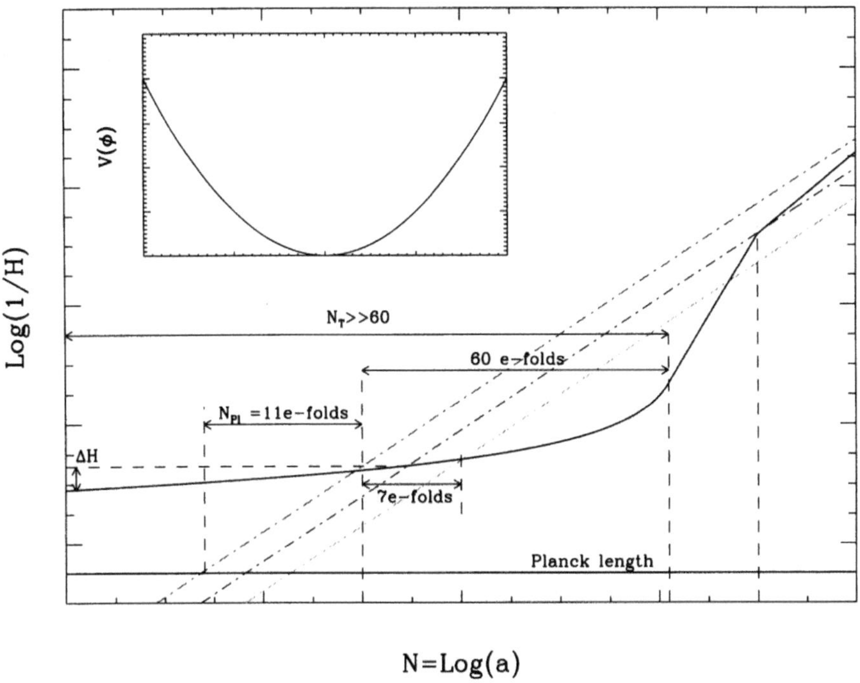

Fig. 1. Evolution of the various scales discussed in the text during inflation and the subsequent radiation and matter dominated epochs. In particular, it is apparent that the CMBR measurements only probe the inflationary model when the modes of astrophysical interest today crossed out the horizon during inflation. The small window shows a typical inflationary potential, see (21)

$$\left\langle \frac{\mathrm{d}\rho}{\mathrm{d}t} \right\rangle = -\left\langle 6H(\rho - V)\right\rangle \simeq -6H\left\langle \rho - V\right\rangle , \tag{33}$$

where we have used the fact that the Hubble rate does not change during one period of the oscillations. The right hand side of the equation above can be evaluated as

$$\left\langle \rho - V\right\rangle \equiv \frac{1}{T}\int_0^T (\rho - V)\mathrm{d}t = \left[\int_{-\varphi_\mathrm{m}}^{\varphi_\mathrm{m}} \sqrt{\rho - V(\varphi)}\mathrm{d}\varphi\right]\left[\int_{-\varphi_\mathrm{m}}^{\varphi_\mathrm{m}} \frac{\mathrm{d}\varphi}{\sqrt{\rho - V(\varphi)}}\right]^{-1} , \tag{34}$$

where φ_m is the value of the scalar field at the maximum of its oscillations. In order to obtain the previous relation, one has also utilized that $\mathrm{d}t = \mathrm{d}\varphi/\sqrt{2(\rho - V)}$. Then, one uses that over one period, $\rho \simeq V(\varphi_\mathrm{m}) \equiv V_\mathrm{m}$ is a constant and one obtains

$$\left\langle \rho - V\right\rangle \simeq \gamma\rho , \tag{35}$$

where the number γ is defined by

$$\gamma \equiv \left[\int_{-\varphi_{\rm m}}^{\varphi_{\rm m}} \sqrt{1-\frac{V(\varphi)}{V_{\rm m}}}{\rm d}\varphi\right]\left\{\int_{-\varphi_{\rm m}}^{\varphi_{\rm m}} \left[1-\frac{V(\varphi)}{V_{\rm m}}\right]^{-1/2}{\rm d}\varphi\right\}^{-1} = \frac{n}{n+2}\,, \quad (36)$$

the last result being valid for potentials of the form $V(\varphi) \propto \varphi^n$. Let us now turn to the left hand side of (33). The term $\langle {\rm d}\rho/{\rm d}t\rangle$ can be written as $\Delta\rho/T$. It can be expressed as a finite difference expression and one can rewrite it as $\dot{\rho}$. This is valid for time intervals much larger than the period of the oscillations. Therefore, the equation governing the evolution of the energy density of the field (33) can be rewritten as

$$\dot{\rho} = -\frac{6n}{n+2}H\rho \Rightarrow \rho \propto a^{-6n/(n+2)}\,. \quad (37)$$

Then, the scale factor is given by $a(t) \propto t^{(n+2)/(3n)}$. For the massive case, $n=2$, the energy density evolves as in a matter-dominated epoch. This can be easily understood since, in this case, the Klein-Gordon equation is exactly the equation of an harmonic oscillator. In this situation, it is known that $\langle \dot{\varphi}^2/2\rangle = \langle V(\varphi)\rangle$ which implies that the pressure vanishes.

So far, we have not taken into account the effect of particles creation. Phenomenologically, it can be described by adding a term $\Gamma\dot{\varphi}$ in the Klein-Gordon equation which now reads

$$\ddot{\varphi} + 3H\dot{\varphi} + \Gamma\dot{\varphi} + \frac{{\rm d}V(\varphi)}{{\rm d}\varphi} = 0 \quad \Rightarrow \frac{{\rm d}\rho}{{\rm d}t} = -\frac{2n}{n+2}(3H+\Gamma)\rho. \quad (38)$$

If we assume that the particles produced are very light in comparison with the mass of the inflaton, these particles will be very relativistic. This means that the equation of conservation of radiation should also be modified according to

$$\frac{{\rm d}\rho_{\rm r}}{{\rm d}t} = -4H\rho_{\rm r} + \Gamma\rho\,, \quad (39)$$

so that the total energy is still conserved. Equation (38) can be easily integrated and the solution reads

$$\rho(t) = \rho_{\rm osci}\left(\frac{a}{a_{\rm osci}}\right)^{-6n/(n+2)} \exp\left[-\frac{2n}{n+2}\Gamma(t-t_{\rm osci})\right] \quad (40)$$

where $t = t_{\rm osci}$ is the time at which the oscillations start (i.e. the time at which the slow-roll period ends) and $\rho_{\rm osci}$ is the value of the scalar field energy density at that time. The effect of the term $\Gamma\dot{\varphi}$ is to multiply the result (37) by a decreasing exponential factor. Equipped with this solution, we can solve (39) and determine the evolution of $\rho_{\rm r}$. If the scalar field energy density dominates the radiation, as it is the case at the beginning of the reheating period, the solution reads (using the fact that the scale factor is known in this regime, see above)

$$\rho_{\mathrm{r}}(t) = \Gamma t_{\mathrm{osci}}\rho_{\mathrm{osci}}\left(\frac{a}{a_{\mathrm{osci}}}\right)^{-4}\left(\frac{n+2}{2n\Gamma t_{\mathrm{osci}}}\right)^{(n+8)/(3n)}\exp\left(\frac{2n}{n+2}\Gamma t_{\mathrm{osci}}\right) \times\left[\gamma\left(\frac{n+8}{3n},\frac{2n}{n+2}\Gamma t\right)-\gamma\left(\frac{n+8}{3n},\frac{2n}{n+2}\Gamma t_{\mathrm{osci}}\right)\right], \tag{41}$$

where the function $\gamma(\alpha, x)$ is the incomplete gamma function defined by $\gamma(\alpha, x) \equiv \int_0^x \mathrm{e}^{-t}t^{\alpha-1}\mathrm{d}t$. In the above formula, we have assumed that, at the end of the slow-roll epoch $t = t_{\mathrm{osci}}$, $\rho_{\mathrm{r}} \simeq 0$. For small values of the argument x, the incomplete gamma function reduces to $\simeq x^\alpha/\alpha$. We define $t_{\mathrm{RH}} \equiv \Gamma^{-1}$ and then we have $x = 2n\Gamma t/(n+2) = 2n/(n+2)(t/t_{\mathrm{RH}})$ and for times $t > t_{\mathrm{osci}} \ll \Gamma^{-1}$, the argument of the incomplete gamma function is small. In this limit, one obtains

$$\rho_{\mathrm{r}}(t) \simeq \Gamma\rho_{\mathrm{osci}}t_{\mathrm{osci}}^2\frac{3n}{(n+8)t}\left[1-\left(\frac{t}{t_{\mathrm{osci}}}\right)^{-(n+8)/(3n)}\right]. \tag{42}$$

We see that ρ_{r} starts to increase, reaches a maximum and then decreases. When t approaches t_{RH}, the previous approximation breaks down since the argument of the incomplete gamma function is no longer small. However, for order of magnitude estimates, we can try to push this approximation. At $t \simeq t_{\mathrm{RH}}$, we have $\rho_{\mathrm{r}} \simeq \Gamma\rho_{\mathrm{osci}}t_{\mathrm{osci}}^2 3n/[(n+8)t_{\mathrm{RH}}]$ since the second term in the square bracket in (42) is negligible. After thermalization, the energy density of radiation takes the form $\rho_{\mathrm{r}} = g_*\pi^2T^4/30$. Using the fact that $\rho_{\mathrm{osci}} \simeq H_{\mathrm{inf}}^2 m_{\mathrm{Pl}}^2$ (nothing but the Friedman equation) and that $t_{\mathrm{osci}} \simeq H_{\mathrm{inf}}^{-1}$, one can deduce the reheating temperature

$$T_{\mathrm{RH}} \simeq \frac{30^{1/4}}{\sqrt{\pi}}g_*^{-1/4}\left(\frac{3n}{n+8}\right)^{1/4}(\Gamma m_{\mathrm{Pl}})^{1/2}. \tag{43}$$

Then, from this temperature, the universe evolves in a standard radiation dominated era. The remarkable feature of the previous equation is that it does not depend on the scale of inflation H_{inf} but only on the decay rate Γ of the inflaton. This means that whatever the scale of inflation is, the radiation dominated era always starts at the same energy (at fixed decay rate) and that the duration of the period of coherent oscillations can change quite a lot. The number of e-foldings during this epoch can be evaluated as [since during this epoch, the scale factor scales as $\propto t^{(n+2)/(3n)}$]

$$N \simeq \frac{n+2}{3n}\ln\left(\frac{H_{\mathrm{inf}}}{\Gamma}\right). \tag{44}$$

The previous considerations are valid if the life time of the inflaton is bigger than the age of the universe at the end of inflation. Otherwise, there is no period of coherent oscillations. In this case, the vacuum energy $H_{\mathrm{inf}}^2 m_{\mathrm{Pl}}^2$ is directly converted into radiation and the reheating temperature is

$$T_{\rm RH} \simeq \frac{30^{1/4}}{\sqrt{\pi}} g_*^{-1/4} (H_{\rm inf} m_{\rm Pl})^{1/2} . \tag{45}$$

Finally, let us recall that the calculations above assume that the physical quantities are time averaged and therefore that the time scales considered are larger than the period of the oscillations. In Fig. 2, where the evolution of the field versus the number of e-folds is displayed, we have integrated the equations of motion numerically. This plot confirms our analytical estimates: inflation consists of a phase of slow-roll followed by a phase of oscillations. This concludes our study of the inflationary background.

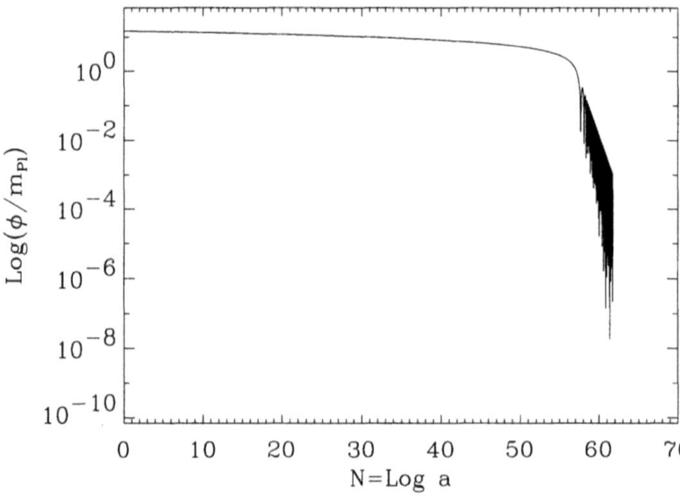

Fig. 2. Evolution of the scalar field during slow-roll inflation and the reheating phase (where the field oscillates) obtained by numerical integration of the equations of motion. The potential is of the type of (21) with $n = 2$ and $\lambda_2 = (8\pi/6) \times 10^{-10}$. The initial conditions are such that $\varphi_{\rm ini} \simeq 3m_{\rm Pl}$ leading to $N_{\rm T} \simeq 60$ as confirmed by the plot

3 Cosmological Perturbations

3.1 General Framework

It is an observational fact that the universe is not isotropic and homogeneous. Therefore, if one wants to have an accurate description, it is clearly mandatory to go beyond the FLRW model. On the other hand, it is also an experimental fact that, in the early Universe, the deviations from the isotropy and from the homogeneity were small (e.g. from the COBE measurement, $\delta T/T \simeq 10^{-5}$). This suggests a perturbative treatment. Therefore, the following metric tensor [8] gives a refined description of our Universe

$$\gamma_{\mu\nu}(\eta, \boldsymbol{x}) = [g_{\mu\nu}(\eta) + \epsilon h_{\mu\nu}(\eta, \boldsymbol{x}) + \epsilon^2 \ell_{\mu\nu}(\eta, \boldsymbol{x}) + \cdots] \mathrm{d}x^\mu \mathrm{d}x^\nu \,, \tag{46}$$

where $g_{\mu\nu}(\eta)$ is the standard FLRW metric introduced previously and represents the "background" (the parameter ϵ in the above equation should not be confused with the first slow-roll parameter; they have nothing to do with each other). The perturbed metric depends on $\boldsymbol{x}$ and this is the signature of the fact that we now go beyond the cosmological principle. In order to be consistent, the same expansion must be performed for the quantities describing matter. For example, if there is a background scalar field $\varphi(\eta)$, a refined description of the scalar field can be expressed as

$$\varphi(\eta, \boldsymbol{x}) = \varphi(\eta) + \epsilon \delta\varphi(\eta, \boldsymbol{x}) + \epsilon^2 \delta^{(2)} \varphi(\eta, \boldsymbol{x}) + \cdots \,. \tag{47}$$

The main goal of the theory of cosmological perturbations is to determine the evolution of the perturbed quantities $h_{\mu\nu}$ and $\delta\varphi$ and, then, to use them in order to calculate observable quantities. To find the behavior of the perturbed quantities, one needs some equations of motion. Naturally, these equations are taken to be the perturbed Einstein equations written order by order (we assume that gravity is described by General Relativity). Therefore, we expand the Einstein tensor and the stress-energy tensor according to

$$G_{\mu\nu} = G^{(0)}_{\mu\nu} + \epsilon G^{(1)}_{\mu\nu} + \epsilon^2 G^{(2)}_{\mu\nu} + \cdots \,, \quad T_{\mu\nu} = T^{(0)}_{\mu\nu} + \epsilon T^{(1)}_{\mu\nu} + \epsilon^2 T^{(2)}_{\mu\nu} + \cdots \,, \tag{48}$$

and then identify the terms of same order to obtain

$$G^{(0)}_{\mu\nu} = \kappa T^{(0)}_{\mu\nu}, \quad G^{(1)}_{\mu\nu} = \kappa T^{(1)}_{\mu\nu}, \quad G^{(2)}_{\mu\nu} = \kappa T^{(2)}_{\mu\nu}, \cdots \,. \tag{49}$$

In the present context, we will restrict ourselves to the linear order in the parameter ϵ.

Let us now try to describe the perturbed metric tensor in more details. For any symmetric two-rank tensor, there is a theorem [9] which states that $h_{\mu\nu}(\eta, \boldsymbol{x})$ can be decomposed as $h_{\mu\nu}(\eta, \boldsymbol{x}) = h^{(\mathrm{S})}_{\mu\nu} + h^{(\mathrm{V})}_{\mu\nu} + h^{(\mathrm{T})}_{\mu\nu}$, where $h^{(\mathrm{S})}_{\mu\nu}$ is constructed only from scalar functions, $h^{(\mathrm{V})}_{\mu\nu}$ is constructed only from three-dimensional vectors with vanishing divergences and $h^{(\mathrm{T})}_{\mu\nu}$ is obtained only from transverse and traceless three-dimensional tensors. These three types of perturbations are the scalar, rotational and tensorial fluctuations respectively. Explicitly, the theorem implies that the unperturbed metric plus the perturbed metric can be expressed as [8]

$$\begin{aligned} \mathrm{d}s^2 = a^2(\eta)\{&-(1+2\phi)\mathrm{d}\eta^2 + 2(\partial_i B - S_i)\mathrm{d}x^i \mathrm{d}\eta + [(1-2\psi)\delta_{ij} + 2\partial_i\partial_j E \\ &+ \partial_j F_i + \partial_i F_j + h^{(\mathrm{T})}_{ij}]\mathrm{d}x^i \mathrm{d}x^j\} \,, \end{aligned} \tag{50}$$

with S_i and F_i being transverse vectors, i.e. $\partial^i S_i = \partial^i F_i = 0$ and $h^{(\mathrm{T})}_{ij}$ being a transverse and traceless tensor, i.e. $\delta^{ij} h_{ij} = 0$, $\partial^j h_{ij} = 0$. We see that the scalar part of the metric depends on four unknown functions: ϕ, B, ψ and E. The vector part depends on two vectors with vanishing divergence, i.e. S_i

and F_i and, finally, the tensor part depends on one transverse and traceless tensor, namely $h_{ij}^{(\mathrm{T})}$. At linear order, each type of perturbations decouple and, as a consequence, can be treated separately.

In the specific case of inflation, one can show that vector perturbations cannot be produced [8]. Hence, in the following, we will consider that only scalar and tensor perturbations are present.

At this point, one should discuss a well-known problem of the theory of cosmological perturbations: the gauge issue. A complete study of this question can be found in [8, 10, 11] but, roughly speaking, it consists in the following. There exist solutions to the perturbed Einstein equations which are coordinates dependent, i.e. which can be removed by performing an infinitesimal change of coordinates. These solutions are fictitious and should not be considered as physical. The following analogy may help to understand the problem [12]. Let us consider the four-dimensional FLRW manifold denoted V_4 in what follows. It can be embedded into a higher dimensional manifold, more precisely into the five-dimensional Minkowski spacetime $E^5_{1,4}$ whose metric is η_{AB} where the indexes A and B runs from 0 to 4. A point in $E^5_{1,4}$ is located by its coordinates z^A. An embedding is a map from V_4 to $E^5_{1,4}$: $z^A = z^A(x^\mu)$. For a spatially flat FLRW spacetime endowed with Cartesian coordinates, the embedding explicitly reads:

$$z^0(\eta, x, y, z) = \frac{1}{2}a(\eta)(x^2 + y^2 + z^2 + 1) + \frac{1}{2}\int^\eta \frac{a^2(\tau)}{a'(\tau)}\mathrm{d}\tau \,, \tag{51}$$

$$z^1(\eta, x, y, z) = \frac{1}{2}a(\eta)(x^2 + y^2 + z^2 - 1) + \frac{1}{2}\int^\eta \frac{a^2(\tau)}{a'(\tau)}\mathrm{d}\tau \,, \tag{52}$$

$$z^2(\eta, x, y, z) = ax, \quad z^3(\eta, x, y, z) = ay, \qquad z^4(\eta, x, y, z) = az \,. \tag{53}$$

Therefore, the FLRW manifold can be viewed as a surface into the higher dimensional spacetime $E^5_{1,4}$. The metric of this surface can be calculated by means of the well-known formula

$$g_{\mu\nu}(\eta, \boldsymbol{x}) = \eta_{AB}\partial_\mu z^A \partial_\nu z^B \,, \tag{54}$$

and we can indeed check that this reproduces the metric of a spatially flat FLRW universe. Let us now try to “deform” this manifold since this is what we have in mind when we consider small perturbations around the background. In the present context, a deformation consists of the following. If we consider a point M in the manifold V_4 located by its coordinates $z^A(x^\mu)$ in $E^5_{1,4}$, deforming the manifold means slightly displacing the point M in $E^5_{1,4}$. This means that the coordinates of this point are no longer z^A but $z^A + \epsilon v^A(x^\mu)$ where ϵ is a small parameter. The vector $v^A(x^\mu)$ characterizes the deformation. As a consequence, the new metric of V_4 calculated by means of (54) reads

$$g_{\mu\nu} = \gamma_{\mu\nu} + 2\epsilon\eta_{AB}\partial_\mu z^A \partial_\nu v^B \,. \tag{55}$$

However, all the vectors $v^A(x^\mu)$ do not represent a genuine deformation. Indeed, if the following relation is satisfied

$$z^A(x^\mu) + \epsilon v^A(x^\mu) = z^A(x^\mu + \epsilon \xi^\mu) , \tag{56}$$

then, clearly, the displacement is within V_4 and does not correspond to a deformation. This is merely a change of coordinates that should not be considered as a physical deformation of V_4. This gauge problem consists of identifying the spurious modes and in removing them from the theory. To conclude this digression, it should be emphasized that the link between the previous approach and the theory of cosmological perturbations has never been worked out in details. Therefore, an important warning is that it may well turn out that the analogy used above cannot be applied completely to the theory studied here.

Having realized that there are non physical modes, the problem is now to find a method to get rid of them. Following Bardeen's seminal paper, an efficient way is to work with combinations of the metric tensor components which are invariant under a general change of coordinates (a "gauge" transformation) and, hence, which cannot contain a spurious mode. For scalar perturbations, the two following combinations [10]

$$\Phi(\eta, \boldsymbol{x}) \equiv \phi + \frac{1}{a}\left[a(B - E')\right]' , \quad \Psi(\eta, \boldsymbol{x}) \equiv \psi - \frac{a'}{a}(B - E') , \tag{57}$$

are gauge invariant. They are called the Bardeen potentials. In what follows, we will see that, in the simple case where matter is described by a scalar field, one has in fact $\Phi = \Psi$. This means that we have reduced the study of the scalar perturbations to the study of a single quantity: the Bardeen potential $\Phi(\eta, \boldsymbol{x})$.

The case of gravitational waves remains to be treated. In fact, it is easy to realize that the gravitational waves are gauge-invariant by definition because one cannot construct an infinitesimal change of coordinates with a tensor. Therefore, one can safely work with the tensor $h_{ij}^{(\mathrm{T})}(\eta, \boldsymbol{x})$ introduced before.

We have identified the gauge invariant variables that describe the gravitational sector. Our next move is to do the same but for the matter sector. This can be done in general [10, 11] but, since we have inflation in mind, we just consider the case of a scalar field. Then, one can show that

$$\delta\varphi^{(\mathrm{gi})}(\eta, \boldsymbol{x}) \equiv \delta\varphi + \varphi' (B - E') , \tag{58}$$

is the gauge-invariant perturbed scalar field.

Finally, we need a last ingredient. Since the spacelike sections are flat and since we study the linear theory, it is very convenient to work in the Fourier space. Indeed, because of the above properties, each Fourier mode evolve independently (the mode coupling appearing at quadratic order only) and it is sufficient to follow their time evolution. Therefore, we Fourier transform the Bardeen potential and the gravitational waves according to

$$\Phi(\eta, \boldsymbol{x}) = \frac{1}{(2\pi)^{3/2}} \int \mathrm{d}\boldsymbol{k} \Phi(\eta, \boldsymbol{k}) \mathrm{e}^{i\boldsymbol{k}\cdot\boldsymbol{x}} , \tag{59}$$

$$h_{ij}^{(\mathrm{T})}(\eta, \boldsymbol{x}) = \frac{1}{(2\pi)^{3/2}} \int \mathrm{d}\boldsymbol{k} \sum_{s=+,\times} p_{ij}^s(\boldsymbol{k}) h_{\mathrm{T}}^s(\eta, \boldsymbol{k}) \mathrm{e}^{i\boldsymbol{k}\cdot\boldsymbol{x}} . \tag{60}$$

In the last equation, $p_{ij}(\boldsymbol{k})$ is the transverse and traceless polarization tensor satisfying the following properties: $p^s_{ij}(\boldsymbol{k})p^{ij\,s'}(\boldsymbol{k}) = 2\delta^{ss'}$. The symbols "+" and "×" denote the two possible states of polarization of the gravitational waves. Of course, we also Fourier transform the perturbed scalar field and work with $\delta\varphi^{(\mathrm{gi})}(\eta, \boldsymbol{k})$.

Having identified what the relevant degrees of freedom are, we now turn to the question of establishing their equation of motion.

3.2 Equations of Motion

Since the Einstein equations are obviously gauge-invariant "by definition", it is clear that it is possible to express them in terms of gauge invariant quantities only. We start with density perturbations. Lengthy but straightforward calculations lead to (for a fixed Fourier mode k)

$$-3\mathcal{H}(\mathcal{H}\Phi + \Psi') - k^2\Psi = \frac{\kappa}{2}\left\{-(\varphi')^2\Phi + \varphi'\left[\delta\varphi^{(\mathrm{gi})}\right]' + a^2\frac{\mathrm{d}V}{\mathrm{d}\varphi}\delta\varphi^{(\mathrm{gi})}\right\}, \quad (61)$$

$$\mathcal{H}\Phi + \Psi' = \frac{\kappa}{2}\varphi'\delta\varphi^{(\mathrm{gi})}, \quad \Phi - \Psi = 0\,, \quad (62)$$

$$(2\mathcal{H}' + \mathcal{H}^2)\Phi + \mathcal{H}\Phi' + \Psi'' + 2\mathcal{H}\Psi' - \frac{1}{3}k^2(\Phi - \Psi) = \frac{\kappa}{2}\left\{-(\varphi')^2\Phi\right.$$
$$\left. + \varphi'\left[\delta\varphi^{(\mathrm{gi})}\right]' - a^2\frac{\mathrm{d}V}{\mathrm{d}\varphi}\delta\varphi^{(\mathrm{gi})}\right\}. \quad (63)$$

As announced, the two Bardeen potentials Φ and Ψ are equal. This is true as long as there is no anisotropic stress. Despite the apparent complexity of this system of equations, straightforward manipulations show that everything can be reduced to the study of a single equation which reads

$$\Phi'' + 2\left(\mathcal{H} - \frac{\varphi''}{\varphi'}\right)\Phi' + \left[k^2 + 2\left(\mathcal{H}' - \mathcal{H}\frac{\varphi''}{\varphi'}\right)\right]\Phi = 0\,. \quad (64)$$

This equation is valid provided $\varphi' \neq 0$. In this case, for which the scalar field plays the role of a cosmological constant, we have no density perturbations at all, $\Phi = 0$. This does not mean that the perturbed scalar field cannot fluctuate in de Sitter spacetime (as a matter of fact, it does) but that, in this case, these fluctuations do not couple to the fluctuations of the metric. Equation (64) can be transformed in order to permit a more transparent physical interpretation. If we consider the variables u and θ defined by

$$u \equiv \frac{4}{3}\frac{a^2\theta}{\mathcal{H}}\Phi\,, \qquad \theta \equiv \frac{1}{a}\left(\frac{\rho}{\rho + p}\right)^{1/2} = \sqrt{3}\frac{\mathcal{H}}{a\varphi'} = \sqrt{\frac{3}{2}}\frac{1}{a\sqrt{\gamma}}\,, \quad (65)$$

then (64) can be expressed as:

$$u'' + \left(k^2 - \frac{\theta''}{\theta}\right)u = 0\,. \quad (66)$$

The above equation can be viewed either as the equation of a parametric oscillator, with a time-dependent frequency given by $\omega^2(\boldsymbol{k},\eta) \equiv k^2 - \theta''/\theta$, or as a Schrödinger equation with the potential θ''/θ. This effective potential contains derivatives of the scale factor up to the fourth order. The typical behavior of the solutions can be easily found. For modes $k^2 \gg \theta''/\theta$, the variable u oscillate, $u \propto \mathrm{e}^{ik\eta}$, while for modes $k^2 \ll \theta''/\theta$ the solution of (66) may be expanded in powers of k^2. At leading order we obtain

$$u(\eta,\boldsymbol{k}) = A_1(k)\theta(\eta)\int^{\eta}\frac{\mathrm{d}\tau}{\theta^2(\tau)} + A_2(k)\theta(\eta)\,. \tag{67}$$

Since $\theta \to \infty$ for $a \to 0$ in general, A_1 is the arbitrary constant in front of the regular (growing) mode and A_2 a constant associated with the singular (decaying) mode. We will see in the following that the variable u is in fact not the most interesting for density perturbations. Quantum-mechanical considerations, among others, will lead us to work with another variable. From the above solution for $u(\eta,\boldsymbol{k})$, we easily deduce the Bardeen potential in the superhorizon regime. One obtains

$$\Phi(\eta,\boldsymbol{k}) \simeq -A_1(k)\frac{\mathcal{H}}{2a^2}\int^{\eta} a^2\gamma(\tau)\mathrm{d}\tau + \frac{A_2(k)}{2k^2}\frac{\mathcal{H}}{a^2}\,. \tag{68}$$

For example, for a power-law behavior of the scale factor, i.e. $a \propto |\eta|^{1+\beta}$ with $\beta \leq -2$ in order to have inflation, the "growing" mode turns out to be constant in time, namely

$$\Phi(\eta,\boldsymbol{k}) \simeq \frac{3}{2}\frac{1+\omega}{5+3\omega}A_1(k)\,. \tag{69}$$

For the de Sitter case, $\omega = -1$ and we recover the fact that there no density perturbations at all.

The equation of motion (64) has a first integral for modes that are much larger than the Hubble scale, i.e. $k/a \ll H$. Following [8] we define

$$\zeta \equiv \frac{2}{3}\frac{\mathcal{H}^{-1}\Phi' + \Phi}{1+\omega} + \Phi\,, \tag{70}$$

which was introduced by Lyth [13] originally. Essentially, the quantity ζ is the perturbation of the intrinsic curvature in the comoving gauge [13] and is written $-\mathcal{R}$ in that reference. The equation of motion for the Bardeen potential can be re-written as an expression for the first derivative of the quantity ζ. One obtains [11]

$$\frac{1}{\mathcal{H}}\frac{\mathrm{d}\zeta}{\mathrm{d}\eta} \propto \left(\frac{k}{\mathcal{H}}\right)^2\Phi\,. \tag{71}$$

Of course, to derive this equation we have not assumed that the equation of state parameter is a constant. Thus, ζ is constant in time for superhorizon modes $k/\mathcal{H} \ll 1$ since then $\dot{\zeta} = 0$.

The importance of the quantity ζ is due to the fact that this is a pure geometrical quantity. Concretely, this means that the conservation law established above in the case of a scalar field is in fact valid for any type of matter (at least, provided that the so-called entropy perturbations do not play an important role). Therefore, ζ can be used as a "tracer" of density perturbations regardless of the type of matter responsible for those fluctuations. In particular, it can be used to propagate the spectrum from the end of inflation (where the Universe is dominated by a scalar field) to the radiation dominated era (where the Universe is dominated by a relativistic fluid) without knowing the details of the reheating process. Let us now study how the calculation works in details. On superhorizon scales, the Bardeen potential is almost constant. If we neglect its time derivation then the equation of motion for Φ and the definition of ζ lead to

$$\Phi \simeq \frac{3(1+\omega_{\rm inf})}{5+3\omega_{\rm inf}}\zeta \simeq \frac{\kappa\varphi'}{2\mathcal{H}}\delta\varphi^{(\rm gi)} . \tag{72}$$

Using the equation of motion of the background, the above formula can be put under the following form

$$\zeta_{\rm inf}(\eta, \boldsymbol{k}) = \frac{5+3\omega_{\rm inf}}{2}\mathcal{H}\left[\frac{\delta\varphi^{(\rm gi)}(\eta, \boldsymbol{k})}{\varphi'}\right] . \tag{73}$$

Now, let us assume that we want to know the Bardeen potential in an era dominated by a fluid with a given equation of state ω (concretely, we have in mind $\omega = 1/3$ or $\omega = 0$ for the radiation or matter dominated epochs, respectively). Writing the constancy of ζ, i.e. $\zeta_{\rm inf} = \zeta_\omega$, one arrives at

$$\Phi_\omega(\eta, \boldsymbol{k}) \simeq 3\mathcal{H}\left(\frac{1+\omega}{5+3\omega}\right)\left[\frac{\delta\varphi^{(\rm gi)}(\eta, \boldsymbol{k})}{\varphi'}\right] , \tag{74}$$

where we have used $\omega_{\rm inf} \simeq -1$. This equation is very important since it links the primordial fluctuations of the scalar field to the fluctuations of the gravitational potential during the subsequent phases of evolution of the Universe.

Let us now turn to gravitational waves. In order to obtain the equation of motion, we must compute the perturbed Ricci or Einstein tensors for the metric given in (50). One finds that $\delta R^0{}_0 = \delta R^0{}_i = 0$. This result is consistent with the fact that the gravitational waves are transverse and traceless since only the trace and/or the derivative of the metric tensor can appear in these components. On the other hand, the component $\delta R^i{}_j$, $i \neq j$ is non vanishing and the leads to

$$\delta R^i{}_j = \frac{1}{2a^2}\left[h^{(\rm T)i}{}_j\right]'' + \frac{a'}{a^3}\left[h^{(\rm T)i}{}_j\right]' - \frac{1}{2a^2}\partial_k\partial^k h^{(\rm T)i}{}_j = \kappa\delta T^i{}_j , \tag{75}$$

where $\delta T^i{}_j$ represents the anisotropic pressure part of the perturbed stress-energy tensor. For a perfect fluid (e.g. for a scalar field), it vanishes. However, it is important to keep in mind that, a priori, the gravitational waves are not "independent" from the matter fluctuations and, hence, that they should be considered on the same footing as density perturbations. This is conceptually important because this means that it would be incorrect to argue that both types of perturbations should be treated differently, in particular with respect to the quantization of the cosmological perturbations.

Then, for the rescaled Fourier amplitude defined by $\mu_{\mathrm{T}}^s(\eta, \boldsymbol{k}) \equiv a(\eta) h^s(\eta, \boldsymbol{k})$, the equation $\delta R^i{}_j = 0$ can be re-written as [14]

$$\left(\mu_{\mathrm{T}}^s\right)'' + \left(k^2 - \frac{a''}{a}\right)\mu_{\mathrm{T}}^s = 0 . \tag{76}$$

This equation can be viewed either as the equation of a parametric oscillator, i.e. an oscillator with a time-dependent frequency, $\omega^2(\eta, \boldsymbol{k}) \equiv k^2 - a''/a$ or as a "time-independent" Schrödinger equation with a potential $U_{\mathrm{T}}(\eta) = a''/a$. Therefore, we obtain the same type of equation as for density perturbations. However, it is also interesting to notice that the effective potential for tensor perturbations involves the scale factor and its derivatives up to second order only. This difference is especially important during the reheating phase.

As it was the case for density perturbations, it is clear that the solution to the equation of motion possesses two regimes. If the wave number k is such that $k^2 \gg U_{\mathrm{T}}$, then the mode function oscillates, i.e. $\mu_{\mathrm{T}} \propto \mathrm{e}^{ik\eta}$. The interaction with the barrier (if any) corresponds to the time $k\eta_{\mathrm{t}} \simeq 1$ since, in the inflationary phase, one has $U_{\mathrm{T}} \simeq 1/\eta^2$ (for a power-law scale factor; this is also the case for slow-roll inflation, see below). This time is also, roughly speaking, the time of Hubble radius exit. Indeed, the wavelength is given by $\lambda = 2\pi a(\eta)/k$ and the Hubble scale is $H^{-1} = a/\mathcal{H}$. The condition $\lambda = H^{-1}$ gives $k\eta_{\mathrm{t}} \simeq 1$ since $\mathcal{H} \simeq 1/\eta$. The second regime is when the wave is below the potential, $k^2 \ll U_{\mathrm{T}}$. An approximate solution is

$$\mu_{\mathrm{T}}^s(\eta, \boldsymbol{k}) \simeq B_1^s(k)a(\eta) + B_2^s(k)a(\eta)\int^{\eta} \frac{\mathrm{d}\tau}{a^2(\tau)} , \tag{77}$$

where $B_1^s(k)$ and $B_2^s(k)$ are two constants which are a priori free. The first term in (77) is the growing mode whereas the second term is the decaying mode. This can be seen, for example, if we consider scale factors of the form $a(\eta) = \ell_0|\eta|^{1+\beta}$. In this case, $\mu_{\mathrm{T}}^s \simeq B_1^s(k)|\eta|^{1+\beta} + B_2^s(k)|\eta|^{-\beta}$ and for $\beta \simeq -2$, the first term goes to infinity while the conformal time goes to zero at the end of inflation. Therefore, this is indeed the growing mode. In terms of the amplitude $h^s(\eta, \boldsymbol{k})$ itself, one sees that the growing mode corresponds in fact to a constant and hence is conserved on large scales. Somehow, $h^s(\eta, \boldsymbol{k})$ plays for gravitational waves the same role as ζ for density perturbations.

Let us end this section with a comparison between density perturbations and gravitational waves. Using the results obtained before, the tensor to scalar amplitudes ratio today is given by

$$\frac{h_{\omega}}{\Phi_{\omega}} \sim (1 + \omega_{\mathrm{inf}})\frac{h_{\mathrm{inf}}}{\Phi_{\mathrm{inf}}} . \tag{78}$$

Therefore, if we assume that $h_{\rm inf} \simeq \Phi_{\rm inf}$, which is the case if the perturbations are of quantum-mechanical origin, then we have $h_\omega/\Phi_\omega \ll 1$, i.e. scalar fluctuations dominates over tensor fluctuations, because during inflation $\omega_{\rm inf} \simeq -1$.

The previous considerations also illustrate the limitations of the classical approach. Without a theory of the initial conditions, i.e. without a prescription to choose the k-dependent constants $A_{1,2}(k)$ for density perturbations and $B_{1,2}(k)$ for gravitational waves, we cannot really go further. This will be one of the main advantage of the quantum-mechanical version of the previous theory: a natural choice for $A_{1,2}(k)$ and $B_{1,2}(k)$.

3.3 The Sachs-Wolfe Effect

The production and the amplification of small inhomogeneities in the early Universe described above has several observational consequences. In this review, we focus on one of them: the presence of small angular anisotropies in the temperature of the Cosmic Microwave Background Radiation (CMBR), at the level of $\delta T/T \simeq 10^{-5}$, detected for the first time by the COBE satellite in 1992 [15]. These anisotropies are of utmost importance for the theory of inflation because they allow us to check the predictions of this scenario and/or to constrain the physics of the early Universe. We now turn to a rapid discussion of this effect, a complete presentation being available in [16].

The Sachs-Wolfe effect [17] links the angular variations of the temperature on the celestial sphere to the presence of cosmological fluctuations in the early Universe. We have to calculate the change in the energy of the photons propagating from the last scattering surface to Earth. This energy is given by $E = -\gamma_{\mu\nu}u^\mu k^\nu$, where k^μ is the wave vector of the photon and u^μ the velocity of the observer. Let us first investigate this relation for the background.

Fundamental observers are observers who move with the cosmological flow. A trajectory is given by the set $x^\mu = x^\mu(s)$, where s is a affine parameter along the line. The velocity along this curve is given by $u^\mu \equiv \mathrm{d}x^\mu/\mathrm{d}s$ and satisfies $u^\mu u_\mu = -1$. For a fundamental observer, one has $u^i = 0$ by definition and the normalization of the four velocity implies that $u^\mu = (1/a, 0)$ and $u_\mu = (-a, 0)$. Let us now study the propagation of a photon. If $k^\mu \equiv \mathrm{d}x^\mu/\mathrm{d}\lambda$ is the wave vector of a photon, then the path followed by this photon is such that

$$\frac{\mathrm{d}k^\mu}{\mathrm{d}\lambda} + \Gamma^\mu_{\nu\rho} k^\nu k^\rho = 0\,. \tag{79}$$

The solution of this equation is the trajectory of the photon: $x^\mu = x^\mu(\lambda)$. In addition, we have the constrain $k^\mu k_\mu = 0$, expressing the fact that the photon follows a null geodesic. At zeroth order, this constraint gives $\delta_{ij}k^i k^j = \left(k^0\right)^2$. In the non perturbed universe, (79) possesses the solutions $k^0 = C^0/a^2$, where C^0 is a constant and $k^i = -C^0 e^i/a^2$, where e^i is a three vector such that $\mathrm{d}e^i/\mathrm{d}\lambda = 0$. Taking the ratio of the wave vector components, we deduce that $\mathrm{d}x^i/\mathrm{d}\eta = -e^i$. Finally, integrating this relation, we find the equation of the trajectory in the unperturbed Universe

$$x^i = -e^i(\eta - \eta_{\rm D}) + x^i_{\rm D} \,, \tag{80}$$

where $(\eta_{\rm D}, x^i_{\rm D})$ are the coordinates at detection of the photon. It does not come as a surprise that the photons propagate along a straight line. On the other hand, the energy is given by

$$E(\eta) = \frac{C^0}{a(\eta)} \,, \tag{81}$$

and, in fact, we just recover the well-known time evolution of the temperature (which, as expected, does not depend on space for the unperturbed Universe).

Let us now turn to the Sachs-Wolfe effect itself. Essentially, this consists in computing the energy of the photons at first order. In a perturbed Universe, the most general observer possesses a velocity given by $u^\mu + \delta u^\mu$, where u^μ denotes the velocity of a fundamental observer calculated above and where we assume that the components of δu^μ are small with respect to this fundamental velocity. The fact that the total velocity is normalized to -1 implies that $\delta u^0 = -\phi/a$. We also write δu^i as $\delta u^i \equiv v^i/a$, from which we deduce that $\delta u_i = a v_i + a\partial_i B$. The trajectory of the photons can also be expanded as $x^\mu + \delta x^\mu$, where x^μ is the path of the photon in an unperturbed Universe determined before and δx^μ are the small corrections around the background trajectory due to the presence of the fluctuations. In the same manner as we did for the four-velocity, we can expand the wave vector of the photon according to $k^\mu + \delta k^\mu$, where $\delta k^\mu \equiv {\rm d}(\delta x^\mu)/{\rm d}\lambda$. At first order, the variation of energy can be expressed as

$$\delta E = -h_{\mu\nu} u^\mu k^\nu - g_{\mu\nu}\delta u^\mu k^\nu - g_{\mu\nu} u^\mu \delta k^\nu \,, \tag{82}$$

which can be re-written as $\delta E = C^0\phi/a + C^0 e^i(\partial_i B + v_i)/a + a\delta k^0$. In this equation, the only unknown quantity is δk^0 and we now establish its expression. Integrating the perturbed version of (79), one finds that

$$\begin{aligned} \delta k^0 &= -\frac{C^0}{a^2(\eta)} \int_{\eta_{\rm E}}^{\eta} {\rm d}\tau \Big\{ \phi' - 2e^i\partial_i\phi - e^i e^j \partial_i\partial_j B + \frac{1}{2}\Big[-2\psi\delta_{ij} + 2\partial_i\partial_j E \\ &\quad + h_{ij}^{(\rm T)}\Big]' e^i e^j \Big\} \,, \end{aligned} \tag{83}$$

where $\eta_{\rm E}$ is the conformal time at emission. Let us stress again that the integration is performed along the unperturbed path of the photon. Putting everything together, we finally obtain

$$\begin{aligned} \frac{E_{\rm D}}{E_{\rm E}} &= \frac{a_{\rm E}}{a_{\rm D}} \Big\{ 1 + \left[\phi + e^i\left(\partial_i B + v_i\right)\right]_{\rm E}^{\rm D} - \int_{\eta_{\rm E}}^{\eta_{\rm D}} {\rm d}\tau \Big[\phi' - 2e^i\partial_i\phi - e^i e^j \partial_i\partial_j B \\ &\quad + \frac{1}{2}\left(-2\psi\delta_{ij} + 2\partial_i\partial_j E + h_{ij}^{(\rm T)}\right)' e^i e^j \Big]\Big\} \,. \end{aligned} \tag{84}$$

The above expression depends on the coordinates of emission and detection of the photons. To go further, it is necessary to specify the conditions of emission, that is to say the characteristics of the last scattering surface. At zeroth order, the surface of last scattering has coordinates $\eta_{\rm E} = \eta_{\rm lss}$, $x^i_{\rm E} = -e^i(\eta_{\rm lss} - \eta_{\rm D}) + x^i_{\rm D}$ where $\eta_{\rm lss}$ is fixed and corresponds to the redshift $z_{\rm lss} \simeq 1100$. The only dependence is now the vector e^i and this corresponds to different directions on the celestial sphere. However, in presence of perturbations, emission occurs at different times and at different positions. In other words, the time of emission is given by $\eta_{\rm E} = \eta_{\rm lss} + \delta\eta(\eta_{\rm lss}, x^i_{\rm E})$. The quantity $\delta\eta(\eta_{\rm lss}, x^i_{\rm E})$ depends on our definition of the surface of emission. Let us assume that this surface is such that the density of photons, ρ_γ, is constant. Writing this condition at first order gives $\delta\rho_\gamma(\eta_{\rm lss}, x^i_{\rm E}) + \rho'_\gamma(\eta_{\rm lss})\delta\eta(\eta_{\rm lss}, x^i_{\rm E}) = 0$. Using the conservation equation which implies that $\rho'_\gamma = -4\mathcal{H}\rho_\gamma$, we arrive at

$$\delta\eta(\eta_{\rm lss}, x^i_{\rm E}) = \frac{1}{4\mathcal{H}(\eta_{\rm lss})}\frac{\delta\rho_\gamma(\eta_{\rm lss}, x^i_{\rm E})}{\rho_\gamma(\eta_{\rm lss})} . \tag{85}$$

Therefore, the term $a(\eta_{\rm E})$ in (84) should be written as

$$a(\eta_{\rm E}) = a(\eta_{\rm lss}) + \mathcal{H}(\eta_{\rm lss})\delta\eta(\eta_{\rm lss}, x^i_{\rm E}) = a(\eta_{\rm lss}) + \frac{1}{4}\delta_\gamma(\eta_{\rm lss}, x^i_{\rm E}) , \tag{86}$$

where $\delta_\gamma \equiv \delta\rho_\gamma/\rho_\gamma$ is the density contrast. In the same manner, if we say that detection takes place on a surface such that the baryons energy density is constant, the factor $a^{-1}(\eta_{\rm R})$ should be written as $a^{-1}(\eta_{\rm D}) = a^{-1}(\eta_0)[1 - (1/3)\delta_{\rm b}(\eta_0, x^i_{\rm D})]$ (the factor $1/3$ comes from the equation of conservation but now written for a fluid whose equation of state vanishes). Finally, (84) takes the form

$$\begin{aligned}\frac{E_{\rm D}}{E_{\rm E}} = \frac{a(\eta_{\rm lss})}{a(\eta_0)}\biggl\{&1 + \frac{1}{4}\delta_\gamma(\eta_{\rm lss}, x^i_{\rm E}) - \frac{1}{3}\delta_{\rm b}(\eta_0, x^i_{\rm D}) + \left[\phi + e^i(\partial_i B + v_i)\right]^{\rm D}_{\rm E} \\ &- \int_{\eta_{\rm lss}}^{\eta_0}{\rm d}\tau\biggl[\phi' - 2e^i\partial_i\phi - e^ie^j\partial_i\partial_j B + \biggl(-\psi\delta_{ij} + \partial_i\partial_j E \\ &+ \frac{1}{2}h^{\rm (T)}_{ij}\biggr)' e^ie^j\biggr]\biggr\} .\end{aligned} \tag{87}$$

Having established this important relation, we must now show that this expression is gauge-invariant. For this purpose, it is sufficient to express the ratio of the energies at emission and detection only in terms of gauge-invariant quantities. We have already described the gauge-invariant variables for the gravity sector. For the variables describing matter, we only need the gauge-invariant density contrast $\delta_{\rm g} \equiv \delta + \rho'/\rho(B - E')$. Finally, one has to decompose the three-velocity as $v_i = \partial_i v$ and the gauge-invariant velocity can be expressed as $v^{\rm (gi)} \equiv v + E'$. Let us also notice that the spatial derivatives can be expressed in terms of time derivatives. Indeed, along a trajectory, one has ${\rm d}f/{\rm d}\eta = \partial_\eta f - e^i\partial_i f$ from which we find $e^i\partial_i f = \partial_\eta f - {\rm d}f/{\rm d}\eta$. Then, straightforward calculations show that

$$\frac{E_{\rm D}}{E_{\rm E}} = \frac{a(\eta_{\rm lss})}{a(\eta_0)}\left\{1 + \frac{1}{4}(\delta_\gamma)_{\rm g}(\eta_{\rm lss}, x^i_{\rm E}) - \frac{1}{3}(\delta_{\rm b})_{\rm g}(\eta_0, x^i_{\rm D}) + \Phi(\eta_{\rm lss}, x^i_{\rm E})\right.$$
$$- \Phi(\eta_0, x^i_{\rm D}) + e^i\partial_i v^{(\rm gi)}(\eta_0, x^i_{\rm D}) - e^i\partial_i v^{(\rm gi)}(\eta_{\rm lss}, x^i_{\rm E})$$
$$\left. + \int_{\eta_{\rm lss}}^{\eta_0} {\rm d}\tau \left[\Phi' + \Psi' - \frac{1}{2}h^{(\rm T)\prime}_{ij}e^i e^j\right]\right\} . \tag{88}$$

We have thus proved the gauge invariance of the ratio $E_{\rm D}/E_{\rm E}$ [18].

The Sachs-Wolfe effect is frequency independent. This means that the shape of the black body is preserved at the perturbed level and this is why a perturbed temperature is still a meaningful concept. If we define $\delta T/T \equiv [\delta T_{\rm D} - T_{\rm D}]/T_{\rm D}$ with $T_{\rm D} = T_{\rm E}a(\eta_{\rm lss})/a(\eta_0)$, we arrive at the final form of the Sachs-Wolfe effect, namely

$$\frac{\delta T}{T} = \left(\frac{\delta T}{T}\right)^{(\rm D)} + \left(\frac{\delta T}{T}\right)^{(\rm S)} + \left(\frac{\delta T}{T}\right)^{(\rm T)} , \tag{89}$$

with,

$$\left(\frac{\delta T}{T}\right)^{(\rm D)} = e^i\partial_i v^{(\rm gi)}(\eta_0, x^i_{\rm D})$$
$$\left(\frac{\delta T}{T}\right)^{(\rm S)} = \frac{1}{4}(\delta_\gamma)_{\rm g}(\eta_{\rm lss}, x^i_{\rm E}) + \Phi(\eta_{\rm lss}, x^i_{\rm E}) - e^i\partial_i v^{(\rm gi)}(\eta_{\rm lss}, x^i_{\rm E})$$
$$+ \int_{\eta_{\rm lss}}^{\eta_0} {\rm d}\tau\, (\Phi' + \Psi') ,$$
$$\left(\frac{\delta T}{T}\right)^{(\rm T)} = -\frac{1}{2}\int_{\eta_{\rm lss}}^{\eta_0} {\rm d}\tau\, \frac{\partial}{\partial\eta}h^{(\rm T)}_{ij}e^i e^j . \tag{90}$$

Several comments are in order here. Firstly, we have discarded the terms $\Phi(\eta_0, x^i_{\rm D})$ and $\delta^{(\rm gi)}_{\rm b}(\eta_0, x^i_{\rm D})/3$ since they do not depend on the vector e^i. Secondly, the first term $[\delta T/T]^{(\rm D)}$ has its e^i dependence fixed. This is just the dipole term due to our motion with respect to the frame of the CMBR. Thirdly, the other terms are genuine fluctuations of primordial origin. As already mentioned, they have been discovered in 1992 by the COBE satellite.

Finally, let us conclude this section by establishing the expression of the Sachs-Wolfe effect due to density perturbations on large scales. On these scales, the Doppler term is negligible. The integrated Sachs-Wolfe effect is also negligible because, on superhorizon scales, the Bardeen potential is approximatively constant, hence its derivative vanishes (see before). Therefore, only the first two terms remain. One can show that they combine such that

$$\left(\frac{\delta T}{T}\right)^{(\rm S)} \simeq \frac{1}{3}\Phi(\eta_{\rm lss}, x^i_{\rm E}) . \tag{91}$$

This equation permits to compute the angular power spectrum in the COBE regime, i.e. for large angular scales.

4 Quantization of Cosmological Perturbations

We start this section with a discussion of the quantization of a free scalar field. This constitutes the prototype of methods used in the sequel for the cosmological perturbations.

4.1 Quantization of a Free Scalar Field

We consider the question of quantizing a (massless) scalar field in curved space-time. The starting point is the following action

$$S = -\frac{1}{c}\int \mathrm{d}^4x\sqrt{-g}g^{\mu\nu}\frac{1}{2}\partial_\mu\Phi\partial_\nu\Phi\,, \tag{92}$$

which, in a FLRW Universe, reads

$$S = \frac{1}{2c}\int \mathrm{d}^4x a^2(\eta)\left(\phi'^2 - \delta^{ij}\partial_i\Phi\partial_j\Phi\right)\,. \tag{93}$$

It follows immediately that the conjugate momentum to the scalar field can be expressed as

$$\Pi(\eta,\boldsymbol{x}) = \frac{a^2}{c}\Phi'(\eta,\boldsymbol{x})\,. \tag{94}$$

It is convenient to Fourier expand the field $\Phi(\eta,\boldsymbol{x})$ over the basis of plane waves (therefore, here, we use explicitly the fact that the spacelike hypersurfaces are flat). This gives

$$\Phi(\eta,\boldsymbol{x}) = \frac{1}{a(\eta)}\frac{1}{(2\pi)^{3/2}}\int \mathrm{d}\boldsymbol{k}\mu_{\boldsymbol{k}}(\eta)\mathrm{e}^{i\boldsymbol{k}\cdot\boldsymbol{x}}\,. \tag{95}$$

We have chosen to define the Fourier component with a factor $1/a(\eta)$ for future convenience. Since the scalar field is real, this last relation allows us to write $\mu_{\boldsymbol{k}}^* = \mu_{-\boldsymbol{k}}$. The next step consists in inserting the expression of $\Phi(\eta,\boldsymbol{x})$ into the action. This gives

$$\begin{aligned} S = \frac{1}{2c}\int \mathrm{d}\eta\int_{R^{3+}} \mathrm{d}^3\boldsymbol{k}\Bigg[&\mu_{\boldsymbol{k}}'^*\mu_{\boldsymbol{k}}' + \mu_{\boldsymbol{k}}'\mu_{\boldsymbol{k}}'^* - 2\frac{a'}{a}\left(\mu_{\boldsymbol{k}}'\mu_{\boldsymbol{k}}^* + \mu_{\boldsymbol{k}}'^*\mu_{\boldsymbol{k}}\right) \\ &+\left(\frac{a'^2}{a^2} - k^2\right)\left(\mu_{\boldsymbol{k}}\mu_{\boldsymbol{k}}^* + \mu_{\boldsymbol{k}}^*\mu_{\boldsymbol{k}}\right)\Bigg]\,. \end{aligned} \tag{96}$$

Notice that the integral over the wavenumbers is calculated in half of the space in order to sum over independent variables only. Equipped with the Lagrangian in the momentum space (that, in the following, we denote by $\bar{\mathcal{L}}$), we can now go to the Hamiltonian formalism. The conjugate momentum to $\mu_{\boldsymbol{k}}$ is defined by the formula

$$p_{\boldsymbol{k}} \equiv \frac{\delta\bar{\mathcal{L}}}{\delta\mu_{\boldsymbol{k}}'^*} = \frac{1}{c}\left(\mu_{\boldsymbol{k}}' - \frac{a'}{a}\mu_{\boldsymbol{k}}\right)\,. \tag{97}$$

One can check that the definitions of the conjugate momenta in the real and Fourier spaces are consistent in the sense that they are linked by the (expected) expression

$$\Pi(\eta, \boldsymbol{x}) = \frac{a(\eta)}{(2\pi)^{3/2}} \int \mathrm{d}\boldsymbol{k} p_{\boldsymbol{k}} \mathrm{e}^{i\boldsymbol{k}\cdot\boldsymbol{x}} . \tag{98}$$

We see that the definition of the conjugate momentum $p_{\boldsymbol{k}}$ as the derivative of the Lagrangian in the Fourier space with respect to $\mu_{\boldsymbol{k}}'^*$ and not to $\mu_{\boldsymbol{k}}'$ is consistent with the expression of the momentum in the real space. Otherwise the momentum $\Pi(\eta, \boldsymbol{x})$ in real space would have been expressed in terms of $p_{\boldsymbol{k}}^*$ instead of $p_{\boldsymbol{k}}$.

One can also check that the Lagrangian leads to the correct equation of motion. Since we have $\delta\bar{\mathcal{L}}/\delta\mu_{\boldsymbol{k}}^* = 1/(2c)[-2\mathcal{H}\mu_{\boldsymbol{k}}' + 2(\mathcal{H}^2 - k^2)\mu_{\boldsymbol{k}}']$, the Euler-Lagrange equation $\mathrm{d}[\delta\bar{\mathcal{L}}/\delta\mu_{\boldsymbol{k}}'^*]/\mathrm{d}\eta - \delta\bar{\mathcal{L}}/\delta\mu_{\boldsymbol{k}}^* = 0$ reproduces the correct equation of motion for the variable $\mu_{\boldsymbol{k}}$, namely

$$\frac{\mathrm{d}^2\mu_{\boldsymbol{k}}}{\mathrm{d}\eta^2} + \left(k^2 - \frac{a''}{a}\right)\mu_{\boldsymbol{k}} = 0 , \tag{99}$$

which is indeed the well-known result.

We are now in a position where we can go to the Hamiltonian formalism. The Hamiltonian density, $\bar{\mathcal{H}}$, is defined by

$$\bar{\mathcal{H}} \equiv p_{\boldsymbol{k}}\mu_{\boldsymbol{k}}'^* + p_{\boldsymbol{k}}^*\mu_{\boldsymbol{k}}' - \bar{\mathcal{L}} , \tag{100}$$

and we obtain

$$\bar{\mathcal{H}} = c\left(p_{\boldsymbol{k}}p_{\boldsymbol{k}}^* + \frac{k^2}{c^2}\mu_{\boldsymbol{k}}\mu_{\boldsymbol{k}}^*\right) + \frac{a'}{a}\left(p_{\boldsymbol{k}}\mu_{\boldsymbol{k}}^* + p_{\boldsymbol{k}}^*\mu_{\boldsymbol{k}}\right) . \tag{101}$$

One can check that the Hamilton equations

$$\frac{\mathrm{d}\mu_{\boldsymbol{k}}^*}{\mathrm{d}\eta} = \frac{\partial\bar{\mathcal{H}}}{\partial p_{\boldsymbol{k}}} = cp_{\boldsymbol{k}}^* + \frac{a'}{a}\mu_{\boldsymbol{k}}^* , \quad \frac{\mathrm{d}p_{\boldsymbol{k}}^*}{\mathrm{d}\eta} = -\frac{\partial\bar{\mathcal{H}}}{\partial\mu_{\boldsymbol{k}}} = -\frac{a'}{a}p_{\boldsymbol{k}}^* - \frac{k^2}{c}\mu_{\boldsymbol{k}}^* , \tag{102}$$

lead to the correct equation of motion given by (99).

As a preparation to canonical quantization, we now introduce the normal variable $\alpha_{\boldsymbol{k}}$ [19] defined by

$$\alpha_{\boldsymbol{k}}(\eta) \equiv N(k)\mu_{\boldsymbol{k}} + ic\frac{M(k)}{k}p_{\boldsymbol{k}} , \tag{103}$$

where, for the moment, the functions $N(k)$ and $M(k)$ are free but will be specified later on. In terms of the normal variables, the scalar field and its conjugate momentum can be expressed as

$$\Phi(\eta, \boldsymbol{k}) = \frac{1}{a(\eta)}\frac{1}{(2\pi)^{3/2}}\int\frac{\mathrm{d}\boldsymbol{k}}{2N(k)}\left[\alpha_{\boldsymbol{k}}(\eta)\mathrm{e}^{i\boldsymbol{k}\cdot\boldsymbol{x}} + \alpha_{\boldsymbol{k}}^*(\eta)\mathrm{e}^{-i\boldsymbol{k}\cdot\boldsymbol{x}}\right] , \tag{104}$$

$$\Pi(\eta, \boldsymbol{x}) = \frac{a(\eta)}{(2\pi)^{3/2}}\int\mathrm{d}\boldsymbol{k}\frac{k}{2icM(k)}\left[\alpha_{\boldsymbol{k}}(\eta)\mathrm{e}^{i\boldsymbol{k}\cdot\boldsymbol{x}} - \alpha_{\boldsymbol{k}}^*(\eta)\mathrm{e}^{-i\boldsymbol{k}\cdot\boldsymbol{x}}\right] . \tag{105}$$

We are now ready to quantize the system. So far, we were dealing with a relativistic field theory and only the constant c appeared in the equations. Now, the constant $\hbar$, which fixes the amplitude of the fluctuations shows up. Concretely, the quantization is carried out by requiring that $\Phi(\eta, \boldsymbol{x})$ and $\Pi(\eta, \boldsymbol{x})$ become quantum operators satisfying the usual commutation relation, namely

$$[\hat{\Phi}(\eta, \boldsymbol{x}), \hat{\Pi}(\eta, \boldsymbol{y})] = i\hbar\delta^3(\boldsymbol{x} - \boldsymbol{y}) . \tag{106}$$

The normal variable $\alpha_{\boldsymbol{k}}(\eta)$ is promoted to an operator $c_{\boldsymbol{k}}(\eta)$. We choose the commutation relation to be $[c_{\boldsymbol{k}}(\eta), c_{\boldsymbol{p}}^{\dagger}(\eta)] = C\delta(\boldsymbol{k} - \boldsymbol{p})$. In the last expression, C is a free dimensionless constant. Notice that the commutation relation is time-independent. Then, the expressions of $N(k)$ and $M(k)$ are fully determined. Let us see in more details how the calculation proceeds. The commutator is given by

$$[\hat{\Phi}(\eta, \boldsymbol{x}), \hat{\Pi}(\eta, \boldsymbol{y})] = \frac{iC}{4c(2\pi)^3} \int \mathrm{d}^3\boldsymbol{k} \frac{k}{N(k)M(k)} \left[\mathrm{e}^{i\boldsymbol{k}(\boldsymbol{x}-\boldsymbol{y})} + \mathrm{e}^{-i\boldsymbol{k}(\boldsymbol{x}-\boldsymbol{y})}\right] . \tag{107}$$

We see that, in order to produce a Dirac function $\delta^3(\boldsymbol{x} - \boldsymbol{y})$ which is necessary in order to reproduce the relation given by (106) by integration of the exponentials, the term $k/(NM)$ must be k-independent. The link between the functions $N(k)$ and $M(k)$ is therefore determined. Let us call D the term $k/(NM)$. Then the result reads

$$[\hat{\Phi}(\eta, \boldsymbol{x}), \hat{\Pi}(\eta, \boldsymbol{y})] = \frac{iC}{4c} \times 2D\delta^3(\boldsymbol{x} - \boldsymbol{y}) . \tag{108}$$

As a consequence we have $CD = 2\hbar c$. As expected, the normalization is given by a combination of $\hbar$ and c. In the following, we will adopt the convenient choice $C = 1$.

Everything has been fixed but the function $N(k)$. This function is chosen by means of the following considerations. The energy of a scalar field is given by the formula

$$\hat{E} = \int \mathrm{d}^3\boldsymbol{x}\sqrt{-{}^{(3)}g}\hat{\rho} = \int \mathrm{d}^3\boldsymbol{x}\sqrt{-{}^{(3)}g}\frac{1}{2a^2}\left(\hat{\Phi}'^2 + \delta^{ij}\partial_i\hat{\Phi}\partial_j\hat{\Phi}\right) \tag{109}$$

$$= \int \mathrm{d}^3\boldsymbol{x}\sqrt{-{}^{(3)}g}\frac{1}{2a^2}\left[\left(\frac{c}{a^2}\hat{\Pi}\right)^2 + \delta^{ij}\partial_i\hat{\Phi}\partial_j\hat{\Phi}\right] , \tag{110}$$

where we have used the expression of the conjugate momentum. In this expression, the determinant of the metric is the determinant of the spatial part of the metric (including the factor a). We can now insert the expression of the operators $\hat{\Phi}$ and $\hat{\Pi}$ in the above equation giving $\hat{E}$. One finds

$$\hat{E} = \frac{1}{2a}\int \mathrm{d}^3\boldsymbol{k}\frac{1}{4}\left[\frac{k^2}{M^2(k)}\left(-c_{\boldsymbol{k}}c_{-\boldsymbol{k}} + c_{\boldsymbol{k}}c_{\boldsymbol{k}}^{\dagger} + c_{\boldsymbol{k}}^{\dagger}c_{\boldsymbol{k}} - c_{\boldsymbol{k}}^{\dagger}c_{-\boldsymbol{k}}^{\dagger}\right)\right.$$

$$\left. + \frac{k^2}{N^2(k)}\left(c_{\boldsymbol{k}}c_{-\boldsymbol{k}} + c_{\boldsymbol{k}}c_{\boldsymbol{k}}^{\dagger} + c_{\boldsymbol{k}}^{\dagger}c_{\boldsymbol{k}} + c_{\boldsymbol{k}}^{\dagger}c_{-\boldsymbol{k}}^{\dagger}\right)\right] . \tag{111}$$

Our criterion is to put "half of a quanta in each mode". Technically, this means that we would like the energy to take the following suggestive form

$$\hat{E} = \int \mathrm{d}^3\boldsymbol{k} \frac{\hbar\omega(\eta)}{2} \left(c_{\boldsymbol{k}} c_{\boldsymbol{k}}^{\dagger} + c_{\boldsymbol{k}}^{\dagger} c_{\boldsymbol{k}} \right) , \tag{112}$$

where $\omega(\eta) = kc/a(\eta)$ is the physical frequency. We see that the only way to cancel the unnecessary terms in (111) is to have $N(k) = M(k)$. Together with the relation established previously, $D = 2\hbar c$, this gives $N^2(k) = k/(2\hbar c)$. As a consequence, The scalar field operator now reads

$$\hat{\Phi}(\eta, \boldsymbol{x}) = \frac{\sqrt{\hbar c}}{a(\eta)} \frac{1}{(2\pi)^{3/2}} \int \frac{\mathrm{d}\boldsymbol{k}}{\sqrt{2k}} \left[c_{\boldsymbol{k}}(\eta) \mathrm{e}^{i\boldsymbol{k}\cdot\boldsymbol{x}} + c_{\boldsymbol{k}}^{\dagger}(\eta) \mathrm{e}^{-i\boldsymbol{k}\cdot\boldsymbol{x}} \right] . \tag{113}$$

Everything is now fixed. The expression of the scalar field operator contains no unspecified factor. Even the amplitude is fixed and is given by the factor $\sqrt{\hbar c}$.

We can now calculate the Hamiltonian operator. Using (101) one obtains

$$\hat{\mathbf{H}} = \frac{1}{2} \int_{R^3} \mathrm{d}^3\boldsymbol{k} \left[\hbar k \left(c_{\boldsymbol{k}} c_{\boldsymbol{k}}{}^{\dagger} + c_{-\boldsymbol{k}}{}^{\dagger} c_{-\boldsymbol{k}} \right) - i\hbar \frac{a'}{a} \left(c_{\boldsymbol{k}} c_{-\boldsymbol{k}} + c_{-\boldsymbol{k}}{}^{\dagger} c_{\boldsymbol{k}}{}^{\dagger} \right) \right] , \tag{114}$$

where it is important to notice that the integral is calculated in R^3 and not in R^{3+}. Let us analyze this Hamiltonian. The first term is the standard one and represents a collection of harmonic oscillators. The most interesting part is the second term. This term is responsible for the quantum creation of particles in curved spacetime. It can be viewed as an interacting term between the scalar field and the classical background. The coupling function ia'/a is proportional to the derivative of the scale factor and therefore vanishes in flat spacetime. From the structure of the interacting term, i.e. in particular the product of two creation operators for the mode $\boldsymbol{k}$ and $-\boldsymbol{k}$, we can also see that we have creation of pairs of quanta with opposite momenta during the cosmological expansion.

We can now calculate the time evolution of the quantum operators (we are here in the Heisenberg picture). Everything is known if we can determine what the temporal behavior of the creation and annihilation behavior is. The temporal behavior is given by the Heisenberg equations which read

$$\frac{\mathrm{d}c_{\boldsymbol{k}}}{\mathrm{d}\eta} = -\frac{i}{\hbar}[c_{\boldsymbol{k}}, \hat{\mathbf{H}}] , \quad \frac{\mathrm{d}c_{\boldsymbol{k}}{}^{\dagger}}{\mathrm{d}\eta} = -\frac{i}{\hbar}[c_{\boldsymbol{k}}{}^{\dagger}, \hat{\mathbf{H}}] . \tag{115}$$

Inserting the expression of the Hamiltonian derived above, we arrive at the equations

$$\frac{\mathrm{d}c_{\boldsymbol{k}}}{\mathrm{d}\eta} = kc_{\boldsymbol{k}}^{s} + i\frac{a'}{a} c_{-\boldsymbol{k}}{}^{\dagger} , \quad \frac{\mathrm{d}c_{\boldsymbol{k}}{}^{\dagger}}{\mathrm{d}\eta} = -kc_{\boldsymbol{k}}^{s}{}^{\dagger} - i\frac{a'}{a} c_{-\boldsymbol{k}}^{s} . \tag{116}$$

This system of equations can be solved by means of a Bogoliubov transformation and the solution can be written as

$$c_{\boldsymbol{k}}(\eta) = u_k(\eta)c_{\boldsymbol{k}}(\eta_{\rm ini}) + v_k(\eta)c_{-\boldsymbol{k}}{}^{\dagger}(\eta_{\rm ini}) \,, \tag{117}$$

$$c_{\boldsymbol{k}}{}^{\dagger}(\eta) = u_k^*(\eta)c_{\boldsymbol{k}}{}^{\dagger}(\eta_{\rm ini}) + v_k^*(\eta)c_{-\boldsymbol{k}}(\eta_{\rm ini}) \,, \tag{118}$$

where $\eta_{\rm ini}$ is a given initial time and where the functions $u_k(\eta)$ and $v_k(\eta)$ satisfy the equations

$$i\frac{{\rm d}u_k(\eta)}{{\rm d}\eta} = ku_k(\eta) + i\frac{a'}{a}v_k^*(\eta) \,, \quad i\frac{{\rm d}v_k(\eta)}{{\rm d}\eta} = kv_k(\eta) + i\frac{a'}{a}u_k(\eta) \,. \tag{119}$$

In addition, these two functions must satisfy $|u_k|^2 - |v_k|^2 = 1$ such that the commutation relation between the creation and annihilation operators is preserved in time. A very important property is the initial values of the two functions are fixed and, from the Bogoliubov transformation, read

$$u_k(\eta_{\rm ini}) = 1 \,, \quad v_k(\eta_{\rm ini}) = 0 \,. \tag{120}$$

At this point, the next move is to establish the link between the formalism exposed above and the classical picture. For this purpose, it is interesting to establish the equation of motion obeyed by the function $u_k + v_k^*$. Straightforward manipulations from (119) lead to

$$(u_k + v_k^*)'' + \left(k^2 - \frac{a''}{a}\right)(u_k + v_k^*) = 0 \,. \tag{121}$$

Therefore, the function $u_k + v_k^*$ obeys the same equation as the variable $\mu_{\boldsymbol{k}}$. This is to be expected since, using the Bogoliubov transformation, the scalar field operator can be re-written as

$$\hat{\Phi}(\eta, \mathbf{x}) = \frac{\sqrt{\hbar c}}{a(\eta)}\frac{1}{(2\pi)^{3/2}}\int \frac{{\rm d}\boldsymbol{k}}{\sqrt{2k}}\Big[(u_k + v_k^*)(\eta)c_{\boldsymbol{k}}(\eta_{\rm ini}){\rm e}^{i\boldsymbol{k}\cdot\boldsymbol{x}} + (u_k^* + v_k)(\eta)c_{\boldsymbol{k}}^{\dagger}(\eta_{\rm ini}){\rm e}^{-i\boldsymbol{k}\cdot\boldsymbol{x}}\Big] \,. \tag{122}$$

If we are given a scale factor, we can now calculate completely the time evolution of the perturbations by means of the formalism presented above. Let us stress again that the quantization procedure has completely fixed the overall amplitude of the field. Indeed, the field is normalized to $\sqrt{\hbar c}$ while the "mode function" $u_k + v_k^*$ has initially an amplitude of one.

Let us now calculate the two-point correlation function in the vacuum state. One gets

$$\left\langle 0 \left| \hat{\Phi}(\eta, \boldsymbol{x})\hat{\Phi}(\eta, \boldsymbol{x} + \boldsymbol{r}) \right| 0 \right\rangle = \frac{\hbar c}{4\pi^2}\int_0^{+\infty} \frac{{\rm d}k}{k}\frac{\sin kr}{kr}k^2\left|\frac{u_k + v_k^*}{a(\eta)}\right|^2 \,. \tag{123}$$

If we assume that the scale factor is given by a power-law of the conformal time, $a(\eta) = \ell_0(-\eta)^{1+\beta}$, where $\beta \le -2$ is a free a parameter and ℓ_0 a constant with the dimension of a length, then the solution of (121) with the initial conditions given by (120) reads

$$(u_k + v_k^*)(\eta) = \sqrt{\frac{\pi}{2}} e^{i(k\eta_{\rm ini} - \pi\beta/2)} \sqrt{-k\eta} H^{(1)}_{-\beta-1/2}(-k\eta) \ , \tag{124}$$

where $H^{(1)}$ is a Hankel function of first kind. From, this solution, it is easy to calculate the spectrum on large angular scales ($k\eta \to 0$)

$$\frac{\hbar c}{4\pi^2} k^2 \left| \frac{u_k + v_k^*}{a(\eta)} \right|^2 = \frac{\hbar c}{4\pi^2} \frac{f(\beta)}{\ell_0^2} k^{4+2\beta} \ , \tag{125}$$

where $f(\beta) \equiv \pi^{-1} \left[2^{-1-\beta} \Gamma(-\beta - 1/2)\right]^2$. In particular, if $\beta = -2$, this case corresponding to de Sitter spacetime for which the Hubble constant is strictly constant, one has $\ell_0 = c/H_{\rm inf}$ and the spectrum reads

$$\frac{\hbar}{c} \left(\frac{H_{\rm inf}}{2\pi} \right)^2 \tag{126}$$

i.e. is scale-invariant (which means that it does not depend on the wavenumber). Of course, if $\beta \neq -2$ then the spectrum is scale dependent.

The above result leads us to a first attempt to quantize cosmological perturbations [20]. Using (73) and taking into account the fact that $\omega_{\rm inf} \simeq -1$ and $\mathcal{H}/\varphi' \simeq \kappa/(2\epsilon)$ (recall that ϵ is the first slow-roll parameter), one obtains for the spectrum of the "tracer" ζ_k

$$P_\zeta \equiv k^3 \zeta_k^2 \simeq \frac{\kappa}{2\epsilon} k^3 \left[\delta\varphi_k^{(\rm gi)}\right]^2 . \tag{127}$$

The question is now how should we calculate $\left[\delta\varphi_k^{(\rm gi)}\right]^2$? Historically, the idea was to consider that the matter fluctuations (i.e. fluctuations in the scalar field) are quantized while the fluctuations in the gravitational field remain classical. Based on this guess, one can used the trick which consists in replacing

$$\left[\delta\varphi^{(\rm gi)}\right]^2 \to \left\langle 0 \left| \left[\delta\hat{\varphi}^{(\rm gi)}\right]^2 \right| 0 \right\rangle \tag{128}$$

or, in the Fourier space, $\left[\delta\varphi_k^{(\rm gi)}\right]^2 \to \hbar H_{\rm inf}^2/(4\pi^2 c)$. This gives for the spectrum of density perturbations

$$P_\zeta \equiv k^3 \zeta_k^2 \simeq \frac{\hbar G}{c^5} \frac{H_{\rm inf}^2}{\pi\epsilon} . \tag{129}$$

As expected the three fundamental constants, G, c and $\hbar$ participate to the final expression. We have kept them in order to be able to trace back their origin. The combination which appears here is the Planck time squared as it has to be since ζ_k is a dimensionless quantity. In natural units, the above spectrum is just $H_{\rm inf}^2/(\pi\epsilon m_{\rm Pl}^2)$. Several remarks are in order at this point. Firstly, as we will see, this trick provides us with the exact result. Secondly, it seems is that there is no way to rigorously justify the replacement (128). The reason is that matter, i.e. $\delta T_{\mu\nu}$, is treated quantum-mechanically, while geometry, i.e. $\delta G_{\mu\nu}$, is still considered to be classical, despite the fact that both are linked by the perturbed Einstein equations, $\delta G_{\mu\nu} = \kappa\delta T_{\mu\nu}$. One could think that a semi-classical equation

$$\delta G_{\mu\nu} = \kappa \left\langle 0 \left| \delta\hat{T}_{\mu\nu} \right| 0 \right\rangle , \tag{130}$$

could do the job but in fact one easily realizes that this cannot be the case because $\delta T_{\mu\nu}$ being linear in $\delta\varphi$ (it is of course quadratic in the scalar fields, but at linear order we have terms like $\varphi'\delta\varphi'$), we have in fact $\langle 0|\delta\hat{T}_{\mu\nu}|0\rangle = 0$ due to $\langle 0\,|\delta\hat{\varphi}|\,0\rangle = 0$. Therefore, (130) is in fact inconsistent in the present context. Thirdly, it would be dangerous to base the physical interpretation of (129) on the above method arguing that it gives the correct result. Here, we emphasize that a convincing physical interpretation should be based on a consistent framework. We can try the following analogy. The correct equation for the energy levels of an Hydrogen atom, $E_n \propto 1/n^2$, has been obtained for the first time by means of the so-called Bohr's model. This model was developed before a consistent framework for Quantum Mechanics become available. But, it is clear that, today, nobody would try to use Bohr's framework to interpret the formula for E_n. We are of the opinion that the situation for the cosmological perturbations is similar. Fourthly, the correct way to proceed is to treat the fluctuations in the geometry and in the scalar field on an equal footing. This amounts to "quantize" both sides of the Einstein equations and to write [8]

$$\delta\hat{G}_{\mu\nu} = \kappa\delta\hat{T}_{\mu\nu} \,. \tag{131}$$

The consequence is of course of utmost importance: the metric operator $h_{\mu\nu}$ should now be considered as a quantum operator, $h_{\mu\nu} \to \hat{h}_{\mu\nu}$. In other words, we have now to deal with the quantum-mechanical nature of the gravitational field, i.e. with quantum gravity (at the linearized level). We now turn to this question.

4.2 Quantization of Density Perturbations

The total action of the system is given by

$$S = -\frac{c^3}{16\pi G}\int \mathrm{d}^4x\sqrt{-g}R - \frac{1}{c}\int \mathrm{d}^4x\sqrt{-g}\left[\frac{1}{2}g^{\mu\nu}\partial_\mu\varphi\partial_\nu\varphi + V(\varphi)\right] . \tag{132}$$

If we perturb this action up to second order in the metric perturbations and in the scalar field fluctuations (this is necessary if we want the variation of this action to reproduce the first order equations of motion) one finds, despite a very long and tedious calculation, that the result is delightfully simple, namely [8]

$$ {}^{(2)}\delta S = \frac{1}{2c}\int \mathrm{d}^4x\left[(v')^2 - \delta^{ij}\partial_i v\partial_j v + \frac{z_\mathrm{S}''}{z_\mathrm{S}}v^2\right], \tag{133} $$

with

$$ v(\eta,\boldsymbol{x}) \equiv a\left[\delta\varphi^{(\mathrm{gi})} + \frac{\varphi'}{\mathcal{H}}\Phi\right]. \tag{134} $$

This is nothing but the action for a scalar field with a time-dependent mass. The constant G does not appear explicitly in the above action because it has been absorbed via the background Einstein equations. It is interesting to notice that the natural variable is not Φ neither ζ but v. It is not a surprise that the system is characterized by a single quantity since gravitational fluctuations are described by Φ and matter fluctuations by $\delta\varphi^{(\mathrm{gi})}$ but are linked by the perturbed Einstein equations. Therefore, only one degree of freedom is left. The link between v and the "tracer" ζ is given by

$$ \zeta = \sqrt{\frac{\kappa}{2}}\frac{v}{a\sqrt{\epsilon}}. \tag{135} $$

Finally, the quantity z_s is given by $z_\mathrm{s} = \sqrt{\kappa/2}a\varphi'/\mathcal{H} = a\sqrt{\epsilon}$ because, from the background Einstein equations, one has $\kappa(\varphi')^2 = 2\mathcal{H}^2\epsilon$.

At this point, the procedure of quantization follows exactly the one presented in the last subsection. The quantity $v(\eta,\boldsymbol{x})$ becomes a quantum operator $\hat{v}(\eta,\boldsymbol{x}) = a\left[\delta\hat{\varphi}^{(\mathrm{gi})} + (\varphi'/\mathcal{H})\hat{\Phi}\right]$, the expression of which can be written as

$$ \hat{v}(\eta,\boldsymbol{x}) = \frac{\sqrt{\hbar c}}{(2\pi)^{3/2}}\int\frac{\mathrm{d}^3\boldsymbol{k}}{\sqrt{2k}}\Big[(u_k+v_k^*)(\eta)c_{\boldsymbol{k}}(\eta_\mathrm{ini})\mathrm{e}^{i\boldsymbol{k}\cdot\boldsymbol{x}} \\ + (u_k^*+v_k)(\eta)c_{\boldsymbol{k}}^\dagger(\eta_\mathrm{ini})\mathrm{e}^{-i\boldsymbol{k}\cdot\boldsymbol{x}}\Big]. \tag{136} $$

As announced, the fluctuations of the metric tensor are now quantized: technically, the Bardeen potential $\Phi(\eta,\boldsymbol{k})$ is now a quantum operator $\hat{\Phi}(\eta,\boldsymbol{x})$ which explicitly appears into the expression of $\hat{v}(\eta,\boldsymbol{x})$. Notice also that the uncertainty principle has completely fixed the overall amplitude of the quantum perturbations since the initial conditions are fixed by $u_k(\eta_\mathrm{ini}) = 1$ and $v_k(\eta_\mathrm{ini}) = 0$. The equation of motion reads

$$ (u_k+v_k^*)'' + \left(k^2 - \frac{z_\mathrm{S}''}{z_\mathrm{S}}\right)(u_k+v_k^*) = 0. \tag{137} $$

The physical meaning of the initial conditions are as follows: initially, we choose the state which is empty of "particles" from the point of view of a local comoving observer at the initial time η_ini. This state $|0\rangle$ is defined by

$c_{\boldsymbol{k}}|0\rangle = 0$. Since, due to the time dependence of the background, there is a nontrivial mixing between positive and negative frequencies, this state is in general not the vacuum at later times.

We are now in a position to calculate the power spectrum of the quantum operator $\hat{\zeta}$. One gets

$$\left\langle 0 \left| \hat{\zeta}(\eta, \boldsymbol{x}) \hat{\zeta}(\eta, \boldsymbol{x}+\boldsymbol{r}) \right| 0 \right\rangle = \frac{\hbar c \kappa}{8\pi^2 z_{\mathrm{S}}^2} \int_0^{+\infty} \frac{\mathrm{d}k}{k} \frac{\sin kr}{kr} k^2 |u_k + v_k^*|^2 \,, \quad (138)$$

from which we easily deduce the expression of the power spectrum

$$k^3 P_\zeta = \frac{\hbar c \kappa}{8\pi^2} k^2 \left| \frac{u_k + v_k^*}{z_{\mathrm{S}}(\eta)} \right|^2 . \quad (139)$$

In order to compare this result with (129), we can evaluate the spectrum for power-law inflation where an exact solution of the equation of motion is available. Indeed, for $a(\eta) = \ell_0 (-\eta)^{1+\beta}$, the function ϵ is a constant, hence one has $z_{\mathrm{S}}''/z_{\mathrm{S}} = a''/a$. This means that the Hankel function of (124) is also solution of (137). Then, straightforward calculations show that

$$k^3 P_\zeta = \frac{1}{\pi \epsilon} \frac{\hbar G}{c^3 \ell_0^2} f(\beta) \, k^{2\beta+4} . \quad (140)$$

If β is close to -2 then $\ell_0 \simeq c/H_{\mathrm{inf}}$ and one recovers exactly the result of (129). As expected, the Planck length "naturally" appears in the above result.

4.3 Quantization of Gravitational Waves

The quantization of gravitational waves proceeds exactly along the same lines as before. Therefore, in this subsection, we only review briefly the main results. The starting point is the Einstein-Hilbert action that we expand to the second order. One gets (in natural units) [8]

$${}^{(2)}\delta S = \frac{m_{\mathrm{Pl}}^2}{64\pi} \int \left[(h^i{}_j)'(h^j{}_i)' - \partial_k (h^i{}_j) \partial^k (h^j{}_i) \right] a^2(\eta) \mathrm{d}^4 x \,. \quad (141)$$

In fact, the action can be re-written as

$${}^{(2)}\delta S_2 = -\frac{m_{\mathrm{Pl}}^2}{16\pi} \sum_{s=+,\times} \int \mathrm{d}^4 x \frac{1}{2} g^{\mu\nu} \partial_\mu h^s \partial_\nu h^s \,, \quad (142)$$

where the quantity $h^s(\eta, \boldsymbol{x})$ is defined by

$$h^s(\eta, \boldsymbol{x}) \equiv \frac{1}{a(\eta)} \frac{1}{(2\pi)^{3/2}} \sum_{s=+,\times}^{2} \int \mathrm{d}\boldsymbol{k} \mu_{\mathrm{T}}^s(\eta, \boldsymbol{k}) \mathrm{e}^{i\boldsymbol{k}\cdot\boldsymbol{x}} \,. \quad (143)$$

Therefore, the action of gravitational waves is equivalent to the action of two decoupled scalar fields (corresponding to the two states of polarization). One can then follow the method presented before. The quantum perturbed metric operator can be written as

$$\hat{h}_{ij}(\eta, \boldsymbol{x}) = \frac{4\sqrt{\pi}}{m_{\rm Pl} a(\eta)} \frac{1}{(2\pi)^{3/2}} \sum_{s=+,\times} \int \frac{\mathrm{d}\boldsymbol{k}}{\sqrt{2k}} p_{ij}^{s}(\boldsymbol{k}) \Big[(u_k^s + v_k^{s*})(\eta) c_{\boldsymbol{k}}^s(\eta_{\rm ini}) \mathrm{e}^{i\boldsymbol{k}\cdot\boldsymbol{x}} + (u_k^{s*} + v_k^s)(\eta) c_{\boldsymbol{k}}^{s\dagger}(\eta_{\rm ini}) \mathrm{e}^{-i\boldsymbol{k}\cdot\boldsymbol{x}} \Big] , \tag{144}$$

where the function $(u_n^s + v_n^{s*})(\eta)$ obeys (76). Finally, the two-point correlation function of the perturbed metric operator can be expressed as

$$\left\langle 0 \left| \hat{h}_{ij}(\eta, \boldsymbol{x}) \hat{h}^{ij}(\eta, \boldsymbol{x}+\boldsymbol{r}) \right| 0 \right\rangle = \frac{16}{\pi m_{\rm Pl}^2 a^2(\eta)} \int_0^{+\infty} \frac{\mathrm{d}k}{k} \frac{\sin kr}{kr} k^2 |u_k^s + v_k^{s*}|^2 , \tag{145}$$

from which we deduce that the power spectrum of the gravitational waves is given by

$$k^3 P_h(k, \eta) = \frac{16}{\pi m_{\rm Pl}^2} k^2 \left| \frac{u_k^s + v_k^{s*}}{a(\eta)} \right|^2 . \tag{146}$$

If we had decided to keep the standard units, the factor $1/m_{\rm Pl}^2$ in the above result would have obviously read $\hbar G/c^3$, i.e. the Planck length squared.

4.4 The Power Spectra in the Slow-roll Approximation

We have established the expression of the scalar and tensor power spectra and calculated these quantities for power-law inflation. However, as discussed at the beginning of this review article, the most interesting physical situation occurs when the slow-roll approximation is valid. As discussed previously, the only thing we need to do in order to compute the spectrum is to solve the equation of a parametric oscillator,

$$\mu'' + \left[k^2 - U(\eta)\right] \mu = \mu'' + \left[k^2 - \frac{z''}{z}(\eta)\right] \mu = 0 , \tag{147}$$

where μ is $u_k + v_k^*$ either for scalar or tensor perturbations and the effective potential $z_{\rm S}''/z_{\rm S}$ or a''/a, i.e. $z = z_{\rm S}$ or $z = a(\eta)$. As already mentioned, on subhorizon or superhorizon scales, this equation can be solved regardless of the detailed form of the scale factor. The solutions are $\exp(-ik\eta)$ and $z(\eta) + z(\eta)\int^{\eta} \mathrm{d}\tau z^{-2}(\tau)$ respectively. However, in order to obtain a reliable solution, one also needs to know the form of the solution in the regime $k^2 \simeq U(\eta)$, that is to say when the corresponding scales crossed out the horizon during inflation.

Let $N_*(\lambda)$ be the number of e-folds before the end of inflation at which the scale λ exits the horizon. We have

$$N_*(\lambda) \simeq \ln\left(\frac{\lambda}{\ell_{\rm H}}\right) + \left[\log_{10}\left(\frac{H_{\rm inf}}{m_{\rm Pl}}\right) - \log_{10}\left(\frac{T_{\rm RH}}{m_{\rm Pl}}\right) + 29\right] \times \ln 10\,. \quad (148)$$

If we take the fiducial values $H_{\rm inf} \simeq 10^{14}$GeV and $T_{\rm RH} \simeq M_{\rm inf} \simeq 10^{16.5}$GeV then $N_* \simeq 60$ for the Hubble scale today, i.e. $\lambda \simeq \ell_{\rm H}$, see also Fig. 1. A scale characterized by its wave-number k corresponds to an angle θ on the celestial sphere of about $k \simeq 1/(2\ell_{\rm H}\theta)$. Given the present CMBR experiments, this means that we probe in fact the scales $\ell_{\rm H} < \lambda < 10^{-3}\ell_{\rm H}$. The smallest scale in this interval crossed out the horizon $\simeq 46$ e-folds before the end of inflation. This means that the time taken by the the scales of astrophysical interest today to cross the horizon during inflation corresponds to $\Delta N \simeq 7$. Therefore, we need an accurate description of the effective potential $U(\eta)$ only during 7 e-folds, see Fig. 1.

In the slow-roll approximation, the effective potentials for scalar and tensor read at linear order

$$U_{\rm S}(\eta) = \frac{2+6\epsilon-3\delta}{\eta^2}\,, \quad U_{\rm T}(\eta) = \frac{2+3\epsilon}{\eta^2}\,. \quad (149)$$

Moreover, the equations of motion for ϵ and δ can be written as:

$$\frac{{\rm d}\epsilon}{H{\rm d}t} = \frac{{\rm d}\epsilon}{{\rm d}N} = 2\epsilon(\epsilon-\delta)\,, \quad \frac{{\rm d}\delta}{H{\rm d}t} = \frac{{\rm d}\delta}{{\rm d}N} = 2\epsilon(\epsilon-\delta) - \xi\,. \quad (150)$$

From these equations, one sees that typically $\mathcal{O}\left(\epsilon^2\right)\Delta N \ll 1$ for $\Delta N \simeq 7$ and, therefore, the slow-roll parameters can be considered as constant during the exit of the physical modes. This simplifies the problem drastically since then the equations of motion in the regime $k^2 \simeq U(\eta)$ can be solved in terms of Bessel functions whose orders depend on the slow-roll parameters. A detailed calculation can be found in [21] and, here, we just give the result

$$k^3 P_\zeta = \frac{H^2}{\pi\epsilon m_{\rm Pl}^2}\left[1 - 2(C+1)\epsilon - 2C(\epsilon-\delta) - 2(2\epsilon-\delta)\ln\frac{k}{k_*}\right]\,, \quad (151)$$

$$k^3 P_h = \frac{16H^2}{\pi m_{\rm Pl}^2}\left[1 - 2(C+1)\epsilon - 2\epsilon\ln\frac{k}{k_*}\right]\,, \quad (152)$$

where C is a numerical constant, $C \simeq -0.73$ and k_* a scale called the "pivot scale". We see that the amplitude of the scalar power spectrum is given by a scale-invariant piece, $H^2/(\pi\epsilon m_{\rm Pl}^2)$ that we had already guessed before, plus logarithmic corrections the amplitude of which is controlled by the slow-roll parameters, i.e. by the microphysics of inflation. It is important to notice that H is the value of the Hubble parameter during the 7 e-folds where the scales of astrophysical interest crossed out the horizon, see Fig. 1. As already mentioned at the end of Sec. (2.4) this can be different from the value of the Hubble parameter at the beginning of inflation. The above remarks are also valid for tensor perturbations. The ratio of tensor over scalar is just given by

$$\frac{k^3 P_h}{k^3 P_\zeta} = 16\epsilon \,. \tag{153}$$

This means that the gravitational are always sub-dominant and that, when we measure the CMBR anisotropies, we essentially see the scalar modes. This is rather unfortunate because this implies that one cannot measure the energy scale of inflation since the amplitude of the scalar power spectrum also depends on the slow-roll parameter ϵ. Only an independent measure of the gravitational waves contribution could allow us to break this degeneracy. On the other hand, the spectral indexes are given by

$$n_{\mathrm{S}} = \left.\frac{\ln k^3 P_\zeta}{\mathrm{d}\ln k}\right|_{k=k_*} = 1 - 4\epsilon + 2\delta \,, \quad n_{\mathrm{T}} = \left.\frac{\ln k^3 P_h}{\mathrm{d}\ln k}\right|_{k=k_*} = -2\epsilon \,. \tag{154}$$

As expected, the power spectra are always close to scale invariance and the deviation from it is controlled by the magnitude of the two slow-roll parameters. Finally, at the next-to-leading order there is no running of the spectral indexes since they are in fact second order in the slow-roll parameters.

5 Comparison with Observations

In this section, we briefly discuss the impact of the recent Wilkinson Microwave Anisotropy Probe (WMAP) measurements on inflation [22]. We have seen previously that the presence of cosmological perturbations causes anisotropies in the CMBR (the Sachs-Wolfe effect) and we have established the link between $\delta T/T$ and the metric fluctuations, see (89) and (91). The fact that the metric fluctuations are described by a quantum operator has an immediate consequence: $\delta T/T$ should be considered as a quantum operator as well. It is convenient to expand this operator on the celestial sphere, i.e. on the basis of spherical harmonics

$$\frac{\widehat{\delta T}}{T}(\boldsymbol{e}) = \sum_{\ell=2}^{+\infty} \sum_{m=-\ell}^{m=\ell} \hat{a}_{\ell m} Y_{\ell m}(\theta, \varphi) \,. \tag{155}$$

The next step is to calculate the two-point correlation function of temperature fluctuations. One gets

$$\left\langle 0 \left| \frac{\widehat{\delta T}}{T}(\boldsymbol{e}_1) \frac{\widehat{\delta T}}{T}(\boldsymbol{e}_2) \right| 0 \right\rangle = \sum_{\ell=2}^{+\infty} \frac{(2\ell+1)}{4\pi} C_\ell P_\ell\left(\cos\gamma\right) \,, \tag{156}$$

where P_ℓ is a Legendre polynomial and γ is the angle between the two vectors $\boldsymbol{e}_1$ and $\boldsymbol{e}_2$. The C_ℓ's are the multipole moments and have been measured with great accuracy by the WMAP experiment [22].

A remark in passing is in order at this point. As a matter of fact, what has been measured by the WMAP satellite is the correlation function

$$\left\langle \frac{\delta T}{T}(\boldsymbol{e}_1)\frac{\delta T}{T}(\boldsymbol{e}_2)\right\rangle , \qquad (157)$$

where the bracket denotes spatial average over the celestial sphere and not ensemble average as in (156). Going from one to another is not trivial and, in fact, involves profound questions which can even go as further as problems linked to the interpretation of Quantum Mechanics! (another related question is the problem of the "classicalization" of the quantum perturbations, see [23]). In order to check that the predictions of (156) are verified or not, one should repeat the measurement of the CMBR map many times and see whether the result converges toward the theoretical prediction. However, one cannot do that because we only have at our disposal one realization, i.e. one Universe or one CMBR map. Facing this situation, the usual strategy is to construct an unbiased estimator of the quantity that we want to measure (the correlation function or the multipole moments) with the minimum possible variance so that it is very probable that the outcome of one realization is closed to the mean value [24]. Unfortunately, the variance cannot be zero (in this case only one realization would be enough to estimate the result) and one can show that this is linked to the fact that a stochastic process on a sphere cannot be ergodic [24]. This variance is called the "cosmic variance" and is generally large on large scales. More details on this question can be found for instance in [24].

On large scales, i.e. for small ℓ, one can use (91) to find an explicit expression of the multipole moments. One gets

$$C_\ell = \frac{4\pi}{25}\int_0^{+\infty}\frac{\mathrm{d}k}{k} j_\ell^2(k)k^3 P_\zeta\,, \quad \ell \ll 20\,, \qquad (158)$$

where j_ℓ is a spherical Bessel function of order ℓ. Using (151) for density perturbations (since they are dominant) and neglecting the logarithmic corrections (which amounts to consider that the spectrum is scale-invariant), we obtain

$$C_\ell = \frac{2H^2}{25\epsilon m_{\mathrm{Pl}}^2}\frac{1}{\ell(\ell+1)}\,, \quad \ell \ll 20\,. \qquad (159)$$

Therefore, a scale invariant spectrum implies that, on large scales, the quantity $\ell(\ell+1)C_\ell$ is a constant. In order to calculate the inflationary multipole moments C_ℓ for any ℓ one must use a numerical code, for instance the CAMB code [25]. Typically, one gets a plateau and then acoustic oscillations. Here we do not treat this question but the details can be found in [16].

The satellites COBE and WMAP have measured the quantity $Q/T \equiv \sqrt{5C_2/(4\pi)}$ where $T \simeq 2.7\mathrm{K}$ and have found $Q \simeq 18\times 10^{-6}\,\mathrm{K}$. Moreover, recent analysis [26] of the WMAP data have been able to put a constraint on the value of the slow-roll parameter ϵ. It was found that $\epsilon < 0.032$. This allows us to put a constraint on the Hubble parameter at horizon crossing. One finds

$$\frac{H_{\rm inf}^2}{m_{\rm Pl}^2} = 60\pi\epsilon\frac{Q^2}{T^2} \Rightarrow \frac{H_{\rm inf}}{m_{\rm Pl}} < 1.6\times 10^{-5}\,. \tag{160}$$

This also puts a constraint on the amount of gravitational waves. In [26], the following result has been obtained

$$\frac{C_{10}^{\rm T}}{C_{10}^{\rm S}} < 0.3\,, \tag{161}$$

that is to say the contribution of gravitational waves is already constrained to be less than 30% of the total contribution.

We conclude this part by a summary of the main observational predictions of single field inflation: (i) The universe is spatially flat: $\Omega_0 = 1 \pm 10^{-5}$; (ii) The spectrum of density perturbations is scale invariant (Harrison–Zeldovich spectrum) plus logarithmic corrections which are model dependent, i.e. $n_{\rm S} = 1 + \mathcal{O}(\epsilon, \delta)$; (iii) There is a nearly scale invariant background of gravitational waves, i.e. $n_{\rm T} = \mathcal{O}(\epsilon)$; (iv) The statistical properties of the CMB anisotropies are Gaussian, i.e. everything is characterized by the power spectrum and we have the following properties

$$\left\langle 0\left|\left(\frac{\delta\hat{T}}{T}\right)^3\right|0\right\rangle = 0,\quad \left\langle 0\left|\left(\frac{\delta\hat{T}}{T}\right)^4\right|0\right\rangle - 3\left\langle 0\left|\left(\frac{\delta\hat{T}}{T}\right)^2\right|0\right\rangle^2 = 0,\quad \text{etc}\,\ldots\,. \tag{162}$$

This conclusion comes from the fact that the quantum state of the perturbations is the vacuum, the "wave function" of which is a Gaussian; (v) Gravitational waves are sub-dominant and there exists a consistency check relating the importance of gravitational waves with respect to scalar density on one hand to the tensor spectral index on the other hand. This relation reads

$$\frac{C_2^{\rm T}}{C_2^{\rm S}} \simeq -f_2(h, \Omega_{\rm cdm}, \Omega_\Lambda, \cdots)n_{\rm T}\,, \tag{163}$$

where the function f_2 is $f_2 \simeq 5$ for the concordance model (i.e. the cold dark matter model plus dark energy which seems to fit best the data at the time of writing); (vi) There are oscillations in the power spectrum. Although this conclusion is also based on the physics of the transfer function, the fact that the perturbations are generated in a coherent manner plays a crucial role for the survival of the acoustic peaks, see [27].

6 The Trans-Planckian Problem of Inflation

We have seen that the CMBR anisotropies are, if the inflation theory turns out to be correct, an observable signature of quantum gravity. However, as it is clear from the previous considerations, the CMBR anisotropies originate from a regime where the quantization of the gravitational field is carried out in the standard manner. In fact, the situation is similar to the Hawking radiation. In this last case, we have a quantum field living in a classical background. In the present context, we also have a field $\hat{h}_{\mu\nu}(\eta, \boldsymbol{x})$ living in the classical FLRW Universe. Of course, the main difference is that, in the case of inflation, the quantized test field is the perturbed metric, i.e. is the gravitational field itself (at least the small excitations of the gravitational field around a classical background) contrary to the Hawking effect where the field is just a scalar field: this is why, conceptually, the Hawking effect does not involve quantum gravity while the theory of cosmological perturbations does. Nevertheless, from the pure technical point of view, we have just used the techniques of ordinary quantum field theory in curved space-time. In this section, we suggest that the CMBR anisotropies could also carry some signatures of quantum gravity but, this time, originating from the non perturbative regime [28]. Obviously, the price to pay is that the following considerations are much more speculative than the rest of this review article but the hope is to learn about quantum gravity, maybe in the non-linear regime. Therefore, it seems that the potential reward is worth the speculation.

The inflationary trans-Planckian issue is based on a very simple remark [28]. If we assume a model, for instance a potential of the type given by (21) (here, we choose $n = 4$ to be concrete), then one can calculate the coupling constant λ_n. For this purpose, it is convenient to express everything in terms of N_*, the number of e-folds before the end of inflation at which the modes crossed out the Hubble radius, see (148). The corresponding value of the inflaton field is given by $\varphi_*^2 = m_{\rm Pl}^2(N_* + 1)/\pi$. Therefore, the Hubble parameter can be expressed as $H_*^2 = \lambda_4 m_{\rm Pl}^2(\varphi_*/m_{\rm Pl})^4 = \lambda_4 m_{\rm Pl}^2(N_* + 1)^2/\pi^2$. Finally, since the slow-roll parameter ϵ is given by $\epsilon = (N_* + 1)^{-1}$, one arrives at

$$\frac{H_*^2}{\epsilon m_{\rm Pl}^2} = \frac{1}{\pi^2}\lambda_4(N_* + 1)^3 \ . \tag{164}$$

The scale of inflation only enters the above equation through N_* and the corresponding dependence is logarithmic, see (148) hence very mild. One can thus use this formula to determine the coupling constant almost independently of $H_{\rm inf}$. Using (159) for $\ell = 2$ and the link between Q and C_2, one finds that $\lambda_4 \simeq 10^{-13}$, where we have used $N_* \simeq 60$. As already mentioned, this means that the total number of e-folds is huge, $N_{\rm T} \simeq 4.9 \times 10^8$. As a result, the Hubble radius today, $\ell_{\rm H} = 10^{61}\ell_{\rm Pl}$ $(h = 0.5)$, where $\ell_{\rm Pl}$ is the Planck length, was equal to $\simeq {\rm e}^{-10^8}\ell_{\rm Pl} \simeq 10^{-4.7\times 10^7}\ell_{\rm Pl}$ at the beginning of inflation, i.e., very well below the Planck length!

One can view the problem differently and ask how many e-folds before the end of inflation a given scale was equal to the Planck length. The answer can be easily calculated from (148) and reads

$$N_{\rm Pl}(\lambda) = N_*(\lambda) - \log_{10}\left(\frac{H_{\rm inf}}{m_{\rm Pl}}\right) \times \ln 10 \tag{165}$$

$$\simeq \ln\left(\frac{\lambda}{\ell_{\rm H}}\right) + \left[29 - \log_{10}\left(\frac{T_{\rm RH}}{m_{\rm Pl}}\right)\right] \times \ln 10\,. \tag{166}$$

If one takes the fiducial values $H_{\rm inf} \simeq 10^{14}$GeV, $T_{\rm RH} = M_{\rm inf} \simeq 10^{16.5}$GeV, one finds that the Planckian region was reached only 11 e-folds before the modes crossed out the horizon during inflation, see Fig. 1. For instance, for the mode $\lambda = \ell_{\rm H}$, this means 70 e-folds before the end of inflation. Of course, if the scale of inflation is smaller, then the number of e-folds before the exit of the Planckian region and the exit of the horizon can be bigger.

The following point should also be emphasized. At the time at which the modes of astrophysical interest today exit the Planckian region, the value of the Hubble parameter is generically well-below the Planckian mass. This means that the use of a classical FLRW background is well justified. The trans-Planckian problem concerns only the fluctuations and has to do with the fine structure of the Universe or with the "Planckian foam" but does necessitate a full quantum gravity description of the evolution of the underlying manifold (for instance, one does not need quantum cosmology).

Having in mind the above considerations, the trans-Planckian problem of inflation consists in the following [28]. It is likely that the framework of standard quantum field theory described in the previous section and used in order to establish what the predictions of inflation are breaks down when the modes under consideration have a wavelength smaller than the Planck length. Therefore, there is the danger that the so far successful predictions of inflation are in fact based on a theory used outside its domain of validity. In other words, there is the problem that the predictions of inflation could in fact depend on physics on scales shorter than the Planck length, a physics which is clearly largely unknown.

Is it really so? In trying to answer this question we immediately face the problem that the trans-Planckian physics is presently unknown and that, as a consequence, it is a priori impossible to study its influence on the inflationary predictions. To circumvent this difficulty, one studies the robustness of inflationary predictions to ad-hoc ("reasonable") changes in the standard quantum field theory framework supposed to mimic the modifications caused by the actual theory of quantum gravity. If the predictions are robust to some reasonable changes, then there is the hope that they will be robust to the modifications induced by the true theory of quantum gravity. On the other hand, if the predictions are not robust, the knowledge of the exact theory seems to be required in order to predict exactly what the changes are. The next question is of course which kind of modifications can we introduce in the theory in order to test its robustness? Many proposals have been made and discussed recently in the literature [28, 29, 30, 31]. Here, we concentrate on two possibilities: the modified dispersion relation and the so-called "minimal" approach.

6.1 Modified Dispersion Relations

Let us start with the modified dispersion relations. The term k^2 in (147) originates from the use of the standard dispersion relation $\omega_{\rm phys} = k_{\rm phys}$. In condensed matter physics, it is known that the dispersion relation starts departing from the linear relation $\omega = k$ on scales of the order of the atomic separation: the mode feels the granular nature of matter. In the same way, one can expect the dispersion relation to change when the mode starts feeling the discreteness of space-time on scales of the order of the Planck (string) length. Therefore, our method is to replace the linear dispersion relation $\omega_{\rm phys} = k_{\rm phys}$ by a non standard dispersion relation $\omega_{\rm phys} = \omega_{\rm phys}(k)$, this non linear relation having of course the property that $\omega_{\rm phys} \simeq k_{\rm phys}$ for $k \ll k_{\rm C}$ where $k_{\rm C}$ is a new scale introduced in the theory which could be, for instance the string scale. In the context of cosmology, this amounts to replacing the square of the comoving wavenumber k^2 with

$$k^2 \rightarrow k_{\rm eff}^2(k,\eta) \equiv a^2(\eta)\omega_{\rm phys}^2\left[\frac{k}{a(\eta)}\right] . \tag{167}$$

Therefore, this implies that we now deal with a time-dependent dispersion relation, a result first obtained in [28]. As a consequence, the equation of motion (147) now takes the form

$$\mu'' + \left[k_{\rm eff}^2(k,\eta) - \frac{z''}{z}\right]\mu = 0 . \tag{168}$$

The effect of the new physics is to change the time-dependent frequency $\omega(k,\eta)$ of the parametric oscillator. Let us remark that a more rigorous derivation of this equation, based on a variational principle, has been provided in [30].

Then, the only question is whether the fact that we now have a new time-dependent frequency can modify the spectrum $k^3|\mu|^2$ or not? As we now demonstrate, this depends on whether the evolution of the modes is adiabatic or not in the trans-Planckian region. Indeed, if the dynamics is adiabatic throughout (in particular if the z''/z term is negligible), the WKB approximation holds and the solution is always given by

$$\mu(\eta) \simeq \frac{1}{\sqrt{2k_{\rm eff}(k,\eta)}}\exp\left[-i\int_{\eta_{\rm ini}}^{\eta} k_{\rm eff}(k,\tau){\rm d}\tau\right] , \tag{169}$$

where $\eta_{\rm ini}$ is some initial time. Therefore, if we start with a positive frequency solution only and uses this solution, one finds that no negative frequency solution appears. Deep in the region where $k_{\rm eff} \simeq k$, i.e. for $k \ll k_{\rm C}$, the solution becomes

$$\mu(\eta) \simeq \frac{1}{\sqrt{2k}}\exp\left[-ik\eta - i\int_{\eta_{\rm ini}}^{\eta_1} k_{\rm eff}(k,\tau){\rm d}\tau\right] , \tag{170}$$

where η_1 is the time at which $k_{\rm eff} \simeq k$. Up to an "accumulated" phase which will disappear when we calculate the modulus $|\mu|^2$, we recover the standard vacuum solution $\mathrm{e}^{-ik\eta}/\sqrt{2k}$ and hence the standard spectrum. We have thus identified the criterion which controls whether the spectrum will be changed or not: in order to get a modification, the dispersion relation in the trans-Planckian region must be such that the WKB approximation is violated. This constrains the shape of the modified dispersion relation. It is possible to give the conditions for violation of the WKB approximation. Given an equation of the form $\mu'' + \omega^2\mu = 0$ (in the present context, one has $\omega^2 = k_{\rm eff}^2 - z''/z$), the WKB approximation is valid if the following quantity is small in the trans-Planckian region [32]

$$\left|\frac{Q}{\omega^2}\right| \ll 1 , \tag{171}$$

where Q is defined by the following expression $Q = 3(\omega')^2/(4\omega^2) - \omega''/(2\omega)$. Then, one can insert in the previous expression one's favorite dispersion relation ans see whether this leads to a new spectrum. This has been done recently in the literature, see [29]. For instance, one can show that the dispersion relations introduced in [33, 34] do not lead to any modification. An example where modifications are present has been studied in [30]. However, it remains to be studied whether this can be made compatible with other studies on the subject, in particular those using astrophysical observations to constraint the deviations from the law $\omega = k$ [35]. Rather than studying these examples in great details, we now turn to a new way of modeling the trans-Planckian regime.

6.2 The Minimal Approach

Modifying the dispersion relation is equivalent to changing the form of the equation of motion for the perturbations. The minimal approach consists in working with the same equation of motion (with a standard dispersion relation hence the name "minimal approach") but with modified initial conditions. For a given Fourier mode, the initial conditions are fixed when the mode emerges from the trans-Planckian region, i.e. when its wavelength becomes equal to a new fundamental characteristic scale $\ell_{\rm C} = 1/k_{\rm C}$. The time η_k of mode "appearance" with comoving wavenumber k, can be computed from the condition

$$\lambda(\eta_k) = \frac{2\pi}{k}a(\eta_k) = \ell_{\rm C} \equiv \frac{2\pi}{M_{\rm C}}, \tag{172}$$

which implies that η_k is a function of k. This has to be compared with the standard inflationary calculations where the initial time is taken to be $\eta_k = -\infty$ for any Fourier mode k and where, in a certain sense, the initial time does not depend on k. Then, a crucial question is in which state the Fourier mode is created at the time η_k (here, we cannot take the limit $k\eta \to -\infty$ anymore). The only requirement is that, if we send the new scale $M_{\rm C}$ to infinity (i.e.

if there is no trans-Planckian region), then one must recover the standard WKB vacuum. Therefore, the most general parametrization of these initial conditions read

$$\mu(\eta_k) = \mp \frac{c_k + d_k}{\sqrt{2\omega_{\mathrm{S,T}}(\eta_k)}} \frac{4\sqrt{\pi}}{m_{\mathrm{Pl}}}\,, \quad \mu'(\eta_k) = \pm i \sqrt{\frac{\omega_{\mathrm{S,T}}(\eta_k)}{2}} \frac{4\sqrt{\pi}(c_k - d_k)}{m_{\mathrm{Pl}}}\,. \quad (173)$$

where the coefficients c_k and d_k are a priori two arbitrary complex numbers satisfying the condition $|c_k|^2 - |d_k|^2 = 1$ and which can be expanded as

$$c_k = 1 + y\sigma_0 + \cdots \qquad d_k = x\sigma_0 + \cdots\,, \quad (174)$$

where $\sigma_0 \equiv H/M_{\mathrm{C}}$. When M_{C} is sent to infinity then $\sigma_0 \to 0$, $c_k = 1$, $d_k = 0$ and, indeed, we recover the standard vacuum. Since there are two energy scales in the problem, namely the Hubble parameter H during inflation and the new scale M_{C}, it is natural that the final result is expressed in terms of their ratio H/M_{C}, which is typically a small parameter. The parameters x and y are considered as free parameters that are not fixed by any existing well-established theories except, as already mentioned above, that they should be such that the relation $|c_k|^2 - |d_k|^2 = 1$ is satisfied. One easily shows that this implies $y + y^* = 0$ at leading order in σ_0. Expanding everything in terms of σ_0, one arrives at [31]

$$\begin{aligned}
k^3 P_\zeta &= \frac{H^2}{\pi\epsilon m_{\mathrm{Pl}}^2}\Biggl\{1 - 2(C+1)\epsilon - 2C(\epsilon-\delta) - 2(2\epsilon-\delta)\ln\frac{k}{k_*} - 2|x|\sigma_0 \\
&\quad \times \left[1 - 2(C+1)\epsilon - 2C(\epsilon-\delta) - 2(2\epsilon-\delta)\ln\frac{k}{k_*}\right] \\
&\quad \times \cos\left[\frac{2}{\sigma_0}\left(1 + \epsilon + \epsilon\ln\frac{k}{a_0 M_{\mathrm{C}}}\right) + \varphi\right] \\
&\quad - 2|x|\sigma_0\pi(2\epsilon-\delta)\sin\left[\frac{2}{\sigma_0}\left(1 + \epsilon + \epsilon\ln\frac{k}{a_0 M_{\mathrm{C}}}\right) + \varphi\right]\Biggr\}, \\
k^3 P_h &= \frac{16H^2}{\pi m_{\mathrm{Pl}}^2}\Biggl\{1 - 2(C+1)\epsilon - 2\epsilon\ln\frac{k}{k_*} - 2|x|\sigma_0\left[1 - 2(C+1)\epsilon - 2\epsilon\ln\frac{k}{k_*}\right] \\
&\quad \times \cos\left[\frac{2}{\sigma_0}\left(1 + \epsilon + \epsilon\ln\frac{k}{a_0 M_{\mathrm{C}}}\right) + \varphi\right] \\
&\quad - 2|x|\sigma_0\pi\epsilon\sin\left[\frac{2}{\sigma_0}\left(1 + \epsilon + \epsilon\ln\frac{k}{a_0 M_{\mathrm{C}}}\right) + \varphi\right]\Biggr\}\,,
\end{aligned} \quad (175)$$

where φ is the argument of the complex number x, i.e $x \equiv |x|e^{i\varphi}$. These expressions should be compared with (151) and (152). The effect of the trans-Planckian corrections is clear: superimposed oscillations in the power spectra have appeared. The magnitude of the trans-Planckian corrections are linear in the parameter σ_0 and their amplitude is given by $|x|\sigma_0$. The wavelength of the oscillations can be expressed as $\Delta k/k = \sigma_0\pi/\epsilon$.

The above calculation provides us with an explicit example where the observational predictions of inflation are modified by the trans-Planckian physics. Let us now study this question in more details. Using (91), one can evaluate the modifications of the multipoles moments caused by the trans-Planckian corrections. In the limit $\epsilon/\sigma_0 \gg \ell$, one gets [36, 37]

$$\ell(\ell+1)C_\ell \simeq \frac{2H^2}{25\epsilon m_{\mathrm{Pl}}^2}(1-2\epsilon)\left\{1+\sqrt{\pi}\frac{|x|\sigma_0\ell(\ell+1)}{(\epsilon/\sigma_0)^{5/2}}\right.$$
$$\left.\times\ \cos\left[\pi\ell+\frac{2}{\sigma_0}\left(1+\epsilon\ln\frac{\epsilon/\sigma_0}{a_0 M_{\mathrm{C}} r_{\mathrm{lss}}}\right)+\varphi-\frac{\pi}{4}\right]\right\}. \quad (176)$$

This expression should be compared with (159). The oscillations in the power spectra are transfered to the multipole moments, at least at relatively small scales. At large ℓ, or for not too small values of σ_0, the above equation quickly becomes invalid and an accurate estimation can be made only with the help of numerical calculations. The result in plotted in Fig. 3 for the temperature fluctuations but also for the polarization, for details see [36, 37]. In those references, a detailed comparison of the trans-Planckian signal with the recently released high accuracy WMAP data has been performed. The main result is that, with the oscillations taken into account, it is possible to decrease the χ^2 significantly. Instead of $\chi^2 \simeq 1431$ for 1342 degrees of freedom for the standard slow-roll power spectra, one now obtains $\chi^2 \simeq 1420$ for 1340 degrees of freedom, i.e. $\Delta\chi^2 \simeq 10$ compared to WMAP one. The reason for such an important improvement of the χ^2 is due to the presence of the oscillations which permit a better fit of the cosmic variance outliers at small scales. The main question is of course the statistical significance of this result. In [36, 37], the so-called F-test has been used and indicates that the result is significant. However, it is clear that other statistical tests, a complete exploration of the parameter space and, of course, new data, should be used before one can really conclude that superimposed oscillations are really present in the CMBR multipole moments. A fair description of the present situation is that there seems to be a hint for an interesting feature in the CMBR data and that, maybe, this feature is a signature of very high energy physics (it is clear that the oscillations, if their presence is confirmed, could have another physical origin).

Finally, we would like to conclude by a comment on the back-reaction problem. This question is crucial for the consistency of the approach used before. It is clear that the energy density of the perturbations must be smaller or equal than that of the inflationary background. This leads to the condition $|x| \le \sqrt{3\pi}m_{\mathrm{Pl}}/M_{\mathrm{C}}$ which amounts to

$$|x|\sigma_0 \le 10^4 \times \frac{\sigma_0^2}{\sqrt{\epsilon}} \,. \quad (177)$$

It is important to emphasize that the above constraint is only a sufficient condition, but by no means, unless proved otherwise, a necessary condition.

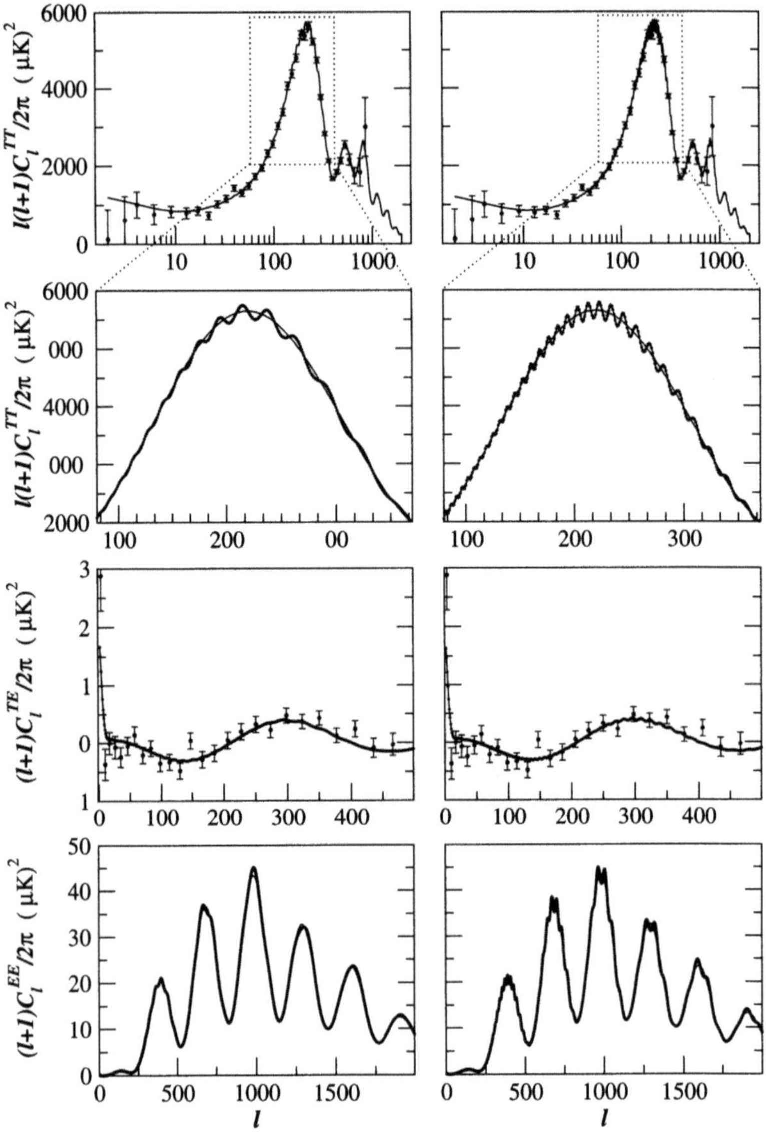

Fig. 3. Angular TT, TE and EE power spectra for two different trans-Planckian models, one with low frequency (LF) superimposed oscillations, the other with high frequency (HF) oscillations, for details see [36, 37]. A zoom of the temperature multipole moments in the first Doppler peak region is also shown (*black curve*) and compared with the standard slow-roll prediction (*blue curve*) calculated with the same cosmological parameters

In general, this constraint is difficult to satisfy. Some of the best fits described above suffer from this back-reaction problem, see [30, 36, 37]. In fact, the above formula expresses a generic difficulty of the trans-Planckian question, this difficulty being present regardless of the approach used in order to model the new physics. The presence of trans-Planckian corrections means the presence of particles (with respect to the standard vacuum) the energy density of which is very easily of the order of the background energy density. On the other hand, if we try to satisfy the back-reaction constraint then the signal very easily becomes tiny and, hence, non observable. A major advance, which would allow us to escape the previous vicious circle, would be to calculate explicitly the effect of the back-reaction. Unfortunately, for the moment, this is still an open question and more work is required to tackle this very important task.

Acknowledgments

It is a pleasure to thank G. Amelino-Camelia and J. Kowalski-Glikman for inviting me to lecture at this school and for their hospitality. I am very grateful to all the participants for very interesting discussions. I also thank P. Brax and M. Lemoine for careful reading of the manuscript.

References

1. A. Guth, Phys. Rev. D **23**, 347 (1981).
2. A. Linde, Phys. Lett. B **108**, 389 (1982); A. Albrecht and P. J. Steinhardt, Phys. Rev. Lett. **48**, 1220 (1982); A. Linde, Phys. Lett. B **129**, 177 (1983).
3. J. Martin, Proceedings of the XXIV Brazilian National Meeting on Particles and Fields, Caxambu, Brazil, (2004), `astro-ph/0312492`.
4. D. Lyth and A. Riotto, Phys. Rept. **314**, 1 (1999), `hep-ph/9807278`.
5. A. Linde, Talk at the Nobel Symposium "Cosmology and String Theory", (2003), `hep-th/0402051`.
6. M. Turner, Phys. Rev. D **28**, 1243 (1983).
7. L. Kofman, A. Linde and A. Starobinsky, Phys. Rev. D **56**, 3258 (1997), `hep-ph/9704452`.
8. V. F. Mukhanov, H. A. Feldman, and R. H. Brandenberger, Phys. Rep. **215**, 203 (1992).
9. J. Stewart, Class. Quantum Grav. **7**, 1169 (1990).
10. J. A. Bardeen, Phys. Rev. D **22**, 1882 (1980).
11. J. Martin and D. J. Schwarz, Phys. Rev. D **57**, 3302 (1998), `gr-qc/970449`.
12. R. Kerner, Gen. Rel. Grav. **9**, 257 (1978).
13. D. H. Lyth, Phys. Rev. D **31**, 1792 (1985).
14. L. P. Grishchuk, Zh. Eksp. Teor. Fiz **67**, 825 (1974).
15. G. F. Smooth, et al., Astrophys. J. **396**, L1 (1992).
16. P. de Bernardis, lectures given at the School.
17. R. K. Sachs and A. M. Wolfe, Astrophys. J. **147**, 73 (1967).
18. M. Panek, Phys. Rev. D **49**, 648 (1986).

19. C. Cohen-Tannoudji, J. Dupont-Roc and G. Grynberg, "Photons et atomes, Introduction a l'électrodynamique quantique", Editions du CNRS, (1987).
20. E. Kolb and M. Turner, "The Early Universe", Frontier in Physics, Perseus Publishing, (1990).
21. J. Martin and D. J. Schwarz, Phys. Rev. D **62**, 103520 (2000), astro-ph/9911225; J. Martin, A. Riazuelo and D. J. Schwarz, Astrophys. J. **543**, L99 (2000), astro-ph/0006392.
22. C. L. Bennet et al., Astrophys. J. Suppl. **148**, 1 (2003), astro-ph/0302207; G. Hinshaw et al., Astrophys. J. Suppl. **148**, 135 (2003), astro-ph/0302217; L. Verde et al., Astrophys. J. Suppl. **148**, 195 (2003), astro-ph/0302218; H. V. Peiris et al., Astrophys. J. Suppl. **148**, 213 (2003), astro-ph/0302225.
23. A. Albrecht, P. Ferreira, M. Joyce and T. Prokopec, Phys. Rev. D **50**, 4807 (1994); D. Polarski and A. Starobinsky, Class. Quantum Grav. **13**, 377 (1996).
24. L. P. Grishchuk and J. Martin, Phys. Rev. D **56**, 1924 (1997), astro-ph/9702018.
25. A. Lewis, A. Challinor and A. Lasenby, Astrophys. J. **538**, 473 (2000), astro-ph/9911177, http://camb.info; S. Leach, http://astronomy.sussex.ac.uk/sleach/inflation/camb-inflation.html.
26. S. Leach and A. Liddle, Phys. Rev. **D 68**, 123508 (2003), astro-ph/0306305.
27. S. Dodelson, hep-ph/0309057.
28. J. Martin and R. H. Brandenberger, Phys. Rev. **D 63**, 123501 (2001), hep-th/0005209; R. H. Brandenberger and J. Martin, Mod. Phys. Lett. **A 16**, 999 (2001), astro-ph/0005432.
29. J. C. Niemeyer, Phys. Rev. **D 63**, 123502 (2001), astro-ph/0005533; A. Kempf, Phys. Rev. **D 63**, 083514 (2001), astro-ph/0009209; R. Easther, B. R. Greene, W. H. Kinney and G. Shiu, Phys. Rev. **D 64**, 103502 (2001), hep-th/0104102; R. Easther, B. R. Greene, W. H. Kinney and G. Shiu, Phys. Rev. **D 67**, 063508 (2003), hep-th/0110226; A. Kempf and J. C. Niemeyer, Phys. Rev. **D 64**, 103501 (2001), astro-ph/0103225; R. H. Brandenberger and P. M. Ho, Phys. Rev. **D 66**, 023517 (2002), hep-th/0203119; S. F. Hassan and M. S. Sloth, hep-th/0204110; F. Lizzi, G. Mangano, G. Miele and M. Peloso, JHEP **0206**, 049 (2002), hep-th/0203099; U. H. Danielsson, Phys. Rev. **D 66**, 023511 (2002), hep-th/0203198; R. Easther, B. R. Greene, W. H. Kinney and G. Shiu, Phys. Rev. **D 66**, 023518 (2002), hep-th/0204129; G. L. Alberghi, R. Casadio and A. Tronconi, gr-qc/0303035; C. Armendáriz-Picón and E. A. Lim, hep-th/0303103.
30. M. Lemoine, M. Lubo, J. Martin and J. P. Uzan, Phys. Rev. **D 65**, 023510 (2002), hep-th/0109128.
31. J. Martin and R. H. Brandenberger, Phys. Rev. **D 68**, 063513 (2003), hep-th/0305161.
32. J. Martin and D. J. Schwarz, Phys. Rev. D **67**, 083512 (2003), astro-ph/0210090.
33. W. Unruh, Phys. Rev. **D 51**, 2827 (1995).
34. S. Corley and T. Jacobson, Phys. Rev. **D 54**, 1568 (1996), hep-th/9601073; S. Corley, Phys. Rev. **D 57**, 6280 (1998), hep-th/9710075.
35. T. Jacobson, S. Liberati, D. Mattingly, this volume.
36. J. Martin and C. Ringeval, Phys. Rev. **D 69**, 064406 (2004), astro-ph/0310382.
37. J. Martin and C. Ringeval, astro-ph/0402609.

CPT Violation and Decoherence in Quantum Gravity

N.E. Mavromatos

King's College London, Department of Physics, Strand, London WC2R 2LS, U.K.
Nikolaos.Mavromatos@kcl.ac.uk

In these lectures I review, in as much pedagogical way as possible, various theoretical ideas and motivation for violation of CPT invariance in some models of Quantum Gravity, and discuss the relevant phenomenology. Since the subject is vast, I pay particular emphasis on the CPT Violating decoherence scenario for quantum gravity, due to space-time foam. In my opinion this seems to be the most likely scenario to be realised in Nature, should quantum gravity be responsible for the violation of this symmetry. In this context, I also discuss how the CPT Violating decoherence scenario can explain experimental "anomalies" in neutrino data, such as LSND results, in agreement with the rest of the presently available data, without enlarging the neutrino sector.

1 Introduction and Summary

Next year, Special Relativity celebrates a century of enormous success, having passed many stringent experimental tests, in both its classical and quantum versions (relativistic quantum field theories in flat space times). Unfortunately, the same is not true for its curved-space counterpart, General Relativity. A consistently quantized theory of gravity, that is a dynamical theory of curved geometries themselves, still remains a mystery. Despite the enormous effort invested for this purpose on behalf of the scientic community over the past ninety years, Quantum Gravity is still far from being understood as a physical theory.

Of course, elegant and mathematically consistent models, such as string or, better, brane theory [1], have been developed to a great detail from a mathematical viewpoint. Nevertheless there are still many fundamental issues and questions which remain unresolved. For instance, the complete process of evaporation of a black hole, or the inverse process of collapsing matter to form a Black Hole, are not completely understood in string theory. The counting of microstates and verification of the Hawking-Bekenstein entropy/area law have been understood mathematically only in specific cases of extremal black

N.E. Mavromatos: *CPT Violation and Decoherence in Quantum Gravity*, Lect. Notes Phys. **669**, 245–320 (2005)
www.springerlink.com

holes, and probably this is the only case that can be studied rigorously in such a framework. Other issues, like the possible existence of space-time foam, that is microscopic singular fluctuations of the (quantum) geometry, which give the space time a “foamy”, topologically non trivial and possibly non-continuous structure at Planck scales (10^{-35} m), still remain far from being resolved in the context of string theory.

In [2] it was suggested that a consistent mathematical framework for dealing with such issues in the context of string theory was the Liouville non-critical srting theory approach, involving strings propagating in non-conformal space-time backgrounds. This violation of conformal symmetry, which lies at the cornerstone of critical string theory, is remedied by the non-decoupling of the Liouville mode, which enters as a whole new target space dimension. In certain models of stringy foam, this extra dimension has time-like signature, and hence it can be identified with a target time, thereby giving the time coordinate a fundamentally irreversible nature, as a result of specific properties of the Liouville dynamics. Indeed, the latter acts as a local renormalization-group scale on the world-sheet of the string, and as such is irreversible. This fundamental irreversibility of non-critical string theory makes it analogous to non-equilibrium systems in field theory. From this point of view, then, critical strings are viewed as asymptotic “equilibrium points” in string theory space.

Alternative approaches to Quantum Gravity, on the other hand, such as the loop gravity approach [3], which has the ambition of formulating a space-time background independent quantum theory of Gravity, have only relatively recently began to deal with non-flat space times (such as those with cosmological constant) or highly curved ones (black holes etc.), and hence their full potential in dealing with the above issues is still not explored [4]. These are very elegant theories from a geometrical viewpoint, which are based on the analogy of gravity to non Abelian gauge theories. Understanding the rôle of matter in such gravity theories is a pressing task, in order to give such mdoels phenomenological relevance. In addition to loop gravity, non commutative geometry [5] is another mathematically elegant route that would certainly prove to be relevant for a dynamical quantum theory of space time at Planck scales, where space time may be discrete. This approach, although existing for some time, has only recently started to be paid attention by the bulk of the theoretical physicists, with a plethora of applications, ranging from field theoretic models to string and brane theories.

A theoretical model, however, no matter how detailed and elegant it might be, does not become a *physical* theory unless it makes some form of contact with *experiment*. Thus, to understand and be guided in our quest for quantum gravity we need experimentally testable or falsifiable predictions. Critical strings, or other approaches to quantum gravity, which respect all local symmetries of classical General Relativity, did not make any predictions for low-energy theories which could be testable in the foreseeable future. The reason is simple: the coupling constant of gravity, the Newton constant $G_N \propto 1/M_P^2$ (in four dimensions) is very small, and, on account of local Lorentz symmetry and general covariance, quantities of possible experimental interest, such as cross sections and probabilities, would be characterised by quantum gravitational loop corrections which would be proportional to some power of curvature tensors. The latter having dimensions of momentum squared, would imply that such quantities would be suppressed at least by the inverse square (and most likely by higher powers) of the Planck Mass scale. This would make the prospects for detection of such quantum gravity effects difficult, if not impossible, for the foreseeable future. Of course this does not necessarily mean that such approaches are physically incorrect, what it means is that, even if they represent reality, we would have no way of testing them in the foreseeable future, and as such they would remain solely mathematically consistent models.

On the other hand, recently, more and more physicists contemplate the idea that some of the fundamental symmetries or laws that govern classical General and Special Relativity, such as *linear Lorentz symmetry*, or principles such as *the equivalence principle*, may not be valid in a full quantum theory of gravity. If true, then, this would probably imply that the above-mentioned Planck-mass strong suppression factors could be modified in such a way that quantum gravity effects are enhanced, thereby leading to some testable/falsifiable predictions in the near future. For instance, in the non-critical string approach to quantum gravity advocated in [2], deviation from conformal invariance due to peculiar backgrounds in string theory, including foamy ones, imply in some models at an effective low-energy field theory level, *modified dispersion relations for photons or at most for some electrically neutral gauge bosons.* Such modifications dot not occur not for charged probes or in general chiral matter [7], thereby violating a form of the equivalence principle, in the sense of the non-universality of gravity effects. In such models it is a gauge symmetry that protects the dispersion relation of charged or chiral matter probes, which, unlike photons, do not interact with space time defects in the foam, the latter consisting of point-like branes in string theory [8]. The modification to the dispersion relations due to such quantum gravity effects are suppressed only by a single power of Planck Mass [6]. Such *minimal suppression models for photons* are not far from being tested, for instance by future Gamma Ray Burst astronomy [9, 10]. On the other hand, models of quantum gravity foam with universal modified dispersions linearly suppressed by the Planck Mass scale are already excluded by means of astrophysical observations of Synchrotron radiation from Crab Nebula [11, 12], and one is not far from reaching sensitivities quadratic to inverse Planck mass [7].

In this context, interesting "bottom-up" approaches to quantum gravity have been proposed and developed rigorously, such as the Doubly-Special Relativity (DSR) theories [13], which are at the focus of this meeting. According to such approaches, the conventional Lorentz symmetry of flat Minkowski space time is not valid, but instead one has a symmetry under non-linear extensions of the Lorentz transformations. Such non-linear extensions are not unique, and this poses an interesting theoretical challenge for these models.

The basic idea behind such theories is that the Planck scale should be observer independent, and hence such non-linear models are characterised not only by the invariance under frame changes of the dimensionless speed of light in vacuo, but also by the frame-invariance of a dimensionful length scale, the Planck length. For this reason, although at present formulated in flat space times, such non-linear extensions of Lorentz symmetry are viewed as a prelude to more complete models of quantum gravity, where the local group is not the conventional (linear) Lorentz, thereby violating the strong form of the equivalence principle. However it remains to be proven whether such models are viable as candidates for a complete and realistic theory of quantum gravity. In other lectures in this meeting we shall hear more about the mathematical foundations and properties of such theories [14], and their phenomenology [12, 15, 16], where we refer the reader for details.

In all approaches mentioned so far as candidate theories for quantum gravity there is a common feature, associated with the violation of a theorem whose validity characterises all consistent flat-space time relativistic quantum field theories known to date. This is the CPT theorem [17, 18, 19, 20]. The violation of this (discrete) space-time symmetry may have important phenomenological implications for low energy physics, and indeed one is prompted immediately to think that this may be a way of testing or falsifying experimentally various theoretical models of quantum gravity entailing such a violation.

There is a number of fundamental questions, however, that one has to ask before embarking on a study of the phenomenology of CPT Violation: (i) What are the theories which allow for CPT breaking?, (ii) How (un)likely is it that somebody, someday finds CPT Violation in the Laboratory, and why?, (iii) What formalism does one has to adopt? Indeed, since our current phenomenology of particle physics is based on CPT invariance, how can we be sure of observing CPT Violation and not something else? And finally, (iv) there does not seem to be a single "figure of merit" for CPT violation. Then how should we compare various "figures of merit" of CPT tests (e.g. direct mass measurement between matter and antimatter (e.g. $K^0 - \overline{K}^0$ mass difference a la CPLEAR), quantum decoherence effects, modifications to Einstein-Podolsky-Rosen (EPR) states in meson factories, neutrino mixing, electron g-2 and cyclotron frequency comparison, neutrino spin-flavour conversion etc.)

In some of these questions I shall try to give answers in the context of the present set of Lectures. I shall not try to present a complete overview of phenomenological tests of CPT Invariance, however, because the subject is vast, and already occupies a considerable part of the published literature. In these lectures I will place the emphasis on neutrino tests of CPT invariance, because as I will argue below, in many instances neutrinos seem to provide at present the best bounds on possible CPT violation. However, I must stress that, precisely because CPT violation is a highly model dependent feature of some approaches to quantum gravity (QG), there may be models in which the sensitivity of other experiments on CPT violation, such as astrophysical experiments, is superior to that of current neutrino experiments. For this reason I will also give a brief outline of alternative tests of CPT violation.

My lectures will focus on the following three major issues:

(a) *What is CPT Symmetry*: I will give a definition of what we mean by CPT invariance, and under what conditions this invariance holds.

(b) *Wny CPT Violation?*: Currently there are various *Quantum Gravity Models* which may *violate* Lorentz symmetry and/or quantum coherence (unitarity etc.), and through this CPT symmetry:
   (i) space-time foam [21] (local field theories [22], non-critical strings [2] etc.),
   (ii) (non supersymmetric) string-inspired standard model extension with Lorentz Violation [23],
   (iii) Loop Quantum Gravity [3].
   (iv) *CPT violation* may also occur at a *global scale*, *cosmologically* [35], as a result of a cosmological constant in the Universe, whose presence may jeopardize the definition of a standard scattering matrix.

(c) *How can we detect CPT Violation?* : Here is a current list of most sensitive particle physics probes for CPT tests:
   (i) *Neutral Mesons*: Kaons [24, 25], B-mesons, and their entangled states in ϕ and B factories [26, 27, 28].
   (ii) *anti-matter factories*: antihydrogen [29] (precision spectroscopic tests on free and trapped molecules [23, 30, 31]),
   (iii) Low energy atomic physics experiments [30], including ultra cold neutron experiments in the gravitational field of the Earth.
   (iv) Astrophysical Tests (especially Lorentz-Invariance violation tests, via modified dispersion relations of matter probes etc.) [9, 10]
   (v) Neutrino Physics, on which we shall mainly concentrate in these lectures [32].

I shall be brief in my description due to space restrictions. For more details I refer the interested reader to the relevant literature. I will present some elementary proofs of theorems that will be essential for the formalism of CPT Violation and its phenomenology. I will not be complete in reviewing the phenomenology of CPT violation; in my lectures I will place emphasis on a specific type of violation, that through quantum decoherence, which I believe to be the most likely one to charactrise space-time foam theories of quantum gravity; this belief is based on the fact that decoherence may be compatible with fundamental local symmetries of space time, such as Lorentz invariance [33, 34]. For completeness, however, I will also give a brief exposition of alternative ways of CPT violation, and refer the reader to some key references, where more detailed information is provided on those topics. Needless to say that I am fully aware of the vastness of the topic of CPT Violation, which grew enormously in recent years, and I realize that I might not have done a perfect job here; I should therefore apologize beforehand for possible omissions in references, and topics, but this was not intentional. I do hope, however, that I give here a rather satisfactory representation of the current situation regarding this important research topic.

2 Theoretical Motivation for CPT Violation and Formalism

2.1 The CPT Theorem and How It May Be Evaded

The CPT theorem refers to quantum field theoretic models of particle physics, and ensures their invariance under the successive operation (in any order) of the following *discrete* transformations: **C**(harge), **P**(arity=reflection), and **T**(ime reversal). The *invariance* of the Lagrangian density $\mathcal{L}(x)$ of the field theory under the combined action of *CPT* is a property of any quantum field theory in a *Flat* space time which respects: *(i) Locality*, *(ii) Unitarity* and *(iii) Lorentz Symmetry.*

$$\Theta\mathcal{L}(x)\Theta^{\dagger} = \mathcal{L}(-x) \ , \ \Theta = CPT \ , \ \mathcal{L} = \mathcal{L}^{\dagger} \tag{1}$$

The theorem has been suggested first by Lüders and Pauli [17], and also by John Bell [18], and has been put on an axiomatic form, using Wightman axiomatic approach to relativistic (Lorentz invariant) field theory, by Jost [19]. Recently the Lorentz covariance of the Wightmann (correlation) functions of field theories [20] as an essential requirement for a proof of CPT has been re-emphasized in [36], in a concise simplified exposition of the work of Jost. The important point to notice in that proof is the use of flat-space Lorentz covariance, which allows the passage onto a momentum (Fourier) formalism. Basically, the Fourier formalism employs appropriately superimposed plane wave solutions for fields, with four-momentum p_{μ}. The proof of CPT, then, follows by the Lorentz covariance transformation properties of the Wightman functions, and the unitarity of the Lorentz transformations of the various fields.

In curved space times, especially highly curved ones with space-time boundaries, such as space-times in the (exterior) vicinity of black holes, where the boundary is provided by the black hole horizons, or space-time foamy situations, in which one has vacuum creation of microscopic (of Planckian size $\ell_P = 10^{-35}$ m) black-hole horizons [21], such an approach is invalid, and Lorentz invariance, and possibly unitarity, are lost. Hence, such models of quantum gravity violate requirements (ii) & (iii) of the CPT theorem, and hence one should expect its *violation*.

It is worthy of discussing briefly the basic mechanism by which unitarity may be lost in space-time foamy situations in quantum gravity. This is the lecturer's favorite route for possible quantum-gravity induced CPT Violation, which may hold independently of possible Lorentz invariance violations. It is at the core of the induced decoherence by quantum gravity [24, 25].

The important point to notice is that, in general, space-time may be *discrete* and *topologically non-trivial* at Planck scales 10^{-35} m, which might (but this is not necessary [33, 34]), imply Lorentz symmetry Violation (LV), and hence CPT Violation (CPTV). Phenomenologically, at a macroscopic level, such LV may lead to extensions of the standard model which violate both Lorentz and CPT invariance [23].

In addition, there may be an *environment of gravitational* degrees of freedom (d.o.f.) *inaccessible* to low-energy experiments (for example non-propagating d.o.f., for which ordinary scattering is not well defined [25]). This will lead in general to an *apparent information loss* for low-energy observers, who by definition can measure only propagating low-energy d.o.f. by means of scattering experiments. As a consequence, an apparent lack of unitarity and hence CPTV may arise, which is in principle independent of any LV effects. The loss of information may be understood simply by the mechanism illustrated in Fig. 1. In a foamy space time there is an ongoing creation and annihilation of Quantum Gravity *singular* fluctuations (e.g. microscopic (Planck size) black holes etc), which indeed implies that the observable space time is an open system. When matter particles pass by such fluctuations (whose life time is Planckian, of order 10^{-43} s), part of the particle's quantum numbers "fall into" the horizons, and are captured by them as the microscopic horizon disappears into the foamy vacuum. This may imply the exchange of information between the observable world and the gravitational "environment" consisting of degrees of freedom inaccessible to low energy scattering experiments, such as back reaction of the absorbed matter onto the space time, recoil of the microscopic black hole etc. In turn, such a loss of information will imply evolution of initially pure quantum-mechanical states to mixed ones for an asymptotic observer.

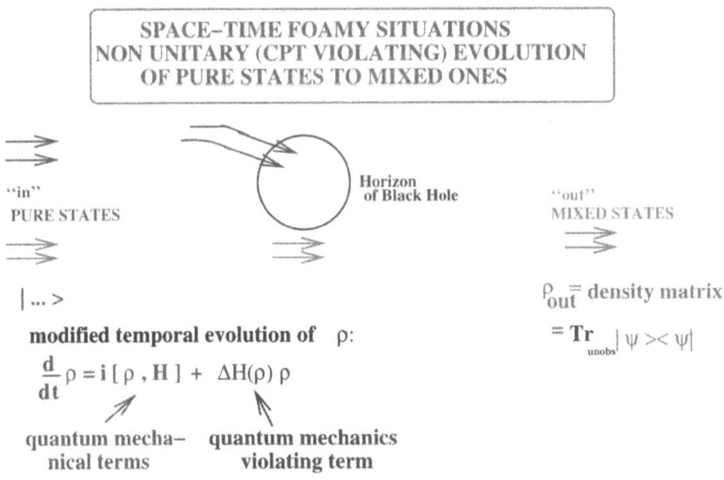

Fig. 1. A basic mechanism for loss of information in a space time foamy situation

As a result, the asymptotic observer will have to use density matrices instead of pure states: $\rho_{out} = \mathrm{Tr}_{\mathrm{unobs}}|out\rangle\langle out| = \$\ \rho_{in}\$ \neq SS^{\dagger}$, with $S = e^{iHt}$ the ordinary scattering materix. Hence, in a foamy situation the concept of the scattering matrix is replaced by that of the *superscattering matrix*, \$, introduced by Hawking [21], which is a *linear*, but *non-invertible* map between "in" and "out" density matrices; in this way, it quantifies the unitarity loss in the effective low-energy theory. The latter violates CPT due to a mathematical theorem by R. Wald, which we describe in the next Subsect. [37].

Notice that this is an effective violation, and indeed the complete theory of quantum gravity (which though is still unknown) may respect some form of CPT invariance. However, from a phenomenological point of view, this effective low-energy violation of CPT is the kind of violation we are interested in here. A word of caution is necessary at this point. Some theorists believe that quantum gravity does not entail an evolution of a pure quantum state to a mixed one, but, as is the case in some quantum mechanical decoherence models of open systems, to be discussed below, the purity of states is maintained during the quantum-gravity induced decoherent evolution. If this is the case, then CPT may be conserved in such models, provided, of course, Lorentz invariance and locality of interactions are respected.

2.2 \$ Matrix and Strong CPT Violation (CPTV)

The theorem of R. Wald states the following [37]: *if* $\$ \neq S\ S^{\dagger}$, *then CPT is violated, at least in its strong form, in the sense that the CPT operator is not well defined.*

For instructive purposes we shall give here an elementary proof. Suppose that CPT is conserved, then there exists a unitary, invertible CPT operator Θ: $\Theta\overline{\rho}_{in} = \rho_{out}$. Since the density matrix acts on a tensor product space between ket and bra vectors by definition, $\rho = \psi \otimes \overline{\psi}$, the action of Θ is defined schematically as: $\Theta = \theta\theta^{\dagger}$, with θ acting on state vectors ψ, and being anti-unitary, i.e. $\theta^{\dagger} = -\theta^{-1}$.

Asuming that such a Θ exists, we have: $\rho_{out} = \$\ \rho_{in} \to \Theta\overline{\rho}_{in} = \$\ \Theta^{-1}\overline{\rho}_{out} \to \overline{\rho}_{in} = \Theta^{-1}\$\ \Theta^{-1}\overline{\rho}_{out}$.

But $\overline{\rho}_{out} = \$\overline{\rho}_{in}$, hence:

$$\overline{\rho}_{in} = \Theta^{-1}\$\Theta^{-1}\$\overline{\rho}_{in}\ . \tag{2}$$

The last relation implies that \$ has an *inverse*

$$\$^{-1} = \Theta^{-1}\$\Theta^{-1}\ , \tag{3}$$

which, however, as we explain now is *impossible*, due to the information loss in case a pure state evolves into a mixed one.

To prove [37] this last statement formally we first notice that from (3) one also obtains the relation:

$$\Theta = \$\Theta^{-1}\$\ . \tag{4}$$

Consider now a pure state density matrix $\rho_{in} = |IN\rangle\langle IN|$, which evolves to the density matrix (mixed state) $\$\rho_{in}$. As a result of (4), the mixed state $\Theta^{-1}\$\rho_{in}$ must evolve to the pure state $\Theta\rho_{in}$. However, suppose we have an out state ψ, which we obtain by the action of \$ on an IN density matrix σ, that is:

$$\$\sigma = \psi \otimes \overline{\psi} \tag{5}$$

where, as mentioned above, $\otimes$ denotes the appropriate tensor product of Hilbert spaces spanned by ket and bra vectors. One may expand σ in terms of its eigenvectors ϕ_i, corresponding physically to a weighted superposition of states that comprise the mixed state σ:

$$\sigma = \sum_i p_i \phi_i \otimes \overline{\phi}_i \ , \tag{6}$$

with p_i positive, and $\sum_i p_i = 1$. Since by definition \$ is a linear map, we have:

$$\sum_i p_i \$(\phi_i \otimes \overline{\phi}_i) = \psi \otimes \overline{\psi} \tag{7}$$

Consider now an OUT state vector χ orthogonal to ψ. Taking the expectation value of (7) in the state χ we obtain:

$$\sum_i p_i \langle \chi | \$(\phi_i \otimes \overline{\phi}_i) | \chi \rangle = 0 \tag{8}$$

Each term in (8) is non negative, due to the positive-definiteness of p_i and the positivity of the density matrix (by definition) $\$(\phi_i \otimes \overline{\phi}_i)$. Therefore, (8) implies

$$\langle \chi | \$(\phi_i \otimes \overline{\phi}_i) | \chi \rangle = 0 \tag{9}$$

for all i and all χ orthogonal to ψ. This implies

$$\$(\phi_i \otimes \overline{\phi}_i) = \psi \otimes \overline{\psi} \tag{10}$$

for all i, i.e. each initial pure state ϕ_i must evolve to the *same* final pure state ψ. In that case, $\theta^{-1}\psi$ must evolve to the final state $\theta\phi_i$ for all i. This is *impossible* if there is more than one ϕ_i, i.e. if the density matrix σ represents a *mixed* state.

Hence, in case where decoherence implies the evolution of a pure state to a mixed one, CPT *must be violated*, at least in its strong form, in the sense of Θ not being a well-defined operator, and the non existence of the inverse of \$, as discussed previoulsy. The non invertibility of \$ should not be considered as a surprise in that case, as a result of the involved loss of information in the problem. CPT symmetry, and also by the same arguments microscopic time reversal [37], fail in a dramatic way in such a case: microscopic time-reversed dynamics does not merely fail to be the same as time evolution forward in time, which would simply mean the non commutativity of the corresponding operators/generators of the symmetry with the hamiltonian of the system under consideration, but *does not exist* at all.

As I remarked before, this is my preferred way of CPTV by Quantum Gravity, given that it may occur in general independently of LV and thus preferred frame approaches to quantum gravity. Indeed, I should stress at this point that the above-mentioned gravitational-environment induced decoherence may be Lorentz invariant [33], the appropriate Lorentz transformations being slightly modified to account, for instance, for the discreteness of space time at Planck length [34]. This is an interesting topic for research, and it is by no means complete. Although the lack of an invertible scattering matrix in most of these cases implies a strong violation of CPT, nevertheless, it is interesting to demonstrate explicitly whether some form of CPT invariance holds in such cases [38]. This also includes cases with non-linear modifications of Lorentz symmetry [13], discussed in this School, which arise from the requirement of viewing the Planck length as an invariant (observer-independent) proper length in space time.

It should be stressed at this stage that, if the CPT operator is not well defined, then this may lead to a whole new perspective of dealing with precision tests in meson factories. In the usual LV case of CPTV [23], the CPT breaking is due to the fact that the CPT operator, which is well-defined as a quantum mechanical operator in this case, does not commute with the effective low-energy Hamiltonian of the matter system. This leads to mass differences between particles and antiparticles. If, however, the CPT operator is not well defined, as is the case of the quantum-gravity induced decoherence [24, 25], then, the concept of the "antiparticle" gets modified [28]. In particular, the antiparticle space is viewed as an independent subspace of the state space of the system, implying that, in the case of neutral mesons, for instance, the anti-neutral meson should not be treated as an identical particle with the corresponding meson. This leads to the possibility of novel effects associated with CPTV as regards entangled states of Einstein-Podolsky-Rosen (EPR) type, which may be testable at meson factories [28]. We shall discuss this in some detail later on.

2.3 CPT Symmetry without CPT Symmetry?

An important issue which arises at this point is whether the above violation of CPT symmetry is actually detectable experimentally. This issue has been examined in [37], where it was proposed that despite the strong CPT violation in cases where decoherence leads to an evolution of a pure state to a mixed one, there is the possibility for a softer (weaker) form of CPT invariance in such cases, compatible with the non-invertibility of $\$$.

The main idea behind such a weak form of CPT invariance is that, although in the full theory CPT is violated in the above sense, nevertheless one can still define asymptotic *pure* scattering IN and OUT states as the CPT inverse of each other. In formal terms, although in the full theory Θ is not well defined, however one can define pure states $\psi \in \mathcal{H}_{IN}$, and $\phi \in \mathcal{H}_{OUT}$ in the respective Hilbert spaces $\mathcal{H}$ of IN and OUT states, such that the following equality between probabilities $\mathcal{P}$ holds:

$$\mathcal{P}(\psi \to \phi) = \mathcal{P}(\theta^{-1}\phi \to \theta\psi) \tag{11}$$

If only pobabilities are measured experimentally, which is certainly our experience so far, then the equality (11) would imply that the strong form of CPT invariance would be undetectable experimentally.

From the point of view of the superscattering matrix \$, the equality (11) implies the following relation [37]:

$$\langle\phi|\$(\psi \otimes \overline{\psi})|\phi\rangle = \langle\theta\psi|\$(\theta^{-1}\phi \otimes \overline{\theta}^{-1}\phi)|\theta\psi\rangle \tag{12}$$

or, equivalently :

$$\$^{\dagger} = \Theta^{-1}\$\Theta^{-1} \tag{13}$$

when the action is considered on pure asymptotic states. Relation (13) is compatible with the non-existence of an inverse of \$, unless the full CPT invariance holds, which would imply unitarity of \$, i.e. $\$^{\dagger} = \$^{-1}$. Wald has argued in favour of this conclusion by considering a simple case of finite-dimension (n) Hilbert spaces of IN and OUT states, and assuming that every pure IN state evolves to the density matrix $1/n\delta^a_b$ in the OUT Hilbert space. It is clear that in this example $\$^{-1}$ does not exist, but for all ψ and ϕ the relation (11) holds, since both sides equal $1/n$.

2.4 Decoherence and Purity of States under Evolution

Since the above result of weak CPT invariance requires the purity of asymptotic scattering states, a natural question to ask is whether there exist concrete models of decoherence where the purity of an initial state vector remains, while time irreversibility holds.

A physically acceptable framework for discussing decohering evolution of an open quantum mechanical system is that of Lindblad or the so-called dynamical semigroup approach [39], which ensures the *complete positivity* of the density matrix $\rho(t)$ at any time moment t during the evolution, and the *conservation of probability* $\mathrm{Tr}\rho = 1$. The Lindblad evolution of open systems [39], with Hamiltonian H, interacting with an environment through operators $D_j, D_j^{\dagger}$, is described as a *linear* evolution in the density matrix ρ:

$$\dot{\rho} = i[\rho, H] + \mathcal{D}[\rho]; \qquad \mathcal{D}[\rho] = \sum_j \left(\{\rho, D_j^{\dagger}D_j\} - 2D_j\rho D_j^{\dagger}\right) \tag{14}$$

where $\{.,.\}$ denotes an anticommutator. The Hamiltonian H in (14) may contain terms from the environmental entanglement which can be expressed as commutators with ρ, and hence it should be understood as an effective

Hamiltonian of the system. The decoherence term $\mathcal{D}[\rho]$, on the other hand, cannot be expressed as such a commutator.

To ensure energy conservation on the average, and monotonic increase of the von-Neumann entropy $S = -\text{Tr}\,(\rho\ln\rho)$, one has to impose self-adjointness of the Lindblad environmental operators

$$D_j^\dagger = D_j \tag{15}$$

and also require that these operators commute with the Hamiltonian

$$[D_j, H] \qquad \text{for} \quad \text{all} \quad j \tag{16}$$

This leads to a double commutator structure of the decoherence terms in (14):

$$\mathcal{D}[\rho] = \sum_j [D_j, [D_j, \rho]] \tag{17}$$

In general, in this type of decoherence one has the evolution of a pure state into a mixed one. However, there exist subclasses of Lindblad evolution, in particular energy-driven simple decoherence models [40, 41], where the purity of state vectors is preserved. A mathematical criterion for this feature is that

$$\rho^2 = \rho, \qquad \text{Tr}\rho = 1\ , \tag{18}$$

during the evolution.

In such models, $D_j = \lambda_j H$, with λ_j c-number constants. Without loss of generality, we can substitute in such a case the sums in (17) by a single environmental operator

$$D = \lambda H, \qquad \lambda^2 = \sum_j \lambda_j^2 \tag{19}$$

This simplifies the situation and will suffice for our purposes in this work.

In this type of decoherence, the density matrix evolution preserves the purity of states, and can be written in terms of stochastic Ito differential equations for the state vectors $|\psi\rangle$ (or equivalently the pure state density matrix $\rho = |\psi\rangle\langle\psi|$):

$$d\rho = -i[H, \rho]dt - \frac{1}{8}[D, [D, \rho]]dt + \frac{1}{2}[\rho, [\rho, D]]dW_t \tag{20}$$

where t is the time, and dW_t is an Ito stochastic differential obeying

$$dW_t^2 = dt, \qquad dtdW_t = 0 \tag{21}$$

which are the equivalent of white noise conditions. Needelss to say that one can generalise the above equation to the case where sums of D_j operators are involved, but as we mentioned above this will not be necessary for our purposes here.

We remark that, in terms of state vectors $|\psi\rangle$, the first term in (20) is nothing but the Schrödinger Hamiltonian term $-iH|\psi\rangle$, while the second term resembles Fokker-Planck stochastic diffusion terms. Unlike the Schrödinger term, the diffusion term is not invariant under the time reversal operation $t \to -t$ and $i \to -i$, and hence *time irreversibility* occurs in the problem, despite the purity of states.

Upon using (19) in (20), one obtains a stochastic equation for this energy-driven decoherence:

$$d\rho = -i[H,\rho] - \frac{\lambda^2}{8}[H,[H,\rho]]dt + \frac{\lambda}{2}[\rho,[\rho,H]]dW_t \tag{22}$$

The double commutator of the Hamiltonian, together with the purity-of-states condition (18), leads to the following order of the decoherence term in such models, obtained by considering the vacuum expectation value of the double commutator term in (22): $\gamma \equiv \langle\langle \mathcal{D}[\rho]\rangle\rangle = \mathrm{Tr}\left(\rho\frac{\lambda^2}{8}[H,[H,\rho]\right)$. Using as a complete orthonormal basis of states energy eigenstates $|m\rangle$, then, it is straightforward to see that the above estimate leads to the square of the energy variance

$$\gamma = \frac{\lambda^2}{8}\left(\Delta H\right)^2 = \langle\langle H^2\rangle\rangle - \left(\langle\langle H\rangle\rangle\right)^2 \tag{23}$$

for this model of decoherence.

In quantum-gravity driven models of decoherence it is natural to assume that $\lambda^2 \propto 1/M_P$, where $M_P \sim 10^{19}$ GeV is the Planck scale, which is expected to be the characteristic scale of quantum gravity. In such models then one obtains the following estimate for the decoherence coefficient γ [41]

$$\gamma \sim \left(\Delta H\right)^2/M_P \tag{24}$$

We shall come back to physical applications of this case later on, when we discuss sensitive probes of quantum mechanics, such as neutral mesons and neutrinos.

Before closing this subsection we should remark that other types of decoherence models, which are not energy driven, but correspond to spontaneous localisation in space, also exist. One such model is the one presented in [42], in which the operator D is taken to be proportional to the spatial coordinate operator q, thereby leading to spatial localisation. In such a case again the decoherence coefficient γ (24) is found to be proportional to the square of the position operator variance $\gamma \propto (\Delta q)^2$, expressing, e.g. spatial separation between centres of wavepackets, resulting for instance from the mass difference.

2.5 More General Case: Dynamical Semi-Group Approach to Decoherence, and Evolution of Pure States to Mixed

In the previous subsection we examined special cases of Lindblad decohering evolution, which preserved the purity of quantum states. The Lindblad approach to decoherence, however, in general has the feature of implying an evolution of a pure state to a mixed one, in the sense of $\mathrm{Tr}\rho(t)^2 \neq \mathrm{Tr}\rho(t)$, thereby leading to a violation of the strong form of CPT, according to the theorem of [37]. The general Lindblad evolution can be formulated in such a way that no detailed knowledge of the underlying microscopic dynamics of the decohering environment is necessary in order to arrive at certain conclusions of phenomenological interest. This is achieved by means of the so-called *dynamical semigroups approach to decoherence* [39], which is a generic formalism to describe a decohering evolution obeying some basic properties. The time irreversibility in this approach is linked to the lack of an inverse of an element in an appropriate semigroup.

Consider the generic case of a decohering (of Lindblad, or even more general, type) evolution for an N-level system, that is a system whose Hamiltonian (energy) eigenstates span an N-dimensional state vector space. The decohering operators, assumed bounded for our purposes here, can be represented by $N \times N$ matrices generated by a basis F_μ, $\mu = 0, 1, \ldots N^2 - 1$, endowed by the scalar product $(F_\mu, F_\nu) = \frac{1}{2}\delta_{\mu\nu}$. For the purposes in this work we shall be dealing explicitly with $N = 2, 3$ level systems, in which cases the basis $\{F_\mu\}$ consists of: (i) the three 2×2 Pauli matrices plus the 2×2 identity matrix I_2 for the $N = 2$ case, and (ii) the 3×3 Gell-Mann matrices Λ_i,$i = 1, \ldots 8$ plus the 3×3 identity matrix I_3 for the $N = 3$ case.

Generically the matrices F_μ satisfy the following commutation relations:

$$[F_i, F_j] = i \sum_k f_{ijk} F_k, \qquad 1 \leq i, j, k \leq 8 \tag{25}$$

where f_{ijk} are the structure constants of the $SU(N)$ group, and we follow the notation that Latin indices run from 1, to $N^2 - 1$, while Greek indices run from $0, 1, \ldots N^2 - 1$.

Expanding the environmental operators, as well as the (effective) Hamiltonian and the density matrix in (14) in terms of the basis $\{F_\mu\}$:

$$H = \sum_\mu h_\mu F_\mu, \qquad \rho = \sum_\mu \rho_\mu F_\mu, \qquad D_j = \sum_{mu} d_\mu^{(j)} F_\mu \tag{26}$$

and imposing the hermiticity of D, which ensures the monotonic increase of the von Neumann entropy $S = -\mathrm{Tr}\rho\ln\rho$, we can write the decoherence term $\mathcal{D}[\rho]$ in (14) as:

$$\mathcal{D}[\rho]_{\mathrm{Lindblad}} = \sum_{\mu,\nu} L_{\mu\nu} \rho_\mu F_\nu \ , \tag{27}$$

where the matrix $L_{\mu\nu}$ is real and symmetric, with the properties:

$$L_{\mu 0} = L_{0\mu} = 0, \qquad L_{ij} = \frac{1}{2} \sum_{k,l,m} (\boldsymbol{d}_m \cdot \boldsymbol{d}_k) f_{iml} f_{ikj} \ , \tag{28}$$

whereby $\boldsymbol{d}_\mu = (d_\mu^{(1)}, \ldots d_\mu^{(N^2-1)})$.

The vanishing of the first row and column is due to entropy increase. Notice that if we do not impose the requirement of energy conservation on the average, then it is not necessary to assume the commutativity of the operators with the Hamiltonian, so in general $[D_j, H] \neq 0$. In fact below we shall examine some examples where energy may be violated due to foam interactions.

The evolution equation (14), then, reads:

$$\dot{\rho} = \sum_{i,j} h_i \rho_j f_{ij\mu} + \sum_{\nu} L_{\mu\nu}\rho_\nu , \qquad \mu, \nu = 0, \dots N^2 - 1 . \tag{29}$$

where the overdot denotes derivative with respect to time t.

Probability conservation $\mathrm{Tr}\rho(t) = 1$ at any time moment t implies that the differential equation for the ρ_0 component decouples, yielding

$$\rho_0(t) = \mathrm{const} \tag{30}$$

The remaining differential equations (29) then can be written in the form:

$$\dot{\rho}_k = \sum_j \left(\sum_i h_i f_{ijk} + L_{kj} \right) \rho_j = \sum_k \mathcal{M}_{kj} \rho_j \tag{31}$$

Representing by $\mathcal{A}$ the matrix that diagonalises the matrix $\mathcal{M}$, and letting $\{\lambda_1, \dots \lambda_{N^2-1}\}$ be the set of eigenavalues of $\mathcal{M}$, and $\{v_1, \dots v_{N^2-1}\}$ be the corresponding set of its eigenvectors, we have for the ij elements of $\mathcal{A}$: $\mathcal{A}_{ij} = (v_i)_j$. The solution of (31), then, can be written as:

$$\rho_i(t) = \sum_{k,j} e^{\lambda_k t} \mathcal{A}_{ik} \mathcal{A}^{-1}_{kj} \rho(0)_j \tag{32}$$

Thus, in the dynamical semigroup approach, we have seen that the imposition of generic properties, such as monotonic entropy increase, probability conservation etc., allows for an apparently complicated decoherence/entanglement problem to be transformed into an algebraic problem of determining the eigenvalues and eigenvectors of finite-dimensional matrices. In general, for N-level systems, with $N \geq 3$, the general form of the decoherence matrix is too complicated to allow for clear physical meaning of all its entries. As we shall discuss below, however, in the context of specific examples, one can make physically meaningful simplicifcations, which allow for physical predictions to be made from such a formalism.

2.6 State Vector Reduction ("Wavefunction Collapse") in Lindblad Decoherence

Decoherence in general is expected to lead to a decay with time t of the off-diagonal elements of the reduced density matrix of an open system, which are in general of the form [44, 45].

$$\rho(x,x',t) \sim \exp\left(-ND(x-x')^2 t\right) , \tag{33}$$

where x, x' denote the spatial locations of the centre of mass of a system of N particles, and D is a generic decoherence parameter. Notice the dependence of the exponent on the square of the distance $|x - x'|^2$, and on the number of particles N, which implies that the larger the $N|x - x'|^2$ the faster the decoherence. Hence, macroscopic bodies (containing, say, at least an Avogadro number of particles) will in general decohere very fast. Such considerations are general, and can also be extended to decoherence models that may have relevance to quantum gravity, such as the wormhole-induced decoherence [44].

I should stress at this point that in general, decoherence does not necessarily solves the problem of measurement, because it cannot explain which one of the diagonal entries of the density matrix is picked up during a "measurement", that is an interaction of the subsystem with a macroscopic environment.

In some models of decoherence, though, especially the ones where the purity of states is preserved during the evolution, like the ones examined above, it is possible to establish a mathematical criterion for the state vector reduction, that is the localisation of the state vector in a given "measurement" channel in state space. It is the point of this subsection to discuss briefly this issue.

First of all we note that the temporal evolution (14) for these specific Lindblad systems can be written in terms of the corresponding state vectors $|\psi\rangle$ via the Ito form [43]:

$$\begin{aligned}|d\psi\rangle &= -iH|d\psi\rangle dt + \sum_j \left(\langle D_j^\dagger\rangle_\psi D_j - \frac{1}{2}D_j^\dagger D_j - \frac{1}{2}\langle D_j^\dagger\rangle_\psi\langle D_j\rangle_\psi\right)|\psi\rangle dt \\ &\quad + \sum_j \left(D_j - \langle D_j\rangle_\psi\right)|\psi\rangle dW_{j,t} \end{aligned}\tag{34}$$

where $dW_{j,t}$ are the stochastic differential random variables satisfying (21).

The state vector reduction, or equivalently "collapse" of the wavefunction that characterises this formalism can be proven as follows [43]: one makes the assumption that the Hamiltonian of the system H can be cast in a *block-diagonal* form in terms of state-space "channels" $\{k\}$, which exist independently of any "measurement" (i.e. interaction with a macroscopic measurement apparatus). This means that, if $\mathcal{P}_k$ denotes the projection operator on channel k, then

$$[H, \mathcal{P}_k] = 0 \tag{35}$$

The state vector reduction is then proven by demonstrating the localisation of $|\psi\rangle$ on a state-space channel k due to the environmental entanglement in (34). A mathematical measure of this localisation is the so-called *Quantum Dispersion Entropy* $\mathcal{K}$ defined as [43]:

$$\mathcal{K} = -\sum_k \langle\mathcal{P}_k\rangle_\psi \ln\langle\mathcal{P}_k\rangle_\psi \tag{36}$$

which, if one uses (34), and the above assumptions, can be shown to have the following monotonic decrease properties:

$$\frac{d}{dt}\left(M\mathcal{K}\right) = -\sum_{k}\frac{1-\langle\mathcal{P}_k\rangle_\psi}{\langle\mathcal{P}_k\rangle_\psi}\sum_{j}|\langle\mathcal{P}_k D_j \mathcal{P}_k\rangle_\psi|^2 \le 0 \tag{37}$$

where M denotes an average over an ensemble of theories. The monotonic decrease (37) implies localisation of the state vector in state space, in a time which depends on the details of the environmental entaglement, and specifically on the so-called effective interaction rates $R_k \equiv \sum_j |\langle\mathcal{P}_k D_j \mathcal{P}_k\rangle_\psi|^2$, which are positive semi-definite quantities, characteristic of the system. This localisation seems therefore a rather generic feature of the Lindblad stochastic decoherence (22). We remark, however, that in some specific cases of environmental entaglement, such a localisation may not be complete, and one may obtain pointer states (i.e. minimum uncertainty coherent states) from decoherence [46]. This is an important topic, which however we shall not dwell upon in these lectures.

2.7 Non-Critical String Decoherence: a Link between Decoherent Quantum Mechanics and Gravity?

There is an interesting connection between the above-mentioned models of decoherence with non-critical string theory. The latter is viewed as a *non-equilibrium* version of string theory, the equilibrium "points" corresponding to the critical strings. In these lectures we shall not describe in detail the corresponding formalism, but we shall rather give a comprehensive outline of the approach, and concentrate on those aspects of the framework that are relevant for our purposes here. For details we refer the interested reader to the literature [2, 47].

The basic idea [2, 48] is the identification of the target time in non-critical strings with a world-sheet renormalization group scale, the Liouville field zero mode. Non-critical strings are described, in a first-quantised framework, by world-sheet sigma models with non-conformal background fields $\{g^i\}$. The corresponding two-dimensional world-sheet action is then given schematically by:

$$S_\sigma = S^* + \int_\Sigma g^i V_i \tag{38}$$

where S^* is a two-dimensional conformal world-sheet action, corresponding to a critical string theory, and the second term on the right hand side of (38) represents deformations from this "conformal point". The operators V_i are the vertex operators, which describe the string excitations corresponding to the background fields g^i, over which the string propagates in target space time. This set may contain gravitons, dilatons, gauge fields, etc., $\{g^i\} = \{G_{\mu\nu}, \Phi, A_\mu \ldots\}$. The important thing to notice is that the background space time fields g^i appear as couplings of the two-dimensional σ-model theory.

The non conformal nature of the backgrounds implies that the world-sheet renormalization group (RG) β-functions $\beta^i = dg^i/d\ln\mu$, where μ is a two-dimensional RG scale, are non zero. For a critical string $\beta^i = 0$, which determines the "consistent" target space backgrounds over which the string propagates. These are the equilibrium "points" in the (infinite dimensional) space of string theories, spanned by the "coordinates" $\{g^i\}$.

For consistency of the world-sheet theory, such non conformal backgrounds require dressing with the Liouville mode, an extra σ-model field, playing the rôle of a target-space coordinate. This field restores conformal invariance, at the cost of enlarging the target space by one extra dimension [47], whose rôle is played by the world sheet zero mode of the Liouville field. Depending on the kind of deformation, the Liouville mode could be space-like or time-like in target space. In these lectures we shall be interested in the time-like Liouville mode case. The Liouville zero mode then can be identified with the target time in a consistent way [2, 47], which in some cases is forced upon us dynamically, due to minimization, upon such an identification, of the effective potential of the low-energy field theory [49]. In this way, the low-energy theory does not have two times.

Since the Liouville mode may be viewed as a world-sheet RG scale $\ln\mu$, we have a situation in which a target time variable is identified with a σ-model RG scale. The irreversibility of the latter has been proven for unitary theories by means of the Zamolodchikov's c-theorem [50], but is expected to hold also for non-unitary ones, due to the presence of a cutoff scale on the world-sheet, which is associated with "loss" of information due to modes with two-dimensional momenta beyond the cutoff [51]. This guarantees a *microscopic time irreversibility*, in a non trivial way.

Formally, in Liouville strings, the world-sheet correlators of vertex operators are identified with well-defined \$-matrix elements rather than scattering amplitudes. The non-factorisability of the \$-matrix into proper S-matrix amplitudes, $\$ \neq SS^\dagger$, is obtained by noting that in Liouville strings, which by definition propagate on non-conformal backgrounds, one may define the Liouville zero mode world-sheet path integral in a steepest-descent fashion by means of the curve indicated in Fig. 2 [2, 47, 48]. Upon the identification of the Liouville zero mode with time, such a curve resembles closed-time-paths in non-equilibrium field theories. It is the short-distance world-sheet singularities (UV) near the origin of the curve of Fig. 2 that cause the aforementioned non factorizability of the \$ matrix. One may link the breathing world sheet, arising from the steepest-descent path of the Liouville mode, to a "bounce" on the infrared (IR, large world-sheet area) limit [48], implying an irreversible RG flow from the ultraviolate to infrared fixed points of the world-sheet system. Details are given in the literature [2, 47], where we refer the interested reader for details. For our purposes we only mention that this property links the time irreversibility of the Liouville mode, stemming from world-sheet RG properties, to fundamental properties of space-time \$-matrix elements, in a similar fashion to the analysis in [37].

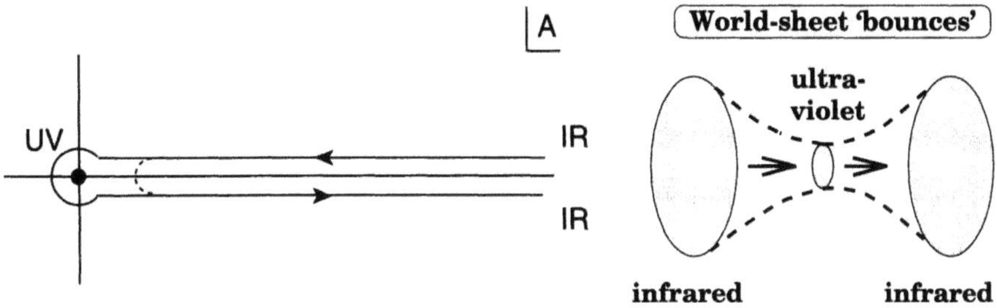

Fig. 2. *Left Picture*: Steepest-descent curve for Liouville zero mode path integration, in the complex plane obtained after complexifying the world-sheet area A. Upon the identification of this mode with target time, such curves resemble closed time paths of non equilibrium field theories, in agreement with the non-equilibrium nature of the Liouville string. *Right Picture*: The "breathing world-sheet", as a result of the path on the *left*. The target-space irreversibility arises from a "bounce" interpretation of this process

The theory space "coordinates"/backgrounds fields g^i become quantum operators upon summing up world-sheet genera [2]; decoherence in this theory space is induced precisely by the non vanishing β-functions, that is the departure from the conformal point [2]. To see this one invokes the principle of world-sheet renormalization group invariance of target-space quantities with physical significance for the string propagating in such non-conformal backgrounds. One such quantity is the density matrix of this string matter ρ_s. The RG invariance implies that $\frac{d}{dt}\rho_s = 0$, where $t \equiv \ln\mu$ is the world-sheet RG scale.

In the quantum theory this equation reads [2, 48]:

$$\dot{\rho}_s = i[\rho_s, H] + :\beta^j \mathcal{G}_{ji}[g^i, \rho_s]: \tag{39}$$

where the overdot denotes partial derivative with respect to t, H is the effective low-energy string Hamlitonian, and $\mathcal{G}_{ij} = z^2\bar{z}^2\langle V_i(z)V_j(0)\rangle$ is the Zamolodchikov's metric in "theory space" [47]. The notation $:\cdots:$ denotes appropriate ordering of the quantum operators.

Equation (39) has similar form to that of a 'decoherent evolution' in the parameter t. Clearly, for critical backgrounds $\beta^i = 0$, and hence the evolution in RG space does not imply any such "decoherence". However, this decoherence would acquire physical significance only if the identification of the scale t with the real target time variable in string theory holds [2]. This is not a trivial issue, and in fact it can be shown that it does not hold for any non conformal deformation. However, as already mentioned, there are physically interesting cases, among which strings in de Sitter space times [35], to be discussed separately in the next subsection, or colliding brane cosmologies [49], which are non conformal backgrounds in string theory, and in which the above-mentioned identification of time with the world-sheet RG scale, that is the Liouville zero mode, occurs due to dynamical reasons, leading to minimization of energy.

Under such an identification, the RG evolution (39) becomes a real temporal evolution for the reduced density matrix of a string interacting with the non conformal background, which leads to the presence of decoherence terms proportional to the RG $\beta^i \neq 0$. Using (39) it can be shown [2] that such Liouville-string decoherence has the following properties:

(i) *Conservation of Probability,*

(ii) *Von-Neumann entropy monotonic increase*: one calculates the relevant rate as:

$$\frac{\partial}{\partial t}\left(\mathrm{Tr}\rho\ln\rho\right) = \beta^i \mathcal{G}_{ij}\beta^j \left(\mathrm{Tr}\rho\ln\rho\right) \tag{40}$$

which is positive semi-definite, since $\beta^i \mathcal{G}_{ij}\beta^j \geq 0$ due to Zamolodchikov's c-theorem for unitary theories or its extension for non-unitary ones [51].

(iii) *Energy conservation* on the average, since

$$\frac{\partial}{\partial t}\langle\langle H\rangle\rangle = \frac{\partial}{\partial t}\left(\frac{\partial \beta^i}{\partial g^i}\right) = 0 \tag{41}$$

due to the fact that there is no explicit RG scale t dependence on the β^i function, due to renormalizability of the σ-model. However, a word of caution should be placed here. In some cases, in particular logarithmic conformal field theories, such as D-particle recoil [48], where the short-distance limit of two deformation operators contain explicit logarithms $V_i(z)V_j(z) \sim \ln z c_{ijk} V_k/|z-z|^2$, there is explicit t dependence in the Operator Product Expansion coefficients appearing in the perturbative expansion of the β-function in powers of coupling constants g^i [52]. For instance, in the recoil problem, the anomalous dimension coefficients are t-dependent [48]. In such cases, the energy conservation on the average may be spoiled.

This type of Liouville-string decoherence leads to "localisation" in theory space g^i [2], which can be seen as follows: the RG β-functions are expressed as a power series in the coordinates/background fields g^i, $\beta^i = C^i_{i_1 \ldots i_n} g^{i_1} \ldots g^{i_n}$. The linear term is the anomalous dimension term. In a weak field expansion, i.e. when g^i are assumed sufficiently weak so that perturbation theory holds, one may assume to a good approximation $\beta^i \simeq y_i g^i$, with y_i the anomalous scaling dimension of the σ-model coupling/background field g^i. Note also that this is an exact result (in terms of a g^i expansion) in some non conformal cosmological backgrounds of string theory, such as de Sitter space, i.e. a space time with a non zero cosmological constant $\Lambda > 0$. In such a case, the graviton β-function, to order $\alpha' = M_s^{-2}$, with M_s the string mass scale, is given by the Ricci tensor: $\beta^{\mu\nu} = \alpha' R_{\mu\nu} = \Lambda g_{\mu\nu}$, and thus is linear in the graviton background. We shall examine this case in some detail in the next subsection.

In such linearised cases, one may choose the antisymmetric quantum ordering prescription which leads to a double commutator structure in the theory space coordinate operators g^i, so that (39) reads:

$$\dot{\rho}_s = i[\rho_s, H] + y_i g_i [g^i, \rho_s] , \tag{42}$$

where we have used the fact that, to leading order in g^i, the Zamolodchikov "metric in theory space" $\mathcal{G}_{ij} \simeq \delta_{ij} + \mathcal{O}(g^2)$, which can always be arranged by an appropriate choice of a Renormalization scheme [47].

Comparing (42) with (14), (17) we observe that we are encountering here exactly an analogous situation, but instead of energy driven or position localisation decoherence models, we have a non-critical string theory induced decoherence. Since g^i are generalised "position vectors" in theory space, the same arguments leading to localisation of the state vector in those models will lead here to "localisation in theory space g^i". From a physical viewpoint this would imply the emergence of the equilibrium target space of string theories in a dynamical way, due to evolution of a non critical string theory to those equilibrium points. Moreover, the double commutator structure in (42) will also lead to variances $(\Delta g^i)^2$ for the background fields g^i, expressing the back reaction of string matter on those backgrounds. In the next subsection we shall examine a concrete and physically interesting example of such a situation, that of a de Sitter space time background. As we shall discuss below, in such a case one also obtains an interesting case of CPT Violation of unconventional form, which may be related to some energy-driven decoherence models mentioned above [40, 41].

2.8 Cosmological CPTV?

One of the reasons that make me prefer the Violation of CPT via the \$ matrix decoherence approach over other approaches to CPT Violation, concerns a novel type of CPT Violation *at a global scale*, which may characterize our Universe. This has been proposed in [35], and was given the name *cosmological CPT Violation*. This type of CPTV is prompted by recent astrophysical Evidence for the existence of a Dark Energy component of the Universe. See Fig. 3 for instance, there is direct evidence for a current era acceleration of the Universe, based on measurements of distant supernovae SnIa [53], which is supported also by complementary observations on Cosmic Microwave Background (CMB) anisotropies (most spectacularly by the recent data of WMAP satellite experiment) [54].

Best fit models of the Universe from such combined data appear at present consistent with a non-zero, positive *cosmological constant* $\Lambda \neq 0$. Such a Λ-universe will eternally accelerate, as it will enter eventually an inflationary (de Sitter) phase again, in which the scale factor will diverge exponentially $a(t) \sim e^{\sqrt{\Lambda/3}t}$, $t \to \infty$. This implies that there exists a *cosmological horizon.*

The existence of such horizons implies incompatibility with a S-matrix: no proper definition of asymptotic state vectors is possible, and there is always an environment of d.o.f. crossing the horizon. This situation may be considered as dual to that of a black hole, depicted in Fig. 1: in that case the asymptotic observer was in the exterior of the black hole horizon, in the cosmological case

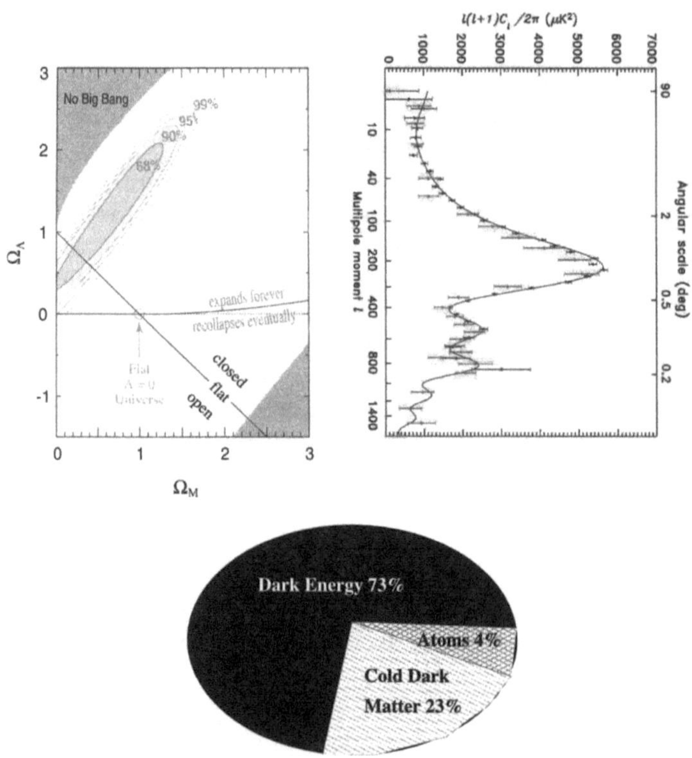

Fig. 3. Recent observational evidence for Dark Energy of the Universe (*upper left figure*: evidence from SnIa [17], *upper right figure*: evidence from CMB measurements [18]) and a pie graph (*lower central figure*) of the energy budget of our world according to these observations

the observer is inside the horizon. However, both situations are characterized by the lack of an invertible scattering matrix, hence the above-described theorem by Wald [37] on $\$$-matrix and CPTV applies [35], and thus CPT is violated, at a global scale, due to a cosmological constant $\Lambda > 0$.

It has been argued in [35] that such a violation is described effectively by a modified temporal evolution of matter in such a Λ-universe, which is given by

$$\partial_t \rho = [\rho, H] + \mathcal{O}(\Lambda M_s)\rho \tag{43}$$

where Λ is a dimensionless cosmological constant in four dimensions, and M_s is the quantum gravity scale (which may be different from the four-dimensional Planck scale, see discussion below).

This form has been derived in the above-described context of non critical strings. Indeed, a de Sitter space time constitutes a non conformal string background, and according to the ideas presented in the previous subsection the temporal evolution of string matter in such a space time is described by the decohering evolution (39). Since, as mentioned previously, the main source of departure from conformal symmetry comes from the graviton $g_{\mu\nu}$ background, whose $\beta^{\mu\nu} = \Lambda g_{\mu\nu}$, one actually has the evolution (42), with the double commutaror structure for the background $g_{\mu\nu}$. The order of the decoherence parameter γ, then, in such a case is:

$$\gamma \sim \Lambda M_s (\Delta g_{\mu\nu})^2 , \tag{44}$$

where M_s is the string scale, and Λ is a dimensionless cosmological constant in the d-dimensional space time the string propagates on. One may use the modern view point that our four dimensional world is actually a string membrane (D-brane), embedded in a ten-dimensional target (bulk) space. The Standard Model matter is localised on such brane worlds. In the bulk, only fields belonging to the gravitational multiplet of the string spectrum are allowed to propagate. From this view point, then, the string scale M_s may be different from the four dimensional Planck scale. However, since string matter is confined on the brane world, it essentially interacts effectively only with the four-dimensional graviton fields, that lie on, or cross, our brane world, and hence one arrives at the estimate (43), with Λ the effective four-dimensional cosmological constant on the brane.

An important issue concerns the order of the variance of the metric fluctuations $(\Delta g_{\mu\nu})^2$. To arrive at the estimate (43) one has to assume that such variances are of order one. However, there are models of space time foam in string/membrane theory [7, 8], where the foam is represented as a gas of D-particle (point-like) defects on three branes, which recoil upon interaction with matter strings. As a result of recoil, there are induced space-time distortions, of the form $g_{0i} \sim u_i$, where i is a spatial three-brane index, and u_i is the recoil velocity of the D-particle. By momentum conservation, $u_i \sim g_s \Delta k_i / M_s$, where Δk_i is the momentum transfer, which is of order of the incident momentum, k, g_s is the (weak) string coupling, and M_s/g_s is the mass of the D-particle. Upon summing world-sheet genera, u_i becomes an operator, which acts on energy eigenstates, yielding appropriate eigenvalues of order $g_s \Delta k_i / M_s$.

Considering the case of a two state system, say neutrinos oscillating between two energy states, with the corresponding energy difference arising from a mass-squared difference Δm^2 in the neutrino Hamiltonian $H \simeq p + m^2/p + \dots$, one has for the model of [8]:

$$(\Delta g_{0i})^2 \sim (g_s \Delta E / M_s)^2 = g_s^2 (\Delta m^2)^2 / E^2 M_s^2 \tag{45}$$

where E is the energy of the low-energy neutrino interacting with the foam.

From (45) and (39), then, we obtain the order of the decoherence coefficient for this case:

$$\gamma \sim \Lambda g_s^2 (\Delta m^2)^2 / E^2 M_s \tag{46}$$

Comparing with (24) we observe that it is of the same form as in the energy-driven decoherence model of [40, 41], provided the decoherence coupling with the environment is of order $\lambda^2 \sim g_s^2 \Lambda / M_s$. In fact, one gets exactly the result (24), if one identifies $M_s/g_s = M_P$, and assumes a $\Lambda \sim 1/g_s$, which could be induced by quantum string loop effects (but, of course, this is too big for a realistic cosmological constant). The equivalence with energy driven decoherence of the D-particle foam model should not have come as a surprise, given that the space-time distortion due to the recoil of the D-particle is driven by the energy content of the matter probe, on account of energy conservation.

For realistic values of $\Lambda \sim 10^{-122}$ in Planck units, the above effects are undetectable in any oscillation experiment. Although the order of the cosmological CPTV effects in this scenario is tiny, if we accept that the Planck scale is the ordinary four-dimensional one $M_P \sim 10^{-19}$ GeV, and hence undetectable in direct particle physics interactions, however, such cosmological-constant induced CPTV may have already been detected indirectly through the (claimed) observational evidence for a current-era acceleration of the Universe ! Of course, the existence of a cosmological constant brings up other interesting challenges, such as the possibility of a proper quantization of de Sitter space as an open system, which are still unsolved.

At this point I should mention that time Relaxation models for Dark Energy, e.g. quintessence model, where eventually the vacuum energy asymptotes (in cosmological time) an equilibrium zero value, are still currently compatible with the data [56]. In such cases it might be possible that there is no cosmological CPTV, since a proper S-matrix can be defined, due to lack of cosmological horizons.

From the point of view of string theory the impossibility of defining a S-matrix in de Sitter space times is very problematic, because critical strings by their very definition depend crucially on such a concept. However, this is not the case of non-critical string theory, which can accommodate in their formalism Λ universes [35]. It is worthy of mentioning briefly that such non-critical (non-equilibrium) string theory cases are capable of accommodating models with large extra dimensions, in which the string gravitational scale M_s is not necessarily the same as the Planck scale M_P, but it could be much smaller, e.g. in the range of a few TeV. In such cases, the CPTV effects in (43) may be much larger, since they would be suppressed by M_s rather than M_P.

It would be interesting to study further the cosmology of such models and see whether the global type of CPTV proposed in [35], which also entails primordial CP violation of similar order, distinct from the ordinary (observed) CP violation which occurs at a later stage in the evolution of the Early Universe, may provide a realistic explanation of the initial matter-antimatter asymmetry in the Universe, and the fact that antimatter is highly suppressed today. In the standard CPT invariant approach this asymmetry is supposed to be due to ordinary CP Violation. In this respect, I mention that speculations about the possibility that a primordial CPTV space-time foam is responsible for the observed matter-antimatter asymmetry in the Universe have also been put forward in [57] but from a different perspective than the one I am suggesting here. In [57] it was suggested that a novel CPTV foam-induced phase difference between a space-time spinor and its antiparticle may be responsible for the required asymmetry. Similar properties of spinors may also characterize space times with deformed Poincare symmetries [58], which may also be viewed as candidate models of quantum gravity. In addition, other attempts to discuss the origin of such an asymmetry in the Universe have been made within the loop gravity approach to quantum gravity [59] exploring Lorentz Violating modified dispersion relations for matter probes, especially neutrinos, which we shall discuss below.

3 Phenomenology of CPT Violation

3.1 Order of Magnitude Estimates of CPTV

Before embarking on a detailed phenomenology of CPTV it is worth asking whether such a task is really sensible, in other words how feasible it is to detect such effects in the foreseeable future. To answer this question we should present some estimates of the expected effects in some models of quantum gravity.

The order of magnitude of the CPTV effects is a highly model dependent issue, and it depends crucially on the specific way CPT is violated in a model. As we have seen cosmological (global) CPTV effects are tiny, on the other hand, quantum Gravity (local) space-time effects (e.g. space time foam) may be much larger.

Naively, Quantum Gravity (QG) has a dimensionful constant: $G_N \sim 1/M_P^2$, $M_P = 10^{19}\,\mathrm{GeV}$. Hence, CPT violating and decohering effects may be expected to be suppressed by E^3/M_P^2 , where E is a typical energy scale of the low-energy probe. This would be practically undetectable in neutral mesons, but some neutrino flavour-oscillation experiments (in models where flavour symmetry is broken by quantum gravity), or some cosmic neutrino future observations might be sensitive to this order: for instance, in models with LV, one expects modified dispersion relations (m.d.r.) which could yield significant effects for ultrahigh energy ($10^{19}\,\mathrm{eV}$) ν from Gamma Ray Bursts (GRB) [60], that could be close to observation. Also in some astrophysical cases, e.g. observations of synchrotron radiation from Crab Nebula or Vela Pulsar, one is able to constraint electron m.d.r. almost near this (quadratic) order [7].

However, resummation and other effects in some theoretical models may result in much larger CPTV effects of order: $\frac{E^2}{M_P}$. This happens, e.g., in some loop gravity models [3], or in some (non-critical) stringy models of quantum gravity involving open string excitations [25]. Such large effects may already be accessible in current experiments, and most of them are excluded by current observations. Indeed, the Crab nebula synchrotron constraint [11] for instance already excludes such effects for electrons. Nevertheless, similar effects for photons are still escaping exclusion at present, and in view of possible violations of the equivalence principle, which might occur in some theoretical models of foam [7], according to which only photons are susceptible to such QG-induced m.d.r., the last word on minimal suppression QG effects has not been spoken yet.

On the other hand, as we discussed previously, in some models of decoherence [41] one may have single Planck mass suppression, $1/M_P$, however the decoherence parameters γ depend on the energy variance, rather than the average energy of the probe, $\Delta E = E_2 - E_1$ between, say, the two energy eigenstates of a two-state system, such as neutral kaon, or two-generation neutrino oscillations in hierarchical neutrino models, $\gamma \sim (\Delta E)^2/M_P$. This will also be undetectable in oscillation experiments in the foreseeable future, despite the minimal Planck scale suppression of the effect in this case.

From the above discussion it is therefore clear that we are in need of guidance by experiment in our quest for the order of decoherence or other non-trivial quantum gravity effects, since theoretically the situation is far from being resolved. Since very little is known about such models, it is important to obtain as much experimental information on bounds of the relevant parameters as possible. Hopefully, this will help us focusing our future research in the phenomenology of quantum gravity on the right track.

3.2 Mnemonic Cubes for CPTV Phenomenology

When CPT is violated there are many possibilities, due to the fact that C,P and T may be violated individually, with their violation independent from one another. This was emphasized by Okun [61] some years ago, who presented a set of mnemonic rules for CPTV phenomenology, which are summarized in Fig. 4. In this figure I also draw a kind of Penrose cube, indicating where the violations of CPT may come from. The diagram has to be interpreted as follows: CPTV may come from violations of special relativity (axis $1/c$), where the speed of light does not have its value, exhibiting some sort of refractive index *in vacuo*, or from departure from quantum mechanics (axis h), or from gravity considerations, where the gravitational constant departs from its value (axis G_N), or finally (and most likely) from quantum gravity considerations, where all such effects may coexist.

3.3 Lorentz Violation and CPT: The Standard Model Extension (SME)

We start our discussion on phenomenology of CPT violation by considering CPTV models in which requirement (iii) of the CPT theorem is violated, that is Lorentz invariance. As mentioned previously, such a violation may be a consequence of quantum gravity fluctuations. In this case Lorentz symmetry

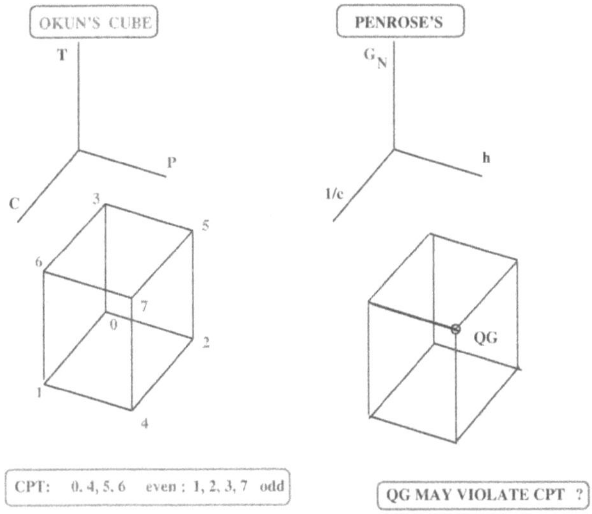

Fig. 4. Mnemonic cubes for CPT Violation: *Left* its phenomenology. *Right*: its possible theoretical origin

is violated (LV) and hence CPT, but there is no necessarily quantum decoherence or unitarity loss. Phenomenologically, at low energies, such a LV will manifest itself as an extension of the standard model in (effectively) flat space times, whereby LV terms will be introduced by hand in the relevant lagrangian, with coefficients whose magnitude will be bounded by experiment [23].

Such SME lagrangians may be viewed as the low energy limit of some string theory vacua, in which some tensorial fields acquire non-trivial vacuum expectation values $\langle A_\mu \rangle \neq 0$, $\langle T_{\mu_1 \dots \mu_n} \rangle \neq 0$. This implies a *spontaneous* breaking of Lorentz symmetry by these (exotic) string vacua [23].

The simplest phenomenology of CPTV in the context of SME is done by studying the physical consequences of a modified Dirac equation for charged fermion fields in SME. This is relevant for phenomenology using data from the recently produced antihydrogen factories [29, 31].

In these lectures I will not cover this part in detail, as I will concentrate mainly on neutrinos within the SME context. It suffices to mention that for free hydrogen H (anti-hydrogen $\overline{H}$) one may consider the spinor ψ representing electron (positron) with charge $q = -|e|(q = |e|)$ around a proton (antiproton) of charge $-q$, which obeys the modified Dirac equation (MDE):

$$(i\gamma^\mu D^\mu - M - a_\mu \gamma^\mu - b_\mu \gamma_5 \gamma^\mu - \\ -\frac{1}{2} H_{\mu\nu} \sigma^{\mu\nu} + i c_{\mu\nu} \gamma^\mu D^\nu + i d_{\mu\nu} \gamma_5 \gamma^\mu D^\nu)\psi = 0 \tag{47}$$

where $D_\mu = \partial_\mu - qA_\mu$, and $A_\mu = (-q/4\pi r, 0)$ is the Coulomb potential. CPT & Lorentz violation is described by terms with parameters a_μ , b_μ , while Lorentz violation only is described by the terms with coefficients $c_{\mu\nu}$, $d_{\mu\nu}$, $H_{\mu\nu}$.

One can perform spectroscopic tests on free and magnetically trapped molecules, looking essentially for transitions that were forbidden if CPTV and SME/MDE were not taking place. The basic conclusion is that for sensitive tests of CPT in antimatter factories frequency resolution in spectroscopic measurements has to be improved down to a range of a 1 mHz, which at present is far from being achieved [31].

Since the presence of LV interactions in the SME affects dispersion relations of matter probes, other interesting precision tests of such extensions can be made in atomic and nuclear physics experiments, exploiting the fact of the existence of a preferred frame where observations take place. The results and the respective sensitivities of the various parameters appearing in SME are summarized in the table of Fig. 5, taken from [30]. As we see, the frame dependence of such LV effects leads to very stringent bounds of the values of the LV parameters in some cases, which are far more superior than the corresponding bounds obtained at present in antihydrogen factories.

LEADING ORDER BOUNDS

EXPER.	SECTOR	PARAMS. (J=X,Y)	BOUND (GeV)
Penning Trap	electron	$\bar{b}_J^e$	5×10^{-25}
Hg–Cs clock comparison	electron	$\bar{b}_J^e$	10^{-27}
	proton	$\bar{b}_J^p$	10^{-27}
	neutron	$\bar{b}_J^n$	10^{-30}
H Maser	electron	$\bar{b}_J^e$	10^{-27}
	proton	$\bar{b}_J^p$	10^{-27}
spin polarized matter	electron	$\bar{b}_J^e / \bar{b}_Z^e$	10^{-29} / 10^{-28}
He–Xe Maser	neutron	$\bar{b}_J^n$	10^{-31}
Muonium	muon	$\bar{b}_J^\mu$	2×10^{-23}
Muon g–2	muon	$\bar{b}_J^\mu$	5×10^{-25} (estimated)

X,Y.Z celestial equatorial coordinates $\bar{b}_J = b_3 - m d_{30} - H_{12}$

(Bluhm, hep–ph/0111323)

Fig. 5. Sensitivities of CPTV and LV parameters appearing in SME/Modified Dirac equation for charged probes, from various atomic and nuclear physics experiments

3.4 Direct SME Tests and Modified Dispersion Relations (MDR)

Many LV Models of Quantum Gravity (QG) predict modified dispersion relations relations (MDR) for matter probes, including neutrinos ν [9, 10, 60]. This leads to one important class of experimental tests using ν: each mass eigenstate of ν has QG deformed dispersion relations, which may, or may not, be the same for all flavours:

$$E^2 = \boldsymbol{k}^2 + m_i^2 + f_i(E, M_{qg}, \boldsymbol{k}) \ , \qquad \text{e.g. } f_i = \sum_{\alpha} C_{\alpha=1,2,\ldots} \boldsymbol{k}^2 \left(\frac{|\boldsymbol{k}|}{M_{qg}}\right)^{\alpha} \tag{48}$$

There are stringent bounds on f_i from oscillation experiments, as we shall discuss below.

It must be stressed that such MDR also characterize SME, although the origin of MDR in the approach of [9, 10, 60] is due to an induced non-trivial microscopic curvature of space time as a result of a back reaction of matter interacting with a stringy space time foam vacuum. This is to be contrasted with the SME approach [23], where the analysis is done exclusively on flat Minkowski space times, at a phenomenological level.

In general there are various experimental tests that can set bounds on MDR parameters, which can be summarized as follows:

(i) astrophysics tests – arrival time fluctuations for photons (model independent analysis of astrophysical GRB data [10]
(ii) Nuclear/Atomic Physics precision measurements (clock comparison, spectroscopic tests on free and trapped molecules, quadrupole moments etc) [30].
(iii) antihydrogen factories (precision spectroscopic tests on free and trapped molecules: e.g. $1S \to 2S$ forbidden transitions) [31],
(iv) Neutrino mixing and spin-flavour conversion, a brief discussion of which we now turn to.

3.5 Neutrinos and SME

The SME formalism naturally includes the neutrino sector. Recently a SME-LV + CPTV phenomenological model for neutrinos has been given in [62]. The pertinent lagrangian terms are given by:

$$\begin{aligned}\mathcal{L}^{\nu}_{SME} \ni & \frac{1}{2} i \overline{\psi}_{a,L} \gamma^{\mu} D_{\mu} \psi_{a,L} - (a_L)_{\mu ab} \overline{\psi}_{a,L} \gamma^{\mu} \psi_{b,L} \\ & + \frac{1}{2} i (c_L)_{\mu\nu ab} \overline{\psi}_{a,L} \gamma^{\mu} D^{\nu} \psi_{b,L} \end{aligned} \tag{49}$$

where a, b are flavour indices. The model has (for simplicity) no ν-mass differences. Notice that the presence of LV induces directional dependence (sidereal effects)!

To analyze the physical consequences of the model, one passes to an Effective Hamiltonian [62]

$$(H_{\text{eff}})_{ab} = |\boldsymbol{p}|\delta_{ab} + \frac{1}{|\boldsymbol{p}|}((a_L)^{\mu}p_{\mu} - (c_L)^{\mu\nu}p_{\mu}p_{\nu})_{ab} \qquad (50)$$

Notice that ν oscillations are now controlled by the (dimensionless) quantities $a_L L$ & $c_L LE$ where L is the oscillation length. This is to be contrasted with the conventional case, where the relevant parameter is associated necessarily with a ν-mass difference Δm: $\Delta m^2 L/E$.

There is an important feature of the SME/ν: despite CPTV, the oscillation probabilities are the same between ν and their antiparticles, if there are no mass differences between ν and $\bar{\nu}$: $P_{\nu_x \to \nu_y} = P_{\bar{\nu}_x \to \bar{\nu}_y}$.

Experimentally, it is possible to bound LV + CPTV SME parameters in the neutrino sector with high sensitivity, if we use data from high energy long baseline experiments [62]. Indeed, from the fact that there is no evidence for $\nu_{e,\mu} \to \nu_\tau$ oscillations, for instance, at $E \sim 100\,\text{GeV}$, $L \sim 10^{-18}$ GeV^{-1} we conclude that $a_L < 10^{-18}\,\text{GeV}$, $c_L < 10^{-20}$.

Similarly for an explanation of the LSND anomaly [63], claiming evidence for oscillations between $(\bar{\nu}_\mu - \bar{\nu}_e)$ but not for the corresponding neutrinos, a mass-squared difference of order $\Delta m^2 = 10^{-19}\,\text{GeV}^2 = 10^{-1}$ eV^2 is required, which implies that $a_L \sim 10^{-18}\,\text{GeV}$, $c_L \sim 10^{-17}$. This would affect other experiments, and in fact one can easily come to the conclusion that SME/ν does not offer a good explanation for LSND, if we accept the result of that experiment as correct, which is not clear at present.

A summary of the Experimental Sensitivities for ν's SME parameters are given in the table of Fig. 6, taken from [62].

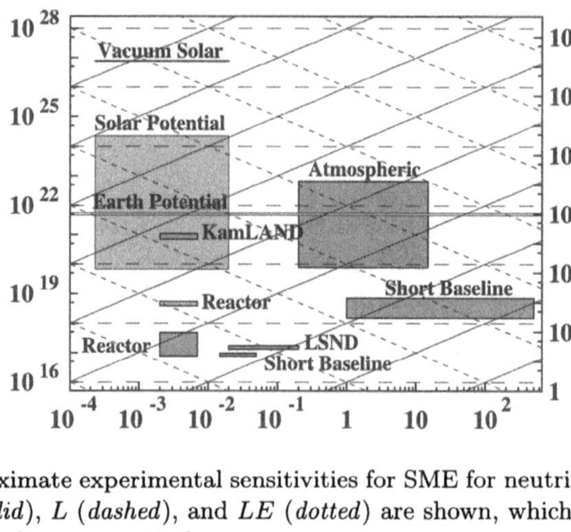

Fig. 6. Approximate experimental sensitivities for SME for neutrinos. Lines of constant L/E (*solid*), L (*dashed*), and LE (*dotted*) are shown, which give sensitivities to Δm^2, a_L, and c_L, respectively

3.6 Lorentz Non-Invariance, MDR and ν-Oscillations

Models of quantum gravity predicting MDR of the type (48) for neutrinos [60, 64], with a leading order E^2/M_{qg} modification, can be severely constrained by a study of the induced oscillations between neutrino flavours, as a result of the departure from the standard dispersion relations provided that the quantum-gravity foam responsible for the MDR breaks flavour symmetry, which however is not always the case [65].

This approach has been followed in [66], where it was shown that if flavour symmetry is not protected in such MDR models, then the extra terms in (48), proportional to E^2/M_{qg} will induce an oscillation length $L \sim 2\pi M_{qg}/(\alpha E^2)$, where α is a phenomenological parameter that controls the size of the effect. This should be contrasted to the Lorentz Invariant case where $L_{\rm LI} \sim 4\pi E/\Delta m^2$, with Δm^2 the square mass difference between neutrino flavours. From a field theoretic view point, terms in MDR proportional to some positive integer power of E^2/M_{qg} may behave as non-renormalizable operators, for instance, dimension five [67] in the case of leading order QG effects suppressed only by a single power of M_{qg}.

The sensitivity of the various neutrino oscillation experiments to the parameter α is shown in Fig. 7 [66]. The conclusion from such analyses, therefore, is that, if the flavour number symmetry is not protected in such MDR foam models with minimal $1/M_P$ suppression in the correction terms, then neutrino observatories and long base-line experiments should have already observed such oscillations. As remarked above, however, not all foam models that lead to such MDR predict such oscillations [65], and hence such constraints are highly foam-model dependent.

EXPER.	STATUS	<E> (GeV)	L (Km)	X (Km)	$\alpha = \frac{L}{X}$
CHORUS	closed 97	26	10^{-2}	0.85	10^{-2}
NOMAD	closed 99	24	10^{-2}	0.94	10^{-2}
SK	operating	1.3	10	$10\text{–}10^{4}$	$1\text{–}10^{-3}$
K2K	operating	1.3	10	250	10^{-2}
SNO	operating	0.008	10^{5}	10^{8}	10^{-3}
MINOS	under constr. (04)	15	0.1	730	10^{-4}
CNGS	starting 05	17	0.1	732	10^{-4}

Fig. 7. Shown for each experiment are: (i) operation status, (ii) mean value of observed ν energy, (iii) the oscillation length L, (iv) typical ν-flight distance X, and the ratio $\alpha = L/X$, which, in models where the foam induces ν flavour oscillations, coincides with the phenomenological parameter that controls the size of MDR effects

3.7 Lorentz Non Invariance, MDR and ν Spin-Flavor Conversion

An interesting consequence of MDR in LV quantum gravity theories is associated with modifications to the well-known phenomenon of spin-flavour conversion in ν interactions [68]. To be specific, we shall consider an example of a MDR for ν provided by a Loop Gravity approach to quantum gravity. According to such an approach, the dispersion relations for neutrinos are modified to [64]:

$$E_{\pm}^2 = A_p^2 p^2 + \eta p^4 \pm 2\lambda p + m^2 \tag{51}$$

where $A_p = 1 + \kappa_1 \frac{\ell_P}{\mathcal{L}}$, $\eta = \kappa_3 \ell_P^2$, $\lambda = \kappa_5 \frac{\ell_P}{2\mathcal{L}^2}$, and $\mathcal{L}$ is a characteristic scale of the problem, which can be either (i) $\mathcal{L} \sim E^{-1}$, or (ii) $\mathcal{L} =$ constant.

It has been noted in [68] that such a modification in the dispersion relation will affect the form of the spin-flavour conversion mechanism. Indeed, it is well known through the Mikheyev-Smirnov-Wolfenstein (MSW) effect [69] that Weak interaction Effects of ν propagating in a medium result in an energy shift $\sqrt{2}G_F(2n_e - n_n)$, where $n_e(n_n)$'s denote electron (neutron) densities. In addition to such effects, one should also take into account the interaction of ν with external magnetic fields, B, via a radiatively induced magnetic moment μ, corresponding to a term in the effective lagrangian: $\mathcal{L}_{\rm int} = \mu\overline{\psi}\sigma^{\mu\nu}F_{\mu\nu}\psi$, with ψ the neutrino fermionic field.

According to the standard theory, the equation for evolution describing the *spin-flavour* conversion phenomenon due to the above-described medium and magnetic moment effects for, say, two neutrino flavours (ν_e, ν_μ) is given by:

$$i\frac{d}{dr}\begin{pmatrix}\nu_{eL}\\ \nu_{\mu L}\\ \nu_{eR}\\ \nu_{\mu R}\end{pmatrix} = \mathcal{H}\begin{pmatrix}\nu_{eL}\\ \nu_{\mu L}\\ \nu_{eR}\\ \nu_{\mu R}\end{pmatrix}, \tag{52}$$

where the effective Hamiltonian $\mathcal{H}$ should be corrected in the loop gravity case to take into account λ-effects, associated with MDR (51):

$$\mathcal{H} = \begin{pmatrix} -\frac{\Delta m^2}{4p}\cos 2\theta - \lambda + \sqrt{2}G_F n_e & \frac{\Delta m^2}{4p}\sin 2\theta & \mu_{ee}B & \mu B \\ \frac{\Delta m^2}{4p}\sin 2\theta & \frac{\Delta m^2}{4p}\cos 2\theta - \lambda + \sqrt{2}G_F n_e & \mu B & \mu_{\mu\mu}B \\ \mu_{ee}B & \mu B & -\frac{\Delta m^2}{4p}\cos 2\theta + \lambda & \frac{\Delta m^2}{4p}\sin 2\theta \\ \mu B & \mu_{\mu\mu}B & \frac{\Delta m^2}{4p}\sin 2\theta & \frac{\Delta m^2}{4p}\cos 2\theta + \lambda \end{pmatrix} \tag{53}$$

where $\mu \equiv \mu_{e\mu}$, $\Delta m^2 = m_2^2 - m_1^2$, and B is the magnetic field. We should notice at this stage that the above formalism refers to Dirac ν; for Majorana ν one should replace: $\nu_{iL} \to \nu_i$, $\nu_{iR} \to \overline{\nu}_i$. Details can be found in [68].

For our purposes we note that the Resonant Conditions for Flavour-Spin-flip are [68]:

$$\nu_{eL} \to \nu_{\mu R}: \quad 2\lambda + \frac{\Delta m^2}{2p}\cos 2\theta - \sqrt{2}G_F n_e(r_{res}) = 0$$
$$\nu_{\mu L} \to \nu_{eR}: \quad 2\lambda - \frac{\Delta m^2}{2p}\cos 2\theta - \sqrt{2}G_F n_e(r_{res}) = 0 \, . \tag{54}$$

One can use the above conditions to obtain bounds for λ, κ_i via the oscillation probabilities for spin-flavour conversion:

$$P_{\nu_{eL}\to\nu_{\mu R}} = \frac{1}{2}(1 - \cos 2\tilde{\theta}\cos 2\theta) \; , \tag{55}$$

where $\tan 2\tilde{\theta}(r) = \frac{4\mu B(r)E}{|\Delta m^2|\cos 2\theta - 4E\lambda + 2\sqrt{2}G_F E n_e(r)}$.

To obtain these bounds the author of [68] made the following physically relevant assumptions: (a) Reasonable profiles for solar $n_e \sim n_0 e^{-10.5r/R_\odot}$, $n_0 = 85 N_A \mathrm{cm}^{-3}$. (b) Also: $\mu \sim 10^{-11}\mu_B$. Then, an upper bound on λ is obtained of order: $\lambda \le \frac{1}{2}\left(10^{-12}e^{-10.5r_{res}/R_\odot}\mathrm{eV} + \frac{|\Delta m^2|}{2E}\right)$.

To obtain bounds on κ we need to distinguish two cases:

(I) $\mathcal{L}$ = *universal constant*: In this case, we already know from photon dispersion tests on GRB and (AGN) [10, 64] that $\mathcal{L} \sim 10^{-18}\,\mathrm{eV}^{-1}$. Then, from best-fit solar ν-oscillations induced by MSW, one may use experimental values of Δm^2, $\sin^2 2\theta$, and obtain the following bound on κ_i: $\kappa_5 < 10^{-25}$. From atmospheric oscillations, in particular LSND experiment [63], $\nu_\mu \to \nu_e$ fits the data with: $|\Delta m^2| \sim eV^2$, $\sin^2 2\theta \sim (0.2 - 3) \times 10^{-3}$, $E_{\max} \sim 10\,\mathrm{MeV}$, then $\kappa_5 < 10^{-17}$.

(II) $\mathcal{L} \sim p^{-1}$ *a mobile scale*: In that case, from SOLAR oscillations, with $p \sim 1$–$10\,\mathrm{MeV}$ one gets $\kappa_5 = \mathcal{O}(1 - 100)$, which is a natural range of values from a quantum-gravity view point. From atmospheric oscillations, for the maximum $\nu E \sim 10\,\mathrm{MeV}$, and $\mathcal{L} \sim E^{-1}$, one obtains $\kappa_5 \sim 10^4$, which is a very weak bound.

The conclusion from these considerations, therefore, is that the experimental data seem to favour case **(II)**, at least from a naturalness point of view.

3.8 ν-Flavour States and Modified Lorentz Invariance (MLI)

An interesting recent idea [70], which we would like to discuss now briefly, arises from the observation of the peculiar way in which flavour ν states experience Lorentz Invariance. Indeed, neutrino flavour states are *a superposition of mass eigenstates* with standard dispersion relations of *different mass*. If one computes the expectation value of the Hamiltonian with respect to flavour states, e.g. in a simplified two-flavour scenario discussed in [70], then one finds:

$$E_e = \langle \nu_e | H | \nu_e \rangle = \omega_{k,1} \cos^2\theta + \omega_{k,2} \sin^2\theta \ ,$$
$$E_\mu = \langle \nu_\mu | H | \nu_\mu \rangle = \omega_{k,2} \cos^2\theta + \omega_{k,1} \sin^2\theta \ , \tag{56}$$

with θ the mixing angle.

One has: $H|\nu_i\rangle = \omega_i|\nu_i\rangle$, $i = 1, 2$, where the $\omega_{k,i} = \sqrt{\boldsymbol{k}^2 + m_i^2}$ is a standard dispersion relation. However, since the sum of two square roots in not in general a square root, one concludes that flavour states do not satisfy the standard dispersion relations. In general this poses a problem, as it would naively imply the introduction of a preferred frame, due to an apparent violation of the standard linear Lorentz symmetry.

The idea of [70], whose validity of course remains to be seen, but which I find rather intriguing, and this is why I decided to include it in these lectures, is to avoid using preferred frames by introducing instead non-linearly modified Lorentz transformations to account for the modified dispersion relations of the flavour states. The idea is formally similar, but physically very different, to the approach of [13], in which, in order to ensure observer independence of the Planck length, viewed as an ordinary length in quantum gravity, and not as a universal coupling constant, one has to modify non linearly the Lorentz transformations. The result is that flavour states satisfy the following MDR:

$$E_i^2 f_i^2(E_i) - \boldsymbol{k}^2 g_i^2(E_i) = M_i^2 \quad i = e, \mu \tag{57}$$

One can determine [70] the $f_i(E_i, \theta, m_i), g_i(E_i, \theta, m_i), M_i(m_i, \theta)$ by comparing with $E_i = E(\omega_i, m_i)$ above (c.f. (56).

Then, in the spirit of [13], one can identify the non-linear Lorentz transformation that leaves the MDR (57) invariant: $U \circ (E, \boldsymbol{k}) = (Ef, \boldsymbol{k}g)$.

The interesting feature is that these ideas can be tested experimentally, e.g. in β-decay experiments: $N_1 \to N_2 + e^- + \bar{\nu}_e$, where e.g. $N_1 = {}^3\mathrm{H}$, $N_2 = {}^3\mathrm{He}$.

Energy conservation in conventional β-decay implies: $E_{N_1} = E_{N_2} + E + E_e$, where E is the energy of e, which would unavoidably introduce a preferred frame. However, in the non-linear LI case for flavour states, where the use of preferred frame is avoided, this relation is modified [70]: $E_{N_1} = E_{N_2} + E + E_e f_e(E_e)$.

These two choices are reflected in different predictions for the endpoint of the β-decay, that is the maximal kinetic energy the electron can carry (c.f. Fig. 8). We refer the interested reader to [70] for further discussion on the experimental set up to test these ideas.

From the point of view of CPTV, which is our main topic of discussion here, I must mention that in such non-linearly modified Lorentz symmetry cases it is not clear what form the CPT theorem, if any, takes. This is currently under investigation [38]. In this sense, the link between CPTV and modified flavour-state dispersion relations, and therefore the interpretation of the associated experiments from this viewpoint, are issues which are not yet clear.

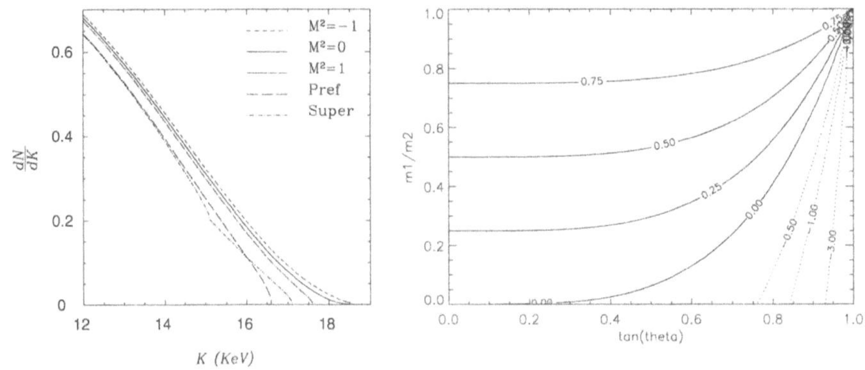

Fig. 8. *Left*: Tail of tritium β-decay spectrum, for massless ν (*solid*) and for LI flavour states (*dashed and dot-long-dashed*). Also plotted is the preferred frame case. *Right*: Likelihood Contours of M^2 (in units of m_2^2) upon which β-decay depends

3.9 CPTV and Departure from Locality for Neutrinos

As another way of violating CPT one can relax the requirement of *locality*. This idea has been pursued in [81], in an attempt to present a concrete model for CPT Violation for neutrinos, with CPTV Dirac masses, in an attempt to explain the LSND anomalous results [63], according to which there is experimental evidence for oscillations in the antineutrino sector, $\overline{\nu}_e \rightarrow \overline{\nu}_\mu$, but not in the corresponding neutrino one. In fact, the idea of invoking CPTV Dirac mass spectra for neutrinos in order to account for the LSND results without invoking a sterile neutrino is due to the authors of [82] (see Fig. 9). However no concrete theoretical model was presented there.

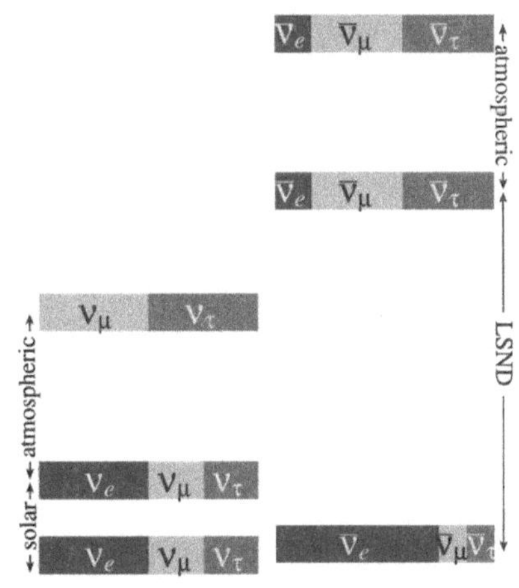

Fig. 9. The CPTV neutrino spectrum proposed by Murayama-Yanagida to explain LSND. One needs $m_\nu^2 - m_{\overline{\nu}}^2 \sim 0.1 \text{ ev}^{-2} = 10^{-19} \text{ GeV}^2$

The model lagrangian of [81] reads:

$$S = \int d^4x \bar{\psi} i \partial_\mu \gamma^\mu \psi + \frac{im}{2\pi} \int d^3x dt dt' \bar{\psi}(t) \frac{1}{t - t'} \psi(t') \tag{58}$$

The on shell equations (in momentum space) for the (Dirac) spinors are:

$$(p_\mu \gamma^\mu - m\epsilon(p_0)) u_\pm(p) = 0 \ , \tag{59}$$

with $\epsilon(p_0)$ the sign function, and

$$\begin{aligned} \psi_+(x) &= u_+(p) e^{-ip\cdot x}, \qquad p^2 = m^2,\ p_0 > 0 \\ \psi_-(x) &= u_-(p) e^{-ip\cdot x}, \qquad p^2 = m^2,\ p_0 < 0 \end{aligned} \tag{60}$$

Notice that on-shell Lorentz invariance is maintained due to the presence of $(\epsilon(p_0))$ but is relaxed.

As remarked in [36], however, the model of [81], although respecting Lorentz symmetry on-shell, has correlation functions (which are in general off-shell quantities) that do violate Lorentz symmetry, in the sense that they transform non covariantly under Lorentz transformations. Therefore, the CPTV in this model is ultimately connected to LV.

The two-generation non-local model of [81] seems to be marginally disfavoured by the current neutrino data, as claimed in [83] (see Fig. 10).

A summary of data and interpretations of current models, including those which entail CPT violation is given in Table 1, taken from the first paper in [83]. In that paper it has also been claimed that the recent WMAP [54] data on neutrinos seem to disfavour 3 + 1 scenaria which conserve CPT invariance. In my opinion one has to wait for future data from WMAP, before definite

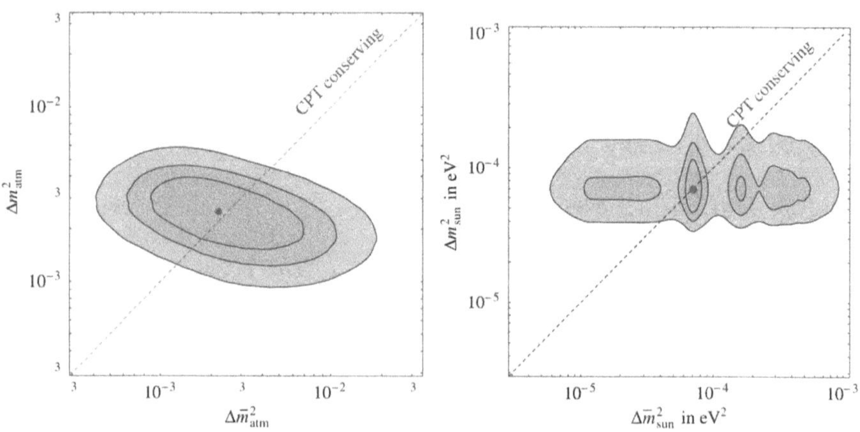

Fig. 10. *Left:* Atmospheric $m_\nu - m_{\bar{\nu}}$ (68, 90, 99%, 2 d.o.f.). *Right:* For solar & reactor data (68, 90, 99%, 2 d.o.f.)

Table 1. Interpretations of solar, atmospheric and LSND data, ordered according to the quality of their global fit. A $\Delta\chi^2 = n^2$ roughly signals an incompatibility at n standard deviations

Model & No. of Free Parameters		$\Delta\chi^2$	Mainly Incompatible with	Main Future Test
ideal fit (no known model)		0		?
$\Delta L = 2$ decay $\bar{\mu} \to \bar{e}\bar{\nu}_\mu\bar{\nu}_e$	6	12	KARMEN	TWIST
3+1: $\Delta m^2_{\text{sterile}} = \Delta m^2_{\text{LSND}}$	9	6+9?	BUGEY + cosmology?	MINIBOONE
3 ν and CPTV (no $\Delta\bar{m}^2_{\text{sun}}$)	10	15	KamLAND	KAMLAND
3 ν and CPTV (no $\Delta\bar{m}^2_{\text{atm}}$)	10	25	SK atmospheric	$\bar{\nu}_\mu$ LBL?
normal 3 neutrinos	5	25	LSND	MINIBOONE
$2+2:\ \Delta m^2_{\text{sterile}} = \Delta m^2_{\text{sun}}$	9	30	SNO	SNO
$2+2:\ \Delta m^2_{\text{sterile}} = \Delta m^2_{\text{atm}}$	9	50	SK atmospheric	ν_μ LBL

conclusions on this issue are reached, given that the current WMAP data are rather crude in this respect. I will not go further into a detailed discussion of this topic, as such summaries of neutrino data and their interpretations can be found in the literature, where I refer the interested reader [84].

Before closing this section, I would like to remark that most of the theoretical analyses for QG-induced CPTV in neutrinos have been done in simplified two-flavour oscillation models. Including all three generations in the formalism may lead to differences in the corresponding conclusions regarding sensitivity (or conclusions about exclusion) of the associated CPTV effects. In this respect the measurements of the mixing angle angle θ_{13} in the immediate future [85], as a way of detecting generic three-flavour effects, will be very interesting. In the current phenomenology, CPT invariance is assumed for the theoretical estimates of this parameter [86].

3.10 Four-Generation ν Models with CPTV

A natural question arises at this point, concerning $(3+1$ or $2+2)$ ν scenaria which violate CPT symmetry. This issue has been studied recently in [87]. These authors postulated a model for CPTV with four generations for neutrinos which leads to different flavor mixing between ν, $\bar{\nu}$: $\nu_a = \sum_{i=1}^4 U^*_{ai}\nu_i$, $\bar{\nu}_a = \sum_{i=1}^4 \bar{U}_{ai}\bar{\nu}_i$, with $U \neq \bar{U}$ due to CPTV. There are various cases to be studied:

- $3+1$ models (see Figs. 11, 12): one ν mass well separated from others, sterile ν couples only to isolated state. The relevant Oscillation probabilities are: $P_{\nu_i \to \nu_i}(|U_{ij}|^2) \neq P_{\bar{\nu}_i \to \bar{\nu}_i}(|\bar{U}_{ij}|^2)$
  Experimentally one may bound $|\bar{U}_{e4}|$ and $U_{\mu 4}$ but there are no tight constraints for $|\bar{U}_{\mu 4}|$, U_{e4}. This is to be contrasted with $(3+1)\nu$ CPT conserving models where $U = \bar{U}$. Hence $(3+1)\nu$ + CPTV seems still viable.

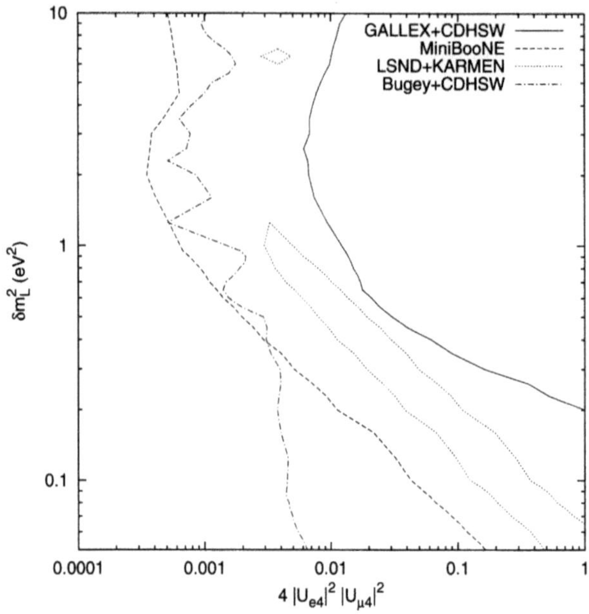

Fig. 11. Upper bound (*solid*) on the $\nu_\mu \to \nu_e$ oscillation amplitude $4|U_{e4}|^2|U_{\mu 4}|^2$ from the GALLEX limit on $|U_{e4}|$ and the CDHSW limit on $|U_{\mu 4}|$ (90% C. L. results are used in both cases). The *dot-dashed line* is the 99% C. L. upper bound from Bugey and CDHSW if *CPT* is conserved. Also shown are the expected sensitivity (*dashed*) of the MiniBooNE experiment and, for comparison, the allowed region (within the *dotted lines*) for $4|\bar{U}_{e4}|^2|\bar{U}_{\mu 4}|^2$ from a combined analysis of LSND and KARMEN data, both at the 90% C. L

- 2 + 2 models (see Fig. 13): sterile ν couples to solar and atmospheric ν oscillations. This structure is only permitted in $\bar{\nu}$ sector. Even with CPT Violation, however, 2 + 2 models are strongly disfavoured by data.

Although the introduction of a fourth neutrino generation with CPTV within conventional field theory seems to be consistent with the current neutrino data, however, there seems to be no concrete evidence for a forth generation from any experiment to date, including the most recent astrophysical WMAP data, as we have seen above. This prompts one to examine alternative ways of explaining the current neutrino "anomalous data", like LSND, employing unconventional ways of CPT Violation by means of quantum decoherence, which are in principle independent of mass differences between particle and antiparticles.

In the next subsection we shall be dealing with this topic, reviewing first the relevant phenomenological formalism which allows direct comparison with experiment. I will start with the phenomenology of CPT violation in neutral mesons and neutrons, as a historical introduction to the general reader, and then I will proceed to the neutrino case. I shall argue that minimal decoherence

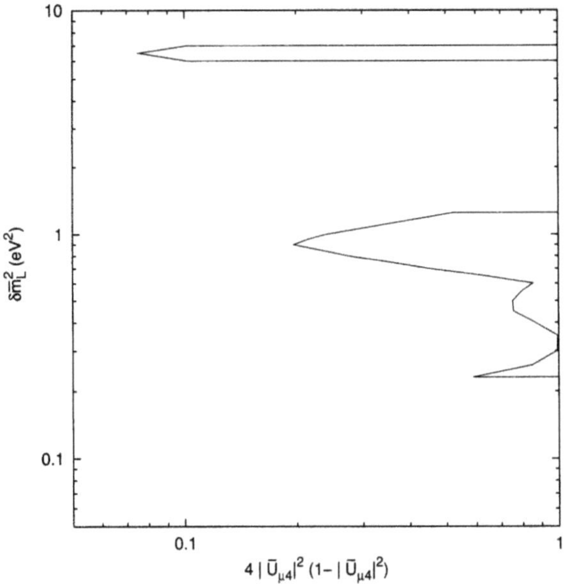

Fig. 12. Lower bounds on $4|\bar{U}_{\mu 4}|^2(1-|\bar{U}_{\mu 4}|^2)$ (the amplitude for atmospheric $\bar{\nu}_\mu$ survival at the LSND mass scale) from the Bugey limit on $\bar{\nu}_e$ disappearance and the $\bar{\nu}_\mu \to \bar{\nu}_e$ oscillation amplitude indicated by LSND and KARMEN (90% C. L. results are used in both cases)

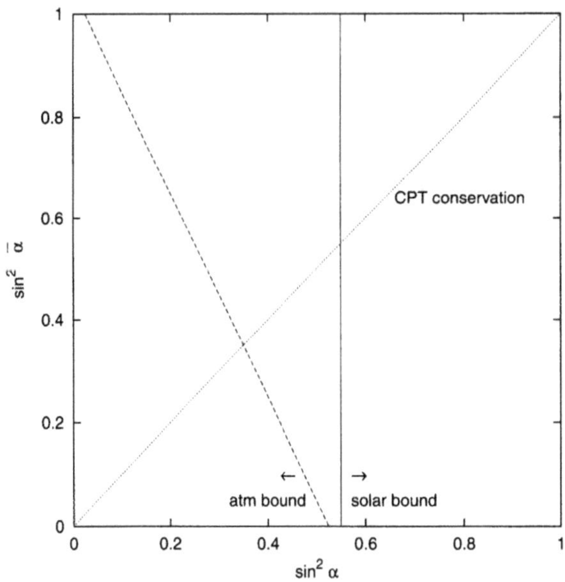

Fig. 13. Constraints on sterile neutrino mixing angles α and $\bar{\alpha}$ from (*solid line*) and (*dashed line*) data. The *dotted line* is the prediction if CPT is conserved

models with CPTV differences in the decoherence parameters between particle and antiparticle sectors, but not CPTV mass differences, can account for all existing neutrino data, including LSND results, without the need for enlarging the neutrino sector, that is staying within three generation models.

3.11 CPTV Through QG Decoherence: Neutral Mesons

In this subsection I will discuss CPTV through decoherence, which is my preferred way of QG-induced CPTV. As mentioned above, in this case the matter systems are viewed as open quantum mechanical or quantum-field theoretic systems interacting with a gravitational "environment", consisting of degrees of freedom inaccessible to low-energy scattering experiments. The presence of such an environment leads to modified quantum evolution, which however is *not necessarily Lorentz Violating* [33]. Thus, such an approach to CPTV should in principle be studied separately, and indeed it is possible for the CPTV decoherence effects to be disentangled experimentally from the LV ones, due to the frame dependence of the latter.

Currently, the most sensitive particle physics probes of such a modification from quantum mechanical behavior (often called "quantum mechanics violation" QMV [24, 25]) are: (i) neutral kaons and B-mesons [24, 25] and ϕ-, B-factories [26, 27, 28] (ii) neutron interferometry [24], (iii) ultracold (slow) neutrons in Earth's gravitational field, and (iv) Neutrino flavour mixing, which is induced independently of masses and mass differences between neutrino species, as we shall discuss below. In these lectures I will discuss briefly (i),(iii) and (iv).

Let us start with the neutral Kaon case. This is a typical two-state system of decoherence. One could follow the Lindblad parametrization [27], in which the requirement of complete positivity would imply a single decoherence parameter γ. The requirement of energy conservation on the average in such models would then imply the double commutator structure (17) for the decoherence term, which however would depend on the square of the energy variance between the two energy-eigenstates of the neutral Kaon system, as in (24). This would be too small to be detected experimentally in neutral meson experiments and factories in the foreseeable future.

However, as argued in [71], complete positivity may not be valid in generic models of quantum gravity, such as the non-critical string decoherence models (39). Indeed, in that case, the decoherence terms contain the Zamolodchikov metric $\langle V_i V_j \rangle$ and as such are non-linear in the probe state density matrix ρ, given that $\langle \dots \rangle = \mathrm{Tr}(\rho \dots)$ depends on it. Complete positivity for non-linear effective theories (e.g. Hartree-Fock type, mean field approaches) is in general a non well defined concept [72].

In fact, in the original parametrization [24] of the QG-induced decoherence effects for the neutral Kaon system this requirement has not been imposed. In such a paremetrization, which has also been followed in more recent, and more complete, phenomenological analyses of this system [25, 26], in addition to the basic principle of entropy increase, one also imposes the requirement of conservation of strangeness by the quantum gravity interactions. This follows from the so-called $\Delta S = \Delta Q$ rule which seems to characterise the leading-order Kaon weak-interaction physics, which in general violates strangeness, but the charge transfer in the neutral Kaon physics is a much more subleading effect than the dominant CP violation effects. This feature is assumed to be obeyed by the quantum gravity interactions [24], which are thus assume to conserve strangeness to leading order.

According to our general discussion in Sect. 2 on the dynamical-semigroup approach to decoherence, on which the formalism of [24] is based, for the neutral-Kaon two-level system the non-Hamiltonian decoherence term in the evolution equation for ρ can be parametrized by a 4×4 matrix $\delta \not{H}_{\alpha\beta}$, where the indices $\alpha, \beta, \ldots$ enumerate the Hermitian σ-matrices $\sigma_{0,1,2,3}$, which we represent in the so-called $K_{1,2}$ basis, defined as $|K^{1,2}\rangle = \frac{1}{\sqrt{2}}\left(|K^0\rangle \pm |\overline{K}^0\rangle\right)$. In Neutral Kaons, the CP eigenstates are not energy (physical) eigenstates, thereby leading to mixing. We refer the reader to the literature [24, 25] for details of this description, noting here the following forms for the neutral kaon Hamiltonian

$$H = \begin{pmatrix} M - \frac{i}{2}\Gamma - \mathrm{Re}M_{12} + \frac{i}{2}\mathrm{Re}\Gamma_{12} & \frac{1}{2}\delta M - \frac{i}{4}\delta\Gamma - i\mathrm{Im}M_{12} - \frac{1}{2}\mathrm{Im}\Gamma_{12} \\ \frac{1}{2}\delta M - \frac{i}{4}\delta\Gamma + i\mathrm{Im}M_{12} - \frac{1}{2}\mathrm{Im}\Gamma_{12} & M - \frac{i}{2}\Gamma + \mathrm{Re}M_{12} - \frac{i}{2}\mathrm{Re}\Gamma_{12} \end{pmatrix} \tag{61}$$

in the $K_{1,2}$ basis, or

$$H_{\alpha\beta} = \begin{pmatrix} -\Gamma & -\frac{1}{2}\delta\Gamma & -\mathrm{Im}\Gamma_{12} & -\mathrm{Re}\Gamma_{12} \\ -\frac{1}{2}\delta\Gamma & -\Gamma & -2\mathrm{Re}M_{12} & -2\mathrm{Im}M_{12} \\ -\mathrm{Im}\Gamma_{12} & 2\mathrm{Re}M_{12} & -\Gamma & -\delta M \\ -\mathrm{Re}\Gamma_{12} & -2\mathrm{Im}M_{12} & \delta M & -\Gamma \end{pmatrix} \tag{62}$$

in the σ-matrix basis. Above, M denotes mass parameters, Γ denotes widths, and $\delta(\ldots)$ denotes CPTV differences between particle and antiparticle sectors, which are due to quantum mechanical effects, such as LV etc.

As mentioned previously, we assume that the dominant violations of quantum mechanics conserve strangeness, so that $\delta \not{H}_{1\beta} = 0$, and that $\delta \not{H}_{0\beta} = 0$ so as to conserve probability. Since $\delta \not{H}_{\alpha\beta}$ is a symmetric matrix, it follows that also $\delta \not{H}_{\alpha 0} = \delta \not{H}_{\alpha 1} = 0$. Thus, we arrive at the general parametrization

$$\delta \not{H}_{\alpha\beta} = \begin{pmatrix} 0 & 0 & 0 & 0 \\ 0 & 0 & 0 & 0 \\ 0 & 0 & -2\alpha & -2\beta \\ 0 & 0 & -2\beta & -2\gamma \end{pmatrix} \tag{63}$$

where, as a result of the positivity of the hermitian density matrix ρ [24]

$$\alpha, \gamma > 0, \qquad \alpha\gamma > \beta^2 \,. \tag{64}$$

We recall [25] that the decoherence terms violate CP, given that the latter transformation can be expressed as a linear combination of $\sigma_{2,3}$ in the $K_{1,2}$ basis : CP $= \sigma_3 \cos\theta + \sigma_2 \sin\theta$, for some choice of phase θ. It is apparent that none of the non-zero terms $\propto \alpha, \beta, \gamma$ in $\delta \not{H}_{\alpha\beta}$ (63) commutes with this CP transformation. In other words, each of the three parameters α, β, γ violates CP. Moreover, in the problem there is evolution of pure to mixed states, as we shall discuss below, leading, according to the theorem of [37], described above, also to a strong form of CPT Violation. Thus, the decoherent CPTV evolution in the neutral Kaon system leads to a much richer phenomenology than in conventional CPT Violations within a quantum mechanical framework, in the absence of decoherence, where the CPT may be violated only through differences in masses δM and widths $\delta\Gamma$ between particles and antiparticles. This is because the symmetric $\delta\not{H}$ matrix has three parameters in its bottom right-hand 2×2 submatrix, whereas the h matrix appearing in the time evolution within quantum mechanics has only one complex CPT-violating parameter δ,

$$\delta = -\frac{1}{2}\frac{\frac{1}{2}\delta\Gamma + i\delta M}{\frac{1}{2}|\Delta\Gamma| + i\Delta m} , \tag{65}$$

where δM and $\delta\Gamma$ violate CPT, but do not induce any mixing in the time evolution of pure state vectors [25]. The parameters $\Delta m = M_L - M_S$ and $|\Delta\Gamma| = \Gamma_S - \Gamma_L$ are the usual differences between mass and decay widths, respectively, of the long-lived K_L and short-lived K_S energy (physical) eigenstates. For more details we refer the reader to the literature [25]. The above results imply that the experimental constraints [90] on CPT Violation have to be rethought. As we shall discuss later on, there are essential differences between quantum-mechanical CPT Violation and the non-quantum-mechanical CPT violation induced by the effective parameters α, β, γ [24].

Useful observables are associated with the decays of neutral kaons to 2π or 3π final states, or semileptonic decays to $\pi l \nu$. In the density matrix formalism introduced above, their values are given by expressions of the form [24, 25]

$$\langle O_i \rangle = \mathrm{Tr}\,[O_i \rho] , \tag{66}$$

where the observables O_i are represented by 2×2 hermitian matrices. For instructive purposes we give their expressions in the $K_{1,2}$ basis

$$O_{2\pi} = \begin{pmatrix} 0 & 0 \\ 0 & 1 \end{pmatrix} , \qquad O_{3\pi} \propto \begin{pmatrix} 1 & 0 \\ 0 & 0 \end{pmatrix} , \tag{67}$$

$$O_{\pi^- l^+ \nu} = \begin{pmatrix} 1 & 1 \\ 1 & 1 \end{pmatrix} , \qquad O_{\pi^+ l^- \bar{\nu}} = \begin{pmatrix} 1 & -1 \\ -1 & 1 \end{pmatrix} . \tag{68}$$

which constitute a complete hermitian set. As we discuss in detail in [25], it is possible to measure the interferecne between $K_{1,2}$ decays into $\pi^+\pi^-\pi^0$ final states with different CP properties, by restricting one's attention to part of the phase space Ω, *e.g.*, final states with $m(\pi^+\pi^0) > m(\pi^-\pi^0)$. In order to separate this interference from that due to $K_{S,L}$ decays into final states with identical CP properties, due to CP Violation in the $K_{1,2}$ mass matrix or in decay amplitudes, we consider the difference between final states with $m(\pi^+\pi^0) > m(\pi^-\pi^0)$ and $m(\pi^+\pi^0) < m(\pi^-\pi^0)$. This observable is represented by the matrix

$$O^{\rm int}_{3\pi} = \begin{pmatrix} 0 & \mathcal{K} \\ \mathcal{K}^* & 0 \end{pmatrix} \tag{69}$$

where

$$\mathcal{K} \equiv \frac{\left[\int_{m(\pi^+\pi^0)>m(\pi^-\pi^0)} d\Omega - \int_{m(\pi^+\pi^0)<m(\pi^-\pi^0)} d\Omega\right] A_2(I_{3\pi}=2)A_1(I_{3\pi}=1)}{\int d\Omega |A_1(I_{3\pi}=1)|^2} \tag{70}$$

where $\mathcal{K}$ is expected to be essentially real, so that the $O^{\rm int}_{3\pi}$ observable provides essentially the same information as $O_{\pi^-l^+\nu} - O_{\pi^+l^-\bar{\nu}}$.

In this formalism, pure K^0 or $\bar{K}^0$ states, such as the ones used as initial conditions in the CPLEAR experiment [89], are described by the following density matrices

$$\rho_{K^0} = \tfrac{1}{2}\begin{pmatrix} 1 & 1 \\ 1 & 1 \end{pmatrix}, \qquad \rho_{\bar{K}^0} = \tfrac{1}{2}\begin{pmatrix} 1 & -1 \\ -1 & 1 \end{pmatrix}. \tag{71}$$

We note the similarity of the above density matrices (71) to the semileptonic decay observables in (68), which is due to the strange quark (s) content of the kaon $K^0 \ni \bar{s} \to \bar{u}l^+\nu, \bar{K}^0 \ni s \to ul^-\bar{\nu}$, and our assumption of the validity of the $\Delta S = \Delta Q$ rule.

One can apply the above formalism to compute the time evolution of certain quantities that are of relevance to experiment [89], being directly observable. These are asymmetries associated with decays of an initial K^0 beam as compared to corresponding decays of an initial $\bar{K}^0$ beam

$$A(t) = \frac{R(\bar{K}^0_{t=0} \to \bar{f}) - R(K^0_{t=0} \to f)}{R(\bar{K}^0_{t=0} \to \bar{f}) + R(K^0_{t=0} \to f)}, \tag{72}$$

where $R(K^0 \to f) \equiv {\rm Tr}\left[O_f\rho(t)\right]$, denotes the decay rate into the final state f, given that one starts from a pure K^0 at $t=0$, whose density matrix is given in (71), and $R(\bar{K}^0 \to \bar{f}) \equiv {\rm Tr}\left[O_{\bar{f}}\bar{\rho}(t)\right]$ denotes the decay rate into the conjugate state $\bar{f}$, given that one starts from a pure $\bar{K}^0$ at $t=0$. One considers the following set of asymmetries: (i) *identical final states*: $f = \bar{f} = 2\pi$: $A_{2\pi}$, $A_{3\pi}$, (ii) *semileptonic* : A_T (final states $f = \pi^+l^-\bar{\nu} \neq \bar{f} = \pi^-l^+\nu$), A_{CPT} ($\bar{f} = \pi^+l^-\bar{\nu}$, $f = \pi^-l^+\nu$), $A_{\Delta m}$.

Typically, for instance when the final states are 2π, one has a time evolution of the decay rate $R_{2\pi}$: $R_{2\pi}(t) = c_S\, e^{-\Gamma_S t} + c_L\, e^{-\Gamma_L t} + 2c_I\, e^{-\Gamma t}\cos(\Delta m t - \phi)$, where S = short-lived, L = long-lived, I = interference term, $\Delta m = m_L - m_S$, $\Gamma = \frac{1}{2}(\Gamma_S + \Gamma_L)$. One may define the *decoherence parameter* $\zeta = 1 - \frac{c_I}{\sqrt{c_S c_L}}$, as a measure of quantum decoherence induced in the system. For larger sensitivities one can look at this parameter in the presence of a regenerator [25]. In our decoherence scenario, it can be shown [25] that ζ depends primarily on β, hence the best bounds on β can be placed by implementing a regenerator.

Let us illustrate the formalism by two explicit examples. We may compute the asymmetry for the case where there are identical final states $f = \bar{f} = 2\pi$, in which case the observable is given in (67). We obtain

$$A_{2\pi} = \frac{\mathrm{Tr}\,[O_{2\pi}\bar{\rho}(t)] - \mathrm{Tr}\,[O_{2\pi}\rho(t)]}{\mathrm{Tr}\,[O_{2\pi}\bar{\rho}(t)] + \mathrm{Tr}\,[O_{2\pi}\rho(t)]} = \frac{\mathrm{Tr}\,[O_{2\pi}\Delta\rho(t)]}{\mathrm{Tr}\,[O_{2\pi}\Sigma\rho(t)]}\,, \tag{73}$$

where we have defined: $\Delta\rho(t) \equiv \bar{\rho}(t) - \rho(t)$ and $\Sigma\rho(t) \equiv \bar{\rho}(t) + \rho(t)$. We note that in the above formalism we make no distinction between neutral and charged two-pion final states. This is because we neglect, for simplicity, the effects of ϵ'. Since $|\epsilon'/\epsilon| \lesssim 10^{-3}$, this implies that our analysis of the new quantum-mechanics-violating parameters must be refined if magnitudes $\lesssim \epsilon'|\Delta\Gamma| \simeq 10^{-6}|\Delta\Gamma|$ are to be studied [26].

In a similar spirit to the identical final state case, one can compute the asymmetry A_{T} for the semileptonic decay case, where $f = \pi^+l^-\bar{\nu} \neq \bar{f} = \pi^-l^+\nu$. The formula for this observable is

$$A_{\mathrm{T}}(t) = \frac{\mathrm{Tr}\,[O_{\pi^-l^+\nu}\bar{\rho}(t)] - \mathrm{Tr}\,[O_{\pi^+l^-\bar{\nu}}\rho(t)]}{\mathrm{Tr}\,[O_{\pi^-l^+\nu}\bar{\rho}(t)] + \mathrm{Tr}\,[O_{\pi^+l^-\bar{\nu}}\rho(t)]}\,. \tag{74}$$

Other observables are discussed in [25], where a complete phenomenological description of CPTV decohering effects is presented.

To determine the temporal evolution of the above observables, which is crucial for experimental fits, it is necessary to know the equations of motion for the components of ρ in the $K_{1,2}$ basis. These are [25][1]

$$\dot{\rho}_{11} = -\Gamma_L\rho_{11} + \gamma\rho_{22} - 2\mathrm{Re}\,[(\mathrm{Im}M_{12} - i\beta)\rho_{12}]\,, \tag{75}$$
$$\dot{\rho}_{12} = -(\Gamma + i\Delta m)\rho_{12} - 2i\alpha\mathrm{Im}\,\rho_{12} + (\mathrm{Im}M_{12} - i\beta)(\rho_{11} - \rho_{22})\,, \tag{76}$$
$$\dot{\rho}_{22} = -\Gamma_S\rho_{22} + \gamma\rho_{11} + 2\mathrm{Re}\,[(\mathrm{Im}M_{12} - i\beta)\rho_{12}]\,, \tag{77}$$

where for instance ρ may represent $\Delta\rho$ or $\Sigma\rho$, defined by the initial conditions

$$\Delta\rho(0) = \begin{pmatrix} 0 & -1 \\ -1 & 0 \end{pmatrix}, \qquad \Sigma\rho(0) = \begin{pmatrix} 1 & 0 \\ 0 & 1 \end{pmatrix}. \tag{78}$$

In these equations $\Gamma_L = (5.17 \times 10^{-8}\sigma)^{-1}$ and $\Gamma_S = (0.8922 \times 10^{-10}\sigma)^{-1}$ are the inverse K_L and K_S lifetimes, $\Gamma \equiv (\Gamma_S + \Gamma_L)/2$, $|\Delta\Gamma| \equiv \Gamma_S - \Gamma_L = (7.364\pm0.016)\times10^{-15}$ GeV, and $\Delta m = 0.5351\times10^{10}\sigma^{-1} = 3.522\times10^{-15}$ GeV is the $K_L - K_S$ mass difference. Also, the CP impurity parameter ϵ is given by

$$\epsilon = \frac{\mathrm{Im}M_{12}}{\frac{1}{2}|\Delta\Gamma| + i\Delta m}\,, \tag{79}$$

[1] Since we neglect ϵ' effects and assume the validity of the $\Delta S = \Delta Q$ rule, in what follows we also consistently neglect $\mathrm{Im}\,\Gamma_{12}$ [88].

which leads to the relations

$$\operatorname{Im} M_{12} = \frac{1}{2}\frac{|\Delta\Gamma||\epsilon|}{\cos\phi}, \quad \epsilon = |\epsilon|e^{-i\phi} \quad : \quad \tan\phi = \frac{\Delta m}{\frac{1}{2}|\Delta\Gamma|}, \tag{80}$$

with $|\epsilon| \approx 2.2 \times 10^{-3}$ and $\phi \approx 45°$ the "superweak" phase [88].

These equations are to be compared with the corresponding quantum-mechanical equations, which are reviewed in [25]. The parameters δM and β play similar roles, although they appear with different relative signs in different places, because of the symmetry of δH as opposed to the antisymmetry of the quantum-mechanical evolution matrix H. These differences are important for the asymptotic limits of the density matrix, and its impurity. In our approach, one can readily show that, at large t, ρ decays exponentially to [25]:

$$\rho_L \approx \begin{pmatrix} 1 & (|\epsilon| + i2\widehat{\beta}\cos\phi)e^{i\phi} \\ (|\epsilon| - i2\widehat{\beta}\cos\phi)e^{-i\phi} & |\epsilon|^2 + \widehat{\gamma} - 4\widehat{\beta}^2\cos^2\phi - 4\widehat{\beta}|\epsilon|\sin\phi \end{pmatrix}, \tag{81}$$

where we have defined the following scaled variables

$$\widehat{\alpha} = \alpha/|\Delta\Gamma|, \quad \widehat{\beta} = \beta/|\Delta\Gamma|, \quad \widehat{\gamma} = \gamma/|\Delta\Gamma|. \tag{82}$$

Conversely, if we look in the short-time limit for a solution of the (75) to (77) with $\rho_{11} \ll \rho_{12} \ll \rho_{22}$, we find [25]

$$\rho_S \approx \begin{pmatrix} |\epsilon|^2 + \widehat{\gamma} - 4\widehat{\beta}^2\cos^2\phi + 4\widehat{\beta}|\epsilon|\sin\phi & (|\epsilon| + i2\widehat{\beta}\cos\phi)e^{-i\phi} \\ (|\epsilon| - i2\widehat{\beta}\cos\phi)e^{i\phi} & 1 \end{pmatrix}. \tag{83}$$

These results are to be contrasted with those obtained within conventional quantum mechanics

$$\rho_L \approx \begin{pmatrix} 1 & \epsilon^* \\ \epsilon & |\epsilon|^2 \end{pmatrix}, \qquad \rho_S \approx \begin{pmatrix} |\epsilon|^2 & \epsilon \\ \epsilon^* & 1 \end{pmatrix}, \tag{84}$$

which, as can be seen from their *vanishing determinant*, correspond to pure K_L and K_S states respectively.

This is an important difference of the decoherence approach of [24] from others, as it implies an evolution of pure states to mixed. Indeed, a pure state will remain pure as long as $\operatorname{Tr}\rho^2 = (\operatorname{Tr}\rho)^2 = \operatorname{Tr}\rho = 1$, or equivalently if $\rho^2 = \rho$ as operator relations, as discussed in Sect. 2 (the normalisation $\operatorname{Tr}\rho = 1$ expresses conservation of probability). In the case of 2×2 matrices $\operatorname{Tr}\rho^2 = (\operatorname{Tr}\rho)^2 - 2\det\rho$, and therefore the purity condition is equivalently expressed as $\det\rho = 0$. In contrast, ρ_L, ρ_S in (81, 83) describe *mixed states*. Even in the limit of the imposition of complete positivity, which according to the analysis of [27], would imply $\alpha = \beta = 0$, $\gamma > 0$, there is a non vanishing determinant for the above matrices, indicating the difference of the decoherence model of [24, 25] from others in the literature where purity of states has been maintained during the evolution [40, 41].

As mentioned above, the maximum possible order of magnitude for the decoherence parameters $|\alpha|, |\beta|$ or $|\gamma|$ that we could expect theoretically is $\mathcal{O}(E^2/M_{Pl}) \sim \mathcal{O}((\Lambda_{\rm QCD} \text{ or } m_s)^2/M_{Pl}) \sim 10^{-19}$ GeV in the neutral kaon system. The fact that the model is different, in general, from the double commutator Lindblad model of decoherence (17), is welcome from a phenomenological view point, given that it avoids the suppression (24) [71]. Such unsuppressed models may characterise, for instance, the Liouville-string decoherence [2], described above, which are thus subject to direct experimental tests in the near future.

To make a consistent phenomenological study of the various asymmetries discussed above, in particular to determine their time profiles and compare them with experiment [89], it is essential to solve the coupled system of equations (75) to (77) for intermediate times. This requires approximations in powers of the decoherence parameters in order to get analytic results [25], which we shall not describe here. Below we shall only outline the results briefly by demonstrating the time profiles of the asymmetries $A_{2\pi}$ and A_T, as well as the asymmetry $A_{\Delta m}$ used in the CPLEAR experiment [89]. The relevant results are outlined in Figs. 14, 15, 16.

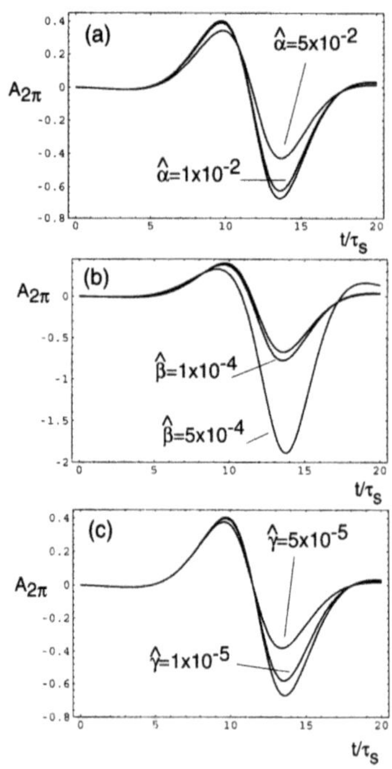

Fig. 14. The time-dependent asymmetry $A_{2\pi}$ for various choices of the CPT-violating parameters: (**a**) dependence on $\widehat{\alpha}$, (**b**) dependence on $\widehat{\beta}$, (**c**) dependence on $\widehat{\gamma}$. The unspecified parameters are set to zero. The curve with no labels corresponds to the standard quantum-mechanical case ($\widehat{\alpha} = \widehat{\beta} = \widehat{\gamma} = 0$)

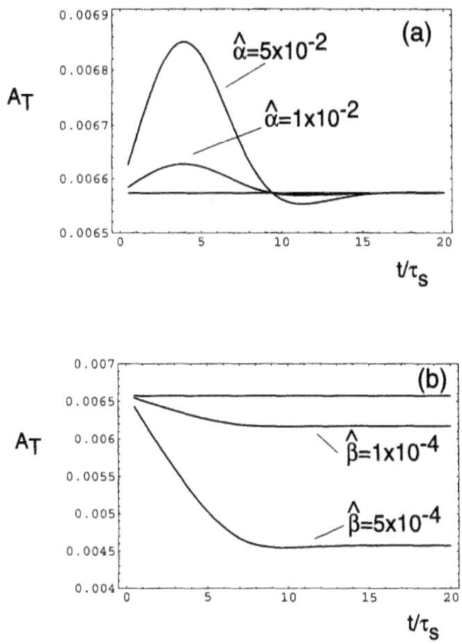

Fig. 15. The time-dependent asymmetry A_T for representative choices of (**a**) $\widehat{\alpha}$ ($\widehat{\beta} = 0$) and (**b**) $\widehat{\beta}$ ($\widehat{\alpha} = 0$). The dependence on $\widehat{\gamma}$ is negligible. The flat line corresponds to the standard case

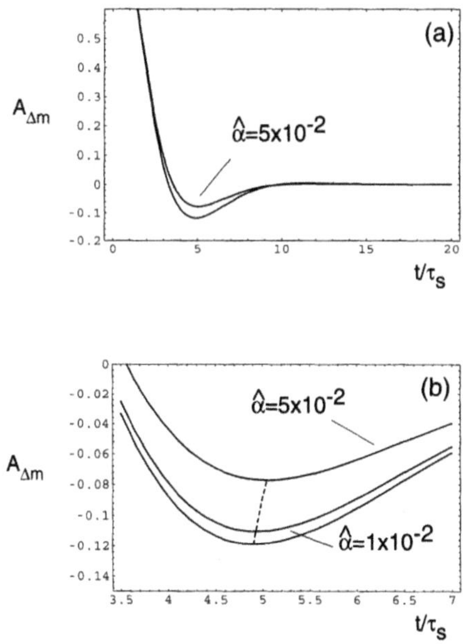

Fig. 16. The time-dependent asymmetry $A_{\Delta m}$ for representative choices of $\widehat{\alpha}$ ($\widehat{\beta} = \widehat{\gamma} = 0$). This asymmetry depends most sensitively only on $\widehat{\alpha}$. In both panels, the *bottom curve* corresponds to the standard case. In the detail (**b**), the *dashed line* indicates the location of the minimum as $\widehat{\alpha}$ is varied

The important point in such an analysis is that CPTV due to decoherence in neutral mesons can be disentangled from CPTV within quantum mechanics, for instance due to Lorentz Violation a lá SME [23]. The experimental tests (decay asymmetries) that can be performed in order to disentangle decoherence from quantum mechanical CPT violating effects are summarized in Table 2. Experimentally, the best available bounds to date for the neutral meson case come from CPLEAR measurements [89] $\alpha < 4.0 \times 10^{-17}$ GeV, $|\beta| < 2.3. \times 10^{-19}$ GeV, $\gamma < 3.7 \times 10^{-21}$ GeV, which are not much different from theoretically expected values in some models, $\alpha, \beta, \gamma = O(\xi \frac{E^2}{M_P})$.

Table 2. Qualitative comparison of predictions for various observables in CPT-violating theories beyond (QMV) and within (QM) quantum mechanics. Predictions either differ ($\neq$) or agree ($=$) with the results obtained in conventional quantum-mechanical CP violation. Note that these frameworks can be qualitatively distinguished via their predictions for A_{T}, A_{CPT}, $A_{\Delta m}$, and ζ

Process	QMV	QM
$A_{2\pi}$	$\neq$	$\neq$
$A_{3\pi}$	$\neq$	$\neq$
A_{T}	$\neq$	$=$
A_{CPT}	$=$	$\neq$
$A_{\Delta m}$	$\neq$	$=$
ζ	$\neq$	$=$

Before closing this section it is worthy of mentioning that above we have considered the same set of decoherence parameters α, β, γ in both particle and antiparticle sectors. However, in view of the induced CPTV in the strong form [37], it is not clear that the order of these two sets of parameters is the same between particle and antiparticle sectors. Although we have no concrete theoretical models at present, nevertheless, one may envisage cases where the strength of the interaction with the foam is different between matter and antimatter. An example of such a case will be seen later on, in the context of neutrino physics. As we shall see there, minimal models of QG-induced decoherence, with the latter being dominant only in the antiparticle sector, will be capable of explaining current neutrino anomalous data, such as LSND reasults [63], in a way consistent with all the other data.

3.12 EPR Entangled Neutral Meson States and Novel Decoherence-Induced CPT Violating Effects

In experiments involving multiparticle states, such as those produced in a ϕ or B factory, the fact that CPT may not be a well defined operation, as a result of decoherence induced by quantum gravity [37], could imply novel effects [28], which may affect the properties of the entangled states, and as such are unique to such situations, and absent in single particle experiments.

In conventional formulations of *entangled* meson states [91] one imposes the requirement of *Bose statistics* for the state $K^0\overline{K}^0$ (or $B^0\overline{B}^0$), which implies that the physical neutral meson-antimeson state must be *symmetric* under the combined operation $C\mathcal{P}$, with C the charge conjugation and $\mathcal{P}$ the operator that permutes the spatial coordinates. Specifically, assuming *conservation* of angular momentum, and a proper existence of the *antiparticle state* (denoted by a bar), one observes that, for $K^0\overline{K}^0$ states which are C-conjugates with $C = (-1)^\ell$ (with ℓ the quantum number), the system has to be an eigenstate of $\mathcal{P}$ with eigenvalue $(-1)^\ell$. Hence, for $\ell = 1$, we have that $C = -$, implying $\mathcal{P} = -$. As a consequence of Bose statistics this ensures that for $\ell = 1$ the state of two identical bosons is forbidden [91]. As a result, the initial entangled state $K^0\overline{K}^0$ produced in a ϕ factory can be written as:

$$|i\rangle = \frac{1}{\sqrt{2}}\left(|K^0(\boldsymbol{k}), \overline{K}^0(-\boldsymbol{k})\rangle - |\overline{K}^0(\boldsymbol{k}), K^0(-\boldsymbol{k})\rangle\right) \tag{85}$$

This is the starting point of all formalisms known to date, either in the K-system [91] or in the B-system, including those [26] where the evolution of the entangled state is described by non-quantum mechanical terms, in the formalism of [24]. In fact, in all these works it has been claimed that the expression in (85) is actually independent of any assumption about CP, T or CPT symmetries.

However, as has been alluded above, the assumptions leading to (85) may not be valid if CPT symmetry is violated. In such a case $\overline{K}^0$ cannot be considered as identical to K^0, and thus the requirement of $C\mathcal{P} = +$, imposed by Bose-statistics, is relaxed. As a result, the initial entangled state (85) can be parametrised in general as [28]:

$$\begin{aligned}|i\rangle = &\frac{1}{\sqrt{2}}\left(|K^0(\boldsymbol{k}), \overline{K}^0(-\boldsymbol{k})\rangle - |\overline{K}^0(\boldsymbol{k}), K^0(-\boldsymbol{k})\rangle\right)\\ &+ \frac{\omega}{\sqrt{2}}\left(|K^0(\boldsymbol{k}), \overline{K}^0(-\boldsymbol{k})\rangle + |\overline{K}^0(\boldsymbol{k}), K^0(-\boldsymbol{k})\rangle\right)\end{aligned} \tag{86}$$

where $\omega = |\omega|e^{i\Omega}$ is a *complex* CPTV parameter, associated with the non-identical particle nature of the neutral meson and antimeson states. This parameter describes a *novel* phenomenon, not included in previous analyses.

Notice that an equation such as the one given in (86) could also be produced as a result of deviations from the laws of quantum mechanics during the initial decay of the ϕ or Υ states. Thus, (86) could receive contributions from two different effects, and can be thought off as simultaneously parametrizing both of them.

In terms of physical (energy) eigenstates, $|K_{S,L}\rangle$, the state (86) is written as (we keep linear terms in the small parameters ω, δ, i.e. in the following we ignore higher-order terms $\omega\delta$, δ^2 etc.)

$$|i\rangle = C\Big[\left(|K_S(\boldsymbol{k}), K_L(-\boldsymbol{k})\rangle - |K_L(\boldsymbol{k}), K_S(-\boldsymbol{k})\rangle\right) + \omega \left(|K_S(\boldsymbol{k}), K_S(-\boldsymbol{k})\rangle - |K_L(\boldsymbol{k}), K_L(-\boldsymbol{k})\rangle\right)\Big] \quad (87)$$

with $C = \frac{\sqrt{(1+|\epsilon_1|^2)(1+|\epsilon_2|^2)}}{\sqrt{2}(1-\epsilon_1\epsilon_2)} \simeq \frac{1+|\epsilon^2|}{\sqrt{2}(1-\epsilon^2)}$. Notice again the presence of combinations $K_S K_S$ and $K_L K_L$ states, proportional to the novel CPTV parameter ω.

Such terms become important when one considers decay channels. Specifically, consider the decay amplitude $A(X,Y)$, corresponding to the appearance of a final state X at time t_1 and Y at time t_2, as illustrated in Fig. 17.

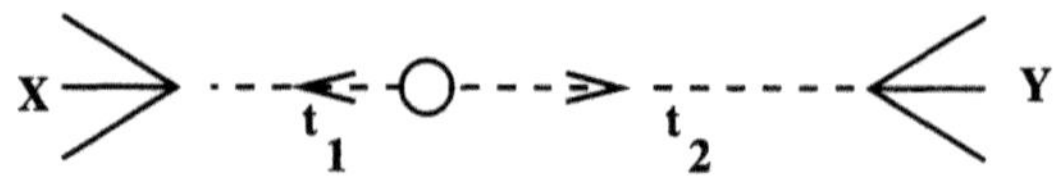

Fig. 17. A typical amplitude corresponding to the decay of, say, a ϕ state into final states X, Y; $t_i, i = 1,2$ denote the corresponding time scales for the appearance of the final products of the decay

One assumes (87) for the initial two-Kaon system, after the ϕ decay. The time is set $t = 0$ at the moment of the decay. Next, one integrates the square of the amplitude over all accessible times $t = t_1 + t_2$, keeping the difference $\Delta t = t_2 - t_1$ as constant. This defines the "intensity" $I(\Delta t)$ [28]:

$$I(\Delta t) \equiv \frac{1}{2}\int_{|\Delta t|}^{\infty} dt\, |A(X,Y)|^2 \quad (88)$$

In what follows we concentrate on identical final states $X = Y = \pi^+\pi^-$, because as we shall argue they are the most sensitive channels to probe the novel effects associated with the CPTV parameter ω. Indeed [90], the amplitudes of the CP violating decays $K_L \to \pi^+\pi^-$ are suppressed by factors of order $\mathcal{O}(10^{-3})$, as compared to the principal decay mode of $K_S \to \pi^+\pi^-$. In the absence of CPTV ω, (85), due to the $K_S K_L$ mixing, such decay rates would be suppressed. This would not be the case, however, when the CPTV ω (86) parameter is non zero, due to the existence of a separate $K_S K_S$ term in that case (87). This implies that the relevant parameter for CPT violation in the intensity is ω/η_X, where $\eta_X = \langle X|K_S\rangle / \langle X|K_L\rangle$ which enhances the potentially observed effect.

The effects of the CPTV ω on such intensities $I(\Delta t)$ are indicated in Fig. 18. We next comment on the distinguishability of the ω effect from conventional background effects. Specifically, the mixing of the initial state due to the non-identity of the antiparticle to the corresponding particle state has similar form to that induced by a non-resonant background with $C = +$ [91]. This latter effect is known to have a small size; estimates based on unitarity

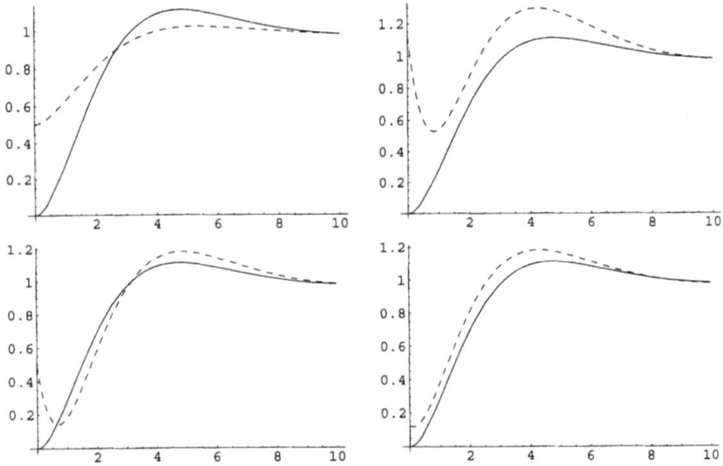

Fig. 18. Characteristic cases of the intensity $I(\Delta t)$, with $|\omega| = 0$ (*solid line*) vs $I(\Delta t)$ (*dashed line*) with (*from top left to right*): (i) $|\omega| = |\eta_{+-}|$, $\Omega = \phi_{+-} - 0.16\pi$, (ii) $|\omega| = |\eta_{+-}|$, $\Omega = \phi_{+-} + 0.95\pi$, (iii) $|\omega| = 0.5|\eta_{+-}|$, $\Omega = \phi_{+-} + 0.16\pi$, (iv) $|\omega| = 1.5|\eta_{+-}|$, $\Omega = \phi_{+-}$. Δt is measured in units of τ_S (the mean life-time of K_S) and $I(\Delta t)$ in units of $|C|^2|\eta_{+-}|^2|\langle\pi^+\pi^-|K_S\rangle|^4\tau_S$

bounds give a size of many orders of magnitude smaller than the $C = -$ effect in the ϕ decays [88, 91]. Terms of the type $K_S K_S$ (which dominate over $K_L K_L$) coming from the ϕ-resonance as a result of CPTV can be distinguished from those coming from the $C = +$ background because they interfere differently with the regular $C = -$ resonant contribution (i.e. (87) with $\omega = 0$). Indeed, in the CPTV case, the $K_L K_S$ and $\omega K_S K_S$ terms have the same dependence on the center-of-mass energy s of the colliding particles producing the resonance, because both terms originate from the ϕ-particle. Their interference, therefore, being proportional to the real part of the product of the corresponding amplitudes, still displays a peak at the resonance. On the other hand, the amplitude of the $K_S K_S$ coming from the $C = +$ background has no appreciable dependence on s and has practically vanishing imaginary part. Therefore, given that the real part of a Breit-Wigner amplitude vanishes at the top of the resonance, this implies that the interference of the $C = +$ background with the regular $C = -$ resonant contribution vanishes at the top of the resonance, with opposite signs on both sides of the latter. This clearly distinguishes experimentally the two cases.

We continue with a brief discussion concerning the distinguishability of the ω effect (86, 87) from non-quantum mechanical effects associated with the evolution, as in [24]. The ω effect can be distinguished from those of the QG-decohering evolution parameters α, β, γ, when the formalism is applied to the entangled states ϕ [26, 75]. A non-quantum mechanical evolution of the entangled Kaon state with $\omega = 0$ has been considered in [26]. In such a case the resulting density-matrix ϕ state $\tilde{\rho}_\phi = \mathrm{Tr}|\phi\rangle\langle\phi|$ can be written as

$$\begin{aligned}\tilde{\rho}_\phi &= \rho_S \otimes \rho_L + \rho_L \otimes \rho_S - \rho_I \otimes \rho_{\overline{I}} - \rho_{\overline{I}} \otimes \rho_I \\ &- \frac{2\beta}{d}(\rho_I \otimes \rho_S + \rho_S \otimes \rho_I) - \frac{2\beta}{d^*}(\rho_{\overline{I}} \otimes \rho_S + \rho_S \otimes \rho_{\overline{I}}) \\ &+ \frac{2\beta}{d}(\rho_{\overline{I}} \otimes \rho_L + \rho_L \otimes \rho_{\overline{I}}) + \frac{2\beta}{d^*}(\rho_I \otimes \rho_L + \rho_L \otimes \rho_I) \\ &- \frac{i\alpha}{\Delta M}(\rho_I \otimes \rho_I - \rho_{\overline{I}} \otimes \rho_{\overline{I}}) - \frac{2\gamma}{\Delta \Gamma}(\rho_S \otimes \rho_S - \rho_L \otimes \rho_L)\end{aligned}$$

where the standard notation $\rho_S = |S\rangle\langle S|$, $\rho_L = |L\rangle\langle L|$, $\rho_I = |S\rangle\langle L|$, $\rho_{\overline{I}} = |L\rangle\langle S|$ has been employed, $d = -\Delta M + i\Delta\Gamma/2$, and an overall multiplicative factor of $\frac{1}{2}\frac{(1+2|\epsilon|^2)}{1-2|\epsilon|^2\cos(2\phi_\epsilon)}$ has been suppressed. On the other hand, the corresponding density matrix description of the ϕ state (87) in our case reads:

$$\begin{aligned}\rho_\phi &= \rho_S \otimes \rho_L + \rho_L \otimes \rho_S - \rho_I \otimes \rho_{\overline{I}} - \rho_{\overline{I}} \otimes \rho_I \\ &- \omega(\rho_I \otimes \rho_S - \rho_S \otimes \rho_I) - \omega^*(\rho_{\overline{I}} \otimes \rho_S - \rho_S \otimes \rho_{\overline{I}}) \\ &- \omega(\rho_{\overline{I}} \otimes \rho_L - \rho_L \otimes \rho_{\overline{I}}) - \omega^*(\rho_I \otimes \rho_L - \rho_L \otimes \rho_I) \\ &- |\omega|^2(\rho_I \otimes \rho_I + \rho_{\overline{I}} \otimes \rho_{\overline{I}}) + |\omega|^2(\rho_S \otimes \rho_S + \rho_L \otimes \rho_L)\end{aligned}$$

with the same multiplicative factor suppressed. It is understood that the evolution of both $\tilde{\rho}_\phi$ and ρ_ϕ is governed by the rules given in [24, 25, 26]. As we can see by comparing the two equations, the terms linear in ω in our case are *antisymmetric* under the exchange of particle states 1 and 2, *in contrast* to the *symmetry* of the corresponding terms linear in β in the case of [26]. Similar differences characterize the terms proportional to $|\omega|^2$, and those proportional to α and γ, which involve $\rho_I \otimes \rho_I$, $\rho_{\overline{I}} \otimes \rho_{\overline{I}}$, $\rho_S \otimes \rho_S$, $\rho_L \otimes \rho_L$. Such differences are therefore important in disentangling the ω CPTV effects proposed here from non-quantum mechanical evolution effects [24, 25, 26, 27].

Finally we close this subsection with a comment on the application of this formalism to the B factories. Although, formally, the situation is identical to the one discussed above, however the sensitivity of the CPTV ω effect for the B system is much smaller. This is due to the fact that in B factories there is no particularly "good" channel X (with $X = Y$) for which the corresponding η_X is small. The analysis in that case may therefore be performed in the equal sign dilepton channel, where the branching fraction is more important, and a high statistics is expected.

3.13 CPTV Decoherence and Ultra Cold Neutrons

Before commencing a discussion on QG-induced decoherence in neutrinos we would like to discuss briefly the application of the decoherence formalism of [24] on another interesting experiment, which attracted some attention recently, that of ultracold neutrons in the gravitational field of Earth[2]. The arrangement of this experiment is demonstrated in Fig. 19.

[2] The results in this section have been derived in collaboration with Elias Gravanis.

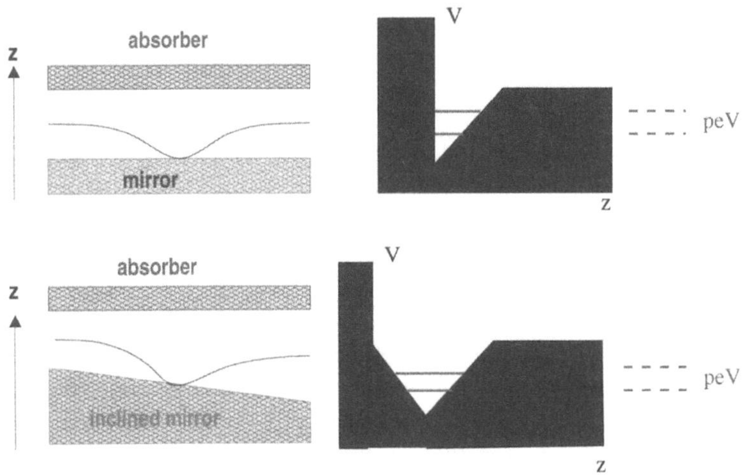

Fig. 19. Inclined mirror ensures Parity invariance of QG modifications and hence formalism similar to neutral kaons. A few (two here) energy states (peV energy differences between levels) are inside the Earth's potential well

The neutrons find themselves on a quantum-mechanical potential which is affected by the gravitational potential of Earth, due to their masses. A few energy states, separated by peV $\sim 10^{-15}$ eV energy differences, lie inside the Earth's potential well. The quantum trajectories of the neutrons are affected by this gravitational potential in the way indicated in the figure. The neutrons are reflected on the mirrors and are collected at the detection point. This has already been demonstrated experimentally, measuring for the first time gravitational effects together with quantum mechanical effects [92].

Consider for our purposes the case where two such energy states find themselves inside the potential well. This constitutes a two-level system, and one may think of applying the two-state decoherence formalism to study QG induced effects in such a situation, which would modify the results concering the probabilities of finding the neutrons in one of the two available energy states at the detection point. Like any two-state oscillation system, the respective probabilities should be equal to $1/2$ in the absence of any decoherence effects, although in the presence of decoherence one would expect a slight bias in the probabilities.

The quantum number which is conserved here is Parity, which however is the case only if the mirror is inclined, so as to eliminate the effects of parity violation induced by the presence of the Gravitational field. The Probability of finding the neutrons in either state at the detection point indicated in the figure can be computed following the same formalism as the two-state parametrization of the neutral kaon system in [24], with the replacement of the strangeness conservation by that of parity. The results read, to leading order in the small decoherence parameters:

$$\mathrm{Tr}(\rho' \varrho_{1,2}) = \frac{1}{2} \pm \frac{1}{2} e^{-\frac{\alpha+\gamma}{2}t} \sin(\Delta E t) \ , \qquad \Delta E = \mathcal{O}(\mathrm{peV}) \tag{89}$$

where t is the time.

We next remark that, if Lorentz invariance is violated by Quantum Gravity, then the decoherence parameters $\alpha, \gamma \simeq \frac{E_{\mathrm{kin}}^2}{M_P}$, where $E_{\mathrm{kin}} = \mathcal{O}(\mathrm{peV})$ is the kinetic energy of the neutrons. This is too small to be detected in this kind of experiment; However, in case QG decoherence respects Lorentz Symmetry, which as mentioned above is possible [33, 34], then $\alpha, \gamma \simeq \frac{m_n^2}{M_P}$. For the duration of the experiment, which is or order $t \sim$ msec, then, we observe that the decoherence effects are much larger. However, at present, there seems to be no significant sensitivity from this type of experiment, as compared with other available tests of decoherence. Nevertheless, one cannot exclude the possibility of a significant improvement in sensitivity in similar experiments in the foreseeable future, and this is the reason why I included this case briefly in the present set of lectures.

3.14 CPTV Through QG Decoherence for Neutrinos: the Most Sensitive Probe to Date

Two-Generation Models

We now come to discuss quantum-gravity decoherence in neutrinos, whose sensitivity in this respect is far more superior than that of neutral kaons, assuming of course a universal nature of QG. This latter assumption, though, requires some second thoughts, given that, as mentioned above, there are theoretical models of quantum space-time foam [7], in which QG effects interact differently with various particle species.

With this in mind we next remark that, QG may induce oscillations between neutrino flavours independently of ν-masses [73, 74, 75, 76]. We begin with the simplified case of two-neutrino generations, that is a two-state system, which makes the formalism very similar to the neutral kaon case described above. In similar spirit to the Kaon case, the energy (physical) eigenstates of neutrinos are not flavour eigenstates, and one has mixing.

The basic formalism for decoherence-induced neutrino oscillations is described by a QMV evolution for the density matrix of the ν, which parallels that of neutral kaon in the case of two generations of neutrinos:

$$\partial_t \rho = i[\rho, H] + \delta H \rho \tag{90}$$

where [24]

$$\delta H_{\alpha\beta} = \begin{pmatrix} 0 & 0 & 0 & 0 \\ 0 & -2\alpha & -2\beta & 0 \\ 0 & -2\beta & -2\gamma & 0 \\ 0 & 0 & 0 & 0 \end{pmatrix}$$

for energy and lepton number conservation, and

$$\delta H_{\alpha\beta} = \begin{pmatrix} 0 & 0 & 0 & 0 \\ 0 & 0 & 0 & 0 \\ 0 & 0 & -2\alpha & -2\beta \\ 0 & 0 & -2\beta & -2\gamma \end{pmatrix}$$

if energy and lepton number are violated, but flavour is conserved (the latter associated formally with the σ_1 Pauli matrix).

Positivity of ρ, but not complete positivity, requires: $\alpha, \gamma > 0,\ \alpha\gamma > \beta^2$. The parameters α, β, γ violate CP, and CPT in general, as discussed previously.

The relevant oscillation probabilities, describing the evolution of a neutrino flavour ν_α, created at time $t = 0$, to a neutrino flavour ν_β at time t, are determined by means of the dynamical semigroup approach (32):

$$P_{\nu_\alpha \to \nu_\beta}(t) = \mathrm{Tr}\left(\rho_\alpha(t)\rho^\beta\right) \tag{91}$$

For our problem we have a two state system, and the computation of the eigenvalue problem is easy. For the two cases above, we obtain after some straightforward algebra [73]:

(A) *For the flavour conserving case*:

As a simplified example, consider the oscillation $\nu_e \to \nu_x \quad (x = \mu, \tau$ or sterile):

$$P_{\nu_e \to \nu_x} = \frac{1}{2} - \frac{1}{2} e^{-\gamma L} \cos^2 2\theta_v - \frac{1}{2} e^{-\alpha L} \sin^2 2\theta_v \cos\left(\frac{|m^2_{\nu_1} - m^2_{\nu_2}|}{2E_\nu} L\right) \tag{92}$$

Here L is the oscillation length and θ_v the mixing angle.

In the mass basis one has: $|\nu_e\rangle = \cos\theta_v|\nu_1\rangle + \sin\theta_v|\nu_2\rangle$, $|\nu_\mu\rangle = -\sin\theta_v|\nu_1\rangle + \cos\theta_v|\nu_2\rangle$. Note that in this case the mixing angle $\theta_v = 0$ if and only if the neutrinos are massless. From the above considerations, however, it is clear that there are flavour oscillations even in the massless case, due to a non-trivial QG parameter γ, compatible with flavour conserving formalism: $\langle\nu_e|\sigma_1|\nu_e\rangle = -\langle\nu_\mu|\sigma_1|\nu_\mu\rangle = 2\sin\theta_v\cos\theta_v$.

(B) *For Energy and Lepton number conserving case*:

Again, we consider a two-flavour example: $\nu_e \to \nu_x \quad (x = \mu, \tau$ or sterile). The relevant oscillation probability in this case is calculated to be [73]:

$$P_{\nu_e \to \nu_x} = \frac{1}{2} \sin^2 2\theta_v \left(1 - e^{-(\alpha+\gamma)L} \cos\left(\frac{|m^2_{\nu_1} - m^2_{\nu_2}|}{2E_\nu} L\right)\right) \tag{93}$$

where we assumed for simplicity, and illustrative purposes, that $\alpha, \beta, \gamma \ll \frac{|m^2_{\nu_1} - m^2_{\nu_2}|}{2E_\nu}$. The reader is invited to contrast this result with case **(A)** above.

One can use the results in the cases **(A)** and **(B)** to bound experimentally $\xi \equiv \{\alpha, \beta, \gamma\}$. At this stage the reader is invited to recall that there exist two kinds of theoretical estimates/predictions for the order of magnitude of the parameters α, β, γ: An optimistic one [25], according to which

$\xi \sim \xi_0(\frac{E}{\text{GeV}})^n, n = 0, 2, -1$, and this has a chance of being falsified in future experiments, if the effect is there, and a pessimistic one [41], which depends on the square of the neutrino mass-squared difference (24), $\xi \sim \frac{(\Delta m^2)^2}{E^2 M_{qg}}$, ($M_{qg} \sim M_P \sim 10^{19}$ GeV), which is much smaller, and probably cannot be accessed by immediate future neutrino oscillation experiments.

We now mention that in some models of QG-induced decoherence, complete positivity of $\rho(t)$ for composite systems, such as ϕ or B mesons, may be imposed [75] (however, I must stress once more that the necessity of this requirement, especially in a QG context where non-linear effects may be present [25], remains to be proven). This results in an ideal Markov environment, with: $\alpha = \beta = 0, \gamma > 0$.

If this model is assumed for ν oscillations induced by QG decoherence [74], then the following phenomenological parametrization can be made: $\gamma = \gamma_0(E/\text{GeV})^n$, $n = 0, 2, -1$. with E the neutrino energy.

From Atmospheric ν data one is led to the following bounds for the QG-decoherence parameter γ (c.f. Figs. 20, 21) [74]:

(a) $n = 0$, $\gamma_0 < 3.5 \times 10^{-23}$ GeV.
(b) $n = 2$, $\gamma_0 < 0.9 \times 10^{-27}$ GeV.
(c) $n = -1$, $\gamma_0 < 2 \times 10^{-21}$ GeV.

Especially with respect to case (b) the reader is reminded that the CPLEAR bound on γ for neutral Kaons was $\gamma < 10^{-21}$ GeV [89], i.e. the ν-oscillation experiments exhibit much higher sensitivity to QG decoherence effects than neutral meson experiments.

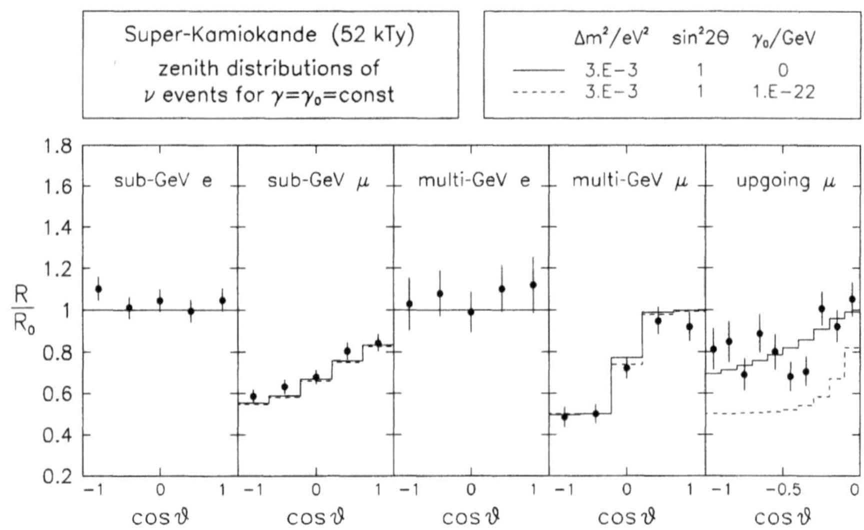

Fig. 20. Effects of decoherence ($\gamma = \gamma_0 = \text{const} \neq 0$) on the distributions of lepton events as a function of the zenith angle ϑ

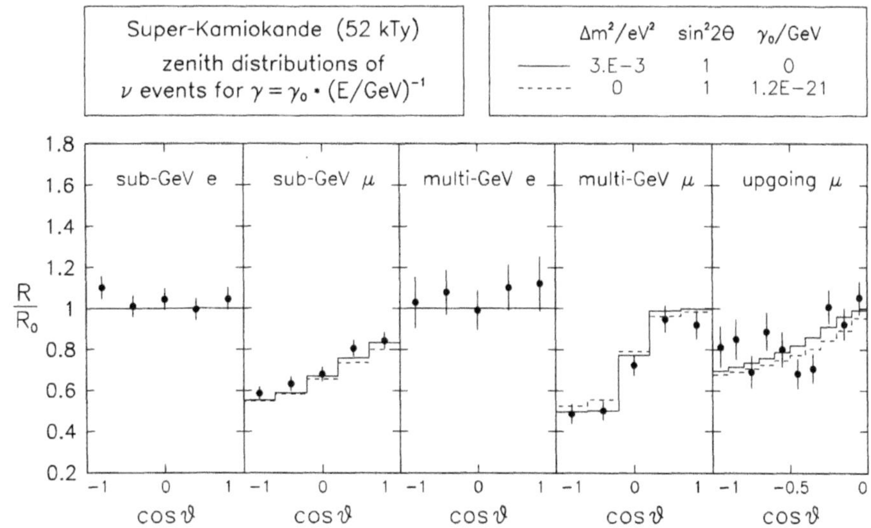

Fig. 21. Best-fit scenarios for pure oscillations ($\gamma = 0$) (*solid line*) and for pure decoherence with $\gamma \propto 1/E$ (*dashed line*)

Finally, I note that in [76] it was remarked that very stringent bounds on α, β and γ (in the lepton number violating QG case) may be imposed by looking at oscillations of neutrinos from astrophysical sources (supernovae and AGN). The corresponding bounds on the γ parameter from oscillation analysis of neutrinos from supernovae and AGN, if QG induces such oscillations, are very strong: $\gamma < 10^{-40}$ GeV from Supernova 1987a, using the observed constraint [77] on the oscillation probability $P_{\nu_e \to \nu_\mu,\tau} < 0.2$, and $\gamma < 10^{-42}$ GeV from AGN, which exhibit sensitivity to order higher than E^3/M_{qg}^2, with $M_{qg} \sim M_P \sim 10^{19}$ GeV! Of course, the bounds from AGN do not correspond to real bounds, awaiting the observation of high energy neutrinos from such astrophysical sources. In [76] bounds have also been derived for the QG decoherence parameters by assuming that QG may induce neutrinoless double-beta decay. However, using current experimental constraints on neutrinoless double-beta decay observables [78] one arrives at very weak bounds for the parameters α, β, γ.

One also expects stringent bounds on decoherence parameters, but also on deformed dispersion relations, if any, for neutrinos, from future underwater neutrino telescopes, such as ANTARES [79], and NESTOR [80][3].

[3] As far as I understand, but I claim no expertise on this issue, the NESTOR experiment has an advantage with respect to detection of very high energy cosmic neutrinos, which may be more sensitive probes of such quantum gravity effects.

Three-Generation Models: Decoherence and the LSND Result

As we discussed above, two-generation CPTV mass models for neutrinos within quantum mechanics are excluded by global fits of available data, especially solar neutrino models. This situation is not expected to change by the inclusion of a third generation, although I must stress that, as far as I am aware of, complete three-generations analyses of this kind have not been performed as yet. Moreover, although four generation CPTV neutrino models are still consistent with experimental data, nevertheless there seems to be no experimental evidence for a forth generation, especially after the recent WMAP astrophysical results. On the other hand, as we have just seen, two-generation neutrino analysis of decoherence effects did not show any spectacular results, apart from the imposition of stringent bounds on the relevant parameters.

This prompts one to think that the extension of the decoherence formalism to three generations of neutrinos, which from a mathematical view point is a problem with considerable increase in technical complexity, is a futile task, with no physical importance whatsoever. However, there are the "anomalous" results provided by the LSND collaboration [63] on the evidence for $\overline{\nu}_e \to \overline{\nu}_\mu$ oscillations, through $\overline{\nu}_e$ disappearance, but not for the cooresponding oscillations in the neutrino sector, which call for an explanation, if one, of course, takes them seriously into account. These effects, as we have seen, cannot be explained by conventional quantum field theoretic analyses, even if CPT is assumed violated.

It is the point of this subsection to point out that, if one extends the decoherence analysis to three generations of neutrinos and allows for CPT Violation among the decoherence parameters, it is possible [93] to fit all the currently availble neutrino data, including the LSND results, by simple decoherence models, in which the dominant decoherence parameters occur in the antineutrino sector. It is important that in such "asymmetric" decoherence models there is no need for enlarging the neutrino sector by a fourth generation, neither for introducing CPTV mass parameters. If the LSND results are confirmed by future experiments, then this would be a significant result, as it would provide for the first time a clear experimental evidence for a CPTV decoherence event, which would be directly related to quantum gravity effects.

Let us briefly present the arguments leading to these results. Formally, the extension of the completely positive decoherence scenario to the standard three-generation neutrino oscillations case is straightforward, and it was described in section two. One adopts a three-state Lindblad problem, and, following the standard procedure outlined there, one determines the corresponding eigenvectors and eigenvalues, as in the two-level case examined in the previous subsection. It is only a considerable increase in mathematical complexity, and obscurity in the precise physical meaning of all the non-trivial entries of the decoherence matrix that one encounters here.

The relativistic neutrino Hamiltonian $H_{\rm eff} \sim p^2 + m^2/2p$, with m the neutrino mass, is used as the effective Hamiltonian of the subsystem in the evolution equation (14). In terms of the generators $\mathcal{J}_\mu$, $\mu = 0, \dots 8$ of the SU(3) group, H_{eff} can be expanded as [94]: $\mathcal{H}_{\text{eff}} = \frac{1}{2p}\sqrt{2/3}\left(6p^2 + \sum_{i=1}^{3} m_i^2\right)\mathcal{J}_0 + \frac{1}{2p}(\Delta m_{12}^2)\mathcal{J}_3 + \frac{1}{2\sqrt{3}p}\left(\Delta m_{13}^2 + \Delta m_{23}^2\right)\mathcal{J}_8$, with the obvious notation $\Delta m_{ij}^2 = m_i^2 - m_j^2$, $i, j = 1, 2, 3$.

The analysis of [94] assumed *ad hoc* a diagonal form for the 9×9 decoherence matrix $\mathcal{L}$ in (27):

$$[\mathcal{L}_{\mu\nu}] = \text{Diag}\,(0, -\gamma_1, -\gamma_2, -\gamma_3, -\gamma_4, -\gamma_5, -\gamma_6, -\gamma_7, -\gamma_8) \tag{94}$$

in direct analogy with the two-level case of complete positivity [74, 75]. As we have mentioned already, there is no strong physical motivation behind such restricted forms of decoherence. This assumption, however, leads to the simplest possible decoherence models, and, for our *phenomenological* purposes in this work, we will assume the above form, which we will use to fit all the available neutrino data. It must be clear to the reader though, that such a simplification, if proven to be successful (which, as we shall argue below, is the case here), just adds more in favor of decoherence models, given the restricted number of available parameters for the fit in this case. In fact, any other non-minimal scenario will have it easier to accommodate data because it will have more degrees of freedom available for such a purpose.

Specifically we shall look at transition probabilities (91), which can be computed in a straightforward manner within the dynamical-semigroups approach outlined previously [94]:

$$\begin{aligned} P(\nu_\alpha \to \nu_\beta) &= \text{Tr}[\rho^\alpha(t)\rho^\beta] \\ &= \frac{1}{3} + \frac{1}{2}\sum_{i,k,j} e^{\lambda_k t}\mathcal{D}_{ik}\mathcal{D}_{kj}^{-1}\rho_j^\alpha(0)\rho_i^\beta \end{aligned} \tag{95}$$

where $\alpha, \beta = e, \mu, \tau$ stand for the three neutrino flavors, and Latin indices run over $1, \dots 8$. The quantities λ_k are the eigenvalues of the matrix $\mathcal{M}$ appearing in the evolution (27), after taking into account probability conservation, which decouples $\rho_0(t) = \sqrt{2/3}$, leaving the remaining equations in the form: $\partial\rho_k/\partial t = \sum_j \mathcal{M}_{kj}\rho_j$. The matrices $\mathcal{D}_{ij}$ are the matrices that diagonalize $\mathcal{M}$ [39]. Explicit forms of these matrices, the eigenvalues λ_k, and consequently the transition probabilities (95), are given in [94].

The important point to stress is that, in generic models of oscillation plus decoherence, the eigenvalues λ_k depend on both the decoherence parameters γ_i and the mass differences Δm_{ij}^2. For instance, $\lambda_1 = \frac{1}{2}[-(\gamma_1 + \gamma_2) - \sqrt{(\gamma_2 - \gamma_1)^2 - 4\Delta_{12}^2}]$, with the notation $\Delta_{ij} \equiv \Delta m_{ij}^2/2p$, $i, j = 1, 2, 3$. Note that, to leading order in the (small) squared-mass differences, one may replace p by the total neutrino energy E, and this will be understood in what follows.

We now note that it is a generic feature of the λ_k to depend on the quantities Ω_{ij} which are given by [93, 94]

$$\Omega_{12} = \sqrt{(\gamma_2 - \gamma_1)^2 - 4\Delta_{12}^2}$$
$$\Omega_{13} = \sqrt{(\gamma_5 - \gamma_4)^2 - 4\Delta_{13}^2}$$
$$\Omega_{23} = \sqrt{(\gamma_7 - \gamma_6)^2 - 4\Delta_{23}^2} \qquad \text{etc.} \tag{96}$$

From the above expressions for the eigenvalues λ_k, it becomes clear that, when decoherence and oscillations are present simultaneously, one should distinguish two cases, according to the relative magnitudes of Δ_{ij} and $\Delta\gamma_{kl} \equiv \gamma_k - \gamma_l$: (i) $2|\Delta_{ij}| \geq |\Delta\gamma_{k\ell}|$, and (ii) $2|\Delta_{ij}| < |\Delta\gamma_{k\ell}|$. In the former case, the probabilities (95) contain trigonometric (sine and cosine) functions, whilst in the latter they exhibit hyperbolic sin and cosine dependence.

Assuming mixing between the flavours, amounts to expressing neutrino flavor eigenstates $|\nu_\alpha\rangle$, $\alpha = e, \mu, \tau$ in terms of mass eigenstates $|\nu_i\rangle$, $i = 1, 2, 3$ through a (unitary) matrix U: $|\nu_\alpha\rangle = \sum_{i=1}^{3} U_{\alpha i}^* |\nu_i\rangle$. This implies that the density matrix of a flavor state ρ^α can be expressed in terms of mass eigenstates as: $\rho^\alpha = |\nu_\alpha\rangle\langle\nu_\alpha| = \sum_{i,j} U_{\alpha i}^* U_{\alpha j} |\nu_i\rangle\langle\nu_j|$. From this we can determine $\rho_\mu^\alpha = 2\mathrm{Tr}(\rho^\alpha \mathcal{J}_\mu)$, a quantity needed to calculate the transition probabilities (95).

The important comment [93] we would like to raise at this point is that, when considering the above probabilities in the antineutrino sector, the respective decoherence parameters $\bar{\gamma}_i$ in general may be different from the corresponding ones in the neutrino sector, as a result of the strong form of CPT violation. In fact, as we shall discuss next, this will be crucial for accommodating the LSND result without conflicting with the rest of the available neutrino data. This feature is totally unrelated to mass differences between flavors.

In [94] a pessimistic conclusion was drawn on the "clear incompatibility between neutrino data and theoretical expectations", as followed by their qualitative tests for decoherence. It is a key feature of the work of [93] to point out that this point of view may not be true at all. In fact, as we shall demonstrate below, if one takes into account all the available neutrino data, including the final LSND results [63], which the authors of [94] did not do, and allows for the above mentioned CPT violation in the decoherence sector, then one will arrive at exactly the opposite conclusion, namely that three-generation decoherence and oscillations can fit the data successfully!

As shown in [93], compatibility of all available data, including CHOOZ [95] and LSND, can be achieved through a set of decoherence parameters γ_j with energy dependences $\gamma_j^0 E$ and γ_j^0/E, with $\gamma_j^0 \sim 10^{-18}, 10^{-24}$ (GeV)^2, respectively, for some j's, and in fact the fit ends up being *significantly better* than the standard one (when LSND results are included) as evidenced by an appropriate χ^2 analysis.

Some important remarks are in order. First of all, in the analysis of [94] pure decoherence is excluded in three-generation scenaria, as in two generation ones, due to the fact that the transition probabilities in the case $\Delta m_{ij}^2 = 0$ (pure decoherence) are such that the survival probabilities in both sectors are equal, i.e. $P(\nu_\alpha \to \nu_\alpha) = P(\overline{\nu}_\alpha \to \overline{\nu}_\alpha)$. From (95) we have in this case [94]:

$$P_{\nu_e \to \nu_e} = P_{\nu_\mu \to \nu_\mu} \simeq \frac{1}{3} + \frac{1}{2} e^{-\gamma_3 t} + \frac{1}{6} e^{-\gamma_8 t} \tag{97}$$

From the CHOOZ experiment [95], for which $L/E \sim 10^3/3$ m/MeV, we have that $\langle P_{\bar{\nu}_e \to \bar{\nu}_e} \rangle \simeq 1$, while the K2K experiment [96] with $L/E \sim 250/1.3$ km/GeV has observed events compatible with $\langle P_{\nu_\mu \to \nu_\mu} \rangle \simeq 0.7$, thereby contradicting the theoretical predictions (97) of pure decoherence.

However, this conclusion is based on the fact that in the antineutrino sector the decoherence matrix is the same as that in the neutrino sector. In general this need not be the case, in view of CPT Violation, which could imply a different interaction of the antiparticle with the gravitational environment as compared with the particle. In fact in models where a pure state evolves to a mixed one, one expects a CPT Violation in the strong form, according to the theorem of [37].

In our tests we took into account this possibility, but pure decoherence can be excluded also in this case, as it is clearly incompatible with the totality of the available data.

In order to check our model, we have performed a χ^2 comparison (as opposed to a χ^2 fit which is still pending) to SuperKamiokande sub-GeV and multi GeV data, CHOOZ data and LSND, for a sample point in the vast parameter space of our extremely simplified version of decoherence models. Since we have *not* performed as yet a χ^2-fit, the point we are selecting (rather visually and not by a proper χ^2 analysis) is not optimized to give the best fit to the existing data. Instead, it must be regarded as one among the many equally good members in this family of solutions, being extremely possible to find another model that fits better the data, through a complete (and highly time consuming) scan over the whole parameter space.

Cutting the long story short, and to make the analysis easier, we have set [93] all the γ_i in the neutrino sector to zero, restricting in this way all the dominant decoherence effects in the antineutrino sector only. For the sake of simplicity we have assumed the form:

$$\bar{\gamma}_i = \bar{\gamma}_{i+1} \text{ for } i = 1,4,6 \qquad \text{and} \qquad \bar{\gamma}_3 = \bar{\gamma}_8 \tag{98}$$

Later on we shall set some of the γ_i's to zero. Furthermore, we have also set the CP violating phase of the NMS matrix to zero, so that all the mixing matrix elements become real.

With these assumptions, the otherwise cumbersome expression (see end of section for more detailed results) for the transition probability for the antineutrino sector takes the form:

$$P_{\bar{\nu}_\alpha \to \bar{\nu}_\beta} = \frac{1}{3} + \frac{1}{2} \left\{ \rho_1^\alpha \rho_1^\beta \cos\left(\frac{|\Omega_{12}|t}{2}\right) e^{-\bar{\gamma}_1 t} \right.$$
$$+ \rho_4^\alpha \rho_4^\beta \cos\left(\frac{|\Omega_{13}|t}{2}\right) e^{-\bar{\gamma}_4 t}$$
$$+ \rho_6^\alpha \rho_6^\beta \cos\left(\frac{|\Omega_{23}|t}{2}\right) e^{-\bar{\gamma}_6 t}$$
$$\left. + e^{-\bar{\gamma}_3 t} \left(\rho_3^\alpha \rho_3^\beta + \rho_8^\alpha \rho_8^\beta\right) \right\}. \tag{99}$$

where the Ω_{ij} were defined in the previous section and are the same in both sectors (due to our choice of γ_i's) and

$$\begin{aligned}
\rho_0^\alpha &= \sqrt{\frac{2}{3}} \\
\rho_1^\alpha &= 2Re(U_{\alpha 1}^* U_{\alpha 2}) \\
\rho_2^\alpha &= -2Im(U_{\alpha 1}^* U_{\alpha 2}) \\
\rho_3^\alpha &= |U_{\alpha 1}|^2 - |U_{\alpha 2}|^2 \\
\rho_4^\alpha &= 2Re(U_{\alpha 1}^* U_{\alpha 3}) \\
\rho_5^\alpha &= -2Im(U_{\alpha 1}^* U_{\alpha 3}) \\
\rho_6^\alpha &= 2Re(U_{\alpha 2}^* U_{\alpha 3}) \\
\rho_7^\alpha &= -2Im(U_{\alpha 2}^* U_{\alpha 3}) \\
\rho_8^\alpha &= \sqrt{\frac{1}{3}} \left(|U_{\alpha 1}|^2 + |U_{\alpha 2}|^2 - 2|U_{\alpha 3}|^2\right)
\end{aligned} \tag{100}$$

where the mixing matrices are the same as in the neutrino sector. For the neutrino sector, as there are no dominant decoherence effects, the standard expression for the transition probability is valid.

It is obvious now that, since the neutrino sector does not suffer from decoherence, there is no need to include the solar data into the fit. We are guaranteed to have an excellent agreement with solar data, as long as we keep the relevant mass difference and mixing angle within the LMA region. As mentioned previously, CPT violation is driven by, and restricted to, the decoherence parameters, and hence masses and mixing angles are the same in both sectors, and selected to be

$$\begin{aligned}
\Delta m_{12}^2 &= \Delta \overline{m_{12}}^2 = 7 \cdot 10^{-5}\,\mathrm{eV}^2\,, \\
\Delta m_{23}^2 &= \Delta \overline{m_{23}}^2 = 2.5 \cdot 10^{-3}\,\mathrm{eV}^2\,, \\
\theta_{23} &= \overline{\theta_{23}} = \pi/4, \theta_{12} = \overline{\theta_{12}} = .45\,, \\
\theta_{13} &= \overline{\theta_{13}} = .05\,,
\end{aligned}$$

as indicated by the state of the art phenomenological analysis in neutrino physics.

For the decoherence parameters we have chosen (c.f. (98))

$$\overline{\gamma_1} = \overline{\gamma_2} = 2 \cdot 10^{-18} \cdot E \quad \text{and} \quad \overline{\gamma_3} = \overline{\gamma_8} = 1 \cdot 10^{-24}/E \ , \tag{101}$$

where E is the neutrino energy, and barred quantities refer to the antineutrinos, given that decoherence takes place only in this sector in our model. All the other parameters are assumed to be zero. All in all, we have introduced only two new parameters, two new degrees of freedom, $\overline{\gamma_1}$ and $\overline{\gamma_3}$, and we shall try to explain with them all the available experimental data.

In order to test our model with these two decoherence parameters in the antineutrino sector, we have calculated the zenith angle dependence of the ratio "observed/(expected in the no oscillation case)", for muon and electron atmospheric neutrinos, for the sub-GeV and multi-GeV energy ranges, when mixing is taken into account. The results are shown in Fig. 22. where, for the sake of comparison, we have also included the experimental data.

As bare-eye comparisons can be misleading, we have also calculated the χ^2 value for each of the cases, defining the atmospheric χ^2 as

$$\chi^2_{\text{atm}} = \sum_{M,S} \sum_{\alpha=e,\mu} \sum_{i=1}^{10} \frac{(R^{\text{exp}}_{\alpha,i} - R^{\text{th}}_{\alpha,i})^2}{\sigma^2_{\alpha i}} \quad . \tag{102}$$

Here $\sigma_{\alpha,i}$ are the statistical errors, the ratios $R_{\alpha,i}$ between the observed and predicted signal can be written as

$$R^{\text{exp}}_{\alpha,i} = N^{\text{exp}}_{\alpha,i} / N^{\text{MC}}_{\alpha,i} \tag{103}$$

(with α indicating the lepton flavor and i counting the different bins, ten in total) and M, S stand for the multi-GeV and sub-GeV data respectively. For the CHOOZ experiment we used the 15 data points with their statistical errors, where in each bin we averaged the probability over energy and for LSND one datum has been included. The results with which we hope all our claims become crystal clear are summarized in Table 3, were we present the χ^2 comparison for the following cases: (a) pure decoherence in the antineutrino sector, (b) pure decoherence in both sectors, (c) mixing plus decoherence in the antineutrino sector, (d) mixing plus decoherence in both sectors, and (e) mixing only – the standard scenario.

From the table it becomes clear that the mixing plus decoherence scenario in the antineutrino sector can easily account for all the available experimental information, including LSND data. It is important to stress once more that our sample point was not obtained through a scan over all the parameter space, but by an educated guess, and therefore plenty of room is left for improvements. At this point a word of warning is in order: although superficially it seems that scenario (d), decoherence plus mixing in both sectors, provides an equally good fit, one should remember that including decoherence effects in the neutrino sector can have undesirable effects in solar neutrinos, especially due to the fact that decoherence effects are weighted by the distance traveled

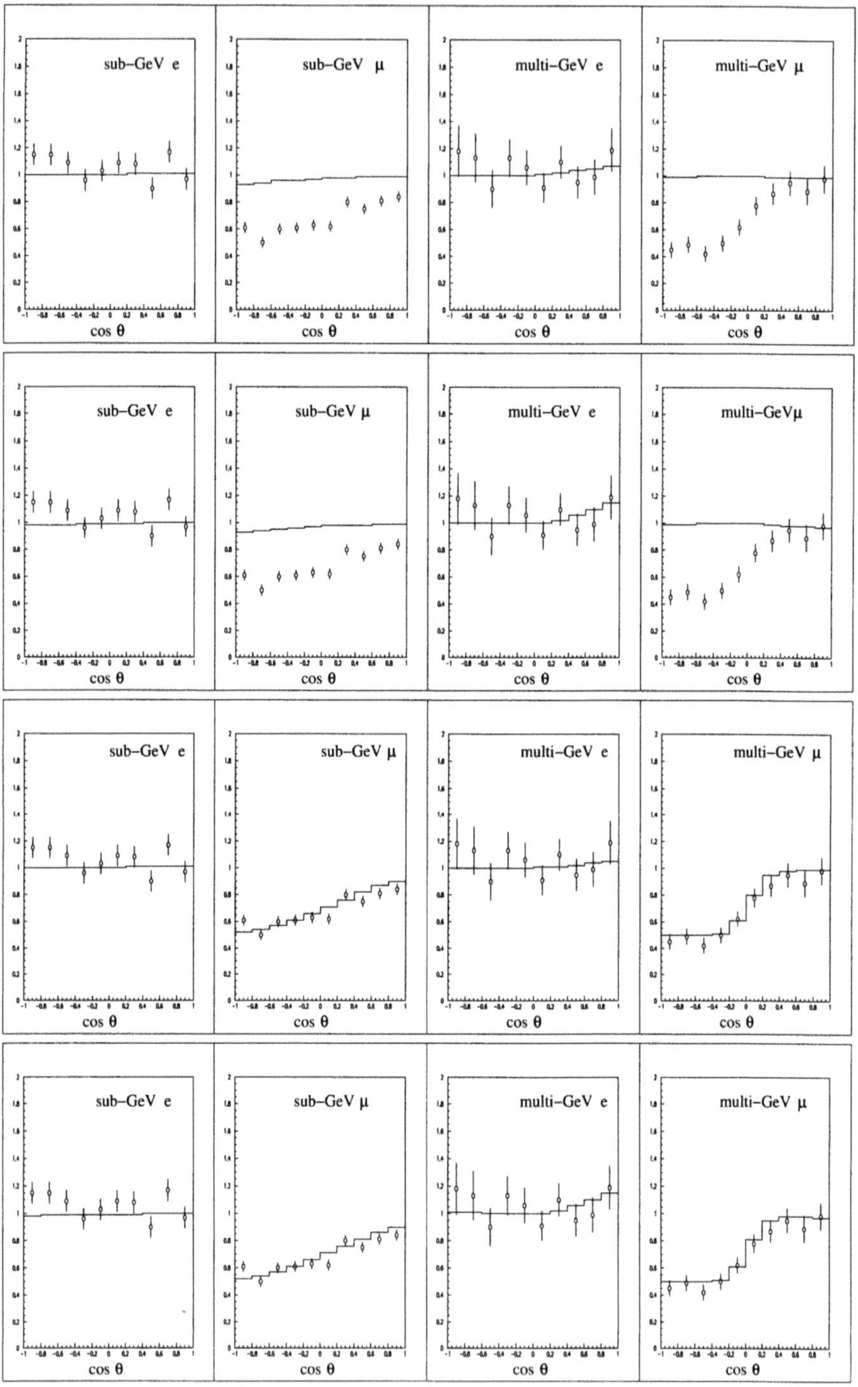

Fig. 22. Decoherence fits, from *top* to *bottom*: (**a**) pure decoherence in antineutrino sector, (**b**) pure decoherence in both sectors, (**c**) mixing plus decoherence in the antineutrino sector only, (**d**) mixing plus decoherence in both sectors. The *dots* correspond to SK data

Table 3. χ^2 obtained for (a) pure decoherence in antineutrino sector, (b) pure decoherence in both sectors, (c) mixing plus decoherence in the antineutrino sector only, (d) mixing plus decoherence in both sectors, (e) standard scenario with and without the LSND result

Model	χ^2 Without LSND	χ^2 Including LSND
(a)	980.7	980.8
(b)	979.8	980.0
(c)	52.2	52.3
(d)	54.4	54.6
(e)	53.9	60.7

by the neutrino, something that may lead to seizable (not observed!) effects in the solar case.

One might wonder then, whether decohering effects, which affect the antineutrino sector sufficiently to account for the LSND result, have any impact on the solar-neutrino related parameters, measured through antineutrinos in the KamLAND experiment [97]. In order to answer this question, it will be sufficient to calculate the electron survival probability for KamLAND in our model, which turns out to be $P_{\bar{\nu}_\alpha \to \bar{\nu}_\beta}|_{\text{KamLAND}} \simeq .63$, in perfect agreement with observations. It is also interesting to notice that in our model, the LSND effect is not given by the phase inside the oscillation term (which is proportional to the solar mass difference) but rather by the decoherence factor multiplying the oscillation term. Therefore the tension between LSND and KARMEN [98] data is naturally eliminated, because the difference in length leads to an exponential suppression.

Having said that, it is now clear that decoherence models (once neutrino mixing is taken into account) are the best (and arguably the only) way to explain all the observations including the LSND result. This scenario, which makes dramatic predictions for the upcoming neutrino experiments, expresses a strong observable form of CPT violation in the laboratory, and in this sense, our fit gives a clear answer to the question as to whether the weak form of CPT invariance (11) is violated in Nature. It seems that, in order to account for the LSND results, we should invoke such a decoherence-induced CPT Violation, which however is independent of any mass differences between particles and antiparticles.

This CPT violating pattern, with equal mass spectra for neutrinos and antineutrinos, if true, will have dramatic signatures in future neutrino oscillation experiments. The most striking consequence will be seen in MiniBooNE [99], According to our picture, MiniBooNE will be able to confirm LSND only when running in the antineutrino mode and not in the neutrino one, as decoherence effects live only in the former. Smaller but experimentally accessible signatures will be seen also in MINOS [85], by comparing conjugated channels (most noticeably, the muon survival probability).

We next remark that fits with decoherence parameters with energy dependences of the form (101) imply that the exponential factors $e^{\lambda_k t}$ in (95) due to decoherence will modify the amplitudes of the oscillatory terms due to mass differences, and while one term depends on L/E the other one is driven by $L \cdot E$, where we have set $t = L$, with L the oscillation length (we are working with natural units where $c = 1$).

The order of the coefficients of these quantities, $\gamma_j^0 \sim 10^{-18}, 10^{-24}$ (GeV)^2, found in our sample point, implies that for energies of a few GeV, which are typical of the pertinent experiments, such values are not far from $\gamma_j^0 \sim \Delta m_{ij}^2$. If our conclusions survive the next round of experiments, and therefore if MiniBOONE experiment [99] confirms previous LSND claims, then this may be a significant result.

Indeed, one would be tempted to speculate that, if the above estimate holds, and the decoherence coefficients are proportional to the neutrino mass-squared differences, this could even indicate that the neutrino mass differences themselves might be due to quantum gravity decoherence, in the sense of environmental contributions to the effective neutrino Hamiltonian appearing in the decoherent evolution (14), which could mascarade themselves as mass terms. Theoretically it is still unknown how the neutrinos acquire a mass, or what kind of mass (Majorana or Dirac) they possess. There are scenaria in which the mass of neutrino may be due to some peculiar backgrounds of string theory for instance. If the above model turns out to be right we might then have, for the first time in low energy physics, an indication of a direct detection of a quantum gravity effect, which disguised itself as an induced decohering neutrino mass difference. Notice that in our sample point only antineutrinos have non-trivial decoherence parameters $\overline{\gamma_i}$, for $i = 1$ and 3, while the corresponding quantities in the neutrino sector vanish. This implies that there is a single cause for mass differences, the decoherence in antineutrino sector, which is compatible with common mass differences in both sectors. This would be very interesting, if true.

Finally, before closing, we would like to remark on extensions of the above phenomenologiocal model for decoherence by including non-diagonal terms in the decoherence matrix $L_{\mu\nu}$. As mentioned above, the physical significance of such extensions is not clear, and indeed it cannot be clear from simple phenomenological analyses like the one presented here. One needs a detailed knoweldge of the QG decoherence effects so as to obtain such an understanding.

Nevertheless one may test the phenomenological efficiency of the simple parametrisation of [93, 94] by comparing the results on the oscillation probabilities versus models where off diagonal terms are included in the decoherence matrix. As a simple example, consider the following form of the decoherence matrix [38]:

$$\mathcal{M} = \begin{pmatrix} L_{11} & -\Delta_{12}+L_{12} & 0 & 0 & 0 & 0 & 0 & 0 \\ \Delta_{12}+L_{12} & L_{22} & 0 & 0 & 0 & 0 & 0 & 0 \\ 0 & 0 & L_{33} & 0 & 0 & 0 & 0 & 0 \\ 0 & 0 & 0 & L_{44} & -\Delta_{13}+L_{45} & 0 & 0 & 0 \\ 0 & 0 & 0 & \Delta_{13}+L_{45} & L_{55} & 0 & 0 & 0 \\ 0 & 0 & 0 & 0 & 0 & L_{66} & -\Delta_{23}+L_{67} & 0 \\ 0 & 0 & 0 & 0 & 0 & \Delta_{23}+L_{67} & L_{77} & 0 \\ 0 & 0 & 0 & 0 & 0 & 0 & 0 & L_{88} \end{pmatrix} \tag{104}$$

where $\mathcal{M}$ is the matrix appearing in the decoherent evolution (31). It is again a straightforward but tedious exercise to determine the matrix which diagonalises $\mathcal{M}$ and find the eigenvalues and eigenvectors of $\mathcal{M}$, which determine the oscillation probabilities (95). Defining Γ_{ij} as

$$\Gamma_{12} \equiv \sqrt{(L_{11}-L_{22})^2 + 4L_{12}^2 - 4\Delta_{12}^2} \tag{105}$$

and similarly for the other elements, and taking notice of the fact that Γ_{ij} are similar to the Ω_{ij} of the diagonal decoherence case (96), with the only difference being an extra positive term (L_{12}^2 etc.) under the sqare root, we can compute the corresponding oscillation probabilities (95). For completeness we give here the relevant expression, which allows the interested reader to derive the diagonal case expressions by setting the off-digonal elements of $L_{\mu\nu}$ equal to zero. We have:

$$\begin{aligned} P_{\nu_\alpha\to\nu_\beta}(t) = \frac{1}{3} + \frac{1}{2} e^{\frac{(L_{11}+L_{22})t}{2}} & \left\{ \left(\rho_1^\alpha\rho_1^\beta + \rho_2^\alpha\rho_2^\beta\right)\left(\frac{e^{\frac{\Gamma_{12}t}{2}} + e^{-\frac{\Gamma_{12}t}{2}}}{2}\right) \right. \\ + & \left[\left(\rho_1^\alpha\rho_1^\beta - \rho_2^\alpha\rho_2^\beta\right)\left(\frac{L_{11}-L_{22}}{\Gamma_{12}}\right)\right) \\ + & \left. \frac{2\rho_2^\alpha\rho_1^\beta(-L_{12}+\Delta_{12}) - 2\rho_1^\alpha\rho_2^\beta(L_{12}+\Delta_{12})}{\Gamma_{12}}\right]\left(\frac{e^{\frac{\Gamma_{12}t}{2}} - e^{-\frac{\Gamma_{12}t}{2}}}{2}\right\} \\ + e^{\frac{(L_{44}+L_{55})t}{2}} & \left\{ \left(\rho_4^\alpha\rho_4^\beta + \rho_5^\alpha\rho_5^\beta\right)\left(\frac{e^{\frac{\Gamma_{13}t}{2}} + e^{-\frac{\Gamma_{13}t}{2}}}{2}\right) \right. \\ + & \left[\left(\rho_4^\alpha\rho_4^\beta - \rho_5^\alpha\rho_5^\beta\right)\left(\frac{L_{44}-L_{55}}{\Gamma_{13}}\right) \right. \\ + & \left. \left. \frac{2\rho_5^\alpha\rho_4^\beta(-L_{45}+\Delta_{13}) - 2\rho_4^\alpha\rho_5^\beta(L_{45}+\Delta_{13})}{\Gamma_{13}}\right]\left(\frac{e^{\frac{\Gamma_{13}t}{2}} - e^{-\frac{\Gamma_{13}t}{2}}}{2}\right)\right\} \\ + e^{\frac{(L_{66}+L_{77})t}{2}} & \left\{ \left(\rho_6^\alpha\rho_6^\beta + \rho_7^\alpha\rho_7^\beta\right)\left(\frac{e^{\frac{\Gamma_{23}t}{2}} + e^{-\frac{\Gamma_{23}t}{2}}}{2}\right) \right. \\ + & \left[\left(\rho_6^\alpha\rho_6^\beta - \rho_7^\alpha\rho_7^\beta\right)\left(\frac{L_{66}-L_{77}}{\Gamma_{23}}\right) \right. \\ + & \left. \left. \frac{2\rho_7^\alpha\rho_6^\beta(-L_{67}+\Delta_{23}) - 2\rho_6^\alpha\rho_7^\beta(L_{67}+\Delta_{23})}{\Gamma_{23}}\right]\left(\frac{e^{\frac{\Gamma_{23}t}{2}} - e^{-\frac{\Gamma_{23}t}{2}}}{2}\right)\right\} \\ + e^{L_{33}t}\rho_3^\alpha\rho_3^\beta + e^{L_{88}t}\rho_8^\alpha\rho_8^\beta & \end{aligned} \tag{106}$$

Assuming Γ_{ij} to be imaginary, as in the diagonal case, taking $\sin\left(\frac{|\Gamma_{ij}|t}{2}\right) \approx 0$, and recalling (100), we observe that, with real values for the elements of the mixing matrix U, one obtains the same form for the oscillation probability as in [93], provided the choice (101) is made for the diagonal elements:

$$\begin{aligned} P_{\nu_\alpha \to \nu_\beta}(t) = \frac{1}{3} + \frac{1}{2} e^{\frac{(L_{11}+L_{22})}{2}} \left(\rho_1^\alpha \rho_1^\beta\right) \cos\left(\frac{|\Gamma_{12}|t}{2}\right) \\ + e^{\frac{(L_{44}+L_{55})}{2}} \left(\rho_4^\alpha \rho_4^\beta\right) \cos\left(\frac{|\Gamma_{13}|t}{2}\right) \\ + e^{\frac{(L_{66}+L_{77})}{2}} \left(\rho_6^\alpha \rho_6^\beta\right) \cos\left(\frac{|\Gamma_{23}|t}{2}\right) \\ e^{L_{33}t} \rho_3^\alpha \rho_3^\beta + e^{L_{88}t} \rho_8^\alpha \rho_8^\beta \end{aligned} \tag{107}$$

the difference being that Γ_{ij}, as noted earlier (105), is of a slightly different form from the respective Ω_{ij} (96), due to the presence of the off-diagonal elements $L_{12} \neq 0$ etc. Notice from (105) that there is a tendency of the off diagonal elements of the decoherence matrix to reduce the effects of the neutrino mass squared difference Δ_{ij}^2. Thus, this sort of extension beyond the diagonal form of the decoherence matrix (94) will affect the magnitude of the oscillation length, as compared to the diagonal case.

It is straightforward to use such parametrizations to obtain bounds on the extra decoherence parameters by comparison with data. We stress again, that, due to CPT Violation, the above probabilities may differ between particles and antiparticles sectors insofar as the order of magnitude of the corresponding decoherence parameters is concerned. Moreover, in view of our comments above on the possible contributions of a decohering environment to the Hamiltonian terms in (14), it is also of great theoretical and phenomenological interest to consider the case of modified dispersion relations for neutrinos simultaneously with the above-described decoherence effects, and compare with current experimental limits. Such modifications may indeed have a common origin with the decohering effects, the interactions with the space time foam. In view of the effects (105) on the oscillation length, analyses like the one in [66], bounding the coefficients of modified dispersion relations by means of their effects on neutrino oscillations, need therefore to be rethought.

4 Conclusions

In these lectures I discussed various theoretical ideas and phenomenological tests of possible CPT Violation induced by quantum gravity. From this exposition it becomes clear, I hope, that CPT Violation may not be an academic issue, and indeed it may characterize a natural theory of quantum gravity.

There are several probes of CPT Violation and there is no single figure of merit for it. Neutrinos seem to provide the most stringent constraints on CPT

Violation through quantum decoherence to date, which in some cases are much stronger than constraints from neutral meson experiments and factories. In this sense neutrinos may provide a very useful guide in our quest for a theory of Quantum Gravity.

Neutrino oscillation experiments provide stringent bounds on many quantum gravity models entailing Lorentz Invariance Violation. There are also plenty of low energy nuclear and atomic physics experiments which yield stringent bounds in models with Lorentz (LV) and CPT Violation (notice that the frame dependence of LV effects is crucial for such high sensitivities). It is my firm opinion that neutrino factories, when built, will undoubtedly shed light on such important and fundamental issues and provide definitive answers to many questions related to LV models of quantum space time.

However, as I repeatedly stressed during these lectures, Quantum Gravity may exhibit Lorentz Invariant (and hence frame independent) CPTV Decoherence. Theoretically, the presence of an environment may be consistent with Lorentz Invariance. This scenario is still compatible with all the existing ν data, including LSND "anomalous" results, within three generation models, and without the need for introducing matter-antimatter mass differences. Of course the order of the decoherence parameters of such models is highly model dependent, and, hence, at present it is the experiment that may guide the theory insofar as properties and estimates of QG decoherence effects are concerned. It is interesting to remark that, in cases where quantum gravity induces neutrino oscillations between flavours or violates lepton number, the sensitivity of experiments looking for astrophysical neutrinos from extragalactic sources may exceed the order of $1/M_P^2$ in the respective figures of merit, and thus is far more superior than the sensitivities of meson factories and nuclear and atomic physics experiments as probes of quantum mechanics.

However, as I remarked previously, the reader should be alert to the fact that there is no single figure of merit for CPT Violation; thus, as we have seen, there may be novel CPTV effects unrelated, in principle, to LV and locality violations, which are associated with modifications of EPR correlations. Such effects may be inapplicable to neutrinos, and thus testable only in meson factories or other situations involving entangled states, e.g. in quantum optics.

Clearly much more work, both theoretical and experimental, is needed before definite conclusions are reached on this important research topic, called phenomenology and theory of CPT Violation. I personally believe that this issue lies at the heart of a complete and realistic theory of quantum gravity. For instance, CPT and its Violation is certainly an issue associated with DSR theories, discussed in this School, and non-commutative geometries, which we did not discuss here, but which, as I mentioned in the beginning of the lectures, is also a very active and rich field of research towards a theory of quantum gravity.

In this respect, I believe firmly that theoretical and phenomenological research on sensitive probes of CPT and quantum mechanics, such as photons from extraglactic sources, neutrinos and neutral mesons, could soon make important contributions to our fundamental quest for understanding the quantum structure of space time. Neutrino research certainly constitutes a very interesting and rapidly developing area of fundamental physics, which already provides fruitful collaboration between astrophysics and particle physics, and which, apart from the exciting results on non-zero neutrino masses which has yielded so far, may still hide even further surprises waiting to be discovered in the near future. But other probes, such as photons and neutral mesons, may also prove invaluable in this respect, especially if QG effects discriminate between particle species, a possibility, which as I mentioned in these lectures, may not be so unrealistic.

Let me close, therefore, these lectures with the wish that by the year 2015, when the physics community will be summoned to celebrate the centennial from the development of General Relativity, the dynamical theory of curved space-time geometries, we shall have obtained some concrete experimental indications on what is going on in Physics near the Planck scale. Let us sincerely hope that this exciting prospect will not remain only a wish for the years to come.

Acknowledgements

It is a real pleasure to thank J. Kowalski-Glikman and G. Amelino-Camelia for the invitation to lecture in this School on *Theory and Phenomenology of Quantum Gravity*, in Poland, and for organizing this very successful and thought-stimulating meeting. I would also like to acknowledge informative discussions with G. Barenboim, J. Bernabeu, D. Binosi, G. Gounaris, C.N. Ktorides, J. Papavassiliou and A. Waldron-Lauda. This work is partly supported by the European Union (contract HPRN-CT-2000-00152).

References

1. M. B. Green, J. H. Schwarz and E. Witten, *Superstring Theory. Vols. 1 & 2* (Cambridge University Press 1987); J. Polchinski, *String Theory. Vols. 1 & 2* (Cambridge University Press 1998). See also, E. Alvarez, lectures in this School.
2. J. Ellis, N.E. Mavromatos and D.V. Nanopoulos, Phys. Lett. **B293** (1992) 37; J. Chaos, Solitons and Fractals **10**, 345 (eds. C. Castro and M.S. El Naschie, Elsevier Science, Pergamon 1999) [hep-th/9805120], and references therein.
3. see for instance L. Smolin, "How far are we from the quantum theory of gravity?," arXiv:hep-th/0303185, and references therein.
4. L. Smolin, Lectures in this volume.
5. J. Kowalski-Glikman, lectures in this School, and references therein.
6. G. Amelino-Camelia, J. R. Ellis, N. E. Mavromatos and D. V. Nanopoulos, "Distance measurement and wave dispersion in a Liouville-string approach to quantum gravity," Int. J. Mod. Phys. A **12** (1997) 607 [arXiv:hep-th/9605211].

7. J. R. Ellis, N. E. Mavromatos and A. S. Sakharov, "Synchrotron radiation from the Crab Nebula discriminates between models of space-time foam," Astropart. Phys. **20** (2004) 669 [arXiv:astro-ph/0308403]. J. R. Ellis, N. E. Mavromatos, D. V. Nanopoulos and A. S. Sakharov, "Space-time foam may violate the principle of equivalence," *Nature, Brief Comm.* in press [arXiv:gr-qc/0312044]; J. R. Ellis, N. E. Mavromatos and M. Westmuckett, arXiv:gr-qc/0405066.
8. J. R. Ellis, N. E. Mavromatos and D. V. Nanopoulos, "Dynamical formation of horizons in recoiling D-branes," Phys. Rev. D **62** (2000) 084019 [arXiv:gr-qc/0006004].
9. G. Amelino-Camelia, J. R. Ellis, N. E. Mavromatos and D. V. Nanopoulos, "Distance measurement and wave dispersion in a Liouville-string approach to quantum gravity," *Int. J. Mod. Phys.* **A12** (1997) 607 [arXiv:hep-th/9605211]; G. Amelino-Camelia, J. R. Ellis, N. E. Mavromatos, D. V. Nanopoulos and S. Sarkar, "Potential Sensitivity of Gamma-Ray Burster Observations to Wave Dispersion in Vacuo," *Nature* **393** (1998) 763 [arXiv:astro-ph/9712103].
10. J. R. Ellis, K. Farakos, N. E. Mavromatos, V. A. Mitsou and D. V. Nanopoulos, "Astrophysical probes of the constancy of the velocity of light," *Astrophys. J.* **535** (2000) 139 [arXiv:astro-ph/9907340]; J. R. Ellis, N. E. Mavromatos, D. V. Nanopoulos and A. S. Sakharov, "Quantum-gravity analysis of gamma-ray bursts using wavelets," *Astron. Astrophys.* **402** (2003) 409 [arXiv:astro-ph/0210124].
11. T. Jacobson, S. Liberati and D. Mattingly, "Lorentz violation and Crab synchrotron emission: A new constraint far beyond *Nature* **424** (2003) 1019[arXiv: astro-ph/0212190].
12. T. Jacobson, Lectures in this volume.
13. G. Amelino-Camelia, "Relativity in space-times with short-distance structure governed by an observer-independent (Planckian) length scale," *Int. J. Mod. Phys.* **D11** (2002) 35 [arXiv:gr-qc/0012051]; J. Magueijo and L. Smolin, "Lorentz invariance with an invariant energy scale," *Phys. Rev. Lett.* **88** (2002) 190403 [arXiv:hep-th/0112090]; "Generalized Lorentz invariance with an invariant energy scale," Phys. Rev. D **67** (2003) 044017 [arXiv:gr-qc/0207085].
14. See lectures by G. Amelino-Camelia and J. Kowalski-Glikman, this volume.
15. A. Grillo, Lectures in this volume.
16. T. Piran, Lectures in this Volume.
17. G. Lüders, *Det. Kong. Danske Videnskabernes Selskab, Mat.-Fys. Medd.* **No 5** (1954) 28; "Proof Of The Tcp Theorem," *Annals Phys.* **2** (1957) 1 [*Annals Phys.* **281** (2000) 1004]. W. Pauli, in *Niels Bohr and the Development of Physics* (Mc Graw-Hill, New York, 1955), p. 30.
18. see: A. Zichichi, in *John Bell and the ten challenges of subnuclear physics* (Bertlmann, R.A. (ed.) et al.: Quantum unspeakables, 2000), p. 429.
19. R. Jost, *Helv. Phys. Acta* **30** (1957), 409; also in *Theoretical Physics in the Twentieth Century* (Interscience, New York, 1960).
20. R. F. Streater and A. S. Wightman, *PCT, Spin & Statistics, and All That* (Benjamin, New York, 1964).
21. see for instance: J. A. Wheeler and K. Ford, *Geons, Black Holes and Quantum Foam: A Life in Physics* (Norton, New York, 1998); S. W. Hawking, "The Unpredictability Of Quantum Gravity," *Commun. Math. Phys.* **87** (1982) 395.
22. L. J. Garay, "Quantum evolution in spacetime foam," Int. J. Mod. Phys. A **14** (1999) 4079 [arXiv:gr-qc/9911002].

23. A. Kostelecky, "Gravity, Lorentz violation, and the standard model," arXiv:hep-th/0312310 and references therein.
24. J. R. Ellis, J. S. Hagelin, D. V. Nanopoulos and M. Srednicki, "Search For Violations Of Quantum Mechanics," *Nucl. Phys.* **B241** (1984) 381.
25. J. R. Ellis, N. E. Mavromatos and D. V. Nanopoulos, "Testing quantum mechanics in the neutral kaon system," *Phys. Lett.* **B293** (1992) 142 [arXiv:hep-ph/9207268]; "CPT violation in string modified quantum mechanics and the neutral kaon system," *Int. J. Mod. Phys.* **A11** (1996) 1489 [arXiv:hep-th/9212057]; J. R. Ellis, J. L. Lopez, N. E. Mavromatos and D. V. Nanopoulos, "Precision tests of CPT symmetry and quantum mechanics in the neutral kaon system," *Phys. Rev.* **D53** (1996) 3846 [arXiv:hep-ph/9505340].
26. P. Huet and M. E. Peskin, "Violation of CPT and quantum mechanics in the K0 - anti-K0 system," Nucl. Phys. B **434** (1995) 3 [arXiv:hep-ph/9403257];
27. F. Benatti and R. Floreanini, "Complete positivity and correlated neutral kaons," Phys. Lett. B **468** (1999) 287 [arXiv:hep-ph/9910508].
28. J. Bernabeu, N. E. Mavromatos and J. Papavassiliou, "Novel type of CPT violation for correlated EPR states," Phys. Rev. Lett. **92** (2004) 131601 [arXiv:hep-ph/0310180].
29. A. Amorreti et al. [ATHENA Coll.], *Nature* **419** (2002) 456; G. Gabrielse et al. [ATRAP Coll.], *Phys. Rev. Lett.* **89** (2002) 213401.
30. R. Bluhm, arXiv:hep-ph/0308281; "Probing the Planck scale in low-energy atomic physics," arXiv:hep-ph/0111323, and references therein.
31. N. E. Mavromatos, "Theoretical and phenomenological aspects of CPT violation," *Nucl. Instrum. Meth.* **B214** (2004) 1 [arXiv:hep-ph/0305215] and references therein. See also: http://ad3-proj-leap03.web.cern.ch/ad3-proj-leap03/
32. N. E. Mavromatos, "Neutrinos and the phenomenology of CPT violation," Proc. Neutrino Oscillation in Venice (NO-VE) 2003 [arXiv:hep-ph/0402005].
33. G. J. Milburn, "Lorentz invariant intrinsic decoherence," arXiv:gr-qc/0308021;
34. F. Dowker, J. Henson and R. D. Sorkin, "Quantum gravity phenomenology, Lorentz invariance and discreteness," arXiv:gr-qc/0311055.
35. N. E. Mavromatos, "On CPT symmetry: Cosmological, quantum-gravitational and other possible violations and their phenomenology," arXiv:hep-ph/0309221, Proc. *Beyond the Desert 2003* (Castle Ringberg, Tegernsee, Germany, 9-14 Jun 2003) in press.
36. O. W. Greenberg, "Why is CPT fundamental?," arXiv:hep-ph/0309309; O. W. Greenberg, "CPT violation implies violation of Lorentz invariance," Phys. Rev. Lett. **89** (2002) 231602 [arXiv:hep-ph/0201258].
37. R. M. Wald, "Quantum Gravity And Time Reversibility," *Phys. Rev.* **D21** (1980) 2742.
38. N. E. Mavromatos and Alison Waldron-Lauda, in preparation.
39. G. Lindblad, Comm. Math. Phys. **48** (1976) 119; R. Alicki and K. Lendi, Lect. Notes Phys. **286** (Springer-Verlag, Berlin (1987)).
40. L. P. Houghston, Proc. Roy. Soc. London **A452** (1996) 953; I.C. Percival, Proc. Roy. Soc. London **A447** (1994) 189.
41. S. L. Adler, "Comment on a proposed Super-Kamiokande test for quantum gravity induced decoherence effects," *Phys. Rev.* **D62** (2000) 117901 [arXiv:hep-ph/0005220].
42. G. C. Ghirardi, A. Rimini and T. Weber, Phys. Rev. **D34** (1986) 470; L. Diósi, Phys. Lett. **129A** (1988) 419; *ibid.* **132A** (1988) 233; G.C. Ghirardi, P. Pearle and A. Rimini, Phys. Rev. **A42** (1990) 78.

43. N. Gisin and I. C. Percival, "Quantum state diffusion, localization and quantum dispersion entropy," J. Phys. A **26** (1993) 2233.
44. J. R. Ellis, S. Mohanty and D. V. Nanopoulos, "Quantum Gravity And The Collapse Of The Wave Function," Phys. Lett. B **221** (1989) 113; "Wormholes Violate Quantum Mechanics In Squids," Phys. Lett. B **235** (1990) 305.
45. see, for instance: D.F. Walls and G.J. Millburn, Phys. Rev. A **31** (1985), 2403.
46. W. H. Zurek, "Quantum Measurements And The Environment Induced Transition From Quantum To Classical," NSF-ITP-88-197; "Decoherence and the transition from quantum to classical," Phys. Today **44** (1991) 36.
47. N. E. Mavromatos, "String cosmology," Lect. Notes Phys. **592**, 392 (2002) [arXiv:hep-th/0111275].
48. Y. I. Kogan, University of British Columbia preprint UBCTP 91-13 (1991); "Regularization Of The Area Integration And Imaginary Parts Of The Correlation Functions In Liouville Theory," Phys. Lett. B **265** (1991) 269.
49. E. Gravanis and N. E. Mavromatos, "Vacuum energy and cosmological supersymmetry breaking in brane worlds," Phys. Lett. B **547**, 117 (2002) [arXiv:hep-th/0205298].
50. A. B. Zamolodchikov, "'Irreversibility' Of The Flux Of The Renormalization Group In A 2-D Field Theory," JETP Lett. **43** (1986) 730 [Pisma Zh. Eksp. Teor. Fiz. **43** (1986) 565].
51. D. Kutasov, "Irreversibility of the renormalization group flow in two-dimensional quantum gravity," Mod. Phys. Lett. A **7** (1992) 2943 [arXiv:hep-th/9207064].
52. N. E. Mavromatos and R. J. Szabo, "Gradient flow in logarithmic conformal field theory," Phys. Lett. B **430** (1998) 94 [arXiv:hep-th/9803092].
53. B. P. Schmidt et al., "The High-Z Supernova Search: Measuring Cosmic Deceleration and Global Curvature of the Universe Using Type Ia Supernovae," *Astrophys. J.* **507** (1998) 46 [arXiv:astro-ph/9805200]; S. Perlmutter et al. [Supernova Cosmology Project Coll.], "Measurements of Omega and Lambda from 42 High-Redshift Supernovae," *Astrophys. J.* **517** (1999) 565 [arXiv:astro-ph/9812133].
54. C. L. Bennett et al., "First Year Wilkinson Microwave Anisotropy Probe (WMAP) Observations: Preliminary Maps and Basic Results," *Astrophys. J. Suppl.* **148** (2003) 1 [arXiv:astro-ph/0302207]; D. N. Spergel et al., "First Year Wilkinson Microwave Anisotropy Probe (WMAP) Observations: Determination of Cosmological Parameters," *Astrophys. J. Suppl.* **148** (2003) 175 [arXiv:astro-ph/0302209].
55. Y. J. Ng and H. van Dam, "Comment on Uncertainty in measurements of distance," Class. Quant. Grav. **20**, (2003) 393 [arXiv:gr-qc/0209021]; "Limit to space-time measurement," Mod. Phys. Lett. A **9**, (1994) 335.
56. see talk by S. Bludman, these proceedings and references therein (http://axpd24.pd.infn.it/NO-VE/prog-NOVE.html).
57. D. V. Ahluwalia and M. Kirchbach, "Primordial space-time foam as an origin of cosmological matter-antimatter asymmetry," *Int. J. Mod. Phys.* **D10** (2001) 811 [arXiv:astro-ph/0107246]. See also: D. V. Ahluwalia-Khalilova, "Extended set of Majorana spinors, a new dispersion relation, and a preferred frame," arXiv:hep-ph/0305336.
58. A. Agostini, "Fields and symmetries in kappa-Minkowski noncommutative spacetime," arXiv:hep-th/0312305 and references therein.
59. G. Lambiase and P. Singh, "Matter-antimatter asymmetry generated by loop quantum gravity," *Phys. Lett.* **B565** (2003) 27 [arXiv:gr-qc/0304051].

60. J. R. Ellis, N. E. Mavromatos, D. V. Nanopoulos and G. Volkov, "Gravitational-recoil effects on fermion propagation in space-time foam," *Gen. Rel. Grav.* **32** (2000) 1777 [arXiv:gr-qc/9911055].
61. L. B. Okun, "C, P, T are broken. Why not CPT?," arXiv:hep-ph/0210052.
62. A. V. Kostelecky and M. Mewes, "Lorentz and CPT violation in the neutrino sector," arXiv:hep-ph/0308300; "Lorentz and CPT violation in neutrinos," arXiv:hep-ph/0309025.
63. see for instance: G. Drexlin, "Final Neutrino Oscillation Results From Lsnd And Karmen," *Nucl. Phys. Proc. Suppl.* **118** (2003) 146, and references therein.
64. J. Alfaro, H. A. Morales-Tecotl and L. F. Urrutia, "Quantum gravity corrections to neutrino propagation," *Phys. Rev. Lett.* **84** (2000) 2318 [arXiv:gr-qc/9909079].
65. J. R. Ellis, N. E. Mavromatos and D. V. Nanopoulos, "Non-critical Liouville string escapes constraints on generic models of quantum gravity," *Phys. Rev.* **D65** (2002) 064007 [arXiv:astro-ph/0108295].
66. R. Brustein, D. Eichler and S. Foffa, "Probing the Planck scale with neutrino oscillations," *Phys. Rev.* **D65** (2002) 105006 [arXiv:hep-ph/0106309].
67. R. C. Myers and M. Pospelov, "Experimental challenges for quantum gravity," *Phys. Rev. Lett.* **90** (2003) 211601 [arXiv:hep-ph/0301124].
68. G. Lambiase, "Spin flavor conversion of neutrinos in loop quantum gravity," *Class. Quant. Grav.* **20** (2003) 4213 [arXiv:gr-qc/0302053].
69. L. Wolfenstein, "Neutrino Oscillations In Matter," *Phys. Rev.* **D17** (1978) 2369. "Neutrino Oscillations And Stellar Collapse," *Phys. Rev.* **D20** (1979) 2634; S. P. Mikheev and A. Y. Smirnov, "Resonance Enhancement Of Oscillations In Matter And Solar Neutrino Spectroscopy," *Sov. J. Nucl. Phys.* **42** (1985) 913 [Yad. Fiz. **42** (1985) 1441];
70. M. Blasone, J. Magueijo and P. Pires-Pacheco, "Neutrino mixing and Lorentz invariance," arXiv:hep-ph/0307205.
71. J. R. Ellis, N. E. Mavromatos and D. V. Nanopoulos, "How large are dissipative effects in noncritical Liouville string theory?," Phys. Rev. D **63**, 024024 (2001) [arXiv:gr-qc/0007044].
72. M. Czachor and M. Kuna, "Puzzle of complete positivity in nonlinear frameworks: A case study," Phys. Rev. A **58** (1998) 128 [arXiv:quant-ph/9708029].
73. Y. Liu, L. z. Hu and M. L. Ge, "The Effect of quantum mechanics violation on neutrino oscillation," *Phys. Rev.* **D56** (1997) 6648; C. H. Chang, W. S. Dai, X. Q. Li, Y. Liu, F. C. Ma and Z. j. Tao, "Possible effects of quantum mechanics violation induced by certain quantum gravity on neutrino oscillations," *Phys. Rev.* **D60** (1999) 033006 [arXiv:hep-ph/9809371].
74. E. Lisi, A. Marrone and D. Montanino, "Probing possible decoherence effects in atmospheric neutrino oscillations," *Phys. Rev. Lett.* **85** (2000) 1166 [arXiv:hep-ph/0002053].
75. F. Benatti and R. Floreanini, "Massless neutrino oscillations," *Phys. Rev.* **D64** (2001) 085015 [arXiv:hep-ph/0105303].
76. H. V. Klapdor-Kleingrothaus, H. Paes and U. Sarkar, "Effects of quantum space time foam in the neutrino sector," *Eur. Phys. J.* **A8** (2000) 577 [arXiv:hep-ph/0004123].
77. A. Y. Smirnov, D. N. Spergel and J. N. Bahcall, *Phys. Rev.* **D49** (1994) 1389.
78. [Heidelberg Moscow Coll.], *Phys. Rev. Lett* **83** (1999) 41 ; H. V. Klapdor-Kleingrothaus et al., hep-ph/99010205 [on behalf of GENIUS Coll.].

79. G. D. Hallewell [ANTARES Coll.], "Status Of The Antares Underwater Neutrino Telescope," *Nucl. Instrum. Meth.* **A502** (2003) 138, and references therein.
80. S. E. Tzamarias [NESTOR Coll.], "Nestor: A Deep-Sea Neutrino Telescope," *Nucl. Instrum. Meth.* **A502** (2003) 150, and references therein.
81. G. Barenboim and J. Lykken, "A model of CPT violation for neutrinos," Phys. Lett. B **554** (2003) 73 [arXiv:hep-ph/0210411] and references therein.
82. H. Murayama and T. Yanagida, "LSND, SN1987A, and CPT violation," Phys. Lett. B **520** (2001) 263 [arXiv:hep-ph/0010178].
83. A. Strumia, "Interpreting the LSND anomaly: Sterile neutrinos or CPT-violation or...?," Phys. Lett. B **539** (2002) 91 [arXiv:hep-ph/0201134]; "Neutrino anomalies," arXiv:hep-ex/0304039.
84. see for instance: A. Smirnov, Proc. of Neutrino Oscillations in Venice (NO-VE) 2003 and references therein (http://axpd24.pd.infn.it/NO-VE/prog-NOVE.html).
85. K. Lang [on behalf of MINOS Coll.] "Minos: The Physics Program And The Construction Status," *Int. J. Mod. Phys.* **A18** (2003) 3857, and references therein.
86. see talk by: M. Lindner, Proc. of Neutrino Oscillations in Venice (NO-VE) 2003 and references therein (http://axpd24.pd.infn.it/NO-VE/prog-NOVE.html).
87. V. Barger, D. Marfatia and K. Whisnant, "LSND anomaly from CPT violation in four-neutrino models," Phys. Lett. B **576** (2003) 303. [arXiv:hep-ph/0308299].
88. *The Second DaΦne Physics Handbook* (eds. L. Maiani, G. Pancheri and N. Paver 1995).
89. R. Adler et al. [CPLEAR Coll.], "Test of CPT Symmetry and Quantum Mechanics with Experimental data from CPLEAR," Phys. Lett. B **364** (1995) 239 [arXiv:hep-ex/9511001].
90. K. Hagiwara et al. [Particle Data Group Collaboration], "Review of Particle Physics," Phys. Rev. D **66**, (2002) 010001.
91. H. J. Lipkin, Phys. Rev. **176**, (1968) 1715; "Simple Symmetries In Epr Correlated Decays Of Kaon And B Meson Pairs With CP Violation," Phys. Lett. B **219**, (1989); I. Dunietz, J. Hauser and J. L. Rosner, "An Experiment Addressing CP And Cpt Violation In The K0 Anti-K0 System," Phys. Rev. D **35**, (1987) 2166.
92. V. V. Nesvizhevsky et al., "Measurement of quantum states of neutrons in the Earth's gravitational field," Phys. Rev. D **67** (2003) 102002 [arXiv:hep-ph/0306198]; "Reply to: Comment on 'Measurement of quantum states of neutrons in the Earth's gravitational field," Phys. Rev. D **68** (2003) 108702.
93. G. Barenboim and N. E. Mavromatos, "CPT violating decoherence and LSND: A possible window to Planck scale arXiv:hep-ph/0404014.
94. A. M. Gago, E. M. Santos, W. J. C. Teves and R. Zukanovich Funchal, "A study on quantum decoherence phenomena with three generations of arXiv:hep-ph/0208166.
95. M. Apollonio et al. [CHOOZ Collaboration], "Limits on neutrino oscillations from the CHOOZ experiment," Phys. Lett. B **466**, 415 (1999) [arXiv:hep-ex/9907037].
96. S. H. Ahn et al. [K2K Collaboration], "Detection of accelerator produced neutrinos at a distance of 250-km," Phys. Lett. B **511**, 178 (2001) [arXiv:hep-ex/0103001]; M. H. Ahn et al. [K2K Collaboration], "Indications of neutrino oscillation in a 250-km long-baseline experiment," Phys. Rev. Lett. **90**, 041801 (2003) [arXiv:hep-ex/0212007].

97. K. Eguchi et al. [KamLAND Collaboration], "First results from KamLAND: Evidence for reactor anti-neutrino disappearance," Phys. Rev. Lett. **90**, 021802 (2003) [arXiv:hep-ex/0212021].
98. B. Armbruster et al. [KARMEN Collaboration], "Upper limits for neutrino oscillations anti-nu/mu → anti-nu/e from muon decay at rest," Phys. Rev. D **65**, 112001 (2002) [arXiv:hep-ex/0203021].
99. E. D. Zimmerman [BooNE Collaboration], "BooNE has begun," eConf **C0209101**, TH05 (2002) [Nucl. Phys. Proc. Suppl. **123**, 267 (2003)] [arXiv:hep-ex/0211039].

Quantum Foam and Quantum Gravity Phenomenology

Y. Jack Ng

Institute of Field Physics, Department of Physics and Astronomy, University of North Carolina, Chapel Hill, NC 27599-3255, USA
yjng@physics.unc.edu

1 Introduction

Our understanding of spacetime has undergone some major changes in the last hundred years. Before last century, spacetime was regarded as nothing more than a passive and static arena in which events took place. Early last century, Einstein's general relativity changed that viewpoint and promoted spacetime to an active and dynamical entity. Nowadays, many physicists also believe that spacetime, like all matter and energy, undergoes quantum fluctuations. Following John Wheeler, many of us think that space is composed of an ever-changing arrangement of bubbles called spacetime foam, a.k.a. quantum foam. To understand the terminology, let us follow Wheeler and consider the following simplified analogy which he gave in a gravity conference at the University of North Carolina in 1957. Imagine yourself flying an airplane over an ocean. At high altitude the ocean appears smooth. But as you descend, it begins to show roughness. Close enough to the ocean surface, you see bubbles and foam. Analogously, spacetime appears smooth on a large scale, but on sufficiently small scales, it will appear rough and foamy, hence the term "spacetime foam." Many physicists believe the foaminess is due to quantum fluctuations of spacetime, hence the alternative term "quantum foam." If spacetime indeed undergoes quantum fluctuations, the fluctuations will show up when we measure a distance (or a time duration), in the form of uncertainties in the measurement. Conversely, if in any distance (or time duration) measurement, we cannot measure the distance (or time duration) precisely, we interpret this intrinsic limitation to spacetime measurements as resulting from fluctuations of spacetime itself.

As we will see below, the physics of spacetime foam is intimately connected to that of black holes. It is related to the holographic principle and has bearings on the physics of clocks and computation. As far as (quantum gravity) phenomenology, the theme of this Winter School, is concerned, we can

Y. Jack Ng: *Quantum Foam and Quantum Gravity Phenomenology*, Lect. Notes Phys. **669**, 321–349 (2005)
www.springerlink.com

only say that it is not easy, but by no means impossible, to detect spacetime foam.[1] We encourage the students to find better ways to do so.

Before we proceed, we should mention that the approach to the physics of quantum foam adopted here is very conservative: the only ingredients we use are quantum mechanics and general relativity. Hopefully, by considering only distances (time durations) much larger than the Planck length (time) or energies (momenta) much smaller than Planck energy (momentum), a semi-classical treatment of gravity suffices and a bona fide theory of quantum gravity is not needed.

We should also make it clear at the outset that we make no assumptions on the high energy regime of the ultimate quantum gravity theory. We refrain from speculating on violations or deformations of Lorentz symmetry and the consequent systematically modified dispersion relations, involving a coefficient of *fixed* magnitude and *fixed* sign, which many people believe are unavoidably induced by quantum gravity. (In the terminology of 2, these quantum gravity effects are called "systematic" effects.) The only quantum gravity effects we are concerned with in these lectures are those due to quantum fuzziness – uncertainties involving *fluctuating* magnitudes with *both* $\pm$ signs, perhaps like a fluctuation with a Gaussian distribution about zero. (In the terminology of 2, these effects are called "non-systematic" effects.)

If quantum fluctuations do make spacetime foamy on small spacetime scales, then it is natural to ask: How large are the fluctuations? How foamy is spacetime? Is there any theoretical evidence of quantum foam? And how can we detect quantum foam? In what follows, we address these questions.

The outline of this manuscript is as follows:

- *Section 2: Quantum fluctuations of spacetime.*
  By analysing a gedanken experiment for spacetime measurement, we show, in Subsec. 2.1, that spacetime fluctuations scale as the cube root of distances or time durations. In Subsect. 2.2, we show that this cube root dependence is consistent with the holographic principle. Subsection 2.3 is devoted to a comparison of this peculiar dependence on distances or time durations with the well-known random-walk problem and other quantum gravity models. In Subsect. 2.4, we consider the cumulative effects of individual spacetime fluctuations.
- *Section 3: Clocks, computers, and black holes.*
  We discuss how quantum foam affects the physics of clocks (Subsect. 3.1) and computation (Subsect. 3.2), and show that the physics of spacetime foam is intimately connected to that of black holes (Subsect. 3.3). In particular, the same underlying physics governs the computational power of black hole quantum computers. In Subsect. 3.3, we give the results for arbitrary spacetime dimensions.
- *Section 4: Energy-momentum uncertainties.*
  Just as there are uncertainties in spacetime measurements, there are also uncertainties in energy-momentum measurements. This topic of energy-

momentum uncertainties is given a brief treatment. Two physical implications are given: dispersion relations are modified, and (as a consequence) energy-dependent speed of light fluctuates around c.
- *Section 5: Spacetime foam phenomenology.*
  Various proposals to detect quantum foam are considered; they include: phase incoherence of light from distant galaxies (Subsect. 5.1), gamma ray bursts (Subsect. 5.2), laser-based interferometry (Subsect. 5.3), and ultra-high energy cosmic ray events (Subsect. 5.4).
- *Section 6: Summary and Conclusions.*

To make the lectures informative and more or less self-contained, "*preparatory remarks*", "*side remarks*", and "*further remarks*", too long for footnotes, are inserted when their additions are warranted. All such remarks are contained inside square brackets [ ]. They are somewhat out of the lectures' main line of development. On notations, the subscript "P" denotes Planck units. Thus $l_P \equiv (\hbar G/c^3)^{1/2} \sim 10^{-33}$ cm is the Planck length, etc.

2 Quantum Fluctuations of Spacetime

The questions are: does spacetime undergo quantum fluctuations? And if so, how large are the fluctuations? To quantify the problem, let us consider measuring a distance l. The question now is: how accurately can we measure this distance? Let us denote by δl the accuracy with which we can measure l. We will also refer to δl as the uncertainty or fluctuation of the distance l for reasons that will become obvious shortly. We will show that δl has a lower bound and will use two ways to calculate it. Neither method is rigorous, but the fact that the two very different methods yield the same result bodes well for the robustness of the conclusion. (Furthermore, the result is also consistent with well-known semi-classical black hole physics. See Sect. 3.)

2.1 Gedanken Experiment

In the first method, we conduct a thought experiment to measure l. The importance of carrying out spacetime measurements to find the quantum fluctuations in the fabric of spacetime cannot be over-emphasized. According to general relativity, coordinates do not have any intrinsic meaning independent of observations; a coordinate system is defined only by explicitly carrying out spacetime distance measurements. Let us measure the distance between point A and point B. Following Wigner [3], we put a clock at A and a mirror at B. Then the distance l that we want to measure is given by the distance between the clock and the mirror. By sending a light signal from the clock to the mirror in a timing experiment, we can determine the distance l. However, quantum uncertainties in the positions of the clock and the mirror introduce an inaccuracy δl in the distance measurement. We expect the clock and the mirror to contribute comparable uncertainties to the measurement. Let us concentrate on the clock and denote its mass by m. Wigner argued that if it has a linear spread δl when the light signal leaves the clock, then its position spread grows to $\delta l + \hbar l (mc\delta l)^{-1}$ when the light signal returns to the clock, with the minimum at $\delta l = (\hbar l/mc)^{1/2}$. Hence one concludes that

$$\delta l^2 \gtrsim \frac{\hbar l}{mc} . \tag{1}$$

Thus quantum mechanics alone would suggest using a massive clock to reduce the jittering of the clock and thereby the uncertainty δl. On the other hand, according to general relativity, a massive clock would distort the surrounding space severely, affecting adversely the accuracy in the measurement of the distance.

Side Remarks

[It is here that we appreciate the importance of taking into account the effects of instruments in this thought-experiment. Usually when one wants to examine a certain a field (say, an electromagnetic field) one uses instruments that are neutral (electromagnetically neutral) and massive for, in that case, the effects of the instruments are negligible. But here in our thought-experiment, the relevant field is the gravitational field. One cannot have a gravitationally neutral yet massive set of instruments because the gravitational charge is equal to the mass according to the principle of equivalence in general relativity. Luckily for us, we can now exploit this equality of the gravitational charge and the inertial mass of the clock to eliminate the dependence on m in the above inequality to promote (1) to a (low-energy) quantum gravitational uncertainty relation.]

To see this, let the clock be a light-clock consisting of a spherical cavity of diameter d, surrounded by a mirror wall of mass m, between which bounces a beam of light (along a diameter). For the uncertainty in distance measurement not to be greater than δl, the clock must tick off time fast enough that $d/c \lesssim \delta l/c$. But d, the size of the clock, must be larger than the Schwarzschild radius $r_S \equiv 2Gm/c^2$ of the mirror, for otherwise one cannot read the time registered on the clock. From these two requirements, it follows that

$$\delta l \gtrsim \frac{Gm}{c^2} . \tag{2}$$

Thus general relativity alone would suggest using a light clock (light as opposed to massive) to do the measurement.

Side Remarks

[This result can also be derived (see the first paper in [4]) in another way. If the clock has a radius $d/2$ (larger than its Schwarzschild radius r_S), then δl, the error in the distance measurement caused by the curvature generated by the mass of the clock, may be estimated by a calculation from the Schwarzschild solution. The result is r_S multiplied by a logarithm involving $2r_S/d$ and $r_S/(l+d/2)$. For $d \gg r_S$, one finds $\delta l = \frac{1}{2} r_S \log \frac{d+2l}{d}$ and hence (2) as an order of magnitude estimate.]

The product of (2) with (1) yields

$$\delta l \gtrsim (l l_P^2)^{1/3} = l_P \left(\frac{l}{l_P} \right)^{1/3} , \tag{3}$$

where $l_P = (\hbar G/c^3)^{1/2}$ is the Planck length. (Note that the result is independent of the mass of the clock and, thereby, one would hope, of the properties of the specific clock used in the measurement.) The end result is as simple as it is strange and appears to be universal: the uncertainty δl in the measurement of the distance l cannot be smaller than the cube root of $l l_P^2$. Reference [4] Obviously the accuracy of the distance measurement is intrinsically limited by this amount of uncertainty or quantum fluctuation. We conclude that there is a limit to the accuracy with which one can measure a distance; in other words, we can never know the distance l to a better accuracy than the cube root of $l l_P^2$. Similarly one can show that we can never know a time duration τ to a better accuracy than the cube root of τt_P^2, i.e.,

$$\delta \tau \gtrsim (\tau t_P^2)^{1/3} , \tag{4}$$

where $t_P \equiv l_P/c \sim 10^{-44}$sec is the Planck time. The spacetime fluctuation translates into a metric fluctuation over a distance l and a time interval τ given by

$$\delta g_{\mu\nu} \gtrsim (l_P/l)^{2/3}, \qquad (t_P/\tau)^{2/3} , \tag{5}$$

respectively. (For a discussion of the related light-cone fluctuations, see [5].)

Because the Planck length is so inconceivably short, the uncertainty or intrinsic limitation to the accuracy in the measurement of any distance, though much larger than the Planck length, is still very small. For example, in the measurement of a distance of one kilometer, the uncertainty in the distance is to an atom as an atom is to a human being. Even for the size of the observable universe ($\sim 10^{10}$ light-years), the uncertainty is only about 10^{-13} cm.

Further Remarks

[*Fluctuations Imply Non-locality?* Fluctuations in spacetime imply that the metrics can be defined only as averages over local regions, and this gives rise to some sort of non-locality. Ahluwalia [4] has observed that spacetime measurements described above alter the spacetime metric in a fundamental manner and that this unavoidable change in the metric destroys the commutativity (and hence locality) of position measurement operators. The gravitationally-induced nonlocality, in turn, suggests a modification of the fundamental commutators.]

Further Remarks

[*On Two Length Scales: An Analogy.* In hindsight it is not too surprising that the uncertainty δl involves two length scales, viz., the fundamental length l_P and the length l itself. There is an analogous result that is relevant for a long thin ruler which can be regarded as a one-dimensional chain of N ions with a spring between successive ions. By a straightforward quantum mechanical calculation [4], one can show that the uncertainty in the length of the ruler scales as $\sqrt{N}$ in the high-temperature limit and as $\sqrt{logN}$ in the zero temperature limit. But $N = l/a$ where l is the length of the ruler and a is the lattice constant (playing the role of l_p in the measurement of distance), so one concludes that the uncertainty of the ruler's length depends on two length scales, viz., l and a.]

Further Remarks

[*Energy Density Fluctuations Associated with Spacetime Fluctuations.* This may be a red herring, but the question has been raised [6] whether the metric fluctuations corresponding to (5) yield an unacceptably large fluctuation in energy density. To see that the associated energy density fluctuation is actually extremely (therefore acceptably) small [7], let us regard metric fluctuations as gravitational waves quantized in a box of volume V (with $\hbar = 1$ and $c = 1$):

$$\delta g(l) = l_P \sum_{\boldsymbol{k}} \frac{A(k)}{\sqrt{2Vk}} coskl \ , \tag{6}$$

with the corresponding energy density fluctuations given by $\delta\rho = V^{-1} \sum A(k)^2 k$ (summation over two different polarizations is understood). In order for (6) to give (5), one needs $A(k) \sim V^{-1/2} l_P^{-1/3} k^{-11/6}$. Replacing the summation over $\boldsymbol{k}$ in $\delta\rho$, in the large volume limit, by an intgral, and using the Planck mass m_P as the upper limit, we get [7]

$$\delta\rho \sim m_P/V \ , \tag{7}$$

an utterly negligible energy density.]

2.2 The Holographic Principle

Alternatively we can estimate δl by applying the holographic principle. Reference [8, 9] but for completeness, let us first recall some physics of black holes and then a heuristic derivation of the holographic principle.

Preparatory Remarks

[*Black Holes.* In our discussion of the gedanken experiment in Subsect. 2.1, we have already used that fact that a chargeless non-rotating black hole of mass m has a size given by

- Size $r_S \sim Gm/c^2$, the Schwarzschild radius. It is also known that a black hole behaves as if it has
- Temperature $\mathcal{T} \sim \frac{\hbar c}{k_B r_S}$ where k_B is the Boltzman's constant;
- Entropy $S \sim k_B \frac{r_S^2}{l_P^2}$, i.e., area in Planck units;
- Finite lifetime $T_{BH} \sim \frac{G^2 m^3}{\hbar c^4}$, first found by Hawking.

Property 3 follows from properties 1 & 2 with the aid of the thermodynamic relation $dS = \frac{dE}{\mathcal{T}}$. For a black hole of one centimeter diameter, its entropy is about 10^{66} bits.

Property 4 can be derived by treating a black hole as a black body for which the emitted power (i.e., $c^2 \frac{dm}{dt}$) per unit area is given by the Stefan's law: $\frac{c^2}{area} \frac{dm}{dt} \sim \sigma \mathcal{T}^4$ where $\sigma \sim k_B^4/(\hbar^3 c^2)$ is the Stefan-Boltzmann constant. The more massive a black hole is, the longer it lasts; a solar-mass black hole is estimated to last 10^{66} years. (By comparison, the present age of the Universe is only about 13.7 billion years.) Unless mini-black holes exist, it will be impossible to directly check Hawking's result for black hole lifetime. But if the physics behind spacetime foam and black holes is the same, as will be shown in Subsect. 3.3, detection of spacetime foam can be taken as an indirect confirmation of Hawking's black hole evaporation process. In passing, we mention that there is increasing evidence that black holes do exist; in particular, supermassive black holes (with mass ranging from a million to a billion times the solar mass) exist at the center of many galaxies, including our own.]

Preparatory Remarks

[*Holographic Principle.* In essence, the holographic principle [10] says that although the world around us appears to have three spatial dimensions, its contents can actually be encoded on a two-dimensional surface, like a hologram. In other words, the maximum entropy of a region of space is given (aside from multiplicative factors of order 1 which we ignore as we have so far) by its surface area in Planck units. This result can be derived by appealing to black hole physics and the second law of theromodynamics as follows. Consider a system with entropy S_0 inside a spherical region Γ bounded by surface area A. Its mass must be less than that of a black hole with horizon area A (otherwise it would have collapsed into a black hole). Now imagine a spherically symmetric shell of matter collapsing onto the original system with just the right amount of energy so that together with the original mass, it forms a black hole which just fills the region Γ. The black hole so formed has entropy $S \sim A/l_P^2$. But according to the second law of thermodynamics, $S_0 \leq S$. It follows immediately that $S_0 \lesssim A/l_P^2$, and hence the maximum entropy of a region of space is bounded by its surface area, as asserted by the holographic principle.]

With the aid of the above preparatory remarks, we are now ready to estimate δl by applying the holographic principle. Reference [8, 9] To be more precise, let us consider a spatial region measuring l by l by l. According to the holographic principle, the number of degrees of freedom that this cubic region can contain is bounded by the surface area of the region in Planck units, i.e., l^2/l_P^2, instead of by the volume of the region as one may naively expect. This principle is strange and counterintuitive, but is supported by black hole physics in conjunction with the laws of thermodynamics (as shown above in the "Preparatory Remarks"), and it is embraced by both string theory and loop gravity, two top contenders of quantum gravity theory. So strange as it may be, let us now apply the holographic principle to deduce the accuracy with which one can measure a distance.

First, imagine partitioning the big cube into small cubes [see Fig. 1]. The small cubes so constructed should be as small as physical laws allow so that intuitively we can associate one degree of freedom with each small cube. In other words, the number of degrees of freedom that the region can hold is given by the number of small cubes that can be put inside that region. But how small can such cubes be? A moment's thought tells us that each side of a small cube cannot be smaller than the accuracy δl with which we can measure each side l of the big cube. This can be easily shown by applying the method of contradiction: assume that we can construct small cubes each of which has sides less than δl. Then by lining up a row of such small cubes along a side of the big cube from end to end, and by counting the number of such small cubes,

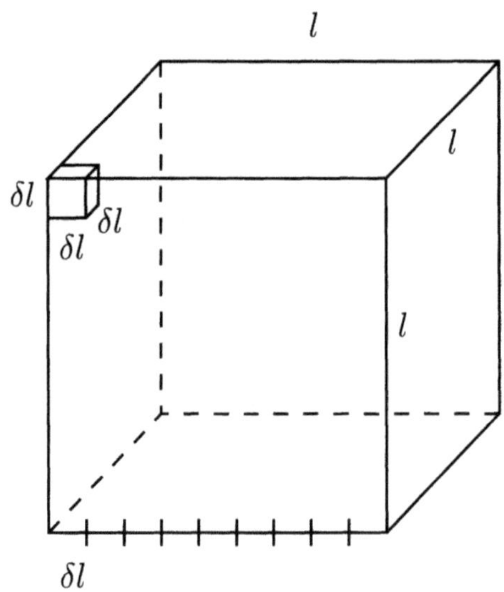

Fig. 1. Partitioning a big cube into small cubes. The big cube represents a region of space measuring l by l by l. The small cubes represent the smallest physically-allowed cubes measuring δl by δl by δl that can be lined up to measure the length of each side of the big cube. Strangely, the size of the small cubes is *not* universal, but depends on the size of the big cube. A simple argument based on this construction leads to the holographic principle

we would be able to measure that side (of length l) of the big cube to a better accuracy than δl. But, by definition, δl is the best accuracy with which we can measure l. The ensuing contradiction is evaded by the realization that each of the smallest cubes (that can be put inside the big cube) indeed measures δl by δl by δl. Thus, the number of degrees of freedom in the region (measuring l by l by l) is given by $l^3/\delta l^3$, which, according to the holographic principle, is no more than l^2/l_p^2. It follows that δl is bounded (from below) by the cube root of ll_P^2, the same result as found above in the gedanken experiment argument. Thus, to the extent that the holographic principle is correct, spacetime indeed fluctuates, forming foams of size δl on the scale of l. Actually, considering the fundamental nature of spacetime and the ubiquity of quantum fluctuations, we should reverse the argument and then we will come to the conclusion that the "strange" *holographic principle has its origin in quantum fluctuations of spacetime.*[1]

Side Remarks

[It is quite possible that the effective dimensional reduction of the the number of degrees of freedom (embodied in the holographic principle) *may* have a dramatic effect on the ultraviolet behaviour of a quantum field theory.]

2.3 Quantum Gravity Models

The consistency of the uncertainties in distance measurements with the holographic principle is reassuring. But the dependence of the fluctuations in distance on the cube root of the distance is still perplexing. To gain further insight into this strange state of affairs, let us compare this peculiar dependence on distance with the well-known one-dimensional random-walk problem. For a random walk of steps of equal size, with each step equally likely to either direction, the root-mean-square deviation from the mean is given by the size of each step multiplied by the square root of the number of steps. It is now simple to concoct a random-walk model [12, 13] for the fluctuations of distances in quantum gravity. Consider a distance l, which we partition into l/l_P units each of length l_P. In the random-walk model of quantum gravity, l_P plays the role of the size of each step and l/l_P plays the role of the number of steps. The fluctuation in distance l is given by l_P times the square root of l/l_P, which comes out to the square root of ll_P. This is much bigger than the cube root of ll_P^2, the fluctuation in distance measurements found above.

The following interpretation of the dependence of δl on the cube root of l now presents itself. As in the random-walk model, the amount of fluctuations in the distance l can be thought of as an accumulation of the l/l_P individual fluctuations each by an amount plus or minus l_P. But, for this

[1] Recently, Scardigli and Casadio [11] claim that the expected holographic scaling seems to hold only in (3+1) dimensions and only for the "generalized uncertainty principle" found above for δl.

case, the individual fluctuations cannot be completely random (as opposed to the random-walk model); actually successive fluctuations must be *entangled* and somewhat *anti-correlated* (i.e., a plus fluctuation is slightly more likely followed by a minus fluctuation and vice versa), in order that together they produce a total fluctuation less than that in the random-walk model. This small amount of anti-correlation between successive fluctuations (corresponding to what statisticians call fractional Brownian motion with self-similarity parameter $\frac{1}{3}$) must be due to quantum gravity effects. Since the cube root dependence on distance has been shown to be consistent with the holographic principle, we will, for the rest of this subsection, refer to this case that we have found (marked by an arrow in Fig. 2) as the holography model.

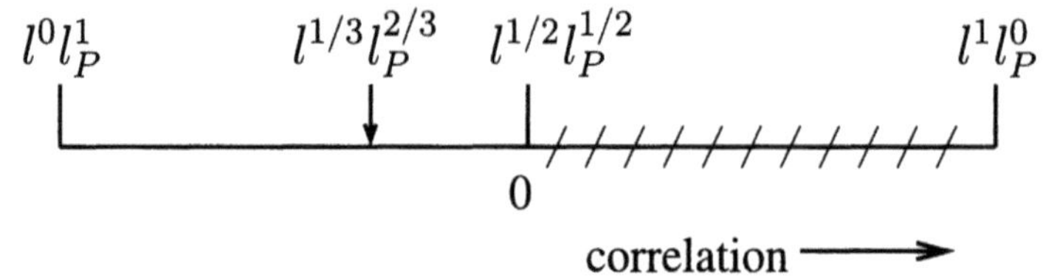

Fig. 2. Lower bounds on δl for the various quantum gravity models. The fluctuation of the distance l is given by the sum of l/l_P fluctuations each by plus or minus l_P. Spacetime foam appears to choose a small anti-correlation (i.e., negative correlation) between successive fluctuations, giving a cube root dependence in the number l/l_p of fluctuations for the total fluctuation of l (indicated by the arrow). It falls between the two extreme cases of complete randomness, i.e., zero (anti-)correlation (corresponding to $\delta l \sim l^{1/2} l_P^{1/2}$) and complete anti-correlation (corresponding to $\delta l \sim l_P$). Quantum gravity models corresponding to positive correlations between successive fluctuations (indicated by the hatched portion) are observationally ruled out. See "Further Remarks" in Subsect. 5.1

Side Remarks

[We leave it as an exercise (albeit a rather non-trivial one) to the students to seek a more microscopic understanding of the holographic principle, at the level of random walk for the random-walk model.]

On the other hand, if successive fluctuations are completely anti-correlated, i.e., a fluctuation by plus l_P is followed by a fluctuation by minus l_P which is succeeded by plus l_P etc. in the pattern $+-+-+-+-+-\ldots$, then the fluctuation of a distance l is given by the minuscule l_P, [14] independent of the size of the distance. Thus the holography model falls between the two extreme cases of complete randomness (square root of $l l_P$) and complete anti-correlation (l_P). For completeness, we mention that *a priori* there are also models with correlating successive fluctuations. But these models yield unacceptably large fluctuations in distance and time duration measurements – we will see below that these models (corresponding to the hatched line to the right of the random-walk model shown in Fig. 2) have already been observationally ruled out.

2.4 Cumulative Effects of Spacetime Fluctuations

Let us now examine the cumulative effects [15] of spacetime fluctuations over a large distance. Consider a distance l, and divide it into l/λ equal parts each of which has length λ. If we start with a fluctuation $\delta\lambda$ from each part, the question is how do the l/λ parts add up to δl for the whole distance l. In other words, we want to find the cumulative factor $\mathcal{C}$ defined by

$$\delta l = \mathcal{C}\, \delta\lambda \,, \tag{8}$$

For the holography model, since $\delta l \sim l^{1/3} l_P^{2/3} = l_P (l/l_P)^{1/3}$ and $\delta\lambda \sim \lambda^{1/3} l_P^{2/3} = l_P(\lambda/l_P)^{1/3}$, the result is

$$\mathcal{C} = \left(\frac{l}{\lambda}\right)^{1/3} . \tag{9}$$

For the random-walk model, the cumulative factor is given by $\mathcal{C} = (l/\lambda)^{1/2}$; for the model corresponding to complete anti-correlation, the cumulative factor is $\mathcal{C} = 1$, *independent* of l. Let us note that, for all quantum gravity models (except for the physically disallowed model corresponding to complete correlation between successive fluctuations), the cumulative factor is *not* linear in (l/λ), i.e., $\frac{\delta l}{\delta\lambda} \neq \frac{l}{\lambda}$. (In general, it is much smaller than l/λ). The reason for this is obvious: the $\delta\lambda$'s from the l/λ parts in l do *not* add coherently. It makes no sense, e.g., to say, for the completely anti-correlating model, that $\delta l \sim \delta\lambda \times l/\lambda \gtrsim l_P l/\lambda$ because it is inconsistent to use the completely anti-correlating model for $\delta\lambda$ while using the completely correlating model for the cumulative factor.

Note that the above discussion on cumulative effects is valid for *any* λ between l and l_P, i.e., it does not matter how one partitions the distance l. In particular, for our holography model, one can choose to partition l into units of Planck length l_P, the smallest physically meaningful length. Then (for $\lambda = l_P$) using $\delta l_P \sim l_P^{1/3} \times l_P^{2/3} = l_P$, one recovers $\delta l \sim (l/l_P)^{1/3} \times l_P = l^{1/3} l_P^{2/3}$, with the dependence on the cube root of l being due to a small amount of anti-correlation between successive fluctuations as noted above. The fact that we can choose λ as small as the Planck length in the partition indicates that, in spite of our earlier disclaimer, it may even be meaningful to consider, in the semi-classical framework we are pursuing, fluctuations of distances close to the Planck length.

Now that we know where the holography model stands among the quantum gravity models, we will restrict ourselves to discuss this model only for the rest of the lectures.

3 Clocks, Computers, and Black Holes

So far there is no experimental evidence for spacetime foam, and, as we will show shortly, no direct evidence is expected in the very near future. In view of this lack of experimental evidence, we should at least look for theoretical corroborations (aside from the "derivation" of the holographic principle discussed in Subsect. 2.2). Fortunately such corroborations do exist – in the sector of black hole physics (this should not come as a surprise to the experts). To show that, we have to make a small detour to consider clocks and computers [16, 17] first.

3.1 Clocks

Consider a clock (technically, a simple and "elementary" clock, not composed of smaller clocks that can be used to read time separately or sequentially), capable of resolving time to an accuracy of t, for a period of T (the running time or lifetime of the clock). Then bounds on the resolution time and the lifetime of the clock can be derived by following an argument very similar to that used above in the analysis of the gedanken experiment to measure distances. Actually, the two arguments are so similar that one can identify the corresponding quantities. [See Table.]

The corresponding quantities in the discussion of distance measurements (first column), time duration measurements (second column), clocks (third column), and computers (fourth column) appear in the same row in the following Table.

Distance Measurements	Time Duration Measurements	Clocks	Computers
$\delta l/c$	$\delta\tau$	t	$1/\nu$
l/c	τ	T	I/ν
$\delta l^2 \gtrsim \hbar l/mc$	$\delta\tau^2 \gtrsim \hbar\tau/mc^2$	$t^2 \gtrsim \hbar T/mc^2$	$I\nu \lesssim mc^2/\hbar$
$\delta l \gtrsim Gm/c^2$	$\delta\tau \gtrsim Gm/c^3$	$t \gtrsim Gm/c^3$	$\nu \lesssim c^3/Gm$
$l/(\delta l)^3 \lesssim l_P^{-2}$ ($\delta l \gtrsim l^{1/3} l_P^{2/3}$)	$\tau/(\delta\tau)^3 \lesssim t_P^{-2}$	$T/t^3 \lesssim t_P^{-2}$	$I\nu^2 \lesssim t_P^{-2} = c^5/\hbar G$

For the discussion of clocks, one argues that at the end of the running time T, the linear spread of the clock (of mass m) grows to $\delta l \gtrsim (\hbar T/m)^{1/2}$. But the position uncertainty due to the act of time measurement must be smaller than the minimum wavelength of the quanta used to read the clock: $\delta l \lesssim ct$, for the entire period T. It follows that [3, 16]

$$t^2 \gtrsim \frac{\hbar T}{mc^2}, \tag{10}$$

which is the analogue of (1). On the other hand, for the clock to be able to resolve time interval as small as t, the cavity of the light-clock must be small enough such that $d \lesssim ct$; but the clock must also be larger than the Schwarzschild radius $2Gm/c^2$ so that the time registered by the clock can be read off at all. These two requirements are satisfied with

$$t \gtrsim \frac{Gm}{c^3}, \tag{11}$$

the analogue of (2). One can combine the above two equations to give [16]

$$T/t^3 \lesssim t_P^{-2} = \frac{c^5}{\hbar G}, \tag{12}$$

which relates clock precision to its lifetime. Numerically, for example, for a femtosecond (10^{-15} sec) precision, the bound on the lifetime of a simple clock is 10^{34} years.

3.2 Computers

Preparatory Remarks

[*Energies Determine the Rate of Computation.* During a logical operation, the bits in a computer go from one state to another. One can use the Heisenberg uncertainty principle in the form $\Delta E \Delta t \geq \hbar$ to show that a quantum state with spread in energy ΔE takes time at least $\Delta t = \pi\hbar/2\Delta E$ to evolve to an orthogonal state. One can further show [18, 19] that it takes a system with average energy E at least the amount of time $\Delta t = \pi\hbar/2E$ to do so. Thus the speed of computation for a computer with total energy E distributed among its various logic gates (labelled by l) is bounded by

$$\sum_l \frac{1}{\Delta t_l} \leq \sum_l \frac{2E_l}{\pi\hbar} = \frac{2E}{\pi\hbar} \sim \frac{E}{\hbar}. \tag{13}$$

That is, energy limits the speed of computation. We will see that a black hole computer can saturate this bound.]

One can easily translate the relations for clocks given in the above subsection into useful relations for a simple computer (technically, it refers to a computer designed to perform highly serial computations, i.e., one that is not divided into subsystems computing in parallel). Since the resolution time t for clocks is the smallest time interval relevant in the problem, the fastest possible processing frequency is given by its reciprocal, i.e., $1/t$. Thus if ν denotes the clock rate of the computer, i.e., the number of operations per bit per unit time, then it is natural to identify ν with $1/t$. To identify the number I of bits of information in the memory space of a simple computer, we recall that the running time T is the longest time interval relevant in the problem. Thus, the maximum number of steps of information processing is given by the running time divided by the resolution time, i.e., T/t. It follows that one can identify the number I of bits of the computer with T/t.[2] In other words, the translations from the case of clocks to the case of computers consist of substituting the clock rate of computation for the reciprocal of the resolution time, and substituting the number of bits for the running time divided by the resolution time. [See Table.] The bounds on the precision and lifetime of a clock given by (10), (11) and (12) are now translated into a bound on the rate of computation and number of bits in the computer, yielding respectively

$$I\nu \lesssim \frac{mc^2}{\hbar}, \qquad \nu \lesssim \frac{c^3}{Gm}, \qquad I\nu^2 \lesssim \frac{c^5}{\hbar G} \sim 10^{86}/\mathrm{sec}^2 \,. \tag{14}$$

The first inequality shows that the speed of computation is bounded by the energy of the computer divided by Planck's constant, in agreement with the result given by (13), found by Margolus and Levitin [18], and by Lloyd [19] (for the ultimate limits to computation). The last bound is perhaps even more intriguing: it requires the product of the number of bits and the square of the computation rate for *any* simple computer to be less than the square of the reciprocal of Planck time, [16] which depends on relativistic quantum gravity (involving c, $\hbar$, and G). This relation links together our concepts of information/computation, relativity, gravity, and quantum uncertainty. The link between information and spacetime foam is perhaps not surprising because, as the above discussion of the holographic principle shows, the maximum amount of information that can be put into a region of space depends on how small the bits are, and they cannot be smaller than the foams of spacetime. So the ultimate power of computation also depends on the structure of spacetime foam. Numerically, the computation bound given by (14) is about seventy-six orders of magnitude above what is available for a current lap-top computer performing ten billion operations per second on ten billion bits, for which $I\nu^2 \sim 10^{10}/\mathrm{sec}^2$.

3.3 Black Holes

Black Hole Lifetime

Now we can apply what we have learned about clocks and computers to black holes. Reference [16, 17] let us consider using a black hole to measure time. It is reasonable to use the light travel time around the black hole's horizon as the resolution time of the clock, i.e., $t \sim \frac{Gm}{c^3} \equiv t_{BH}$, then from (10), one immediately finds that

$$T \sim \frac{G^2 m^3}{\hbar c^4} \equiv T_{BH} \,. \tag{15}$$

We have just recovered Hawking's result for black hole lifetime!

---

[2] One can think of a tape of length cT as the memory space, partitioned into bits each of length ct.

Black Hole Computers

Finally, let us consider using a black hole to do computations. This may sound like a ridiculous proposition. But if we believe that black holes evolve according to quantum mechanical laws, it is possible, at least in principle, to program black holes to perform computations [19] that can be read out of the fluctuations in the Hawking black hole radiation. How large is the memory space of a black hole computer, and how fast can it compute? Applying the results for computation derived above, we readily find the number of bits in the memory space of a black hole computer, given by the lifetime of the black hole divided by its resolution time as a clock, to be

$$I = \frac{T_{BH}}{t_{BH}} \sim \frac{m^2}{m_P^2} \sim \frac{r_S^2}{l_P^2} , \tag{16}$$

where $m_P = \hbar/(t_P c^2)$ is the Planck mass, m and r_S^2 denote the mass and event horizon area of the black hole respectively. This gives the number of bits I as the event horizon area in Planck units, in agreement with the identification of black hole entropy. (Recall that entropy S and the number of bits I are related by $S = k_B I ln 2$.)

Side Remarks

[Recall that the only property of a black hole we have used in the analysis of the gedanken experiment to measure distances (Subsect. 2.1) and in the analysis of clocks (Subsect. 3.1) is that it has a size given by the Schwarzschild radius $r_S \sim Gm/c^2$ (property 1 in first set of "Preparatory Remarks" in Subsect. 2.2). Now we have recovered the results for black hole entropy (property 3) and lifetime (property 4). Actually one can also recover the result for black hole temperature $\mathcal{T} \sim \hbar c/k_B r_S$ (property 2) by using the thermodynamic relation $\mathcal{T} = dE/dS$.]

Furthermore, the number of operations per unit time for a black hole computer is given by

$$I\nu = \frac{T_{BH}}{t_{BH}} \times \frac{1}{t_{BH}} \sim \frac{mc^2}{\hbar} , \tag{17}$$

its energy divided by Planck's constant, in agreement with the result found by Lloyd [19]. It is curious that all the bounds on computation discussed above are saturated by black hole computers. Thus one can even say that once they are programmed to do computations, black holes are the ultimate simple computers.

All these results reinforce the conceptual interconnections of the physics underlying spacetime foam, black holes, and computation. It is intersting that these three subjects share such intimate bonds and are brought together here [see Fig. 3]. The internal consistency of the physics we have uncovered also vindicates the simple (some would say overly simple) arguments we present

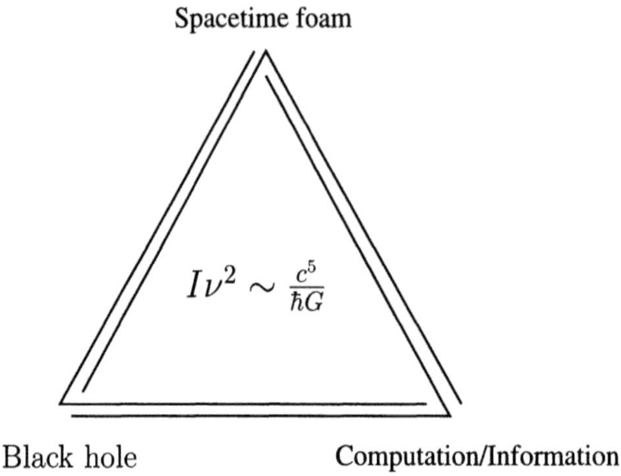

Fig. 3. The quantum foam-black hole-computation/information triangle. At the center of the triangle is the *universal* relation: $I\nu^2 \sim c^5/\hbar G$, where I is the number of bits in the memory space, and ν is the clock rate of computation of a black hole computer. This relation is a combined product of the physics behind spacetime foam, black holes, and computation/information

in Sect. 2 in the derivation of the limits to spacetime measurements. It is as if Nature approves simplicity, and tries to get away with as much simplicity as possible. Perhaps it actually follows Albert Einstein's dictum: Everything should be made as simple as possible, but not simpler.

Side Remarks

[It was John Wheeler who coined the terms "spacetime foam" and "black holes". Also famous for his phrase "its from bits", he was among the first physicists to recognize the importance of quantum information and quantum computation. To honor him for the promotion of these ideas, we should perhaps call the triangle in Fig. 3 the Wheeler Triangle.]

3.4 Results for Arbitrary Dimensions

So far we have been doing $(3 + 1)$-dimensional physics, but it is theoretically interesting to generalize the discussion to arbitrary $(n + 1)$ dimensions. In this subsection we set $c = 1$ and $\hbar = 1$ for convenience. In $(n + 1)$ dimensions, Newton's constant G has the dimension of $[length]^{n-1}$. The corresponding Schwarzschild radius is given by [20] $r_S \sim (mG)^{\frac{1}{n-2}}$. One can carry out a gedankan experiment to measure a distance as described in Subsect. 2.1. Again quantum mechanics imposes the bound $\delta l^2 \gtrsim l/m$; but gravity demands $\delta l^{n-2} \gtrsim mG$. It follows that the uncertainty in distance measurements is given by

$$\delta l \gtrsim (Gl)^{\frac{1}{n}} . \tag{18}$$

Next, following the argument given in Subsect. 2.2, one can alternatively use the holographic principle (that the number of degrees of freedom in an n-dimensional hyper-cube is bounded by l^{n-1}/G) and the fact that the number of small hyper-cubes inside the big hyper-cube is given by $(l/\delta l)^n$, to derive (18).[3]

The discussion given in Subsect. 3.3 for black holes can be duplicated for the case of arbitrary $(n+1)$ dimensions. For a black hole used as a clock, we get

$$t_{BH} \sim (mG)^{\frac{1}{n-2}} , \qquad T_{BH} \sim (m^n G^2)^{\frac{1}{n-2}} , \tag{19}$$

for its resolution time and its total running time respectively. Correspondingly, the number of bits a black hole computer can hold in its memory space and the bound on its rate of computation are respectively given by

$$I_{BH} \sim \frac{r_S^{n-1}}{G} ; \qquad I\nu \sim m . \tag{20}$$

For the rest of the lectures, we go back to $3+1$ dimensions.

4 Energy-Momentum Uncertainties

Just as there are uncertainties in spacetime measurements, there are also uncertainties in energy-momentum measurements due to spacetime foam effects. Thus there is a limit to how accurately we can measure and know the energy and momentum of a system. Reference [4] imagine sending a particle of momentum p to probe a certain structure of spatial extent l so that $p \sim \hbar/l$. It follows that $\delta p \sim (\hbar/l^2)\delta l$. Spacetime fluctuations $\delta l \gtrsim l(l_P/l)^{2/3}$ can now be used to give

$$\delta p - \beta p \left(\frac{p}{m_P c}\right)^{2/3} , \tag{21}$$

where *a priori* $\beta \sim 1$. The corresponding statement for energy uncertainties is

$$\delta E = \gamma E \left(\frac{E}{E_P}\right)^{2/3} , \tag{22}$$

where $E_P = m_P c^2 \sim 10^{19}$ GeV is the Planck energy and a priori $\gamma \sim 1$. We emphasize that all the uncertainties take on $\pm$ sign with equal probability (most likely, a Gaussian distribution about zero). Thus at energy-momentum far below the Planck scale, the energy-momentum uncertainties

3 Note the somewhat different conclusions (for $n \neq 3$) reached in [11] where a different bound from gravity (by following an argument similar to that given in "Side Remarks" after (2)) is used.

are very small, suppressed by a fractional (two-thirds) power of the Planck energy-momentum. (For example, the uncertainty in the energy of a particle of ten trillion electron-volts is about a thousand electron-volts.) A word of caution is in order: while the result for distance and time-interval fluctuations has indirect support from black hole physics (and is thus reasonably trustworthy), the result for energy-momentum fluctuations enjoys no such support and is probably not as reliable.

Side Remarks

[Equations (21) and (22) can also be derived by considering the coupling of the metric to the energy-momentum tensor of a particle: $(g_{\mu\nu} + \delta g_{\mu\nu})t^{\mu\nu} = g_{\mu\nu}(t^{\mu\nu} + \delta t^{\mu\nu})$, where we have noted that the uncertainty in $g_{\mu\nu}$ can be translated into an uncertainty in $t_{\mu\nu}$. Equation (5) and $p \sim \hbar/l$ can then be used to yield (21).

Alternatively, we can consider how δp, the uncertainty of the momentum operator $p = (\hbar/i)(\partial/\partial x)$, is associated with $\delta x = (x l_P^2)^{1/3}$. For any function $f(x)$, $(\delta p)f$ is given by

$$(\delta p)f = \left(\frac{\hbar}{i}\right)\left[\delta x\left(\frac{\partial^2 f}{\partial x^2}\right) + \left(\frac{\partial f}{\partial x}\right)\left(\frac{\partial \delta x}{\partial x}\right)\right] . \tag{23}$$

Taking the function $f(x)$ to be the linear momentum eigenstate $f = \exp(ipx/\hbar)$, we find that the minimum value of $|\ \delta p\ |$ is attained at $x \sim \hbar/p$, yielding (21).

Similar arguments yield (22) for the fluctuations of energy.]

Modified Dispersion Relations

Energy-momentum uncertainties affect both the energy-momentum conservation laws and dispersion relations. Energy-momentum is conserved up to energy-momentum uncertainties due to quantum foam effects, i.e., $\Sigma(p_i^\mu + \delta p_i^\mu)$ is conserved, with p_i^μ being the average values of the various energy-momenta. On the other hand the dispersion relation is now generalized to read

$$E^2 - p^2c^2 - \epsilon p^2 c^2 \left(\frac{pc}{E_P}\right)^{2/3} = m^2c^4 , \tag{24}$$

for high energies with $E \gg mc^2$. A priori we expect $\epsilon \sim 1$ and is independent of β and γ. But due to our present ignorance of quantum gravity, we are not in a position to make any definite statements. In fact, it is possible that $\epsilon = 2(\beta - \gamma)$, which would be the case if the modified dispersion relation is given by $(E + \delta E)^2 - (p + \delta p)^2c^2 = m^2c^4$, with δp and δE given by (21) and (22) respectively.

A Fluctuating Speed of Light

The modified dispersion relation discussed above has an interesting consequence for the speed of light. Reference [21, 22] applying (24) to the massless photon yields

$$E^2 \simeq c^2 p^2 + \epsilon E^2 \left(\frac{E}{E_P} \right)^{2/3} . \tag{25}$$

The speed of (massless) photon

$$v = \frac{\partial E}{\partial p} \simeq c \left(1 + \frac{5}{6} \epsilon \frac{E^{2/3}}{E_P^{2/3}} \right) , \tag{26}$$

becomes energy-dependent and fluctuates around c. For example, a photon of ten trillion electron-volt energy has a speed fluctuating about c by one centimeter per second.

5 Spacetime Foam Phenomenology

Because the Planck length $l_P \sim 10^{-33}$ cm is so minuscule, the Planck time $t_P \sim 10^{-44}$ sec so short, and the Planck energy $E_P \sim 10^{28}$ eV so high, spacetime foam effects, suppressed by Planck scales, are exceedingly small. Accordingly, they are very hard to detect. The trick will be to find ways to amplify the small effects [1].

5.1 Phase Incoherence of Light from Extra-galactic Sources

One way to amplify the minute effects is to add up many such effects, like collecting many small raindrops to fill a reservoir. Consider light coming to us from extragalactic sources. Over one wavelength, the phase of the light-waves advances by 2π; but due to spacetime foam effects, this phase fluctuates by a small amount. The idea is that the fluctuation of the phase over one wavelength is extremely small, but light from distant galaxies has to travel a distance of many wavelengths. It is possible that over so many wavelengths, the fluctuations can cumulatively add up to a detectable level at which point the phase coherence for the light-waves is lost. Loss of phase coherence would mean the loss of interference patterns. Thus the strategy is to look for the blurring of images of distant galaxies in powerful telescopes like the Hubble Space Telescope. This technique to detect spacetime foam was proposed by Lieu and Hillman [23], and elaborated by Ragazzoni and his collaborators [24].

The proposal deals with the phase behavior of radiation with wavelength λ received from a celestial source located at a distance l away. Fundamentally, the wavelength defines the minimum length scale over which physical quantities such as phase and group velocities (and hence dispersion relations) can be defined. Thus, the uncertainty in λ introduced by spacetime foam is the starting point for this analysis. A wave will travel a distance equal to its own wavelength λ in a time $t = \lambda/v_g$ where v_g is the group velocity of propagation, and the phase of the wave consequently changes by an amount

$$\phi = 2\pi \frac{v_p t}{\lambda} = 2\pi \frac{v_p}{v_g} , \tag{27}$$

(i.e., if $v_p = v_g, \phi = 2\pi$) where v_p is the phase velocity of the light wave. Quantum gravity fluctuations, however, introduce random uncertainties into this phase which is simply

$$\delta\phi = 2\pi\, \delta\left(\frac{v_p}{v_g}\right) . \tag{28}$$

Due to quantum fluctuations of energy-momentum [4] and the modified dispersion relations, we obtain

$$\delta\left(\frac{v_p}{v_g}\right) \sim \pm \left(\frac{E}{E_P}\right)^{2/3} = \pm \left(\frac{l_P}{\lambda}\right)^{2/3} , \tag{29}$$

where we have used $v_p = E/p$ and $v_g = dE/dp$, and $E/E_P = l_P/\lambda$. We emphasize that this may be either an incremental advance or a retardation in the phase.

In travelling over the macroscopically large distance, l, from source to observer an electromagnetic wave is continually subjected to random, incoherent spacetime fluctuations. Therefore, by our previous argument given in Subsect. 2.4, the cumulative statistical phase dispersion is $\Delta\phi = \mathcal{C}\delta\phi$ with the cumulative factor $\mathcal{C} = (l/\lambda)^{1/3}$, that is

$$\Delta\phi = 2\pi a \left(\frac{l_P}{\lambda}\right)^{2/3} \left(\frac{l}{\lambda}\right)^{1/3} = 2\pi a \frac{l_P^{2/3} l^{1/3}}{\lambda} , \tag{30}$$

where $a \sim 1$. (This is our fundamental disagreement [15] with Lieu and Hillman who assume that the microscale fluctuations induced by quantum foam into the phase of electromagnetic waves are coherently magnified by the factor l/λ rather than $(l/\lambda)^{1/3}$). Thus even the active galaxy PKS1413 + 135, an example used by Lieu and Hillman, for which $\lambda \simeq 1.6\,\mu\text{m}$ and $l \simeq 1.216$ Gpc, is not far enough to make the light wave front noticeably distorted. A simple calculation [15] shows that, over four billion light years, the phase of the light waves fluctuates only by $\Delta\phi \sim 10^{-9} \times 2\pi$, i.e., only by one billionth of what is required to lose the sharp ring-like interference pattern around the galaxy which, not surprisingly, is observed [25] by the Hubble Telescope. This example illustrates the degree of difficulty which one has to overcome to detect spacetime foam. The origin of the difficulty can be traced to the incoherent nature of the spacetime fluctuations (i.e., the anticorrelations between successive fluctuations).

Further Remarks

[*Ruling Out the Random-Walk Model of Quantum Gravity.* But not all is lost with Lieu and Hillman's proposal. One can check that the proposal can be used to rule out [15], if only marginally, the random-walk model of quantum gravity, which would (incorrectly) predict $\Delta\phi \sim 2\pi(l_P/\lambda)^{1/2}(l/\lambda)^{1/2} = 2\pi(l_P l)^{1/2}/\lambda \sim 10 \times 2\pi$, a large enough phase fluctuation for light from PKS1413 + 135 to lose phase coherence, contradicting evidence of diffraction patterns from the Hubble Telescope observation. It follows that models corresponding to correlating successive fluctuations are also ruled out.]

5.2 High Energy γ Rays from Distant GRB

For another idea to detect spacetime foam, let us recall (26) that, due to quantum fluctuations of spacetime, the speed of light fluctuates around c and the fluctuations increase with energy. Thus for photons (quanta of light) emitted simultaneously from a distant source coming towards our detector, we expect an energy-dependent spread in their arrival times. To maximize the spread in arrival times, we should look for energetic photons from distant sources. High energy gamma rays from distant gamma ray bursts [21] fit the bill. So the idea is to look for a noticeable spread in arrival times for such high energy gamma rays from distant gamma ray bursts. This proposal was first made by G. Amelino-Camelia et al. [21] in another context.

To underscore the importance of using the correct cumulative factor to estimate the spacetime foam effect, let us first proceed in a naive manner. At first sight, the fluctuating speed of light $\delta v \sim c(E/E_P)^{2/3}$ (see (26)) would *seem* to yield [22] an energy-dependent spread in the arrival times of photons of the *same* energy E given by $\delta t \sim \delta v(l/c^2) \sim t(E/E_P)^{2/3}$, where $t = l/c$ is the average overall time of travel from the photon source (distance l away). Furthermore, the modified energy-momentum dispersion relation would seem to predict time-of-flight differences between simultaneously-emitted photons of different energies, E_1 and E_2, given by $\delta t \simeq t(E_1^{2/3} - E_2^{2/3})/E_P^{2/3}$. But these results for the spread of arrival times of photons are *not* correct, because we have inadvertently used $l/\lambda \sim Et/\hbar$ as the cumulative factor instead of the correct factor $(l/\lambda)^{1/3} \sim (Et/\hbar)^{1/3}$. Using the correct cumulative factor, we get a much smaller $\delta t \sim t^{1/3}t_P^{2/3}$ for the spread in arrival time of the photons of the same energy. Thus the result is that the time-of-flight differences increase only with the cube root of the average overall time of travel from the gamma ray bursts to our detector, leading to a time spread too small to be detectable [1].

5.3 Interferometry Techniques

Suppressed by the extraordinarily short Planck length, fluctuations in distances, even large distances, are very small. So, to measure such fluctuations, what one needs is an instrument capable of accurately measuring fluctuations in length over long distances. Modern gravitational-wave interferometers, having attained extraordinary sensitivity, come to mind. The idea of using gravitational-wave interferometers to measure the foaminess of spacetime was proposed by Amelino-Camelia [12] and elaborated by the author and van Dam [8]. Modern gravitational-wave interferometers are sensitive to changes in distances to an accuracy better than 10^{-18} meter. To attain such sensitivity, interferometer researchers have to contend with many different noises, the enemies of gravitational-wave research, such as thermal noise, seismic noise, and photon shot noise. To this list of noises that infest an interferometer, we now have to add the faint yet ubiquitous noise from spacetime foam. In other words, even after one has subtracted all the well-known noises, there is still the noise from spacetime fluctuations left in the read-out of the interferometer.

The secret of this proposal to detect spacetime foam lies in the existence of another length scale [12] available in this particular technique, in addition to the minuscule Planck length. It is the scale provided by the frequency f of the interferometer bandwidth. What is important is whether the length scale $l_P^{2/3}(c/f)^{1/3}$, characteristic of the noise from spacetime foam at that frequency, is comparable to the sensitivity level of the interferometer. The hope is that, within a certain range of frequencies, the experimental limits will soon be comparable to the theoretical predictions for the noise from quantum foam.

The detection of spacetime foam with interferometry techniques is also helped by the fact that the correlation length of the noise from spacetime fluctuations is extremely short, as the characteristic scale is the Planck length. Thus, this faint noise can be easily distinguished from the other sources of noise because of this lack of correlation. In this regard, it will be very useful for the detection of spacetime foam to have two nearby interferometers.

To proceed with the analysis, we recall that the displacement noise due to spacetime foam that involves a time interval t is given by $\delta l(t) \sim l_P^{2/3}(ct)^{1/3}$. Next we decompose the displacement noise in terms of the associated displacement amplitude spectral density [26] $S(f)$ of frequency f. For a frequency-band limited from below by the time of observation t, $\delta l(t)$ and $S(f)$ are related by

$$(\delta l(t))^2 = \int_{1/t} [S(f)]^2 df \, . \tag{31}$$

For the displacement noise due to quantum foam, one can easily check that the amplitude spectral density is given by $S(f) \sim c^{1/3} l_P^{2/3} f^{-5/6}$, inversely proportional to (the 5/6th power of) frequency. So one can optimize the performance of an interferometer at low frequencies. As lower frequency detection is possible only in space, interferometers like the proposed Laser Interferometer Space Antenna [27] may enjoy a certain advantage.

To be specific, let us now compare the predicted spectral density from quantum foam noise with the noise level projected for the Laser Interferometer Gravitational-Wave Observatory. The "advanced phase" of LIGO [28] is expected to achieve a displacement noise level of less than $10^{-20}\,\mathrm{mHz}^{-1/2}$ near 100 Hz; one can show that this would translate into a probe of l_P down to 10^{-31} cm, a mere hundred times the physical Planck length. But can we then conclude that LIGO will be within striking distance of detecting quantum foam? Alas, the above optimistic estimate is based on the assumption that spacetime foam affects the paths of all the photons in the laser beam coherently. But, in reality, this can hardly be the case. Since the total effect on the interferometer is based on averaging over all photons in the wave front, the incoherent contributions from the different photons are expected to cut down the sensitivity of the interferometer by some fractional power of the number of photons in the beam – and there are many photons in the beams used by LIGO. Thus, even with the incredible sensitivity of modern gravitational-wave interferometers like LIGO, the fluctuations of spacetime are too small to be detected – unless one knows how to build a small beam interferometer of slightly improved power and phase sensitivity than what is projected for the advanced phase of LIGO![4]

For completeness, we should mention that the use of atom interferometers [9, 29] and optical interferometers [30] to look for effects of spacetime fluctuations has also been suggested. A recent proposal to build a matter-wave interferometric gravitational-wave observatory [31], using atomic beams emanating from supersonic atomic sources, sounds promising, not only for detecting gravitational radiation, but perhaps also for detecting spacetime foam.

Further Remarks

[*A Suggestion to Use Atom Interferometry Techniques.* Here we propose [9] to use laser-based atom interferometry experiments [29] in the not-too-distant future to detect spacetime fluctuations on the scales of quantum gravity at the level given by (3) and (4). In a laser-based atom interferometer, an atomic beam is split by laser beams into two coherent wave packets which are kept apart before being recombined by laser beams. The phase change of each wave packet is proportional to the proper time along its path, and so the resulting interference pattern depends on the time difference between the two paths. In the absence of spacetime fluctuations, the phase change η over a time interval τ is given by $\eta(\tau) = \Omega\tau$, where $\Omega \equiv mc^2/\hbar$ is the quantum angular frequency associated with the mass m of the atom. Due to spacetime fluctuations (4), there is an additonal fluctuating phase $\delta\eta$ given by

$$\delta\eta \sim \frac{(\tau t_P^2)^{1/3}}{\tau}\eta = (\tau t_P^2)^{1/3}\Omega\,. \tag{32}$$

For example, in 1992, Chu and Kasevich at Stanford University built an atom interferometer which used sodium atoms ($m \sim 4.5 \times 10^{-26}$ kg), and the two

[4] This conclusion is based on the author's discussion with G. Amelino-Camelia and R. Weiss.

wave packets were kept apart for 0.2 sec. Reference [32] for that experiment, one finds that $\eta(\tau) \sim 7 \times 10^{24}$ radians and $\delta\eta \sim 3 \times 10^{-4}$ radians. Thus one needs a precision of about 1 part in 10^{29} to look for spacetime foam (through suppression of the interference pattern), compared with the precision of 1 part in 10^{26} that was then achieved. In other words, it appears that one needs a (mere) thousandfold improvement in noise sensitivity to detect spacetime fluctuations. Though the above argument, a variant of the one given by Percival [29], is necessarily short and perhaps too simplistic and overtly optimistic, hopefully the conclusion is not too far off the mark.]

5.4 Ultra-High Energy Cosmic Ray Events

The universe appears to be more transparent to the ultra-high energy cosmic rays (UHECRs) [33] than expected.[5] Theoretically one expects the UHECRs to interact with the Cosmic Microwave Background Radiation and produce pions. These interactions above the threshold energy should make observations of UHECRs with $E > 5.10^{19}$ eV (the GZK limit) [34] unlikely. Still UHECRs above the GZK limit have been observed. In this subsection, we attempt to explain the UHECR paradox by arguing [22] that energy-momentum uncertainties due to quantum gravity (significant only for high energy particles like the UHECRs), too small to be detected in low-energy regime, can affect particle kinematics so as to raise or even eliminate the energy thresholds, thereby explaining the threshold anomaly.[6] (For similar or related approaches, see [35].)

Relevant to the discussion of the UHECR events is the scattering process in which an energetic particle of energy E_1 and momentum $\mathbf{p}_1$ collides head-on with a soft photon of energy ω in the production of two energetic particles with energy E_2, E_3 and momentum $\mathbf{p}_2$, $\mathbf{p}_3$. After taking into account energy-momentum uncertainties, energy-momentum conservation demands

$$E_1 + \delta E_1 + \omega = E_2 + \delta E_2 + E_3 + \delta E_3 \ , \tag{33}$$

and

$$p_1 + \delta p_1 - \omega = p_2 + \delta p_2 + p_3 + \delta p_3 \ , \tag{34}$$

where δE_i and δp_i $(i = 1, 2, 3)$ are given by (22) and (21),

$$\delta E_i = \gamma_i E_i \left(\frac{E_i}{E_P} \right)^{2/3} , \quad \delta p_i = \beta_i p_i \left(\frac{p_i}{m_P c} \right)^{2/3} , \tag{35}$$

---

[5] For the case of (the not-so-well-established) TeV-γ events, see [1] and references therein.

[6] Unfortunately, we have nothing useful to say about the origins of these energetic particles per se.

and we have omitted $\delta\omega$, the contribution from the uncertainty of ω, because ω is small.[7]

Combining (35) with the modified dispersion relations[8] (24) for the incoming energetic particle ($i = 1$) and the two outgoing particles ($i = 2, 3$), and putting $c = 1$,

$$E_i^2 - p_i^2 - \epsilon_i p_i^2 \left(\frac{p_i}{E_P}\right)^{2/3} = m_i^2 , \tag{36}$$

we obtain the threshold energy equation

$$E_{th} = p_0 + \tilde{\eta} \frac{1}{4\omega} \frac{E_{th}^{8/3}}{E_P^{2/3}} , \tag{37}$$

where

$$p_0 \equiv \frac{(m_2 + m_3)^2 - m_1^2}{4\omega} \tag{38}$$

is the (ordinary) threshold energy if there were no energy-momentum uncertainties, and

$$\tilde{\eta} \equiv \eta_1 - \frac{\eta_2 m_2^{5/3} + \eta_3 m_3^{5/3}}{(m_2 + m_3)^{5/3}} , \tag{39}$$

with

$$\eta_i \equiv 2\beta_i - 2\gamma_i - \epsilon_i . \tag{40}$$

Note that, in (37), the quantum gravity correction term is enhanced by the fact that ω is so small [37] (compared to p_0).

Given that all the β_i's, the γ_i's and the ϵ_i's are of order 1 and can be $\pm$, $\tilde{\eta}$ can be $\pm$ (taking on some unknown Gaussian distribution about zero), but it cannot be much bigger than 1 in magnitude. For positive $\tilde{\eta}$, E_{th} is greater than p_0. The threshold energy increases with $\tilde{\eta}$ to $\frac{3}{2}p_0$ at $\tilde{\eta} = \tilde{\eta}_{max}$, beyond which there is no (real) physical solution to (37) (i.e., E_{th} becomes complex) and we interpret this as *evading* the threshold cut. Reference [22] the cutoff $\tilde{\eta}_{max}$ is actually very small: $\tilde{\eta}_{max} \sim 10^{-17}$. Thus, energy-momentum uncertainties due to quantum gravity, too small to be detected in low-energy regime, can (in principle) affect particle kinematics so as to raise or even eliminate energy thresholds. Can this be the solution to the UHECR threshold anomaly puzzle? On the other hand, for negative $\tilde{\eta}$, the threshold energy is less than p_0, i.e., a negative $\tilde{\eta}$ *lowers* the threshold energy. Reference [2, 38, 39] for example, $\tilde{\eta} \sim -1$ gives $E_{th} \sim 10^{15}$eV. Can this be the explanation of the opening up of the "precocious" threshold in the "knee" region? See Fig. 4. Curiously, the interpolation between the "knee" region and the GKZ limit may even explain the "ankle" region [1].

[7] We should mention that we have not found the proper transformations of the energy-momentum uncertainties between different reference frames. Therefore we apply the results only in the frame in which we do the observations.

[8] The suggestion that the dispersion relation may be modified by quantum gravity first appeared in [36].

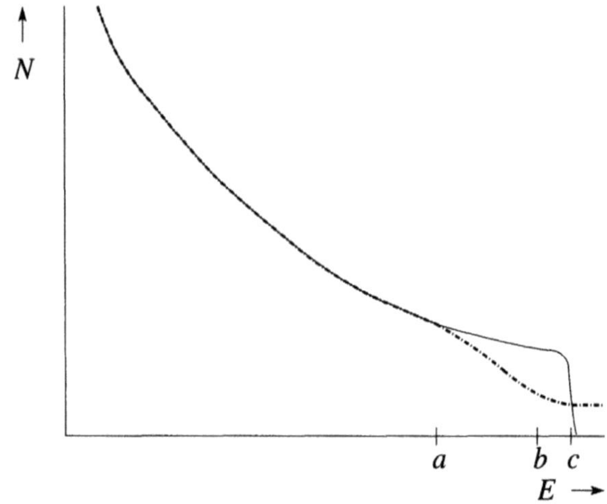

Fig. 4. Schematic plot of the number N of UHECRs versus energy E. Reference [39] the *solid curve* refers to the case of ordinary threshold energy $E_{th} = p_0$. The *dashed-dotted* curve refers to the case of the threshold energy given by (37). The "knee" region is indicated by "a", the "ankle" region by "b", and the GZK limit by "c"

It is far too early to call this a success. In fact there are some problems confronting this particular proposal to solve the astrophysical puzzle. The most serious problem[9] is the question of matter (in) stability [40] because quantum fluctuations in dispersion relations (36) can *lower* as well as raise the reaction thresholds. This problem may force us to entertain one or a combination of the following possibilities: (i) The fluctuations of the energy-momentum of a particle are not completely uncorrelated (e.g, the fluctuating coefficients β, γ, and ϵ in (21), (22), and (24) may be related such that $\eta_i \approx 0$ in (40)); (ii) The time scale at which quantum fluctuations of energy-momentum occur is relatively short[10] (compared to the relevant interaction or decay times); (iii) Both "systematic" and "non-systematic" effects of quantum gravity are present, [2] but the "systematic" effects are large enough to overwhelm the "non-systematic" effects.

On the other hand, *if* (and that is a big "if") the problems, such as the matter instability problem discussed above, can somehow be solved, the proposal suggested by energy-momentum uncertainties will become a rather attractive and simple explanation of the "knee" and "ankle" regions and the threshold anomaly (if indeed there is one) found in the UHECR events.

[9] But this is by no means the only problem; see [2].

[10] Unfortunately, these two scenarios also preclude the possibility that energy-momentum uncertainties are the origin of the threshold anomaly discussed above. On the plus side, the threshold anomaly suggested by the present AGASA data may turn out to be false. Data from the Auger Project are expected to settle the issue.

6 Summary and Conclusions

We summarize by collecting some of the salient points:

- On large scales spacetime appears smooth, but on a sufficiently small scale it is bubbly and foamy (just as the ocean appears smooth at high altitudes but shows its roughness at close distances from its surface).
- Spacetime is foamy because it undergoes quantum fluctuations which give rise to uncertainties in spacetime measurements; spacetime fluctuations scale as the cube root of distances or time durations.
- Quantum foam physics is closely related to black hole physics and computation. The "strange" holographic principle, which limits how densely information can be packed in space, is a manifestation of quantum foam.
- Because the Planck length/time is so small, the uncertainties in spacetime measurements, though much greater than the Planck scale, are still very small.
- It may be difficult to detect the tiny effects of quantum foam, but it is by no means impossible.

Recall that, by analyzing a simple gedanken experiment for spacetime measurements, we arrive at the conclusion that spacetime fluctuations scale as the cube root of distances or time durations. This cube root dependence is strange, but has been shown to be consistent with the holographic principle and with semi-classical black hole physics in general. We think this result for spacetime fluctuations is as beautiful as it is strange. Hopefully it is also true!

But what is really needed is direct detection of quantum foam. Its detection will give us a glimpse of the fabric of spacetime and will help guide physicists to the correct theory of quantum gravity. The importance of direct experimental evidence cannot be over-emphasized.

We hope that the arguments given in these lectures are sufficiently compelling to encourage a determined experimental quest to detect spacetime foam, the ultimate structure of spacetime, for, as Michael Faraday, the discoverer of electromagnetic induction, once observed:

> *Nothing is too wonderful to be true, if it be consistent with the laws of nature, and in such things as these, experiment is the best test of such consistency.*

Acknowledgments

I wish to thank G. Amelino-Camelia and J. Kowalski-Glikman for inviting me to participate in this very stimulating and enjoyable Winter School. This work was supported in part by the US Department of Energy and the Bahnson Fund of the University of North Carolina. Help from A. Giugni, L. L. Ng, and T. Takahashi in the preparation of this manuscript is gratefully acknowledged. I also thank R. Weiss and especially my collaborators H. van Dam and G. Amelino-Camelia, for useful discussions. Thanks are due to B. Schumacher and M. Taqqu for conversations which led to the inclusion of the subsection on the various quantum gravity models.

References

1. Y.J. Ng, gr-qc/0305019, *Mod. Phys. Lett.* **A18**,1073 (2003).
2. G. Amelino-Camelia, Y.J. Ng, H. van Dam, gr-qc/0204077, *Astropart. Phys.* **19**, 729 (2003).
3. E.P. Wigner, *Rev. Mod. Phys.* **29**, 255 (1957); H. Salecker and E.P. Wigner, *Phys. Rev.* **109**, 571 (1958).
4. Y.J. Ng and H. van Dam, *Mod. Phys. Lett.* **A9**, 335 (1994); **A10**, 2801 (1995); in hep-th/9406110, *Proc. of Fundamental Problems in Quantum Theory*, eds. D.M. Greenberger and A. Zeilinger, Ann. New York Acad. Sci. **755**, 579 (1995). Also see F. Karolyhazy, *Nuovo Cimento* **A42**, 390 (1966); T. Padmanabhan, *Class. Quan. Grav.* **4**, L107 (1987); D.V. Ahluwalia, *Phys. Lett.* **B339**, 301 (1994); L.J. Garay, *Int. J. Mod. Phys.* **A10**, 145 (1995); and N. Sasakura, *Prog. Theor. Phys.* **102**, 169 (1999).
5. L.H. Ford, *Phys. Rev.* **D51**, 1692 (1995).
6. L. Diosi and B. Lukas, *Europhys. Lett.* **34**, 479 (1996).
7. Y.J. Ng and H. van Dam, *Europhys. Lett.* **38**, 401 (1997); gr-qc/0209021, *Class. Quant. Grav.* **20**, 393 (2003). See also the second reference in [8].
8. Y.J. Ng and H. van Dam, gr-qc/9906003, *Found. Phys.* **30**, 795 (2000); gr-qc/9911054, *Phys. Lett.* **B477**, 429 (2000).
9. Y.J. Ng, gr-qc/0201022, *Int. J. Mod. Phys.* **D11**, 1585 (2002).
10. G. 't Hooft, in *Salamfestschrift*, edited by A. Ali et al. (World Scientific, Singapore, 1993), p. 284; L. Susskind, *J. Math. Phys.* (N.Y.) **36**, 6377 (1995). Also see J.A. Wheeler, *Int. J. Theor. Phys.* **21**, 557 (1982); J.D. Bekenstein, *Phys. Rev.* **D7**, 2333 (1973); S. Hawking, *Comm. Math. Phys.* **43**, 199 (1975).
11. F. Scardigli and R. Casadio, hep-th/0307174, *Class. Quant. Grav.* **20**, 3915 (2003).
12. G. Amelino-Camelia, *Nature* **398**, 216 (1999).
13. L. Diosi and B. Lukacs, *Phys. Lett.* **A142**, 331 (1989).
14. C.W. Misner, K. S. Thorne, and J. A. Wheeler, *Gravitation* (Freeman, San Francisco, 1973), 1190.
15. Y.J. Ng, W. Christiansen, and H. van Dam, astro-ph/0302372, *Astrophys. J.* **591**, L87 (2003).
16. Y.J. Ng, gr-qc/0006105, *Phys. Rev. Lett.* **86**, 2946 (2001), and (erratum) *ibid* **88**, 139902-1 (2002); Y. J. Ng in hep-th/0010234, *Proc. of OCPA 2000*, eds. N.P. Chang et al. (World Scientific, Singapore, 2002), p. 235.
17. J.D. Barrow, *Phys. Rev.* **D54**, 6563 (1996).
18. N. Margolus and L.B. Levitin, *Physica* **D120**, 188 (1998).
19. S. Lloyd, *Nature* (London) **406**, 1047 (2000).
20. R.C. Myers and M.J. Perry, *Ann. Phys.* **172**, 304 (1986).

21. G. Amelino-Camelia, J. Ellis, N.E. Mavromatos, D.V. Nanopoulos, and S. Sarkar, *Nature* **393**, 763 (1998); B.E. Schaefer, *Phys. Rev. Lett.* **82**, 4964 (1999); S.D. Biller *et al., ibid*, **83**, 2108 (1999).
22. Y.J. Ng, D.S. Lee, M.C. Oh, and H. van Dam, *Phys. Lett.* **B507**, 236 (2001); hep-ph/0010152 and references therein. The arXiv preprint is a more informative version of the paper in Phys. Lett. B.
23. R. Lieu and L.W. Hillman, astro-ph/0301184, *Astrophys. J.* **585**, L77 (2003).
24. R. Ragazzoni, M. Turatto, and W. Gaessler, astro-ph/0303043, *Astrophys. J.* **587**, L1 (2003).
25. E.S. Perlman, et al., 2002, *Astro. J.* **124**, 2401 (2002).
26. V. Radeka, *Ann. Rev. Nucl. Part. Sci.* **38**, 217 (1988).
27. K. Danzmann, *Class. Quant. Grav.* **13**, A247 (1996).
28. A. Abramovici, et. al., Science **256**, 325 (1992).
29. I. Percival, *Phys. World*, March 1997, p.43; F. Benatti and R. Floreanini, quant-ph/0208164.
30. F. Benatti and R. Floreanini, quant-ph/0204094; L.B. Crowell, *Found. Phys.* **16**, 281 (2003).
31. R.Y. Chiao and A.D. Speliotopoulos, arXiv:gr-qc/0312096.
32. M. Kasevich and S. Chu, *Appl. Phys.* **B54**, 321 (1992).
33. M.A. Lawrence et al., *J. Phys.* **G17**, 733 (1991); N.N. Efimov et al., in *22nd Intl. Cosmic Ray Conf.* (Dublin, 1991); D.J. Bird et al., *Astrophys. J.* **441**, 144 (1995); M. Takeda et al., *Phys. Rev. Lett.* **81**, 1163 (1998); A. Watson, in *Proc. Snowmass Workshop*, 126 (1996).
34. K. Greisen, *Phys. Rev. Lett.* **16**, 748 (1966); G.T. Zatsepin and V.A. Kuz'min, *JETP Lett.* **41**, 78 (1966).
35. C.J. Cesarsky, *Nucl. Phys.* (Proc. Suppl.) **B28**, 51 (1992); L. Gonzalez-Mestres, physics/9704017; R. Aloisio, P. Blasi, P.L. Ghia, and A.F. Grillo, astro-ph/0001258; O. Bertolami and C.S. Carvalho, *Phys. Rev.* **D61**, 103002 (2000); H. Sato, astro-ph/0005218; T. Kifune, *Astrophys. J. Lett.* **518**, L21 (1999); W. Kluzniak, astro-ph/9905308; S. Coleman and S.L. Glashow, *Phys. Rev.* **D59**, 116008 (1999); D. Colladay and A. Kostelecky, *Phys. Rev.* **D55**, 6760 (1997); R. Lieu, *ApJ* **568**, L67 (2002); F.W. Stecker, astro-ph/0304527; M. Jankiewicz, R.V. Buniy, T.A. Kephart, and T.J. Weiler, hep-ph/0312221.
36. G. Amelino-Camelia, J. Ellis, N.E. Mavromatos, and D.V. Nanopoulos, *Int. J. Mod. Phys.* **A12**, 607 (1997).
37. G. Amelino-Camelia and T. Piran, hep-th/0006210; astro-ph/0008107.
38. R. Aloisio, P. Blasi, A. Galante, P.L. Ghia, and A.F. Grillo, *Astropart. Phys.* **19**, 127 (2003).
39. Y.J. Ng, talk given in the Huntsville Workshop 2002 (unpublished).
40. R. Aloisio, P. Blasi, A. Galante, and A.F. Grillo, astro-ph/0304050; R. Le Gallou, astro-ph/0304560.

Gamma-Ray Bursts as Probes for Quantum Gravity

T. Piran

Racah Institute for Physics, The Hebrew University, Jerusalem, Israel
tsvi@phys.huji.ac.il

1 Introduction

Gamma ray bursts (GRBs) are short and intense pulses of γ-rays arriving from random directions in the sky. Several years ago Amelino-Camelia et al. [1] (see also [2]) pointed out that a comparison of time of arrival of photons at different energies from a GRB could be used to measure (or obtain a limit on) possible deviations from a constant speed of light at high photons energies. I review here our current understanding of GRBs and reconsider the possibility of performing these observations (see also Norris, Bonnell, Marani, & Scargle [3] for a review of the same topic). I begin (in Sect. 2) with a brief discussion of the motivation to consider an energy dependent variable speed of light. I turn (in Sect. 3) to a general discussion of the detectability of deviations from a constant speed of light via time-lag measurments. I derive constraints on the Energy range, the distance to the sources and the needed temporal resolution of the sources and the detectors. I then turn (in Sect. 4) to a short description of our current understanding of GRBs. This section is included as a background material as for the rest of the discussion GRBs are just cosmological sources of high energy photons and we don't really care how are these photons they produced. In Sect. 5 I return to the subject of the talk and I describe the temporal structure and spectral properties of GRBs. These are the key issues that are relevant for the observations of a variable speed of light. I conclude (in Sect. 6) by confronting the observations needed for determination of (or obtaining a limit on) a variable speed of light with the properties of GRBs. I discuss some recent attempts to obtain limits on Quantum Gravity effects [4, 5, 6, 7] and prospects for future improvements.

2 An Energy Dependent Speed of Light

An energy dependent speed of light arises in a variety of Quantum Gravity models, ranging from critical or noncritical string theories, via noncommutative

T. Piran: *Gamma-Ray Bursts as Probes for Quantum Gravity*, Lect. Notes Phys. **669**, 351–362 (2005)
www.springerlink.com

geometry, to canonical quantum gravity. These models, which involve a breakdown or a modification of Lorentz invariance at high energies, have been discussed extensively in other lectures in this school and are reviewed elsewhere in this volume. I focus here on a simple linear velocity-energy relation (see (1) below) that arises in models for the breakup of Lorentz symmetry proposed by Amelino-Camelia et al., [1]. It appears that a similar analysis is also applicable to the case of "DSR deformation" of Lorentz symmetry, since the same time-of-flight studies are considered in that framework [9, 10, 11, 12]. In fact I would expect that this simple linear velocity-energy relation (1) would be valid, to a leading order, in many other models.

On the phenomenological side an energy dependent speed of light was suggested as a possible resolution of the GZK paradox [13, 14]: The observations of UHECRs (Ultra High Energy Cosmic Rays) above the expected (GZK) threshold for interaction of such cosmic rays with the Cosmic microwave background [15, 16, 17, 18, 19, 20]. Such energy dependence could be related to a threshold violation at very high energies. Another possible indication for this phenomenon is the observation of TeV photons from distant sources [21, 22, 23]. Such photons are expected to be annihilated due to the interaction with the IR background. Again threshold anomalies (that would be associated with an energy dependent speed of light) could resolve this problem [20, 24, 25, 26]. In fact Amelino-Camelia and Piran [20] have pointed out that a simple Lorentz invariance deformation with parameters of the order expected in various quantum gravity theories (namely $\eta \sim 1$ in the notations used below) could resolve both paradoxes.

3 On the Detection of Energy Dependent Time Lags Due to an Energy Dependent Speed of Light

In this short review I will not discuss the theoretical or the phenomenological motivations for an energy dependent speed of light. Instead I focus on the detectability of this phenomenon. I stress that the deviations that I discuss here are drastically different from those that arise from appearance of a photon mass. The effects of a photon mass are most pronounced at low energies. However, the deviations considered here depend on E/M_{pl} and are relevant only at very high energies.

Amelino-Camelia et al. [1] (see also [8] and other talks in this volume) pointed out that even a small variations in the speed of photons with different energies could lead to observable energy dependent time of arrival lags for photons arriving from a cosmological source. Following Amelino-Camelia et al. [1], I consider a linear energy dependence of the form:

$$v = c\left(1 - \frac{E}{\eta M_{pl}}\right) , \tag{1}$$

where M_{pl} is the Planck mass and η is a dimensionless constant. Quantum gravity effects that cause the deviation in the speed of light are expected to take place around the Planck energy, M_{pl}. I characterize the exact energy in which these take effect as $E_{QG} \equiv \eta M_{pl}$. The sign of η determines the direction of these changes.

One can easily generalize the discussion and consider a more general velocity-energy dependance, such as: $v = c[1 - (E/\eta M_{pl})^\alpha]$ [20, 17]. However, for $\alpha > 2$ and for the relevant energy range and for $\eta \approx 1$ the resulting time delays will be so short that I don't discuss this case here.

This velocity law (1) leads to a time lag between a photon at energy E and a very low energy photon of:

$$\delta t(E) \approx 10 \text{ msec } \eta^{-1} d_{Gpc} E_{GeV} \ , \tag{2}$$

where d_{Gpc} is the distance to the source in units of Gpc and E_{GeV} is the photon's energy in GeV. Ellis et al., [6] provide an exact expression as a function of the redshift of the source. However, the approximate expression given above is sufficient for the purpose of this work. The dotted lines in Fig. 1 depict the relation between d and E for different values of $\eta\delta t$. A detection, for a given value of $\eta\delta t$, is possible only above the corresponding line. The value of δt is the minimal time delay that can be detected in the particular source.

It is clear from (2) that we need a very high energy source. However for these sources, because of the enormous energy that each of the photons carries the rate of arrival of high energy photons, $R(E)$, is very often too small. I call these sources which are limited by a too small rate of arrival of photons: *photon starved sources*. This has to be taken into account as the low photon rate limits the shortest possible detectable temporal variation as:

$$\frac{1}{R(E)} = \frac{4\pi d^2 E}{AL(E)} = 180 \text{ msec } \frac{d^2_{Gpc} E_{GeV}}{A_4 L_{50}(E)} \le \delta t_{min} \ , \tag{3}$$

where $L(E)$ is the luminosity at energy E and where I have ignored for simplicity cosmological correction factors. $L_{50}(E)$ is the luminosity at the relevant energy interval in units of 10^{50} ergs/sec and A_4 is the area in units of m^2. Again δt_{min} is the minimal time scale that can be detected in the particular source.

The exact limit that the combination of (3) and (2) imply depends on the spectral shape, on the overall luminosity and on the variability time scale at the source. Quite generally these conditions lead to an *upper limit* on the distance from which the effect could be measured and to a *lower limit* on the energy. As I show in Sect. 6 this limit is important for GRBs. As an example the solid lines in Fig. 1 correspond to equal values of $AL\delta t_{min}$, for the case when the luminosity per decade is constant (i.e. the spectral index is -2) and for the case that the inequality (3) is satisfied as an equality. The dashed line on this figure depicts the same graph for $L(E) \propto E^{-1/2}$

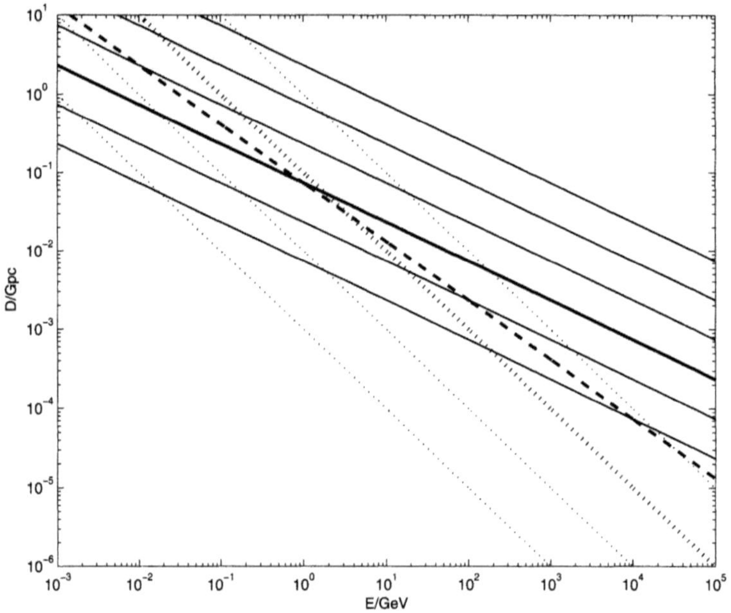

Fig. 1. Lines of a constant values of $\delta t_{ms} L_{50} A_4$ (*solid lines*) for $\delta t_{ms} L_{50} A_4 = 0.01, 0.1, 1, 10, 100, 1000$. δt_{ms} is in units of msec. The canonical value $\delta t_{ms} L_{50} A_4 = 1$ is marked by a *thicker line*. Detection is possible only below a given *solid line*. The single *dashed line* corresponds to $L(E) \propto E^{-1/2}$ and is normalized so that $\delta t_{ms} L_{50}(1\,\mathrm{GeV})\mathrm{A}_4 = 1$. The *dotted lines* mark lines of constant values of $\delta_{ms}\eta = 0.01, 0.1, 1, 10$, where again δt_{ms} is in units of msec. The canonical value of $\delta_{ms}\eta = 1$ is marked by a *thicker line*. Detection is possible only above a given *dotted line*. The combination of both constraints yields an allowed wedge with a maximal distance and minimal energy. Note that the vertical scale of distances ranges from cosmological distances at the *top* ($d_{Gpc} > 1$) to local (galactic) distances at the *bottom* ($d_{Gpc} < 10^{-5}$)

which is normalized so that the luminosity per decade of energy at 1 GeV is 10^{50} ergs/sec. A detection is possible only below these lines. For a given combination of $\eta\delta t$ and $LA\delta t$ a detection is possible only within a wedge outlined by the corresponding solid line and dotted line. Namely, for a given set of parameters there is a **maximal** distance and a **minimal** energy for which the time-lag can be detected. This suggests that in some cases (but not in the general case) a local (galactic) source with a strong very high energy signal might be advantageous over a weak source at a cosmological source. Indeed this was used by Kaaret [27] to obtain a meaningful limit on $\eta > 1.310^{-4}$ using the emission from the Crab pulsar which is only at 2.2 kpc

It is clear from (2) that a cosmological distance and a high energy are needed for a significant δt. However, the interaction of high energy photons with the cosmic IR background limits the distance that high energy photons can travel. For $E \sim 100\,\mathrm{GeV}$ the optical depth to $z = 0.5$ is unity [28][1]. Thus, we must consider photons with $E < 100\,\mathrm{GeV}$. This, in turn gives an upper limit of $\sim 6/\eta$ sec to possible magnitude of the time delay between photons of different energies. This is independent of the source of the emitted photons. It immediately follows that to observe this phenomenon we need cosmological sources of $\sim$ GeV photons with a rapid and detectable variability on the time scale of seconds or less. Amelino-Camelia et al. [1] point out that Gamma-ray bursts are the natural candidates for this task, and indeed several groups obtained lower limits on η using GRBs [4, 5, 6, 7].

4 Gamma-Ray Bursts

GRBs are short and intense pulses of γ-rays that are located at cosmological distances. As such GRBs are ideal sources for the effect that we are looking for. For the purpose of this work GRBs are just a cosmological source of high energy photons. Their exact nature is unimportant for our ability to use the photons to test the predictions of quantum gravity. However, it is worthwhile, for completeness, to review briefly our current understanding of this phenomenon. I refer the readers to several extensive reviews [29, 30, 31, 32, 33, 34, 35, 36, 37] for more details.

It is generally accepted that GRBs are described by the internal-external shocks model [38, 39, 40, 41]. According to this model GRBs are produced when the kinetic energy of an ultra-relativistic flow is dissipated. Internal shocks within the relativistic flow produce the GRB. These shocks take place at a distance of $\sim 10^{13} - 10^{15}\,\mathrm{cm}$ from the center. The short observed time scales (which violates the simple naive rule of $\delta t < R/c$) arises because of the relativistic motion of the flow (with a Lorentz factor $\gamma \geq 100$) towards us. Subsequent interaction of the relativistic outgoing flow with the surrounding matter leads to the production of an afterglow (in x-ray, optical and radio) that lasts days, weeks, months and in some cases even years. This takes place at distances of $\sim 10^{16} - 10^{18}\,\mathrm{cm}$ from the center. The flow is slowed down due to this interaction and eventually it becomes Newtonian.

It is worthwhile to mention what is the validity of this model. Indirect determination of the size of the afterglow of GRB 970508 [42] and direct measurement of the size of the afterglow of GRB 030329 [43] confirmed the predicted relativistic motion . Additionally there is a good agreement between the observed spectra and light curves of the afterglows and the predictions of the relativistic shock synchrotron model. There is also good observational evidence for the "internal-external" shocks transition. On the other hand, little is known about the details of the "inner engine" and the details how does the collapsing core produce the required relativistic jet.

[1] Different authors make different assumptions on the IR background and find different estimates for the optical depth. These quantitative differences are not important for the purpose of this work.

The discovery of long lasting x-ray, optical and radio afterglow enabled the determination of the redshifts and the positions of some bursts. The identification of bursts within star forming regions and the identification of Supernovae (SNe) signatures (SNe bumps) in the afterglow of some bursts (most notably GRB 980425 and GRB 030329) revealed that long[2] bursts are associated with type Ic Supernovae. As the rate of SNe Ic is much larger than the rate of GRBs it is clear that not all Supernovae are associated with GRBs. Jet-breaks detected in the afterglow of many bursts revealed that the bursts are beamed into cones of a few degrees and that their total energy is rather constant $\sim 10^{51}$ ergs [45, 46].

The GRB-SNe association is explained according to the Collapsar model [47], which is a model for the "inner engine". According to this model a black hole – accretion disk system forms during the core collapse. This system produces a relativistic jet that manages to punch a hole in the supernova envelope. The burst and the afterglow are produces along the internal-external shocks model, once the relativistic jet has emerged from the envelope.

5 GRB Observations and Testing of a Variable Speed of Light

The possibility of observing the energy dependent time-lags depend on four factors the distances to the sources, their temporal structure, their spectrum and their intensity. I discuss these three features here:

- **Distances** It is established that the bursts arise from cosmological distances. The identified redshift record is 4.5 (GRB 000131) but it is likely that more distant bursts has been observed but their redshift is unknown [48].
- **Temporal Structure** The bursts durations vary lasting from a few milliseconds to a thousand seconds. The paucity of bursts with a duration around two seconds suggest a classification of the bursts to two groups according to their durations – long bursts with durations longer than 2 seconds and short one with a durations shorter than 2 seconds.
  What is most important for our purpose is that most bursts show a highly variable light curve (see for example Fig. 2). Nakar and Piran [49], for example analyzed the TTE (high resolution data of the short bursts and of the first two seconds of long ones) find in many burst sub-pulses on a time scale of 10 ms (which was about the minimal possible temporal resolution). with sub-pulses on a scale as short as a fraction of a millisecond [50].
- **Spectrum** The bursts's spectrum usually peaks around a few hundred keV. Recently a subgroup of bursts, x-ray flashes, that emits most of their

[2] As afterglow was seen so far only from long burst it is not clear if short bursts are also associated with Supernovae. In fact there are some theoretical considerations that suggest that they are not related.

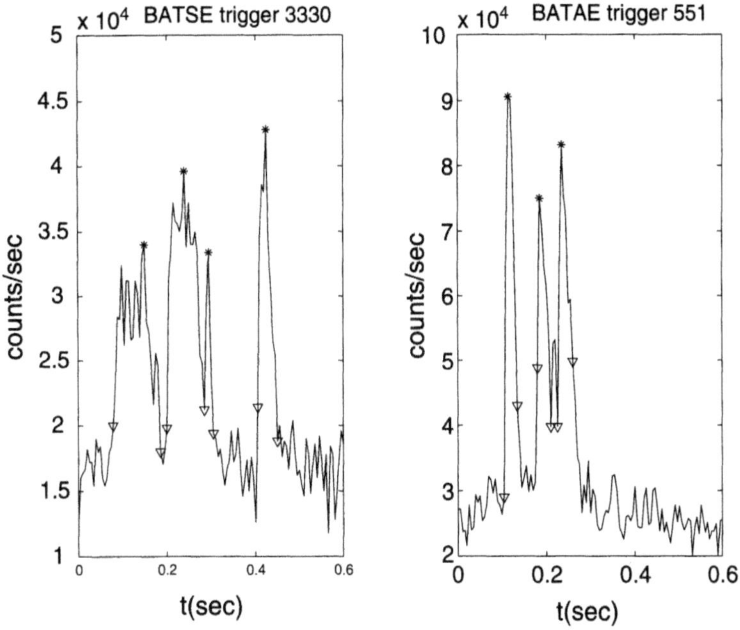

Fig. 2. (*Left*) The beginning of BATSE trigger 3330: a long bright burst with $T_{90} = 62\,\text{sec}$. (*Right*) The whole light curve of BATSE trigger 551: a bright short burst with $T_{90} = 0.25\,\text{sec}$. The peaks are marked by stars and the triangles mark the pulses' width. The figure demonstrates similar short time scale structure in these bursts (at a 5 msec resolution). From [49]

energy in X-ray was discovered. In many cases a high energy tail, with photon energies from 100 MeV to 18 GeV has been observed [51]. The TeV detector, Milagrito, discovered (at a statistical significance of 1.5e-3 or so, namely at 3σ) a TeV signal coincident with GRB 970417 [52, 53]. However no further TeV signals were discovered so far from other 53 bursts observed by Milagrito [52] or from several bursts observed by the more sensitive Milagro [54]. One should recall however, that due to the attenuation of the IR background TeV photons could not be detected from $z > 0.1$. Thus even if most GRBs emit TeV photons those photons won't be detected on Earth. Similarly these photons are too energetic for our purpose.

- **Intensity** The last factor that is important in our consideration is the intensity of the signals. This is important because a significant number of photons is needed to determine exactly the timing of a pulse. The strongest observed bursts have a fluence of $10^{-4}\,\text{ergs/cm}^2$ corresponding to 1000 (100 keV photons)/cm^2. The peak photon flux (on the BATSE[3] 64 msec channel) is $\sim$180 photons/cm^2/sec. With typical detectors' area of several

[3] BATSE, the Burst and Transient Source Experiment on board on NASA's Compton-GRO, is the largest GRB detector flown so far.

square meters this leads to a (100 keV range) photon rate of more than a photon per µsec that in principle could be used to determine the temporal structure down to a very short time scales.

The situation looks at first promising. GRBs are highly variable bright cosmological sources providing γ-ray photons at the right distances. Equation (2) reveals that energies higher than 100 MeV are needed to produce a time delay of a few millisecond and many GRBs have such photons. At the same time many GRBs show variability on such a time scale. However, as we see in the next section one should proceed with cation before concluding that GRB signals could provide a real measure of a variable speed of light.

6 Caveat, Past Observations and Future Prospects

A careful look at the properties of GRBs uncovers, however, problems. The main problem is that it is not clear that the high and low energy photons seen from GRBs are emitted simultaneously. In fact the current understanding is just the opposite. The highest energy (18 GeV) photons discovered by EGREAT (a detector on Compton – GRO), were observed more than an hour after the main burst [55, 56]. Similarly, when Gonzalez et al. [57] combined the BATSE (30 keV–2 Mev) data with the EGRET data they discovered in GRB 941017 a high energy tail that extended up to 200 MeV. This high energy component appeared 10–20 sec after the beginning of the burst and displayed a roughly constant flux up to 200 sec, while the main lower energy burst decayed after several dozen seconds.

One may hope that this non-simulteneity appears only in a "global" sense and that on a short time scale high energy photons are emitted simultaneously with the low energy ones. While there is not enough information on the generic time lag between very high (100 MeV and higher) and low energy (100 keV) GRB photons there is a lot of "alarming" information on lack of simultaneity within the BATSE band (25 keV to 2 MeV). Already in 1992 Fishman et al., [58] (see also Link et al., [59]) noticed that the duration of GRB pulses depend on their energy and that at lower energy the pulses are wider. Band [60] classifies this as a hard to soft evolution. Later Norris et al., [61] noticed that this evolution corresponds to a time lag between pulses at different energies. Typically the higher energy pulses peak before the corresponding low energy ones. For a sample of 174 bright bursts Norris et al., [61] find typical lags between channel 1 (25–50 keV) to channel 4 (300 keV–2 Mev) of the order of 0.1–0.2 sec with a maximal lag of 5 sec. A small fraction (~5%) of the bursts have negative lags of the order of less than 0.1 sec. These lags are larger by a factor of 2–3 than the lags suggested in (2) for a GeV photon!

For the bursts with a known redshift Norris et al. [61] find interesting anti-correlation between the time lags and the peak luminosities of the bursts. This correlation, for which there is no clear theoretical explanation, has been used by Norris et al. [61] to estimate the luminosity of other bursts. It is in a general agreement with other luminosity indicators such as the variability of the bursts [62]. While this correlation is not of interest for the purpose of this work the existence of intrinsic time lag between photons of different energies may jeopardize the whole prospect of detection of energy dependent time of arrival lags arising from an energy dependent travel time. It is clear that such an observation requires a simultaneous emission of photons at different energies.

Ellis et al. [6] suggest to use the redshift dependence of the velocity induced time lags to distinguish them from the intrinsic lags that are produced at the source. By plotting the time lags for several BATSE bursts with a known redshifts they obtain a limit $E_{QG} > 6.9 \cdot 10^{15}$ GeV or in our notations $\eta > 6.9 \cdot 10^{-4}$. As the highest energy photons used are of ~1 MeV, this corresponds, according to (2) above to the conclusion that the redshift dependent time lags are less than ~0.1 sec, which is comparable with the intrinsic time lags of these bursts [61]. Given the time resolution this limit ($\eta > 10^{-3}$ seem to be (see in (2)) the best that can be done using "low" energy (~ MeV) photons.

However, there is another observational factor that appears here. Norris et al. [65] describe the tendency for wide pulses to be more asymmetric, to peak later at lower energy and to be spectrally softer, while narrow bursts are harder, more symmetric, and nearly simultaneous. This implies that the narrowest peaks, those that are most interesting for this experiment have a chance of being simultaneous in both low and high energies. Schafer [4] uses, along these lines, the observations of one of BATSE's brightest bursts, GRB 930131 with 30 keV and 80 MeV photons to obtain a limit of $E_{QG} > 8.3 \cdot 10^{16}$ GeV (or $\eta > 8.3 \cdot 10^{-3}$). Also along this line Boggs et al., [7] analyze GRB 021206. They used the Reuven Ramaty High Energy Solar Spectroscopic Imager (RHESSI) with an energy range of (3 keV to 17 MeV). They noticed that while the lower energy (<2 MeV) light curve of the burst is rather irregular at higher energies the light curve exhibits a single sharp pulse of photons extending to energies above 10 MeV with a duration of 15 msec. This enables Boggs et al., [7] to set a limit of $\delta t/E = 0.0 \pm 0.34\,\mathrm{sec\,GeV^{-1}}$ from which they obtain $E_{QG} > 1.8 \cdot 10^{17}$ GeV ($\eta > 0.018$). Considering (2) this seems to be the best that can be done with 10 MeV photons. To improve we have to get to lower temporal resolution (which might not be possible) or to higher photon energies.

But here arises a second simple but important problem. In spite of the fact that GRBs are the most luminous objects in the universe at GeV energies they are *photon starved*: the observed flux is simply low. The maximal GRB fluxes at energies of a few hundred keV are of ~100 $\mathrm{photons_{100\,keV}/cm^2/sec}$. With a several square meter detector this corresponds to a flux of 10^6 100 keV photons/sec or to a photon rate of one per μsec. However even if GRBs emit the same energy flux at the GeV range this flux corresponds to a meager $10^{-3}/\mathrm{cm^2/sec}$ GeV-photons or to 10 GeV-photons per second with a square meter class detector. As the minimal temporal resolution is larger than the reciprocal of the rate of observed photons it will be impossible to obtain a temporal resolution of better than ~100 msec at the GeV range. From (2) this corresponds to a limit on η of order unity if all other problems are resolved.

The comparison of (2) and (3) (shown in Fig. 1 for a constant energy per logarithmic interval) yields that to resolve the time lags we need a nearby ($d < 1$ Gpc) very luminous GRB with a significant GeV component. Truly the rather "small" distance will reduce δt. However, only in this way there will be enough photons to obtain a sufficient temporal resolution. The requirement of short distances implies that we won't be able to use the redshift effect to distinguish between intrinsic lags and time of flight lags. However, the fact that we consider only very luminous bursts may resolve this problem as the luminosity-lag correlation indicates that intrinsic lags are smaller for more luminous bursts and they may disappear for the very bright ones. These simple considerations are indeed supported by the present observations. Boggs et al., [7] considered a single very bright burst: GRB 0211206 which was one of the most powerful bursts ever [63] and was most likely at $z \approx 0.3$, and obtained $\eta > 0.018$. This should be compared with $\eta > 0.00069$ obtained by Ellis et al. [6] who considered a family of weaker bursts at cosmological distances $z \geq 1$. One has to recall however, that such a burst occurs once per decade and it is not clear when will the next one take place. Hopefully a suitable GRB detector will be in orbit at that time.

The best prospect to estimate the variable velocity energy dependent effect will be with a single observatory that could observe both the low energy γ-rays as well as the GeV emission. Luckily there are two planed mission that can perform this job.

The Italian **Agile** (Astro-rivelatore Gamma a Immagini LEggero) detector [64] is scheduled to be launched in 2005. It is a GRB detector at the energy range of 30 MeV–50 GeV and a low energy detector at 10–40 keV. Thus it is expected to detect GRBs at both very high and very low energy. The temporal resolution is about 1msec. The expected detection rate is about 10 GRBs per year at energies above 100 MeV. The basic limitation of Agile is its relatively small area (~0.16m^2).

An ideal observatory will be NASA's **GLAST** (Gamma Ray Large Area Space Telescope). GLAST is scheduled for launch in 2007 (Norris et al., [3]). GLAST will include the Large Area Telescope, LAT, which will have an effective area of 8m^2 and will be sensitive to photons in the 20 MeV–300 GeV range and GRM, a Gamma-Ray burst Monitor which will be sensitive to photons in the 10 keV to 25 MeV range. Both the LAT and the GBM provide the arrival time of each photon with a resolution requirements of <10 µsec (with a goal of <10 µsec) and will give energies for each detected photon. One cannot ask for more, in terms of the experimental design needed to study the energy dependent time lag. Thus, if the intrinsic time lags will be resolved or shown to be unimportant in some sub class of pulses or bursts, and this is a very big IF in my mind, we might be able to obtain a limit of η around unity towards the end of this decade.

I thank D. Band, E. Nakar and G. Amelino-Camila for helpful remarks. This research was supported by the US-Israel Binational Science Foundation.

References

1. G. Amelino-Camelia, J. Ellis, N.E. Mavromatos, D.V. Nanopoulos, & S. Sarkar, 1998, Nature, 393, 763
2. R. Gambini and J. Pullin, 1999, Phys. Rev. D59:124021
3. J.P. Norris, J.T. Bonnell, G.F. Marani, & J.D. Scargle, 1999, ArXiv Astrophysics e-printsScargle, astro-ph/9912136
4. B.E. Schaefer, 1999, Phys. Rev. Lett., 82, 4964
5. J. Ellis, K. Farakos, N.E. Mavromatos, V.A. Mitsou, & D.V. Nanopoulos, 2000, Ap. J., 535, 139
6. J. Ellis, N.E., Mavromatos, D.V. Nanopoulos, & A.S. Sakharov, 2003, Astron. & Astrophys., 402, 409
7. S.E. Boggs, C.B. Wunderer, K. Hurley, & W. Coburn, 2003, ArXiv Astrophysics e-prints, astro-ph/0310307
8. G. Amelino-Camelia, 2004, *This Volume.*
9. G. Amelino-Camelia, 2002, Int. J. Mod. Phys. D11:35-60
10. J. Kowalski-Glikman, S. Nowak, 2003, Int. J. Mod. Phys. D12:299-316
11. J. Magueijo and L. Smolin, 2003, Phys. Rev. D67:044017
12. G. Amelino-Camelia, L. Smolin, A. Starodubtsev, 2004, Classical & Quantum Gravity 21, 3095
13. K. Greisen. 1996, Phys. Rev. Lett., 16:748
14. G.T. Zatsepin & V.A. Kuzmin, 1966, Sov. Phys.-JETP Lett., 4:78
15. L. Gonzalez-Mestres, 1997, physics/9704017
16. S. Coleman, S.L. Glashow, 1999, Phys. Rev. D59. 116008.
17. R. Aloisio, P. Blasi, P.L. Ghia, A.F. Grillo, 2000, Phys. Rev. D62:053010
18. O. Bertolami and C.S. Carvalho, 2000, Phys. Rev. D61, 103002.
19. H. Sato, astro-ph/0005218.
20. G. Amelino-Camelia, & T. Piran, 2001, Phys. Rev. D., 64, 036005
21. A.I. Nikishov, 1962, Sov. Phys. JETP, 14, 393
22. J. Gould, G. Schreder, 1967, Phys. Reports, 155.5, 1404
23. F.W. Stecker, O.C. De Jager and M.H. Salomon, 1992, Ap. J. Lett., 390, L49
24. T. Kifune, Ap. J. Lett.1999, 518, L21.
25. W. Kluźniak, 1999, Astroparticle Physics, 11, 117
26. R.J. & Meyer, H. 2000, Physics Letters B, 493, 1
27. P. Kaaret, 1999, Astron. & Astrophys., 345, L32
28. J.R. Primack, J.S. Bullock, R.S. Somerville, & D. MacMinn, 1999, Astroparticle Physics, 11, 93
29. G.J. Fishman, and C.A. Meegan, 1995, Ann. Rev. Astron. & Astrophys.33, 415.
30. T. Piran, 1999, Phys. Reports 314, 575.
31. J. van Paradijs, C. Kouveliotou, and R.A.M.J. Wijers, 2000, Ann. Rev. Astron. & Astrophys.38, 379.
32. T. Piran, 2000, Phys. Reports, 333, 529.
33. P. Mészáros, 2001, Science, 291, 79-84
34. K.R. Hurley, Sari, and S.G. Djorgovski, 2002, astro-ph ???
35. P. Mészáros, 2002, Ann. Rev. Astron. & Astrophys.40, 137.

36. T. Galama, and R. Sari, 2002, in Relativistic Flows in Astrophsyics, LNP 589, edited by Axel Guthmann et al., Springer-Verlag, Berlin, 123.
37. T., Piran, 2004, Rev. Modern Phys. in press, astro-ph/0405503.
38. P. Meszaros, & M.J. Rees, 1992, MNRAS, 257, 29P
39. R. Narayan, B. Paczynski, & T. Piran, 1992, Ap. J. Lett., 395, L83
40. M.J. Rees, & P. Meszaros, 1994, Ap. J. Lett., 430, L93
41. R. Sari & T. Piran, 1997, Ap. J., 485, 270
42. D.A. Frail, S.R. Kulkarni, S.R. Nicastro, M. Feroci, & G.B. Taylor, 1997, Nature, 389, 261
43. G.B. Taylor, D.A. Frail, E. Berger, & S.R. Kulkarni, 2004, ArXiv Astrophysics e-prints, astro-ph/0405300
44. J.E. Rhoads, 1999, Ap. J., 525, 737
45. D.A. Frail et al. 2001, Ap. J. Lett., 562, L55
46. A. Panaitescu & P. Kumar, 2001, Ap. J. Lett., 560, L49
47. A.I. MacFadyen and S.E. Woosley, 1999, Ap. J., 524, 262,
48. V. Bromm, & A. Loeb, 2002, Ap. J., 575, 111
49. E. Nakar & T. Piran, 2002, MNRAS, 330, 920
50. J.D. Scargle, J.P. Norris, & J.T. Bonnell, 1998, American Institute of Physics Conference Series, 428, 181
51. B.L. Dingus, & J.R. Catelli, 1998, Abstracts of the 19th Texas Symposium on Relativistic Astrophysics and Cosmology, held in Paris, France, Dec. 14-18, 1998. Eds.: J. Paul, T. Montmerle, and E. Aubourg (CEA Saclay).,
52. R. Atkins et al. 2004, Ap. J. Lett., 604, L25
53. R. Atkins et al. 2000, Ap. J. Lett., 533, L119
54. J. McEnery, 2002, APS Meeting Abstracts, A3007
55. K. Hurley, 1994, Nature, 372, 652
56. M. Sommer et al. 1994, Ap. J. Lett., 422, L63
57. M.M. González, B.L. Dingus, Y. Kaneko, R.D. Preece, C.D. Dermer, & M.S. Briggs, 2003, Nature, 424, 749
58. G.J. Fishman, C.A. Meegan, R.B. Wilson, J.M. Horack, M.N. Brock, W.S. Paciesas, G.N. Pendleton, & C. Kouveliotou, 1992, American Institute of Physics Conference Series, 265, 13
59. B. Link, R.I. Epstein, & W.C. Priedhorsky, 1993, Ap. J. Lett., 408, L81
60. D.L. Band, 1997, Ap. J., 486, 928
61. J.P. Norrisf, G.F. Marani, & J.T. Bonnell, 2000, Ap. J., 534, 248
62. D.E. Reichart, D.Q. Lamb, E.E. Fenimore, E. Ramirez-Ruiz, T.L. Cline, & K. Hurley, 2001, Ap. J., 552, 57
63. E. Nakar, T. Piran, & E. Waxman, 2003, Journal of Cosmology and Astro-Particle Physics, 10, 5
64. M. Tavani et al., 2004, Astron. Astrophys. Suppl. Ser. 138, 569-570
65. J.P. Norris, R.J. Nemiroff, J.T. Bonnell, J.D. Scargle, C. Kouveliotou, W.S. Paciesas, C.A. Meegan, & G.J. Fishman, 1996, Ap. J., 459, 393

Loop Quantum Gravity and planck Scale Phenomenology

L. Smolin

Perimeter Institute for Theoretical Physics Waterloo, Canada N2J 2W9 and Department of Physics, University of Waterloo
lsmolin@perimeterinstitute.ca

1 Introduction

Of the different approaches to quantum gravity, the best developed, from the point of view of addressing the key theoretical questions a quantum theory of gravity must answer, is loop quantum gravity[1]. While string theory appears to better address the issue of unification, at least so far, it fails to provide a background independent approach to the quantum mechanics of spacetime geometry-a necessary condition for any quantum theory of gravity. Moreover, many key conjectures remain unproven, including perturbative finiteness and consistency, which have not been demonstrated for any string theory past second order in perturbation theory[2]. By contrast, loop quantum gravity appears to provide a consistent and finite background independent approach to quantum gravity. There is recent progress on several issues, including accounting for the black hole entropy [8], and giving a precise quantum mechanical description of the earliest phases of the evolution of the universe [9, 10]. Furthermore, it gives unique predictions of novel quantum gravitational phenomena, such as the discreteness of area, volume and other observables.

However, in this new era of quantum gravity phenomenology, a quantum theory of gravity must pass a stricter test to be taken as a serious candidate: it must make unambiguous predictions for the upcoming experiments which probe the Planck scale.

One way such predictions may arise is by modifications of the energy-momentum relations of low energy physics, of the form,

$$E^2 = p^2 + m^2 + al_P E^3 + bl_P^2 E^4 + \dots \tag{1}$$

Related to this may be corrections to other of the basic kinematical formula of special relativity including energy-momentum conservation laws and the

[1] For some reviews, see [1, 3, 4, 6, 7]
[2] For a summary of the status of the main conjectures of string theory, see [6].

L. Smolin: *Loop Quantum Gravity and Planck Scale Phenomenology*, Lect. Notes Phys. **669**, 363–408 (2005)
www.springerlink.com

actions of Lorentz transformations on energy-momentum eigenstates. As described in other papers in this volume, such corrections are in fact amenable to experimental test [11, 12, 13, 14] in present and near future experiments.

In this contribution I would like to do three things. First I will describe the present state of the results relevant for predictions of quantum gravity phenomenology from loop quantum gravity. Second I will give an introduction to some aspects of loop quantum gravity, in particular those having to do with the theory in the presence of a non-zero cosmological constant, Λ[3] Third, I will describe one approach to deriving Planck scale corrections to energy momentum relations from matter fields, of the form of (1). This approach is based on studying the theory for nonzero Λ, and then deriving predictions for zero or very small Λ, corresponding to our universe by taking the limit $\Lambda \rightarrow 0$. For reasons I will explain shortly, it may be very helpful to approach the problem of making predictions for the phenomenology with $\Lambda = 0$ through a limit of this kind.

2 What Should Theory Predict for Phenomenology

There is a key question that any quantum theory of gravity must answer if it is to provide predictions for low energy phenomenology: *What is the fate of Poincaré and Lorentz invariance when Planck scale corrections to the semiclassical limit are taken into account?*

This question arises because Poincaré and *global* Lorentz invariance cannot be symmetries of any quantum theory of gravity. This is because they are not symmetries of general relativity, and any correct quantum theory of gravity must have general relativity as the low energy limit. Poincaré and global Lorentz invariance are only symmetries of one particular solution to general relativity: Minkowski spacetime. It happens that Minkowski spacetime is, under certain choices of asymptotic conditions and with $\Lambda = 0$ the ground state of the classical theory. As such it has the maximal number of symmetries of any spacetime.

But general relativity is a dynamical theory of spacetime, which is background independent, in the sense that no single spacetime-not even Minkowski spacetime-appears in the formulation of the action, equations of motion and Hamiltonian. A particular solution may have symmetries, but those are not symmetries of the dynamics of the theory. From the point of view of the full theory, the symmetries of any classical solution-even the ground state-play no fundamental role.

So in classical general relativity, Poincaré and global Lorentz invariance are only accidental, emergent symmetries that characterize the low energy limit. Such must be true for any quantum theory of gravity that has general relativity as a classical limit.

---

[3] These parts of the paper are cannibalized from a previous review [5].

Because of this, it is possible, in fact necessary to ask, *what symmetries characterize the low energy limit of the correct quantum theory of gravity?*

There appear to be three possibilities:

1. Lorentz and Poincaré invariance emerge exactly as in the classical theory. In this case there are no corrections of order $l_p E$ to relativistic kinematics, a and b in (1) vanish.
2. Lorentz and perhaps Poincaré invariance are broken, so the symmetry is reduced at order $l_P E$ to a smaller algebra. There will then be nonzero corrections and a and b should be computable. Further, there will be a preferred frame of reference, which arises somehow dynamically in the low energy limit.
3. The symmetry is deformed, or non-linearly realized, as in κ-Poincaré symmetry [15] or proposals for doubly special relativity, *DSR* [16, 17, 18, 19]. There is no preferred frame, so that the number of symmetry generators remains 10. However, the algebra of the Poincaré generators may be quantum deformed as in the κ-Poincaré algebra [15]. One symptom of this is that the geometry of Minkowski spacetime can become non-commutative. By the relativity principle, the Lorentz subgroup of Poincaré cannot be deformed, for the group axioms must be satisfied if observers in relative motion are able to transform measurements made among themselves. But its action on energy-momentum eigenstates can become non-linear.
   This scenario leads to non-zero a and b in (1). There are also other corrections to kinematics, for example there must be corrections to the energy and momentum conservation laws so that the modified energy-momentum relations may be true, consistently with the conservation laws, in every inertial frame.

Any background independent theory of quantum gravity must tell us which of the these three scenarios it predicts. Furthermore it must make unique and calculable predictions for a and b for the different particles, as well as for other possible corrections, such as to energy-momentum laws.

Furthermore, it must be emphasized that only a background independent theory could make such a prediction. A background dependent theory, like string theory, cannot in principle, because its definition requires the prior specification of a fixed classical spacetime background. The symmetry of the low energy limit is then put in by hand as the symmetry of the background, it cannot be predicted. Thus, we should beware of "me too" claims for string theory and other background dependent approaches, that assert that they can produce a theory based on a background where any of the three scenarios is realized. This is true, but it is a sign that those theories are not predictive, because the background must be chosen by hand from a (very long) list of possible backgrounds on offer, rather than predicted as the result of a dynamical calculation from the theory.

An example of how such a prediction arises uniquely from a genuine quantum theory of gravity is given by the case of quantum gravity in 2 + 1 dimensions, coupled to point particles. It has long been known that $2+1$ gravity exists, and is an exactly solvable theory [21]. This remains true when coupled to any number of point particles. Thus, the argument I've just made says it must make a unique prediction for which scenario obtains, and for the coefficients a and b.

Does it?

It does [22]. As soon as the question was asked it was realized that the result is already in the literature, indeed the answer was given in a number of different results in separate papers [23]. *The answer is that* $2+1$ *quantum gravity with point particles realizes exactly the third scenario.* Moreover, it predicts that when $\Lambda = 0$ the low energy symmetry is given by the κ-Poincaré algebra. Furthermore, as that algebra is non-linear, the theory must tell us what presentation of the symmetry algebra corresponds to the algebra of the asymptotic energy and momentum operators measured by an observer at the boundary of the spacetime. It does that as well, and by doing so it gives unique predictions for all Planck scale corrections including a and b.

For reasons I will explain, I believe that it is reasonable to conjecture that the same will be true in LQG in $3+1$ dimensions. The specific conjecture is

The symmetry of the ground state of loop quantum gravity when $\Lambda = 0$ *is the* κ*-Poincaré algebra with the invariant energy given by* $M_P = \kappa^{-1}$.

There is a simple argument for this conjecture [24], whose applicability to loop quantum gravity in $3+1$ dimensions will be supported by a result I will describe below.

We first note that it is unlikely that a preferred frame emerges from the low energy limit of loop quantum gravity. This is because the gauge invariances of the theory include the transformations generated by the Hamiltonian constraint. On classical solutions, these generate diffeomorphism corresponding to arbitrary changes in the time coordinate. Hence different slicings of a classical spacetime into a one parameter family of spacelike hypersurfaces are gauge equivalent.

A preferred frame could then only arise by a spontaneous breaking of this gauge invariance. This would require that some vector field v^a acquire a vacuum expectation value. This might occur in a theory coupled to a vector field (that was not a gauge field), but in the absence of such a coupling and a Higgs like dynamics for the vector field, there is no apparent mechanism for spontaneous breaking of local lorentz symmetry to arise.

This would appear to rule out the second possibility, except in special theories where the right kinds of vector fields and interactions were included. A rough argument to prefer the third over the first in the general case, goes as follows.

We recall the classic result that the de Sitter (or anti-de Sitter) algebra $SO(3,2)$ has a contraction to the Poincaré algebra, The generators of $SO(3,2)$ are denoted by M_{AB}, where $A = a, 5$ and $a = 0, \ldots 3$ are the $3+1$ Lorentz indices. If $R = \Lambda^{-1/2}$ is the radius of the universe, then the generators of translations in Minkowski spacetime come from

$$P_a = \sqrt{\Lambda} M_{5a} \tag{2}$$

The Poincaré algebra emerges in the limit $\Lambda \to 0$.

However, there is evidence that for nonzero Λ, the symmetry algebra is quantum deformed to $SO_q(3,2)$ with, for small Λ, and with $z = \ln(q)$

$$z \approx \sqrt{\Lambda} l_p \quad \text{for} \quad d = 2+1 \quad [20,\ 25,\ 26] \tag{3}$$

$$z \approx \Lambda l_p^2 \quad \text{for} \quad d = 3+1 \quad [27,\ 28,\ 29] \tag{4}$$

In the case of 2+1 gravity, the result that the symmetry algebra is quantum deformed when the cosmological constant is turned on is rigorous, a complete argument is given in [26]. For the case of $3+1$ there is good evidence that the local gauge symmetry of the spacetime connection is quantum deformed from $SU(2)$ to $SU_q(2)$ [27, 28, 29]. In [30] an argument is given that this extends to the quantum deformation of the algebra of observables on the boundary of a spacetime with cosmological constant, so that the subgroup of the de Sitter algebra that generates the symmetries of the boundary is quantum deformed. This prompts the conjecture that the algebra of generators that preserve the ground state of $3+1$ quantum gravity with nonzero Λ is quantum deformed.

We now consider taking the contraction of the quantum deformed symmetry algebra. The cosmological constant now occurs in two places, in (2) and in either (3) or (4). As a result, the limit $\Lambda \to 0$ may be no longer the Poincaré algebra. In the case of $2+1$ gravity it is exactly the κ-Poincaré algebra [24]. Indeed this is exactly how the κ-Poincaré algebra was found in the first place [15].

In the case of $3+1$ dimensions, one must take into account an additional renormalization of the energy and momentum generators. This is necessary because, unlike the case of 2+1 dimensions, there are local degrees of freedom, and these will induce a renormalization between the fundamental operators of the theory and the symmetry generators of the low energy limit. This will be proportional to a power of the ratio of the ultraviolet and infrared regulator. Since LQG is known to be ultraviolet finite, the former is the Planck scale. The latter is of course the cosmological constant itself. Thus we have, instead of (2),

$$P_{a,ren} = \left(\frac{1}{\sqrt{\Lambda} l_p} \right)^r \sqrt{\Lambda} M_{5a} \tag{5}$$

It turns out that for $r < 1$ the contraction is the ordinary Poincaré algebra, while for $r = 1$ it is again κ-Poincaré. (For $r > 1$ the contraction does not exist.) This is supported by a result we will describe below.

3 Does Loop Quantum Gravity Make Predictions for Planck Scale Phenomenology?

To make predictions for Planck scale corrections to energy-momentum relations and other laws, a background independent quantum theory of gravity such as loop quantum gravity should ideally have three good features. (1) It should have a unique ground state. (2) It should be possible to derive quantum field theory on Minkowski or (A) de Sitter spacetime (presumably with a finite cutoff) as an approximation to the physics of excitations of that ground stat. (3) There should be a technique for calculating corrections to the predictions of QFT on Minkowski spacetime as a power series in $l_P E$, where E is the energy carried by a particle. The results should allow us to deduce which of the three scenarios above is correct.

There are in the literature a number of proposals for candidates for the ground state, for both vanishing and small positive or negative Λ [31, 32, 33, 34, 35]. The first candidates, known as weave states, for $\Lambda = 0$ were proposed shortly after loop quantum gravity was formulated. In 1989 Kodama proposed a candidate ground state for non-zero Λ, that now goes under his name [35]. It will be discussed in some detail below.

For some of these candidate ground states step two has been carried out, leading to the recovery of quantum field theory on Minkowski or de Sitter spacetimeGP,AMU,chopinlee. In several cases there is also evidence for a Planck scale cutoff. The $l_P E$ corrections have also been computed in several of the cases, and are found to be present. That is the good news. The bad news is that the coefficients a and b turn out to depend on adjustable parameters in the definitions of the candidate ground states.

Before taking these results too seriously, however, it should be noted that there are two ways in which most of these proposals of candidates ground states fall short. The first is that they are not physical states because they are not annihilated by the diffeomorphism and/or the Hamiltonian constraint operators. These are the operators that generate, on the quantum states, the local gauge and diffeomorphism transformations of the classical theory. The states don't satisfy them because they are not gauge invariance, hence they cannot represent physical quantities.

As a caveat to this, it is possible, as pointed out by Ashtekar [37], that it might be possible to construct physical states in a gauge fixed version of the theory. This would be interesting to work out.

The second reservation one must have concerning these results is that there is no hamiltonian which they minimize. This is a shortcoming of the theory, that should be overcome.

It is true that the equivalence principle rules out the existence of a local hamiltonian density in any theory of gravity that implements it. But it is still possible to define a hamiltonian, this requires defining a boundary, and fixing appropriate boundary conditions there. When this is done, a hamiltonian can be defined in general relativity and, under very weak assumptions having mainly to do with the positivity of the energy density of matter fields, shown to be finite.

It would be very useful if a corresponding hamiltonian operator could be defined for the quantum theory of gravity and, proved to be positive definite. Given such a positive definite Hamiltonian operator, the ground state could be found simply minimizing the expectation value of a suitable ansatz for the ground state, as in the rest of physics.

Baring this, results of the kind in the literature show that there is a possibility that quantum gravity may predict corrections to the energy momentum relations. But the results up till now are not reliable or robust enough to count as predictions of the theory.

The only putative ground state that has been studied which is a solution to all the constraints is the Kodama state [35]. As such it is of interest to see if corrections to the energy-momentum relations arise when matter fields propagate on the quantum background defined by this state. We will discuss this in detail in the last sections of this review, and show that the answer is yes. Moreover, we will see that the results support the argument we gave above based on contraction from quantum de Sitter symmetry.

4 The Basic Ideas of Loop Quantum Gravity

Loop quantum gravity is likely the most conservative approach to quantum gravity. This approach is based, to begin with, on the quantization of Einstein's theory of general relativity, using a particular formulation discovered by Sen [38] and completed by Ashtekar [39]. Rather than postulating new degrees of freedom, symmetries or dimensions, loop quantum gravity takes the basic principles of general relativity and quantum field theory seriously, and puts the emphasis on the development of methods that do not compromise either set of principles. These methods highlight the force that the principles of relativity theory, primarily diffeomorphism invariance and the independence from any fixed, non-dynamical background structure have, when treated properly in the context of quantum field theory. Indeed, it turns out that once this is done, the theory admits a wide range of assumptions concerning the fundamental degrees of freedom, symmetries and supersymmetries, as well as the exact dynamical laws. Einstein's equations may be imposed, and to a remarkable extent, solved, quantum mechanically, but other assumptions concerning the fundamental dynamics may also be studied.

Loop quantum gravity has been under development since 1986, and throughout this time there has been continual progress. The various obstacles encountered have in most cases been overcome. As a result it has been possible over time to make increasingly strong claims for this approach to quantum gravity.

We will be concerned here with a subset of the results, those relevant for the case that the cosmological constant is non-zero and positive. For a detailed listing of many of the key results, to date, of loop quantum gravity, see [6]. It must be said that some theoretical physicists find the results gotten using the methods of LQG surprising. It used to be argued that the perturbative non-renormalizability of general relativity implies that quantum gravity requires a modification of the principles of physics such as proposed in different ways in other approaches such as string theory, causal sets, etc. What is perhaps surprising is that loop quantum gravity has been successful, not by being radical, but by sticking rather strictly to the basic principles of general relativity and quantum theory.

Thus, the first question to be addressed is how it is that the theory may exist non-perturbatively, when the theory is nonsense when developed by traditional perturbative methods around fixed backgrounds? The answer has two parts. First the theory is completely *background independent* which means that classical spacetimes play no role in the formulation. This is necessary due to the strongly interacting nature of the Planck scale physics, the fact that the spacetime geometry is represented completely by operators, and the fact that the gauge invariances of the theory include active diffeomorphisms, which are broken by the specification of any given classical background metric[4].

The mistake all background dependent approaches make is to assume that space and spacetime have continuous, classical structures when probed at arbitrarily short wavelengths. Even rough, heuristic arguments, such as those that Wheeler and others used to give based on the uncertainty principle, suggest that this is wrong. The results of loop quantum gravity, based only on the basic principles of quantum theory and general relativity demonstrate conclusively that there is no classical spacetime manifold at Planck scales and shorter[5].

Thus, to arrive at a good quantum theory one must use methods which are background independent, and do not make the incorrect assumption that spacetime is smooth at short distances. To do this means only to take the basic principles of Einstein's theory of general relativity seriously and apply them exactly in the quantum theory. As opposed to other approaches, such as string theory, where new degrees of freedom are posited to move in fixed, classical background spacetimes, loop quantum gravity treats the geometry of spacetime at the quantum mechanical level as Einstein did at the classical level, as a completely dynamical entity. This is in fact required if the gauge invariance of the theory, which is active diffeomorphisms, is to be respected exactly in the quantum theory.

[4] For background on background independence, see [41, 42, 43].

[5] Indeed, the dynamics of Einstein's equations do not come into the derivation of the quantization of area and volume, which tell us that quantum geometry is discrete. These results are consequences only of the canonical commutation relations and the gauge invariances that define the theory. They apply whatever matter the theory is coupled to, and, with appropriate modifications, described in [44] apply also to supergravity.

When diffeomorphism invariance is imposed exactly there is a big payoff, which is that one finds an exact description of the gauge invariant Hilbert space of quantum gravity. These are eigenstates of observables that represent the volume of spacetime and the areas of surfaces in the spacetime. These observables turn out to be finite, when regulated in a manner that respects diffeomorphism invariance. And they have discrete spectra, which demonstrates that quantum geometry is discrete at the Planck scale. These results explain, in detail, why perturbative expansions around smooth backgrounds fail; because they fail to capture any of the structure present in exact solutions of the quantum constraints that impose gauge and diffeomorphism invariance.

The second key to the success of loop quantum gravity is a property of the Einstein equations, which gives force to these general considerations of background independence and diffeomorphism invariance. This is the existence of an intimate connection between the kinematics and dynamics of general relativity and topological field theories.

A topological field theory is a field theory that has only a finite number of degrees of freedom. The few degrees of freedom it has are non-local and generally are measured either at boundaries of spacetime or by measuring phase factors or holonomies associated with loops or surfaces that cannot be shrunk to a point. Topological field theories share some properties with general relativity, such as being invariant under the gauge group of active diffeomorphisms, and being independent of any classical background[6].

At the same time, general relativity is not a topological field theory as it has an infinite number of local degrees of freedom.[7] Even so, there are close connections between general relativity and topological field theory at the classical level, and these are exploited by loop quantum gravity to make a sensible quantization of general relativity. By doing so, results that at first sight are surprising and unexpected, turn out to be not only possible, but accessible with standard methods. This relationship with topological field theory in fact makes possible the key results at both the hamiltonian and path integral level that are the basis for the success of the approach.

These relationships with topological field theory are the subject of the next few sections of this paper, as they are in a good way into the subject. There are in fact three distinct ways that topological field theory enters into quantum gravity, First, the action turns out to be closely related to that of a topological field theory [4]. Second, the natural boundary term in the action, which must be added when a spacetime region with boundary is studied, is a topological field theory [27]. Third, the ground state of the theory with a non-zero Λ is derived from a topological field theory [35].

[6] Some basic references on topological field theory include [45]

[7] There are other diffeomorphism invariant theories that are not topological, even some without a metric, such as Chern-Simons theory for $d \geq 5$ [46] and higher form versions of Chern-Simons theory [47, 48].

Only a few of the results to be described in the next sections are recent. The case of positive cosmological constant has, in fact, been somewhat neglected in the field in recent years. Thus it is worth mentioning some of the reasons to return to it now. These include the fact that a positive cosmological constant appears to be observed (not too mention the fact that it is in any case necessary for inflationary cosmology that the effective cosmological constant at early times be positive, and large). Beyond this, general renormalization group arguments tell us that we cannot make sense of any quantum field theory unless we include all low dimension couplings. This is certainly the case with the cosmological constant; if it is not included in a generic quantum field theory it will be there in the effective theory with a magnitude order one in either the cutoff or the supersymmetry breaking scale. This general expectation is fully born out in all the numerical work done on nonperturbative approaches to quantum gravity. This includes the early work in euclidean dynamical triangulations as well as the recent numerical work of Ambjorn, Loll and collaborators on lorentzian, or causal, dynamical triangulation models [49]. Further the latter calculations show convergence is only possible if $\Lambda > 0$. Indeed, all the evidence we have from explicit renormalization group calculations, at both the perturbative and non-perturbative levels, tells us that the theory cannot have a good low energy limit, leading to the recovery of general relativity, unless the cosmological constant term is included in the formulation of the theory.

We now turn to an introduction of quantum general relativity, for the case of $\Lambda > 0$.

5 Gravity as a Gauge Theory

Let us jump right in and see the power of the connection between gauge theory, gravity and topological field theory uncovered in loop quantum gravity, and then go on to show how this perspective illuminates the geometry of de Sitter spacetime.

A good way into the subject is to begin with the following challenge: *Suppose that you wanted to make a theory of gravity, but you were restricted to using only the fields of an ordinary gauge theory. You are not allowed to assume the existence of a metric, either as a background or as a dynamical field. You have to work only with a gauge field. How close would you come to general relativity?*

The answer is that the simplest guess as to how to do this lands right on the nose on general relativity. Here is how this goes:

It turns out to be most direct to reason in the Hamiltonian language. From this point of view a spacetime is a manifold of the form $\Sigma \times R$, where Σ is a three dimensional manifold which will represent space, at least topologically.

From a hamiltonian perspective the fields we are allowed are a connection, A_a^i, and its conjugate momentum, E_i^a where a is a $3d$ spatial index and i is valued in a lie algebra, G.

Thus, we have the Poisson brackets,

$$\{A_a^i(x), E_j^b(y)\} = \delta_a^b \delta_j^i \delta^3(y,x) \tag{6}$$

We know that in a hamiltonian formulation of a gauge theory there is one constraint for each independent gauge transform [41]. The gauge invariances of a gravitational theory include at least 4 diffeomorphisms, per point. Thus,

$$I^{GR} = \int dt \int_\Sigma E^{ai}\dot{A}_{ai} - N\mathcal{H} - N^a H_a - w_i \mathcal{G}^i - h \tag{7}$$

where $\mathcal{H}_a$ generates the diffeomorphisms of Σ, $\mathcal{H}$ must be the so-called Hamiltonian constraint that generates the rest of the diffeomorphism group of the spacetime (and hence changes of the slicings of the spacetime into spatial slices) while $\mathcal{G}^i$ generates the local gauge transformations. h represents the terms in the hamiltonian that are not proportional to constraints. However, there is a special feature of gravitational theories, which is there is no way *locally* to distinguish the changes in the local fields under evolution from their changes under a diffeomorphism that changes the time coordinate. Hence h is always just a boundary term, in a theory of gravity.

From Yang-Mills theory we know that the constraint that generates local gauge transformations under (6) is just Gauss's law

$$\mathcal{G}^i = \mathcal{D}_a E^{ai} = 0 \tag{8}$$

Note that E_i^a is a vector *density*, so there is no metric used in either the Poisson brackets or Gauss's law[8].

Let us now guess the forms of the other constraints. First there must be three constraints per point that generate the diffeomorphisms of the spatial slice. Infinitesimally these will look like coordinate transformations, hence the parameter that gives the infinitesimal change is a vector field. Hence these constraints must multiply a vector field, without using a metric. Thus these constraints are the components of a one form. It should also be invariant under ordinary gauge transformations, as they commute with diffeomorphisms. We can then ask what is the simplest such beast we can make using A_a^i and E_i^a? The answer is obvious, it is

$$\mathcal{H}_a = E_i^b F_{ab}^i = 0 \tag{9}$$

where F_{ab}^i is the Yang-Mills field strength.

[8] One thing to get used to in this field is that as there is no background metric, while in the quantum theory the metric is a composite operator, one must be completely explicit about all places the metric appears and all density weights.

It is a simple exercise to show that $\mathcal{H}_a$ so defined does in fact generate a spatial diffeomorphism (plus an ordinary gauge transformation) on the fields A_a and E^a.

There remains one constraint per point, which generates changes in the time coordinate, or else in the embedding of Σ in $\mathcal{M} = \Sigma \times R$. This is called the Hamiltonian constraint. Since its action is locally indistinguishable from the effect of changing the time coordinate, it does contain the dynamics.

The Hamiltonian constraint must be gauge invariant and a scalar, since the parameter it multiplies is proportional to the local change in the time coordinate. But it could also be a density, so we have the freedom to find the simplest expression that is a density of some weight. It turns out there are no polynomials in our fields that have density weight zero, without using a metric. But there are simple expressions that have density weight two. The two simplest such terms that can be written, which are lowest order in derivatives, are,

$$\epsilon_{abc} Tr E^a E^b E^c \quad \text{and} \quad Tr E^a E^b F_{ab} \tag{10}$$

where the Tr is in the lie algebra G. If we need to we could go to terms with more derivatives, but such terms will give trouble if we want the theory to have a simple linearization, which will be useful to reproduce Newtonian gravity and gravitational waves.

In fact these two terms already give Einstein's equations, so long as we take the simplest nontrivial choice for G, which is $SU(2)$.

Thus, we take for the Hamiltonian constraint

$$\mathcal{H} = \epsilon_{ijk} E^{ai} E^{bj} \left(F^k_{ab} + \frac{\Lambda}{3} \epsilon_{abc} E^{ck} \right) = 0 \tag{11}$$

There is a place to put a free parameter Λ which indeed will turn out to be the cosmological constant. As far as dimensions are concerned, A_a is a connection and so has dimensions of inverse length. It will turn out that E^a is related to the metric and so we should make the unconventional choice that it is dimensionless.

In fact, what we have here is Euclidean general relativity. If we want the Lorentzian theory, we need only modify what we have by putting an $\imath$ into the commutation relations, so we have instead of (6)

$$\{A^i_a(x), E^b_j(y)\} = \imath G \delta^b_a \delta^i_j \delta^3(y, x) \tag{12}$$

I have also inserted a factor of Newton's constant, G, which is necessary if E^a is dimensionless.

The Einstein's equations of course come from taking Poisson brackets with the Hamiltonian, which is a linear combination of constraints,

$$H(N, v^a) = \int_\Sigma N\mathcal{H} + v^a \mathcal{H}_a \tag{13}$$

here N and v^a are related to the usual lapse and shift, which are in turn related to the time-time and time-space components of the metric, respectively. In fact, noting that $\mathcal{H}$ has density weight two, we see that N must have density weight -1. Hence it is of the form of $g_{00}/\sqrt{detq_{ij}}$, where q_{ij} is the spatial metric. (This may seem pedantic but we will use it to good effect in the next section.)

The simplest way to evolve is with zero shift, which corresponds to the space-time components of the metric vanishing. The equations of motion are then,

$$\dot{A}_{ai} = \{A_{ai}, \textstyle\int N\mathcal{H}\} = N\imath G\epsilon_{ijk}E^{bj}(2F^k_{ab} + \Lambda\epsilon_{abc}E^{ck}) \tag{14}$$

$$\dot{E}^{ai} = \{E^{ai}, \textstyle\int N\mathcal{H}\} = \imath G\epsilon^{ijk}\mathcal{D}_b(NE^a_j E^b_k) \tag{15}$$

These equations, together with the seven constraints make a diffeomorphism invariant field theory whose only degrees of freedom are an $SU(2)$ connection and its conjugate electric field. To see that the theory is consistent one should check the constraint algebra, in fact it is first class. One can then count degrees of freedom and find that there are 2 canonical degrees of freedom per point. If one linearizes, one gets right away the laws for the propagation of spin 2 massless fields.

How can this be, when there is no metric in the world our equations describe? In fact there is one, it is hidden in the gauge fields. The theory we have is general relativity, with the following identifications. E^a_i is related to the three metric q_{ab} by

$$det(q)q^{ab} = E^{ai}E^{bj}\delta_{ij} \tag{16}$$

The determinant is there because the expression is a density of weight two.

The $SU(2)$ connection A_a turns out to be, for solutions, the self-dual part of the spacetime connection. For Lorentzian solutions this is complex, and its real and imaginary parts each have a geometrical interpretation.

$$A_{ai} = \text{3d spin connection}_{ai} + \frac{\imath}{\sqrt{q}}K_{ab}E^b_i \tag{17}$$

where K_{ab} is the extrinsic curvature of the 3 manifold Σ embedded in the spacetime, which in turn is essentially the time derivative of the three metric[9].

6 The de Sitter Solution as a Gauge Field

It is not of course obvious to see that the theory we have constructed is in fact Einstein's theory, or where the correspondences I've just mentioned come from. A bit later we will derive these from an action principle. But for now I want to only show how the de Sitter solution fits into this framework.

[9] For more details on the canonical formulation of GR in Ashtekar-Sen variables, see [38, 39] as well as the books [2].

We begin by noting that a family of solutions to the constraints can be read off immediately, by inspection. These are those that satisfy,

$$F^i_{ab} = -\frac{\Lambda}{3}\epsilon_{abc}E^{ci} \tag{18}$$

It is easy to see that they satisfy all seven[10] constraints. We call these *self-dual solutions* as they have the magnetic fields proportional to the electric fields.

Let us examine the simplest one of these. We can take as an ansatz that A^i_a is proportional to δ^i_a. Of course this breaks gauge invariance but this is what we have to do if we want to write an explicit solution.

As I know the answer, I will put in the right parameters:

$$A_{ai} = \imath\sqrt{\frac{\Lambda}{3}}\ f(t)\ \delta_{ai} \mapsto F_{abi} = -f^2(t)\ \frac{\Lambda}{3}\epsilon_{abi} \tag{19}$$

where $f(t)$ is a function of the time coordinates.

Taking A^i_a to be purely imaginary makes sense in light of (17), it means that we are making an ansatz that the three geometry is flat, so the three dimensional spin connection vanishes. The metric can then be taken to be homogeneous, as must also be its time derivative, which is the imaginary part of A^i_a.

By the self-dual initial conditions we see that

$$E^{ai} = f^2\delta^{ai} \quad \mapsto \quad q_{ab} = f^2\delta_{ab} \tag{20}$$

As we have satisfied the self-dual condition all the constraints are satisfied. We merely have to plug into the equations of motion (14, 15) to find the evolution equations for f. Both equations of motion agree that

$$\dot{f} = N\sqrt{\frac{\Lambda}{3}}f^4 \tag{21}$$

Remembering that N is an inverse density, we should take $\mapsto$ $N \approx det(q)^{-1/2} = f^{-3}$. This gives us

$$\dot{f} = N\sqrt{\frac{\Lambda}{3}}f^4 = \sqrt{\frac{\Lambda}{3}}f \tag{22}$$

so that $f = e^{\sqrt{\frac{\Lambda}{3}}t}$.

With the identifications we have made this gives the de Sitter spacetime in spatially flat coordinates[11]:

$$ds^2 = -dt^2 + e^{2\sqrt{\frac{\Lambda}{3}}\ t}(dx^a)^2 \tag{23}$$

[10] 3 generate spatial diffeo's, three generate $SU(2)$ gauge transformations plus the Hamiltonian constraint.

[11] A good review of the different coordinations of de Sitter spacetime is in [78].

7 Hamilton-Jacobi Theory, de Sitter Spacetime and Chern-Simons Theory

Before we get serious and go back to the action and show why this is really Einstein's theory, there is one more simple trick we can do, which brings to light a connection between de Sitter spacetime and topological field theory.

To see this connection we may begin by asking what insight Hamilton-Jacobi theory may throw on the solutions we have been considering. To use Hamilton-Jacobi theory we assume that there is a Hamilton-Jacobi functional $S(A)$ on the configuration space. As we are studying a gauge theory the configuration space is the space of the connections A_a on the three manifold Σ.

The conjugate electric field must then be the gradient of the Hamiltonian-Jacobi function,

$$E^{ai} = -\frac{\delta S(A)}{\delta A_{ai}} \tag{24}$$

We found that all seven constraints are solved with the self-dual ansatz (18). This means that the Hamilton-Jacobi function must satisfy a first order differential equation,

$$F^i_{ab} = -\frac{\Lambda}{3}\epsilon_{abc}E^{ci} = \frac{\Lambda}{3}\epsilon_{abc}\frac{\delta S(A)}{\delta A_{ci}} \tag{25}$$

This integrates immediately to

$$S_{CS} = \frac{2}{3\Lambda}\int Y_{CS} \tag{26}$$

Here Y_{CS} is the famous Chern-Simons invariant, given by

$$Y_{CS} = \frac{1}{2}Tr\left(A \wedge dA + \frac{2}{3}A^3\right) . \tag{27}$$

It satisfies $\frac{\delta \int Y_{CS}}{\delta A_{ai}} = 2\epsilon^{abc}F^i_{bc}$.

Thus, *the self-dual solutions follow trajectories in configuration space which are gradients of the Chern-Simons invariant.* Not only is de Sitter spacetime one of these, there is the remarkable fact that, while there are an infinite number of self-dual solutions for Euclidean signature, there is only one for Lorentzian signature and it is de Sitter spacetime.

This suggests that the semiclassical state that describes de Sitter is

$$\Psi_K(A) = \mathcal{N}e^{\frac{3}{2\lambda}\int Y_{CS}} \tag{28}$$

$\mathcal{N}$ is a normalization depending only on topology [40].

In fact, *this is an exact quantum state* as was shown in 1990 by Hideo Kodama [35]. We will return to the Kodama state and the physics that may be derived from it. But first we want to go back and find out why what we have been studying is Einstein's general theory of relativity.

8 General Relativity as a Constrained Topological Field Theory

In the last sections a very mysterious fact emerged, which is that when general relativity is written in such a way as to bring it close to gauge theory, in terms of field content and geometry, we fell upon a close relationship between an important set of solutions-the self-dual solutions, and a topological field theory. Given the ease by which topological field theories may be quantized and studied, as well as their remarkable connections with various fields of algebra, representation theory and topology, it is very important to know if this is an accident or if it has its roots in some deep relationship between gravity and topological field theory. In this section we will show that it is indeed no accident and that general relativity and topological field theory are deeply connected at the level of the action principle.

BF Theory

We begin with a four dimensional topological field theory called BF theory [79]. We will work on a four manifold $\mathcal{M} = \Sigma \times R$, where Σ will be the spatial topology. There is no metric, and no other fixed background field.

We introduce now two fields. The first is an $SU(2)$ connection A^i_μ, where μ indices the spatial coordinates (to be suppressed when we use form notation and $i = 1, 2, 3$ label the generators of $SU(2)$. The second field is a two form, $B^i_{\mu\nu}$ which is also valued in the $SU(2)$ generators. To begin with we take them both real.

The action we use is,

$$I^{BF} = \int B^i \wedge F_i + \frac{\Lambda}{2} B^i \wedge B_i \ . \tag{29}$$

It is easy to derive the equations of motion,

$$F^i = -\Lambda B^i \ , \qquad \mathcal{D} \wedge B^i = 0 \tag{30}$$

We see that the curvature is constrained to be proportional to the B field, with Λ the constant of proportionality. B^i is in turn is constrained to be covariantly constant. If one counts one finds there are no local degrees of freedom, hence the theory is topological. It is also invariant under $Diff(\mathcal{M})$, the group of diffeomorphisms of the manifold. Because of the form of the action, this topological field theory is called BF theory.

Self-Dualology

General relativity is in fact closely related to BF theory. To see this, we need first to understand the dynamics of general relativity in 4 spacetime dimensions in terms of self-dual and antiself-dual connections and curvatures.

Let us then have a four dimensional spacetime metric $g_{\mu\nu}$. We will at first take the spacetime to be Euclidean, then we will see how things are modified for the Lorentzian case.

It is convenient to work with frame fields, e^a_μ, with $a = 0,1,2,3 = 0,i$ being four dimensional frame field indices. They are related to the metric by

$$g_{\mu\nu} = e^a_\mu e^b_\nu \eta_{ab} \tag{31}$$

with η_{ab} the flat metric on the tangent space.

Now we need do a little *self-dualology.* Let us consider an antisymmetric tensor A_{ab} in the tangent space. Given the totally antisymmetric ϵ^{abcd} and the metric η^{ab} we may divide A_{ab} into its self-dual and antiself-dual parts

$$A^{\pm}_{ab} = \frac{1}{2}\left(A_{ab} \pm A^*_{ab}\right) \tag{32}$$

where $A^*_{ab} = \frac{1}{2}\epsilon_{ab}{}^{cd}A_{cd}$. We have

$$(A^{\pm}_{ab})^* = \pm A^{\pm}_{ab} \tag{33}$$

Note that these equations are consistent with $** = +1$, which is the case for Euclidean signature.

Among the objects that can be decomposed this way are the spin connection one form A_{ab} and the curvature two form $F_{ab} = dA_{ab} + \frac{1}{2}A_a{}^c A_{bc}$. These are valued in the $SO(4)$ lie algebra. The decomposition of A_{ab} into A^+_{ab} and A^-_{ab} corresponds to the Lie algebra identity $SO(4) = SO(3)_L \oplus SO(3)_R$. There are then three generators (per form index) in A^+_{ab} and they correspond to $SO(3)_L$. These three generators may then be labelled by $i = 1,2,3$ by the correspondence $A^+_i = A^+_{0i} = \frac{1}{2}\epsilon_i{}^{jk}A^+_{jk}$.

It is important to note that F^+_i, being also valued in $SO(3)_L$ is an $SO(3)$ gauge field which is a function only of the $SO(3)_L$ connection A^+_i.

It turns out that not only can the connection and curvature information in a four manifold be decomposed in self-dual and antiself-dual parts, the same is true for the metric information. Given the metric $g_{\mu\nu}$ one can construct three two forms from the self-dual parts of $e^a \wedge e^b$, as

$$\Sigma^i = e^0 \wedge e^i + \epsilon^i{}_{jk} e^j \wedge e^k \tag{34}$$

These forms are self-dual by construction in the internal indices. Each of the three is also *self-dual in the spacetime sense*

$$^*\Sigma^i_{\mu\nu} \equiv \epsilon_{\mu\nu\lambda\sigma} g^{\lambda\alpha} g^{\sigma\beta} \Sigma^i_{\mu\nu} = \Sigma^i_{\alpha\beta} \tag{35}$$

From Self-Dual two forms to General Relativity

The connection of general relativity to BF theory comes about by identifying the $SO(3)$ valued B^i fields, which are three two forms, with the self-dual two forms Σ^i corresponding to some metric $g_{\mu\nu}$. Let us see how this works.

To make the correspondence we cannot just plug

$$B^i = \Sigma^i = e^0 \wedge e^i + \epsilon^i{}_{jk} e^j \wedge e^k \tag{36}$$

into the equations of motion, (30), as the restriction (36) reduces the number of degrees of freedom per point, and there are already zero degrees of freedom per point. But we can plug (36) into the action (29) for BF theory to find that,

$$I^{JSS} = I^{BF}|_{B^i=\Sigma^i} = \int (e^0 \wedge e^i + \epsilon^i{}_{jk} e^j \wedge e^k) \wedge F_i + \frac{\Lambda}{2} \epsilon_{abcd} e^a \wedge e^b \wedge e^c \wedge e^d \tag{37}$$

This is actually an action for general relativity [50]. In fact it is easy to see that it gives the same equations of motion as the Palatini action

$$I^{Palatini} = \int \epsilon_{abcd} \left(e^a \wedge e^b \wedge F^{cd} + \frac{\Lambda}{2} e^a \wedge e^b \wedge e^c \wedge e^d \right) \tag{38}$$

Using the projections into the self-dual and antiself-dual parts of the curvature, our strange looking action (37) can be written as,

$$I^{JSS} = \int \epsilon_{abcd} \left(e^a \wedge e^b \wedge F^{+cd} + \frac{\Lambda}{2} e^a \wedge e^b \wedge e^c \wedge e^d \right) \tag{39}$$

The equations of motion that come from varying the self-dual part of the connection, A_i^+ are

$$(\mathcal{D}^+ \Sigma)^i = 0 \tag{40}$$

These three equations are in fact the self-dual projection of the six equation of motion that corresponds to varying the Palatini action by the full $SO(3) \oplus SO(3)$ connection, A^{ab} to find,

$$\nabla e^a \wedge e^b = 0 \tag{41}$$

It is well known that the solution to this last (41) is that A^{ab} is equal to the $SO(4)$ spin connection, ω^{ab} corresponding to the frame field e^a. The solution to the equations of motion of the modified action (40) are similar, they are that A_i^+ is equal to ω^{+ab}, which is the self-dual part of the spin connection of e^a.

The other equation of motion of the Palatini equation is, with the connection taken to be the spin connection, the Einstein equations,

$$\epsilon_{abcd} \left(e^b \wedge F^{cd} + \Lambda e^a \wedge e^b \wedge e^c \wedge e^d \right) = 0 \,. \tag{42}$$

The equation of motion of the modified equation is instead

$$\epsilon_{abcd} \left(e^b \wedge F^{+cd}(\omega^+) + \Lambda e^a \wedge e^b \wedge e^c \wedge e^d \right) = 0 \tag{43}$$

This differs from the Einstein equation (42) by a single term, which is

$$e_c \wedge F^{cd}(\omega) \,, \tag{44}$$

but this vanishes by the Bianchi identity that sets $R_{\mu[\nu\lambda\sigma]} = 0$.

This establishes the equivalence of (39) to general relativity with Euclidean signature[12].

Back to BF Theory

We are not yet done for as discovered, first by Plebanski [51], and then by Capovilla, Dell and Jacobson [52], we can use what we have just learned to put the action in a form that shows a direct relationship to BF theory. To do this we ask whether there are conditions on the two form fields B^i which are sufficient for there to exist a metric, and hence a frame field, e^a such that B^i are the self-dual two forms of e^a. The answer is yes, these are the five equations

$$B^i \wedge B^j = \frac{1}{3}\delta^{ij} B^k \wedge B_k \tag{45}$$

This is easy to see one way, by plugging in, for the other, see [52].

Why five equations? There are 18 components in the B^i's minus three gauge degrees of freedom for the $SO(3)$ rotations that mix them up, minus five equations yields the 10 components of the metric $g_{\mu\nu}$.

Thus, general relativity is the consequence of varying the BF action with the B^i fields subject to the five constraints, (45). Thus, if we add these constraints times lagrange multipliers to the BF action, we get an action for general relativity in the form,

$$I^{Plebanski} = \int B^i \wedge F_i + \frac{\Lambda}{2} B^i \wedge B_i - \frac{1}{2}\phi_{ij} B^i \wedge B^j \tag{46}$$

so long as ϕ_{ij} itself is constrained to be symmetric and tracefree.

Actually we can incorporate the cosmological constant term by requiring instead that

$$\phi_i{}^i = -\Lambda; \qquad \phi_{[ij]} = 0 \tag{47}$$

so that the action is now,

$$I^{Plebanski} = \int B^i \wedge F_i - \frac{1}{2}\phi_{ij} B^i \wedge B^j \tag{48}$$

Although we will not need it in what follows, it is interesting to note that since the action is quadratic in the B^i these can be integrated out (or solved for) to find an even simpler form for the action

---

[12] Except that, as in the Palatini case, the fact that the action and equations of motion are polynomial means there are solutions when $det(e) = 0$ that would not be non-degenerate solutions of general relativity. Thus the space of solutions has been expanded by the addition of a kind of boundary that includes solutions with degenerate frame fields.

$$I^{CDJ} = \int F^i \wedge F^j (\phi^{-1})_{ij} \tag{49}$$

Thus, we see that an action can be written for general relativity with a non-zero cosmological constant, *in which the metric does not appear at all.* All that appears is the curvature of the left handed part of the spin connection, and a new field ϕ_{ij}, whose trace is constrained by (47) to be the cosmological constant. The metric is instead a *composite field*, which arises only for solutions of the equations of motion.

And what is the physical interpretation of the new field ϕ_{ij}? To answer this we need only look at the field equation gotten from varying B^i in the Plebanski action, (48).

$$F^i = \phi^i{}_j B^i \tag{50}$$

Since we learn by varying ϕ that there exists a metric, whose self-dual two form B^i becomes, we learn that the ϕ_{ij} are just the components of the self-dual half of the curvature two form, when expanded in components of the frame fields, or equivalently, directly in terms of the self-dual two forms of the metric. So the action (49) codes for the metric in the backhanded way that the ϕ_{ij} have to turn out to be the components of the curvatures F^i expanded in frame field components of that metric. Very sneaky, but effective, as we shall see.

The Same Thing with Lorentzian Signature

So far everything was presented assuming the metric has Euclidean signature. But for real physics we need the metric to be Lorentzian.

The same steps yield a connection between Lorentzian general relativity and BF theory, but it is a bit more complicated because all the fields become complex. To understand this it is best to proceed in two steps. First, we go back and fix the definition of self-dual fields. This is necessary because, as may be easily checked, for Lorentzian signature $** = -1$. To accommodate this, we must insert an $\imath$ into the definition of self-dual tensors,

$$A^{\pm}_{ab} = \frac{1}{2}\left(A_{ab} \pm \imath A^*_{ab}\right) \tag{51}$$

Thus, we now have

$$(A^{\pm}_{ab})^* = \pm \imath A^{\pm}_{ab} \tag{52}$$

This means that the self-dual connection and curvature components A^+_{ab} and F^+_{ab} are now complex. That is, the left handed part of an $SO(3,1)$ connection is really a complex one form valued in the complexification of $SO(3)$.

Due to the $\imath$ in (51) the action now has the form,

$$I^{Lorentz} = \imath \int B^i \wedge F_i - \frac{1}{2}\phi_{ij} B^i \wedge B^j \tag{53}$$

One may wonder whether the fact that A^+ and A^- are both complex has doubled the degrees of freedom. The answer depends on whether or not we want the spacetime frame fields e^a to be real. If we don't require the frame fields to be real then we have extended the theory to allow all solutions of Einstein's equations where the metric is complex. In this case we have doubled the number of degrees of freedom. However, if we want the metric, and the frame field components to be real than there is a restriction on the self-dual and antiself-dual components, coming from the fact that the spin connection of the metric is real. Thus, we have

$$\bar{A}_i^- = A_i^+ \tag{54}$$

This is an important difference from the Euclidean theory, in which A^+ and A^- are independent, but both real.

Nevertheless, we can proceed as follows. We can consider the JSS action, (39) for the case of real e^a but complex A_i^+. The equations of motion then still constrain the A_i^+ to be the self-dual parts of the spin connection and since the e^a are assumed real we still get the real Lorentzian Einstein equations from (37). The only difference is there should be an $\imath$ in front of the whole action.

The next stage is to eliminate the metric completely from the action, by going to the Plebanski action (48). Here there is no simple way to incorporate the condition that the metric is real and Lorentzian. The problem is that the ϕ_{ij}'s are complex for real, Lorentzian metrics. The simplest thing to do seems to be to simply consider the actions (48) and (49) for complex fields A^i and ϕ_{ij}. One then gets the complexified Lorentzian Einstein equations. One can then add to the field equations the initial conditions that the frame field or metric components are real. The field equations are complex, but they have the property that restricted to initial data in which the metric is real, the solutions will conserve the reality of the metric.

This may seem a strange thing to do, but from the point of view of the quantum theory it is not so bad to split the field equations into polynomial equations, which are complex, plus reality conditions on the fields. The reason is that in quantum theory the reality conditions become hermiticity conditions on operators, and these are different from the operator equations of motion, in that they involve the inner product. Strictly speaking, in quantum theory one always works with the complexified operator algebra, and imposes reality conditions through the choice of the inner product. So to do the same in quantum gravity is not very much of an innovation[13]. This is the strategy we will take up when we study the Lorentzian theory.

[13] For this reason, in the early days of loop quantum gravity the strategy of expressing the reality conditions on the metric only through the inner product, while working with a complex self-dual connection, seemed a good one as it greatly simplified the dynamics and led to many new results. More recently another alternative was adopted in many calculations, in which one worked with another $SO(3)$ connection, which is real, invented by Barbero [53]. This leads to more complicated constraint equations which, however, Thiemann showed were still

Self-Dual Spacetimes and the de Sitter Solutions

The equations of motion gotten from the Palatini action are,

$$\mathcal{D} \wedge B = 0 \, , \qquad F^i = \phi^i_j B^j \tag{55}$$

We can see immediately the self-dual solutions.

$$\Phi^i_j = -\frac{\Lambda}{3}\delta^i_j \quad \rightarrow \quad F^i = -\frac{\Lambda}{3}B^i \tag{56}$$

This shows us how the de Sitter and self-dual solutions we obtained from the Hamiltonian picture may be obtained directly as solutions to the Euler-Lagrange equations.

Derivation of the Hamiltonian Formalism

Finally, we can briefly sketch how the constraints of the hamiltonian that we guessed in Sect. 5 above are derived from the forms of the action we have just described. It is easiest to work with the CDJ form of the action (49).

We begin by finding the canonical momenta, which is

$$E^{ai} = \epsilon^{abc} F^j_{bc} (\phi^{-1})^i_j \tag{57}$$

with the canonical momenta for A^i_0, E^{0i} of course vanishing.

The action can then be written as

$$I^{CDJ} = \imath \int dt \int_\Sigma E^{ai} \dot{A}_{ai} - A^i_0 \mathcal{G}_i \tag{58}$$

where the $\imath$ is there only for the Lorentzian case. Because of this the Poisson brackets for the Lorentzian case have the form (12) for the Lorentzian case and (6) for the Euclidean theory.

The Gauss's law constraint (8) then holds to preserve the vanishing of E^{0i}. However there are additional constraints, which arise from the fact that ϕ_{ij} is itself subject to constraints, (47), being symmetric and having trace fixed to be the cosmological constant. These must be imposed to recover the equations of motion, because without them the relationship between $\dot{A}_{ai}$ and the momenta cannot be inverted.

It is easy to check that the constraints that arise from the antisymmetric part of ϕ_{ij} vanishing is the diffeo constraint, (9), while the constraint that arises from its trace being fixed is the Hamiltonian constraint (11). Thus we arrive at the hamiltonian formulation we developed by guess work in Sect. 5.

---

manageable [54]. However, this strategy does not help in the case of the results presented here, and so is not adopted in this paper.

9 Boundaries with $\Lambda > 0$ and Chern-Simons Theory

There are several reasons we will want to consider spacetimes with boundaries. These include the important subjects of how we realize the Bekenstein bound, study the entropy of black hole and cosmological horizons and express the holographic principle. Depending on the context these boundaries will be null, as at horizons, or timelike as in the boundary of *AdS* spacetimes or even Euclidean, if we are working in that context. These boundaries may be at infinity, or they may have finite area.

Before we can study the quantum theory with boundaries we have to understand the role boundaries play in the classical theory. Generally when there is a boundary we will not have a sensible variational principle unless the theory is modified to take the boundary into account. Normally these modifications consist of two parts. We have to add a boundary term to the action, which just depends on the fields pulled back into the boundary. And we have to add boundary conditions. Both the boundary action and boundary conditions must be chosen so that the variations of the actions by the fields are pure bulk terms, so that the equations of motion are sensible.

Here we will consider a region of spacetime with topology $\mathcal{M} = \Sigma \times R$ with a boundary $\partial\Sigma = \mathcal{B}$. We will study only one particular class of boundary conditions, which are called *self-dual boundary conditions [27]*.

The basic idea of these boundary conditions is to require that at the boundary some components of the fields satisfy the self-dual relations (18) which define the de Sitter (or with $\Lambda < 0$ anti-de Sitter) spacetime [27]. We cannot require all the components satisfy the self-dual conditions, otherwise only self-dual solutions will be allowed. But we can get interesting boundary conditions by requiring only a subset of the components satisfy the self-dual relations (18) when pulled back into the boundary.

We will consider cases in which the spatial components of the self-dual relation, pulled back into the boundary are satisfied, in at least one spatial slicing of the boundary. Thus, we impose,

$$F^i_{ab}|_{\mathcal{B}} = -\frac{\Lambda}{3}\epsilon_{abc}E^{ci}|_{\mathcal{B}} \tag{59}$$

There may be other components of the boundary condition, imposed on the timelike components of the boundary fields, for details about this in the Euclidean case see [27], in the timelike case see [44] in the null case see [55, 56].

To complement the boundary condition we must add a boundary term to the action. The natural one to add turns out to be the Chern-Simons action of the connection A^i_a pulled back into the boundary [27]. The action then reads,

$$I^{GR} = \epsilon \int_{\mathcal{M}} B^i \wedge F_i + \phi_{ij} B^i \wedge B^j + \frac{\epsilon k}{4\pi}\int_{\partial\mathcal{M}} Y_{CS}(A) \tag{60}$$

where, from now on, $\epsilon = \imath$ for the Lorentzian theory and unity for the Euclidean theory.

There is in both cases a relation between the coupling constant, k of the Chern-Simons theory and the cosmological constant.

$$k = \frac{6\pi}{\lambda}, \qquad \lambda = \hbar G \Lambda \tag{61}$$

This ends our study of the classical physics we need to know to understand the quantum theory of gravity with $\Lambda > 0$. The key lesson of this survey is the connection to topological field theory, which we have seen arises three ways in the classical theory:

- The action for GR has the form of a constrained topological field theory.
- There is a natural class of boundary terms which require that the boundary term added to the action is the Chern-Simons action of the left handed spin connection, pulled back to the boundary.
- The deSitter and other self-dual solutions follow gradients of the Chern-Simons invariant, which can then be taken as the Hamilton-Jacobi function.

10 The Kodama State

We begin with a very brief review of how diffeomorphism invariant theories are to be quantized in the Hamiltonian approach. For more details on the basic approach, see [2, 57, 58, 59]. We do not here describe path integral methods in loop quantum gravity, but they are well developed, see, for example, [60, 61, 62, 63, 64, 65, 66].

10.1 A Brief Review of Quantization

The approach taken here is Dirac quantization. This means that the whole unconstrained configuration space is quantized. This defines a *kinematical state space* $H^{kinematical}$. The constraints are imposed as operator relations on the states, as in

$$\hat{C}|\Psi\rangle = 0 \tag{62}$$

where $\hat{C}$ stands for operators representing all the first class constraints of the theory. The solutions to the constraints define subspaces of the Hilbert space. A physical state must be a simultaneous solution to all the constraints.

Often this is done in two steps. The kernel of the gauge and spatial diffeomorphism constraints is called the diffeo-invariant Hilbert space, and is labelled H^{diffeo}. The simultaneous kernel of all the constraints is called the physical Hilbert space, $H^{physical}$.

Generally, new inner products need to be introduced on these Hilbert spaces, because solutions to the constraints are not normalizable in the inner products on the kinematical Hilbert space.

We will work in this and the next four chapters with the connection representation of quantum gravity [57]. After this we will switch to the loop (or spin network) representation. As in the case of the position and momentum representations in ordinary quantum mechanics these are equivalent, but complementary, in that certain calculations are easier to do in one representation than another.

Both representations are defined as representations of a certain algebra of classical observables.

From a naive point of view, we could take the canonical commutation relations (6) as the basis of the quantization. Thus, we heuristically define the connection representation by the relations

$$\langle A|\Psi\rangle = \Psi(A) \quad E^{ai} = -\hbar G \frac{\delta}{\delta A_{ai}} \tag{63}$$

These satisfy the commutation relations,

$$[A_a^i(x), E_j^b(y)] = \hbar G \delta_a^b \delta_j^i \delta^3(y,x) \tag{64}$$

Note that because there is an $\imath$ in the classical commutation relation (6) no $\imath$ appears here[14]. Unless explicitly mentioned, from now on we are working with the Lorentzian theory.

However, to discuss carefully the regularization of the operators that define the theory, we need to define the quantization in terms of a different set of observables, which are the Wilson loops

$$T[\gamma, A] = TrPe^{\int_\gamma A} \tag{65}$$

in the fundamental, spin 1/2 representation, and the elements of area,

$$A[\mathcal{S}] = \int_{\mathcal{S}} \sqrt{h} \tag{66}$$

where $\mathcal{S}$ is a surface in Σ and h is the determinant of the induced metric in the surface. These have very beautiful Poisson bracket relations,

$$\{T[\gamma, A], A[\mathcal{S}]\} = \hbar G Int[\gamma, \mathcal{S}] T[\gamma, A] \tag{67}$$

where $Int[\gamma, \mathcal{S}]$ is the intersection number of the loop and the surface.

For the definitions of the connection and loop representations in terms of this algebra, see [58]. Here we will work with naive operators and mostly neglect the details of regularization procedures, which can be found in the references. However, it is very important to stress that everything said here does go through when all the details of the regularization procedures are included.

We now need the expressions of the constraints in the connection representation. These are

[14] Were we working instead with the Euclidean theory there would be an $\imath$ here.

– Gauss's law:

$$\mathcal{G}^i\Psi(A) = \mathcal{D}_a \frac{\delta}{\delta A_{ai}} \Psi(A) = 0 \tag{68}$$

– Diffeomorphism constraint

$$\mathcal{H}_a\Psi(A) = F^i_{ab} \frac{\delta}{\delta A_{bi}} \Psi(A) = 0 \tag{69}$$

– Hamiltonian constraint:

$$\mathcal{H}\Psi(A) = \epsilon_{ijk} \frac{\delta}{\delta A_{ai}} \frac{\delta}{\delta A_{bj}} \left(F^k_{ab} + \frac{\lambda}{3} \epsilon_{abc} \frac{\delta}{\delta A_{ck}} \right) \Psi(A) = 0 \tag{70}$$

Note that with the ordering given here, the quantum algebra of constraints can be shown, after a suitable regularization procedure, to be consistent [2, 39, 57, 67]. This means that the commutators give terms proportional to operators, which are of the form of (new operator) × operator constraints. Thus, there are a non-trivial space of states in the simultaneous kernel of all the constraints. In fact, infinite dimensional spaces of simultaneous solutions to all the regularized constraints have been found and studied [54, 58].

For details of the different regularization procedures that can be applied to define these constraints, and the infinite dimensional families of solutions that have been found, see [54, 58, 67].

For the time being, we will be concerned with the case that Σ is compact, and without boundary. A bit later we will show how boundaries are included in the quantum theory.

10.2 The Kodama State

We will be concerned first of all with a particular simultaneous solution to all the constraints, which is the Kodama state we first introduced in Sect. 7. This is the Kodama state, defined by [35]

$$\Psi_K(A) = \mathcal{N} e^{\frac{3}{2\lambda} \int Y_{CS}} \tag{71}$$

To show that this is a solution to all the constraints, one makes use of the identity,

$$\frac{\delta \Psi_K(A)}{\delta A_{ck}} = \frac{3}{2\lambda} \epsilon^{abc} F^i_{ab} \Psi_K(A) \tag{72}$$

Thus, the Kodama state is in the kernel of the operator

$$\mathcal{J}^i_{ab} = F^k_{ab} + \frac{\lambda}{3} \epsilon_{abc} \frac{\delta}{\delta A_{ck}} \tag{73}$$

and satisfies the Hamiltonian constraint because we have chosen an ordering such that

$$\mathcal{H} = \epsilon_{ijk} \frac{\delta}{\delta A_{ai}} \frac{\delta}{\delta A_{bj}} \cdot \mathcal{J}^k_{ab} \tag{74}$$

$\mathcal{J}^i_{ab}$ is of course an operator version of the self-dual condition.

The Kodama state solves the other constraints because it is manifestly invariant under diffeomorphisms of Σ and *small* gauge transformations. (Note that only small gauge transformations are generated by constraints.)

One might think that invariance under large gauge transformations would be achieved because k is an integer. However, this would be wrong, as there is no $\imath$ in the Chern-Simons state.

Invariance under large $SU(2)$ (real) gauge transformations is instead gotten by choosing $\mathcal{N}$ to be a topological invariant also sensitive to them. For details of this, see the paper by Soo [40]. There is a reason for choosing k to be an integer, we will see below.

It will also be important to note that as A_a is complex, so is its Chern-Simons invariant. Hence the Kodama state is complex.

The reader may want to ask a few questions:

- **Does the Kodama state survive the details of a regularization procedure, needed to define the rigorous action of the constraints?** *Yes, for details see [67].*
- **Is the Kodama state normalizable?** *Certainly it is not normalizable in the naive inner product. But this is to be expected, on two grounds. First because solutions to constraints are generally never normalizable in the inner product of the kinematical Hilbert space [2]. Second, because, as we will see below, there are components of the connection that function as a time coordinate on the configuration space [36]. The physical inner product cannot integrate over time, otherwise all energy eigenstates would be non-normalizable. Hence a new physical inner product needs to be chosen, and, given the fact that it satisfies all the physical properties we require of a physical state, it makes sense to take as a condition for the physical inner product that the Chern-Simons state, as well as its perturbations that we see below represent long wave graviton states, are normalizable.*
  We might note that, as has been pointed out by [68, 69], there is an analogue of the Kodama state in Yang-Mills theory, where it fails to be a normalizable state. This does not, however, imply that the Kodama state in quantum gravity must be non-normalizable, for reasons described in [70].
- **Suppose we construct an analogue of the Kodama state for the linearized theory of gravitons on a de Sitter background, by truncating the Chern-Simons invariant to quadratic order in perturbations. Is the result a normalizable state of the linerarized theory?** *The answer is no for the Lorentzian theory, while for the Euclidean theory the state is delta-functional normalizable [70]. Whether this is an indication, however, of a pathology with the state itself, with the linearization, is not yet clear.*

- **But, is the Chern-Simons state really a ground state? Does it really correspond to the vacuum?**
  It is if one can study its weakly coupled excitations and they reproduce quantum field theory in curved spacetime and long wave length gravitons on de Sitter spacetime.
  Note that we need only recover these for low energies in Planck units.
  We show this below for both matter fields and for gravitons.

But first we check that the theory reproduces one of the basic features of quantum theory in de Sitter spacetime, which is that there is a temperature associated with a positive cosmological constant.

11 The Thermal Nature of Quantum Gravity with $\Lambda > 0$

It is well established that quantum field theory on de Sitter spacetime must be interpreted as irreducibly thermal [71], with a temperature given by

$$\mathcal{T} = \frac{1}{2\pi}\sqrt{\frac{\Lambda}{3}} \tag{75}$$

This can be understood as due to the presence of the horizon. Alternatively, one can show that any quantum field on de Sitter spacetime, satisfies the KMS condition for a thermal state. This is that the continuation of any correlation function to imaginary time coordinate $t_E = \imath t$ be periodic, with period $\beta = 1/\mathcal{T}$.

What is not so well known, however, is that in loop quantum gravity the full background independent quantum theory of gravity plus arbitrary matter fields has also an irreducibly thermal nature. This is because it satisfies the KMS condition on the whole configuration space[15].

To apply the KMS condition to quantum gravity we need two things: 1) a definition of a time coordinate on the configuration space and 2) a definition of the continuation to Euclidean time. In a background independent theory we cannot use a time coordinate on a given classical spacetime, as that is just a given classical solution. We need instead a time coordinate on the configuration space of the theory.

We saw above that there is in fact a preferred time coordinate on the configuration space, which is picked out by the semiclassical expansion around the Kodama state. It is equal to the imaginary part of the Chern-Simons invariant, (91). There are other arguments that confirm that (91) is a good time coordinate on the configuration space, for example it is always normal to the gauge directions in the tangent space of the configuration space. For details see [36, 72]. The Chern-Simons time coordinate is dimensionless, as we

[15] The argument of this section is taken from [36].

saw in the last sections when we evaluate it on a given solution we have to scale it appropriately.

It is interesting to note that (Lorentzian) Kodama state can be written

$$\Psi_K(A) = e^{\imath M T_{CS}} e^{\frac{k}{4\pi} \mathcal{R}e \int Y_{CS}(A)} \tag{76}$$

where the dimensionless "energy" is

$$M = \frac{k}{4\pi} = \frac{3}{2\lambda} \tag{77}$$

Now we need a definition of how to continue to Euclidean time. As we are dealing with a theory of spacetime we should continue the whole theory to Euclidean signature. This requires the following changes: The connection A_{ai} becomes a real, $SU(2)$ connection and there is now an $\imath$ in $E^{ai} = \imath\delta/\delta A_{ai}$. As a consequence of which the Chern-Simons state is now,

$$\Psi_K^{Euc}(A) = e^{\frac{\imath k}{4\pi} \int Y_{CS}(A)} \tag{78}$$

The Euclidean time coordinate is then just

$$T_{ECS} = \int Y_{CS}(A) \tag{79}$$

as can be seen directly, or by repeating the derivation from the semiclassical theory. Thus, the Euclidean wavefunction is,

$$\Psi_K^{Euc}(A) = e^{\imath \frac{k}{4\pi} T_{Euc}} \tag{80}$$

This is periodic in T_{Euc}. However, this is not enough to show that the KMS condition is satisfied, for that requires that *every correlation function* be periodic.

Interestingly enough, this is in fact the case whenever Σ is chosen so that $\pi^3(\Sigma)$ is nontrivial. In this case there are large gauge transformations that have the property that

$$\int Y_{CS}(A) \rightarrow \int Y_{CS}(A) + 8\pi^2 n \tag{81}$$

where n is the winding number of the large gauge transformation. This means that $T_{ECS} = \int Y_{CS}(A)$ is actually a periodic function on the configuration space. As a result, *every* correlation function will satisfy the KMS condition in T_{ECS}, no matter what the state. That is, by equating configurations of A_{ai} that differ by a large gauge transformations we reduce the topology of the configuration space to a circle, which is parameterized by T_{ECS}.

As a result of this universal periodicity there is a temperature, given in dimensionless units by $\mathcal{T}_{dimless} = \frac{1}{8\pi^2}$. This dimensionless temperature corresponds to the fact that the time coordinate on the configuration space, T_{CS} is dimensionless.

It is interesting to ask if this dimensionless temperature corresponds to the temperature on de Sitter spacetime. To investigate this we may consider a trajectory in configuration space that corresponds to a slicing of de Sitter spacetime with topology $S^3 \times R$. Such coordinates are given by

$$ds^2 = -\left(1 - \frac{\Lambda r^2}{3}\right) dt^2 + \frac{1}{\left(1 - \frac{\Lambda r^2}{3}\right)} dr^2 + d\Omega^2 \tag{82}$$

To work out the scaling of the coordinate t on the solution with the coordinate T_{CS} on the configuration space, we compute

$$\frac{\partial T_{CS}}{\partial t} = \int_{S^3} N\{T_{CS}(A), \mathcal{H}\} \tag{83}$$

where the (densitized) lapse N is read off from the solution (82). A simple calculation gives

$$\frac{\partial T_{CS}}{\partial t} = 4\pi \sqrt{\frac{\Lambda}{3}} \tag{84}$$

Thus, if the Euclidean continuation T_{ECS} is periodic with period $8\pi^2$, the Euclidean continuation of the time coordinate on the solution must be periodic with period $2\pi\sqrt{\frac{3}{\Lambda}}$. In fact, this is the periodicity of the Euclidean de Sitter solution, in these coordinates! This leads to the temperature of de Sitter spacetime, (75).

Thus, we learn that *the periodicity of the Euclidean de Sitter spacetime is a consequence of that spacetime having an interpretation as a trajectory on the configuration space of $SU(2)$ connections.* The periodicity of the Euclidean Schwarzschild solution is a consequence of the fact that the whole configuration space is periodic due to the action of the large gauge transformations. This is yet another connection between the properties of the gauge theory and the physics of gravitation. Thus, the thermal nature of quantum field theory on de Sitter spacetime is a consequence of a deeper and more general result, which is that the whole quantum theory with $\Lambda > 0$ is thermal.

Finally, we can deduce one more fact from these considerations. For the analysis we have just given to be relevant to the Kodama state, it must be that the Euclidean Kodama state is itself well defined under large gauge transformations. This will only be the case if k is an integer.

12 The Recovery of QFT on de Sitter Spacetime

Now that we have verified that the thermal properties of de Sitter spacetime are extended to the quantum gravity domain, we can go on to examine the properties of the Kodama state. Here we will be mostly concerned with its properties at the semiclassical level.

The first thing we can do to probe the Kodama state is to add matter, and then see what happens if we excite the matter in the presence of the state[16].

Adding matter fields is straightforward. In the language of loop quantum gravity it is simple to add all kinds of matter: gauge fields, fermions, scalars, antisymmetric tensor gauge fields[17]. For what we are doing here we do not need any details, so we will refer to all matter fields as ϕ, their canonical momenta as π and the matter hamiltonian as $H^{matter}(\phi, \pi)$.

All the constraints get new terms in the matter fields. For the Hamiltonian constraint we have

$$\mathcal{H}^{total} = \mathcal{H}^{grav}(A, E) + H^{matter}(A, E, \phi, \pi) \tag{85}$$

We will work in an extended connection representation in which the states are functionals $\Psi[A, \phi]$, π is represented by $-\imath\hbar\delta/\delta\phi$ and so forth.

As in the pure gravity case, the gauge and diffeomorphism constraints, applied to the states, require that the states are gauge invariant and invariant under diffeomorphisms of Σ.

This is straightforward, so we focus here on the hamiltonian constraint.

To study perturbations of the Kodama state,we follow the proposal of Banks [75], which is to study the semiclassical approximation in quantum cosmology by a version of the Born-Oppenheim approximation, in which the gravitational degrees of freedom play the role of the heavy, nuclear degrees of freedom, while the matter degrees of freedom play the role of the light, electron degrees of freedom.

Thus, we consider a product state of the form

$$\Psi(A, \phi) = \Psi_K(A)\chi(A, \phi) \tag{86}$$

The exact Hamiltonian constraint is then of the form

$$\left(H^{grav} + H^{matter}\right)\Psi_K(A)\chi(A, \phi) = 0 \tag{87}$$

The idea is to make an approximation to the exact equations, which is described in terms of quantum matter fields propagating on a classical background spacetime (A^0, E^0). This approximation is gotten by expanding the Wheeler-De Witt equation (87) in a neighborhood of a classical solution. We use the fact that the Kodama state can be understood as a WKB state as well as an exact solution. This tells us that the classical background (A^0, E^0) must be de Sitter spacetime, as it is the unique solution gotten by taking S_{CS} to be the Hamiltonian-Jacobi function, consistent with the requirement that the Lorentzian metric be real.

---

[16] The material in this section comes from [36] to which the reader is referred for more details.

[17] There is also no obstacle to extending the theory to supergravity, so long as $\Lambda \leq 0$ in four dimensions [73, 74]. This has been worked out in some detail up to $N = 2$ in four dimensions. For some partial results on $d = 11$ supergravity, the interested reader can see [48].

We will describe the details of this approximation, for the case of a scalar field, below. As a prelude, we mention here the basic features of the results.

As shown in [36], we find that an approximation to (87) takes the form of a Tomonaga-Schwinger equation:

$$\imath \frac{\delta\chi}{\delta\tau_{CS}} = \frac{1}{\Lambda} H^{matter}_{E^{ai}=(3/\Lambda)\epsilon^{abc}F^{i}_{bc}} \chi + O(l_{Pl}E) \tag{88}$$

In this equation, the matter Hamiltonian is evaluated with classical gravitational fields satisfying the self-dual condition $E^{ai} = (3/\Lambda)\epsilon^{abc}F^{i}_{bc}$. As we just said, the reality conditions then tell us that the background is de Sitter. We have neglected higher order terms in $l_{Pl}E$, where E is the energy of the matter fields measured with respect to the background metric.

The approximation procedure picks out a time coordinate called τ_{CS}, related to the Chern-Simons invariant. It is first of all a coordinate on the configuration space of the theory, defined by

$$\delta\tau_{CS}(x) = \frac{1}{2}\mathcal{I}m\ \epsilon^{abc}F^{i}_{bc}\delta A_{ai}(x) \tag{89}$$

Thus, integrated over the spatial manifold Σ, we have

$$\int_{\Sigma} \delta\tau_{CS}(x) = \delta\mathcal{I}m \int Y_{CS}(A) \tag{90}$$

If we take the integral, we can define

$$T_{CS} = \int_{\Sigma} \tau_{CS} = Im \int_{\Sigma} Y_{CS} \tag{91}$$

This can be argued to provide a provides a good global time parameter on the configuration space [36, 72]. This is because its derivative is always orthogonal, in the tangent space of the configuration space, to both the gauge directions and the directions that parameterize the physical degrees of freedom.

When evaluated on a background solution, this gives rise to a time coordinate on the spacetime. One can then show that, to leading order in λ, $Y_{CS} = \imath\sqrt{det(q)}\mathcal{K} + O(\sqrt{\lambda})$, where $\mathcal{K}$ is the trace of the extrinsic curvature K_{ab}. Thus. this choice of time coordinate agrees, to leading order in λ, with that proposed by York [36]. This time coordinate has been shown to have many good properties that an intrinsic time coordinate should have.

Thus, QFT on de Sitter *is* a good approximation to the physics of $\Psi(A,\phi) = \Psi_K(A)\chi(A,\phi)$ when $\lambda = \hbar G\Lambda$ and $l_{Planck}E$ are small. This stands as a first piece of evidence that $\Psi_K(A)$ may be indeed a good ground state.

There are additional terms in $l_{Planck}E$, where E is the matter energy on the de Sitter background. We will study the effect of these terms in Sect. 14.

13 Gravitons from Perturbations Around the Kodama State

To further probe the properties of the Kodama state we should also investigate its gravitational excitations. To do this we return to the case of pure gravity and consider states of the form

$$\Psi[A] = \mathcal{N} e^{\frac{3}{2\lambda}\int Y_{CS} + \lambda S'(A)} \tag{92}$$

It is not difficult to show that there are solutions of this form, and that they do describe long wavelength gravitons moving on the classical background of de Sitter spacetime[18]. But to do this we first need to know how to recognize gravitons in this language. We then detour to summarize the results in this area,

Linearized Gravity on a de Sitter Background

The quantization of linearized general relativity in Sen-Ashtekar variables of the kind we are using was considered early in the study of loop quantum gravity [77]. There a complete description was obtained of gravitons on a Minkowski spacetime background. It is trivial to extend what was done there to gravitons moving on a de Sitter background. As we want results that hold for small λ, it is convenient to use λ as an expansion parameter.

Thus, we expand classical general relativity around the de Sitter background studied above in Sect. 6,

$$A_{ai} = \imath\sqrt{\Lambda} f(t)\delta_{ai} + \lambda a_{ai}; \quad E^{ai} = f^2\delta^{ai} + \lambda e^{ai} \tag{93}$$

It is trivial then to compute the constraints to linear order to find the linearized constraints satisfied by the a_{ai}'s and e^{ai}'s. To solve them we need to impose 7 gauge fixing conditions. A natural set to impose is[19]

$$a_{[ai]} = a^a_a = \partial_a a^a_i = 0 \tag{94}$$

where indices are raised and lowered by the background metric, q^0_{ab}. The simultaneous solution of the linearized constraints and gauge fixing conditions is

$$\partial_a e^a_i = e_{[ai]} = e^a_a = 0\,. \tag{95}$$

The result is that the theory is reduced to a^r's and e^r's that are tracefree, divergence free and symmetric. These are spin two fields.

The linearized poisson brackets can be derived by applying the full Poisson brackets to the linearized fields. This gives,

[18] More details concerning these results will be reported elsewhere [76].

[19] As an exercise one has to check that these seven gauge fixing conditions do together with the linearized constraints make a second class algebra.

$$\{a^r_{ai}(x), e^{bj}_r(y)\} = \imath P^{bj}_{ai} \delta^3(x,y) \tag{96}$$

where P^{bj}_{ai} is the projection operator onto the symmetric, transverse, tracefree fields.

Finally, we have to construct the linearized hamiltonian. This comes from the quadratic terms in the integral of the hamiltonian constraint, and comes out to be,

$$h(a^r, e^r) = f^{-1}\left[\epsilon^{ajk}(\mathcal{D}^0_a a^r_{bk})e^b_{r\ j} + \Lambda e^{ai}_r e_{r\ ia}\right] \tag{97}$$

It is then straightforward to quantize this theory, yielding a quantum theory of gravitons on the background of de Sitter spacetime.

Linearization of the Exact Quantum Theory Agrees with the Quantization of the Linearized Theory, for Long Wavelength

Now we want to go the other way around and study expansions of exact states in powers of λ around the Kodama state. We consider a product state of the form (92) and solve all seven constraints in a neighborhood of the classical trajectory on the configuration space.

The 6 kinematical constraints give:

$$\int_\Sigma (\mathcal{D}_a w)^i \frac{\delta S'}{\delta A_{ai}} = 0; \quad \int_\Sigma v^a F^i_{ab} \frac{\delta S'}{\delta A_{ai}} = 0 \tag{98}$$

where w^i and v^a are arbitrary functions on Σ. These are linearized around the dS background. They are solved by taking $S' = S'(f, a_{ai})$ with a_{ai} symmetric and transverse with respect to the de Sitter background. Thus,

$$\frac{\delta S'}{\delta A_{ai}} = \frac{\imath}{\sqrt{\Lambda}}\delta_{ai}\frac{\delta S'}{\delta f} + \frac{1}{\lambda}\frac{\delta S'}{\delta a_{ai}} \tag{99}$$

The hamiltonian constraint is,

$$\epsilon_{abc}\epsilon_{ijk}\frac{\delta}{\delta A_{ai}}\frac{\delta}{\delta A_{bj}}\left[\frac{\delta S'}{\delta A_{ck}} e^{\frac{3}{2\Lambda}\int Y_{CS} + S'(A)}\right] = 0 \tag{100}$$

Using (99), this can be expanded to give:

$$\imath\frac{\partial S'}{\partial t} = \hat{H}^2 S' + O(l_{Planck}E) + O(\sqrt{\lambda}) \tag{101}$$

where the free Hamiltonian is

$$\hat{H}^2 = \hat{h}\left(\hat{a}, \hat{e} = \frac{\delta}{\delta a}\right) \tag{102}$$

and E is the energy of the graviton state with respect to the background. Thus we conclude that for long wavelength perturbations, but only so long as $l_{Pl}E \ll 1$, the linearized theory is recovered. However it must be emphasized that we have only obtained a correspondence with the standard linearized theory for low energy and small λ.

14 Corrections to Energy Momentum Relations

Now that we have recovered known physics from the Kodama solution, we may go on to see if the theory makes any predictions beyond the recovery of quantum field theory in the semiclassical limit. To see that it may, let us look in detail at the fundamental (87). For simplicity we consider the case of a massless, non-interacting scalar field, although similar conclusions apply for other matter fields[20]. For this case the form of the matter term in the Hamiltonian constraint is

$$H^{matter}(x) = \frac{G\hbar}{2}\left(\pi^2 + (\partial_a\phi)(\partial_b\phi)E^{ai}E^b_i\right) \tag{103}$$

where π is the canonical momentum of the scalar field. Implemented as a quantum operator this is,

$$\hat{H}^{matter}(x)\Psi_K(A)\chi(A,\phi) = \frac{\hbar G}{2}\left(\pi^2 + (\hbar G)^2(\partial_a\phi)(\partial_b\phi)\frac{\delta}{\delta A_{ai}}\frac{\delta}{\delta A^i_b}\right)\Psi_K(A)\chi(A,\phi) \tag{104}$$

In Sect. 12 we recovered quantum field theory in curved spacetime from an approximation to this last expression. In this approximation we considered only the terms in which the factors of $\hat{E}^{ai} = -\hbar G\delta/\delta A_{ai}$ in the second term act on the Kodama state, giving terms proportional to the background frame field , $\delta_{ai}f^2$. Keeping only these terms we have

$$\hat{H}^{matter}(x)\Psi_K(A)\chi(A,\phi) = \frac{1}{2}\left(\pi^2 + \delta^{ab}f^4(\partial_a\phi)(\partial_b\phi)\right)\chi(A,\phi) \tag{105}$$

which is the hamiltonian for the scalar field on the background spacetime.

To go beyond the semiclassical approximation we may then consider the other terms in (104) in which one or both of the functional derivatives by A_{ai} act directly on the perturbed state $\chi(A,\phi)$. These still give terms linear in χ so they may be interpreted as corrections to the functional Schroedinger equation. We will see that these terms give predictions of new physics.

The interpretation of the new terms is easiest when we can neglect the effect of the cosmological constant, and approximate a region of de Sitter spacetime by a region of flat spacetime. To get predictions for the theory in flat spacetime, we proceed in two steps. First we evaluate the approximate solutions to the Wheeler de Witt equation at the background values of the connection and metric we studied in Sect. 6. These are given by (19, 20). We then approach flat spacetime by neglecting terms such as $k^2\Lambda$, where k is the momentum of a particle, which vanish in the limit $\Lambda \to 0$. This is of course a good approximation in the observed situation in which the cosmological constant is non-zero, but very small.

[20] More details concerning the results of this section will appear in [76].

Evaluating the action of $\delta/\delta A_{ai}$ to leading order on $\Psi_K(A)$, we found that,

$$\hat{E}^{ai}(x)\Psi_K(A) = -\hbar G \frac{\delta \Psi_K(A)}{\delta A_{ai}(x)} = f^2(t)\delta_{ai}\Psi_K(A) \tag{106}$$

Thus, the full action of the functional derivatives gives

$$\begin{aligned}
&\frac{(\hbar G)^2}{2}(\partial_a \phi)(\partial_b \phi)\left(\frac{\delta}{\delta A_{ai}}\frac{\delta}{\delta A_b^i}\right)\Psi_K(A)\chi(A,\phi) \\
&= \Psi_K(A)\left\{\frac{f^4}{2}(\partial_a\phi)(\partial_b\phi)\delta^{ab}\chi(A,\phi)\right. \\
&\quad \left. + (\partial_a\phi)(\partial_b\phi)[-\hbar G f^2 \frac{\delta \chi(A,\phi)}{\delta A_{ab}} + \frac{(\hbar G)^2}{2}\frac{\delta^2 \chi(A,\phi)}{\delta A_{ai}\delta A_b^i}]\right\}
\end{aligned} \tag{107}$$

Above we saw that the time derivative in the functional Schroedinger equation came from terms in which a single derivative in $\frac{\delta}{\delta A_{ai}}$ in the gravity part of the hamiltonian constraint acted on $\chi(A,\phi)$, while the remaining functional derivatives act on $\Psi_K(A)$ giving factors of the background fields through (106). But there are also terms in the gravity part in which two and three functional derivatives act on $\chi(A,\phi)$. This will give additional corrections to the functional Schroedinger equation.

Before writing them all out we have to consider the effect of the gauge and diffeomorphism constraints. By an analysis similar to the one of the last section, they tell us that

$$\chi(A,\phi) = \chi(\tau, a_{ai}, \phi) \tag{108}$$

where as before a_{ai} is transverse and tracefree. The dependence on the a_{ai} describes the couplings to gravitons.

τ is a field that parameterizes the trace part of the background A_{ai} and is given by

$$A_{ai}(x) = \delta_{ai} e^{\sqrt{\frac{\Lambda}{3}}\tau(x)} + \ldots \tag{109}$$

For the background A_{ai} discussed in Sect. 6 we have $\tau(x) = t$. However its important to keep the distinction clear: while t is a coordinate on a particular classical solution, $\tau(x)$ is a field that parameterizes the trace part of A_{ai} on the whole configuration space.

On the solution, τ is related to the Chern-Simon time described above by

$$\tau_{CS}(x) = \left(\frac{\Lambda}{3}\right)^{3/2} e^{3\sqrt{\frac{\Lambda}{3}}\tau(x)} \tag{110}$$

Thus, we have

$$\hat{E}^{ai}\chi(A,\phi) = -\hbar G \frac{\delta \chi(A,\phi)}{\delta A_{ai}} = \frac{\imath \hbar G}{\Lambda}\delta_{ai}\frac{\delta \chi(A,\phi)}{\delta \tau} - \hbar G \frac{\delta \chi(A,\phi)}{\delta a_{ai}} \tag{111}$$

We are interested in finding leading order corrections to the propagation of a free field on the background spacetime. Thus, we can neglect the couplings to gravitons. Doing so gives us corrections to the Tomonaga-Schwinger equation:

$$\begin{aligned} \imath f^4 \frac{\delta\chi(\tau,\phi)}{\delta\tau(x)} &= \frac{1}{2}[\pi^2 + f^4(\partial_a\phi)^2]\chi(\tau,\phi) \\ &+\frac{1}{2}(\partial_a\phi)^2\left[\frac{2\imath\hbar G f^2}{\Lambda}\frac{\delta}{\delta\tau(x)} - \left(\frac{\hbar G}{\Lambda}\right)^2\frac{\delta^2}{\delta\tau^2(x)}\right]\chi(\tau,\phi) \\ &+\left[2\frac{\hbar G f^2}{\Lambda}\frac{\delta^2}{\delta\tau^2(x)} + \imath\left(\frac{\hbar G}{\Lambda}\right)^2\frac{\delta^3}{\delta\tau^3(x)}\right]\chi(\tau,\phi) \qquad (112) \end{aligned}$$

The corrections on the second line come from the action of $\frac{\delta}{\delta A_{ai}}$ on χ from the matter hamiltonian density, while the corrections on the last line come from the higher order terms in $\frac{\delta}{\delta A_{ai}}$ from the gravitational part of the hamiltonian constraint.

To see what the effect of the corrections is on ordinary physics, we have to re-express the Tomonoaga-Schwinger equation in terms of measurable quantities that govern the low energy physics. One way to approach this is the following.

We are interested in extracting quantum field theory on Minkowski spacetime, in the limit $\Lambda \to 0$. For the limit to be non-singular we must rescale the time coordinate, because of the factors of $\hbar G/\Lambda$ in front of the $\delta/\delta\tau$ derivatives. In any case we need to rescale to remove a density factor, as we are interested in expressing the final answer in terms of a Schrodinger equation rather than a Tomonaga-Schwinger type equation. To do this we must replace the functional degree of freedom $\tau(x)$, which we have chosen to represent time by a global coordinate T. This coordinate T is taken to be proportional to τ on a $\tau =$ constant slice. However $\delta/\delta\tau(x)$ and $\partial/\partial T$ have different density weights and dimensions and this must be compensated for.

We accomplish both if we rescale so that on a fixed $\tau =$ constant slice,

$$\frac{\hbar G}{\Lambda}\frac{\delta}{\delta\tau(x)} = \alpha l_{Pl}\sqrt{detq^0_{ab}}\frac{\partial}{\partial T} \qquad (113)$$

where α is a dimensionless parameter. The factor of $\sqrt{detq^0_{ab}}$ is due to the fact that $\delta/\delta\tau(x)$ is a density. This form is required as l_{Pl} is the only dimensional parameter in the theory when $\Lambda \to 0$. We will see shortly how α is fixed.

The next step is to integrate over the spatial manifold, so as to recover the Schroedinger equation. To do this we multiply the whole expression by $1/\sqrt{detq^0_{ab}}$, because the form of the hamiltonian constraint we are using has density weight two, and then integrate. We set $f = 1$ as we are about to take $\Lambda \to 0$ and we note that in the coordinates we are using $det(q^0_{qb}) = 1$. This gives us,

$$\imath \frac{\partial \chi(T,\phi)}{\partial T}\left(\frac{\alpha V \Lambda}{l_{Pl}}\right) = \int_\Sigma \frac{1}{2}[\pi^2 + (\partial_a \phi)^2]\chi(T,\phi)$$
$$+ \int_\Sigma \frac{1}{2}(\partial_a\phi)^2\left[2\imath\alpha l_{Pl}\frac{\partial}{\partial T} - \alpha^2 l_{Pl}^2 \frac{\partial^2}{\partial T^2}\right]\chi(T,\phi)$$
$$+ \left(\frac{\alpha V \Lambda}{l_{Pl}}\right)\left[2\alpha^2 l_{Pl}^2 \frac{\partial}{\partial T} + \imath \alpha^3 l_{Pl}^3 \frac{\partial^2}{\partial T^2}\right]\chi(T,\phi) \quad (114)$$

where V is the volume of the spatial manifold according to the background metric, $V = \int_\Sigma \sqrt{det(q^0)}$. We impose an infrared cutoff so V is finite. We will shortly take $V \to \infty$ as $\Lambda \to 0$.

However before we do this we should take into account the renormalization between the bare fields and the physical fields that enter into the low energy physics. We expect to have to renormalize because there are interactions between the scalar and gravitational fields. However, as there is a cutoff on the spatial resolution in the exact diffeomorphism invariant states, coming from the discreteness of area and volume, we expect the wavefunction renormalization to be finite and to be proportional to powers of the ratio $\frac{L}{l_{Pl}}$, with $V = L^3$ as they represent infrared and ultraviolet cutoffs. Further as we expect relativistic invariance to hold at least up to corrections in l_{Pl}, we expect that $\pi \approx \dot{\phi}$ and $\partial_a \phi$ to renormalize by the same factor, again up to possible corrections in l_{Pl}[21]. Thus we expect

$$\pi = Z\pi_R \, , \quad \partial_a\phi = Z\partial_a\phi_R \quad (115)$$

where Z is a multiplicative renormalization. Let us suppose that $Z = \beta^{1/2}(L/l_{Pl})^{d/2}$. As the background represents de Sitter spacetime, it is natural to scale $\Lambda = \gamma/L^2$ where γ is a factor of order one depending on the topology. Thus we have

$$\imath \frac{\partial \chi(T,\phi)}{\partial T}\frac{\alpha\gamma}{\beta}\left(\frac{l_{Pl}}{L}\right)^{(d-1)} = \int_\Sigma \frac{1}{2}[\pi_R^2 + (\partial_a \phi_R)^2]\chi(T,\phi)$$
$$+ \int_\Sigma \frac{1}{2}(\partial_a\phi_R)^2\left[2\imath\alpha l_{Pl}\frac{\partial}{\partial T} - \alpha^2 l_{Pl}^2 \frac{\partial^2}{\partial T^2}\right]\chi(T,\phi)$$
$$+ \frac{\alpha\gamma}{\beta}\left(\frac{l_{Pl}}{L}\right)^{(d-1)}\left[2\alpha^2 l_{Pl}^2 \frac{\partial^2}{\partial T^2} + \imath \alpha^3 l_{Pl}^3 \frac{\partial^3}{\partial T^3}\right]\chi(T,\phi) \quad (116)$$

We must recover the Schroedinger equation in the limit $L \to \infty$, $l_{Pl} \to 0$. As the renormalized Hamiltonian

$$H_R = \int_\Sigma \frac{1}{2}[\pi_R^2 + (\partial_a\phi_R)^2] \quad (117)$$

should generate evolution in T, we require that the coefficient of $\imath\frac{\partial}{\partial T}$ on the left hand side be unity. This tells us that

[21] In the following we ignore such corrections, but if found by calculations they can be inserted directly in the following expressions.

$$\alpha = \frac{\beta}{\gamma}\left(\frac{L}{l_{Pl}}\right)^{(d-1)} \tag{118}$$

The limit then exists for $d \leq 1$. If $d < 1$ the additional terms disappear, and the usual Lorentz invariant quantum field theory is recovered. But in the case that $d = 1$ we have,

$$\alpha = \frac{\beta}{\gamma} \tag{119}$$

is a factor of order unity. Then our equation is

$$\begin{aligned} \imath\frac{\partial\chi(T,\phi)}{\partial T} &= H_R\chi(T,\phi) \\ &+ \int_\Sigma \frac{1}{2}(\partial_a\phi_R)^2\left[2\imath\alpha l_{Pl}\frac{\partial}{\partial T} - \alpha^2 l_{Pl}^2\frac{\partial^2}{\partial T^2}\right]\chi(T,\phi) \\ &+ \left[2\alpha^2 l_{Pl}^2\frac{\partial^2}{\partial T^2} + \imath\alpha^3 l_{Pl}^3\frac{\partial^3}{\partial T^3}\right]\chi(T,\phi) \end{aligned} \tag{120}$$

Thus, under the assumptions stated, we predict corrections to the Schroedinger equation, of order l_{Pl}, with the finite dimensionless coefficient α determined by the wavefunction renormalization of the scalar field theory interacting with gravity.

Now, to analyze the scalar field theory we can use to a first approximation a regular Fock space quantization in which

$$\phi(x,t) = \int \frac{d^3k}{\sqrt{(2\pi)^3 2\omega}}\left[a_k f_k(x)e^{-\imath\omega t} + a_k^\dagger f_k^*(x)e^{\imath\omega t}\right] \tag{121}$$

where, as $\Lambda \to 0$, $f_k(x) \approx e^{\imath k_a x^a}$.

The Fock space one particle states are not going to be exact solutions to the Wheeler-De Witt equation, but we can search for solutions of the form

$$\chi(T,\phi) = e^{-\imath\omega T}|k\rangle + O(l_{Pl}) \tag{122}$$

where $|k\rangle$ is a one particle Fock state. We do not set $\omega = |k|$ in (121), instead we take the components $\langle k| \ldots |k\rangle$ of the Wheeler-De Witt equation to extract the relation between ω and $|k|$.

We find, after the standard normal ordering

$$\langle k| : H_R : |k\rangle = \frac{\omega^2 + k^2}{2\omega} \tag{123}$$

$$\langle k| : \int_\Sigma (\partial_a\phi_R)^2 : |k\rangle = \frac{k^2}{2\omega} \tag{124}$$

Applying (120) to this state we find

$$\omega^2\frac{(1 + 4\alpha l_{Pl}\omega + 2\alpha^2 l_{Pl}^2\omega^2)}{(1 + \alpha l_{Pl}\omega + \frac{1}{2}\alpha^2 l_{Pl}^2\omega^2)} = k^2 \tag{125}$$

We thus see that there are indeed corrections to the energy momentum relations. However, we see that the precise predictions depend on the renormalization of the matter fields. The renormalization constants are expected to be finite, and computable, but have not yet been computed.

15 Conclusions and Further Developments

The results described here show that there is a good possibility that loop quantum gravity has the capacity to make unambiguous predictions for Planck scale phenomenology. However, it has not yet done so.

I close by briefly summarizing the present situation. Previous results indicated that perturbations of semi-classical states propagate according to modified energy-momentum relations. However, the examples studied in flat space are not definitive because the states are not physical states, hence they may just indicate that there is a preferred frame in the candidate ground state studied. The results are also ambiguous as the exact magnitudes of the effects depend on parameters in the candidate ground state.

The result described in the last section indicates that corrections to energy momentum relations are found even for perturbations of a physical state. However the exact predictions depend on finite renormalization constants that have yet to be computed. There are also open issues regarding the interpretation of the Kodama state, which should be resolved.

We can look forward to further progress along several lines.

- It would be very useful if a quantum hamiltonian operator constructed and its expectation value proved positive definite. In this case one could determine the parameters in ansatzes for the ground state by minimizing the energy.
- It may be possible to investigate propagation of excitations of semiclassical states in a gauge fixed formalism.
- There is always the possibility of an indirect determination of the symmetry of the low energy limit or of a sum rule that would relate different predictions, which could be tested.
- Another domain in which loop quantum gravity may make testable predictions is cosmology. It would be very interesting to see if a connection could be made between predictions in the two domains: cosmology and Planck scale phenomenology.

If progress can be made in any of these directions, while progress continues on the experimental side, it is not impossible that loop quantum gravity may achieve the status of a theory that has been subject to experimental test. So long as this is a possibility its realization should be the first priority of people who work on quantum gravity.

Acknowledgements

I wish first to thank my collaborators on different aspects of the problem of quantum gravity with a cosmological constant, Giovanni Amelino-Camelia, Roumen Borissov, Yi Ling, Jerzy Kowalski-Glikman, Joao Magueijo, Seth Major, Fotini Markopoulou, Carlo Rovelli, Chopin Soo and Artem Starodubstev. I also have learned a lot about this and related subjects from discussions and correspondence with John Baez, Tom Banks, Raphael Bousso, Louis Crane, Laurent Freidel, Lou Kauffman, Renate Loll and Michael Reisenberger. Parts of this work were supported generously by The National Science Foundation and the Jesse Phillips Foundation.

References

1. C. Rovelli, Living Rev. Rel. 1 (1998) 1, gr-qc/9710008.
2. A. Ashtekar, New perspectives in canonical gravity (Bibliopolis, Naples, 1988); Lectures on non-perturbative canonical gravity, Advanced Series in Astrophysics and Cosmology-Vol. 6 (World Scientific, Singapore, 1991).
3. R. Gambini and J. Pullin, Loops, knots, gauge theories and quantum gravity, Cambridge University Press, 1996.
4. L. Smolin: in Quantum Gravity and Cosmology, eds J Perez-Mercader et al., World Scientific, Singapore 1992; "The future of spin networks" gr-qc/9702030 in the Penrose Festschrift.
5. L. Smolin, "Quantum gravity with a positive cosmological constant," hep-th/0209079 .
6. L. Smolin, "How far are we from the quantum theory of gravity?," hep-th/0303185 .
7. A. Ashtekar and J. Lewandowski, "Background independent quantum gravity: a status report," gr-qc/0404018.
8. O. Dreyer, "Ln(3) and Black Hole Entropy," gr-qc/0404055; "Quasinormal Modes, the Area Spectrum, and Black Hole Entropy," Phys. Rev. Lett. 90 (2003) 081301, gr-qc/0211076.
9. Martin Bojowald, "Isotropic Loop Quantum Cosmology", Class. Quant. Grav. 19 (2002) 2717–2742, gr-qc/0202077; "Inflation from Quantum Geometry", gr-qc/0206054; "The Semiclassical Limit of Loop Quantum Cosmology", gr-qc/0105113, Class. Quant. Grav. 18 (2001) L109-L116; "Dynamical Initial Conditions in Quantum Cosmology", gr-qc/0104072, Phys. Rev. Lett. 87 (2001) 121301.
10. S. Tsujikawa, P. Singh, and R. Maartens, "Loop quantum gravity effects on inflation and the CMB," astro-ph/0311015
11. G. Amelino-Camelia et al., Int. J. Mod. Phys. A12:607-624,1997; G. Amelino-Camelia et al. Nature 393:763-765,1998; J. Ellis et al., Astrophys. J. 535:139-151, 2000; J. Ellis, N.E. Mavromatos and D. Nanopoulos, Phys. Rev. D63:124025,2001; ibidem astro-ph/0108295.
12. G. Amelino-Camelia and T. Piran, Phys. Rev. D64 (2001) 036005.

13. Tomasz J. Konopka, Seth A. Major, "Observational Limits on Quantum Geometry Effects", New J. Phys. 4 (2002) 57. hep-ph/0201184; Ted Jacobson, Stefano Liberati, David Mattingly, "TeV Astrophysics Constraints on Planck Scale Lorentz Violation", hep-ph/0112207.
14. Subir Sarkar, "Possible astrophysical probes of quantum gravity", Mod. Phys. Lett. A17 (2002) 1025–1036, gr-qc/0204092.
15. J. Lukierski et al., "Q Deformation Of Poincaré Algebra," Phys. Lett. B264 (1991) 331.
16. G. Amelino-Camelia, Nature 418 (2002) 34.
17. J. Magueijo and L. Smolin, Phys. Rev. Lett. (88) 190403, 2002.
18. J. Magueijo and L. Smolin, gr-qc/0207
19. N.R. Bruno, G. Amelino-Camelia, J. Kowalski-Glikman, Phys. Lett. B522:133-138,2001; J. Kowalski-Glikman and S. Nowak, hep-th/0203040;S. Judes, gr-qc/0205067; M. Visser, gr-qc/0205093; S. Judes, M. Visser, gr-qc/0205067; D. V. Ahluwalia and M. Kirchbach, qr-qc/0207004.
20. Achucarro and Townsend, "A Chern-Simons Action For Three-Dimensional Anti-De Sitter Supergravity Theories," Phys. Lett. B180 (1986) 89; E. Witten, "(2+1)-Dimensional Gravity As An Exactly Soluble System," Nucl. Phys. B311 (1988) 46.
21. A. Ashtekar, V. Husain, C. Rovelli, J. Samuel and L. Smolin, "2+1 quantum gravity as a toy model for the 3+1 theory," Class. and Quantum Grav. L185-L193 (1989); L. Smolin, "Loop representation for quantum gravity in 2+1 dimensions," in the proceedings of the John's Hopkins Conference on Knots, Tolopoly and Quantum Field Theory ed. L. Lusanna (World Scientific, Singapore, 1989) .
22. Laurent Freidel, Jerzy Kowalski-Glikman, Lee Smolin, "2+1 gravity and Doubly Special Relativity," hep-th/0307085, Phys. Rev. D69 (2004) 044001.
23. Hans-Juergen Matschull, Max Welling, "Quantum Mechanics of a Point Particle in 2+1 Dimensional Gravity," gr-qc/9708054, Class. Quant. Grav. 15 (1998) 2981; Hans-Juergen Matschull, "The Phase Space Structure of Multi Particle Models in 2+1 Gravity," gr-qc/0103084 , Class. Quant. Grav. 18 (2001) 3497; F. A. Bais, N. M. Muller, B. J. Schroers, "Quantum group symmetry and particle scattering in (2+1)-dimensional quantum gravity," hep-th/0205021, Nucl. Phys. B640 (2002) 3.
24. Giovanni Amelino-Camelia, Lee Smolin, Artem Starodubtsev, "Quantum symmetry, the cosmological constant and Planck scale phenomenology," hep-th/0306134.
25. J.E. Nelson, R.F. Picken, "Quantum Holonomies in (2+1)-Dimensional Gravity," Phys. Lett. B471 (2000) 367; J.E. Nelson, T. Regge, "Quantisation of 2+1 gravity for genus 2," Phys. Rev. D50 (1994) 5125, gr-qc/9311029
26. E. Buffenoir, K. Noui, P. Roche, "Hamiltonian Quantization of Chern-Simons theory with SL(2,C) Group," hep-th/0202121, Class. Quant. Grav. 19 (2002) 4953; Karim Noui, Philippe Roche, "Cosmological Deformation of Lorentzian Spin Foam Models," gr-qc/0211109, Class. Quant. Grav. 20 (2003) 3175–3214.
27. L. Smolin, "Linking topological quantum field theory and nonperturbative quantum gravity," J. Math. Phys. 36(1995)6417, gr-qc/9505028.
28. J. Baez, "Spin foam models," Class. Quant. Grav. 15 (1998) 1827–1858, gr-qc/9709052; "An introduction to spin foam models of quantum gravity and BF theory," Lect. Notes Phys., 543:2594, 2000.

29. S. Major and L. Smolin, "Quantum deformation of quantum gravity," Nucl. Phys. B473, 267(1996), gr-qc/9512020; R. Borissov, S. Major and L. Smolin, "The geometry of quantum spin networks," Class. and Quant. Grav.12, 3183(1996), gr-qc/9512043.
30. Artem Starodubtsev, "Topological excitations around the vacuum of quantum gravity I: The symmetries of the vacuum," hep-th/0306135.
31. A. Ashtekar, C. Rovelli and L. Smolin, "Weaving a classical geometry with quantum threads," Phys. Rev. Lett. 69 (1992) 237 hep-th/9203079; Luca Bombelli, "Statistical geometry of random weave states," gr-qc/0101080; A. Corichi, J.M. Reyes, "A Gaussian Weave for Kinematical Loop Quantum Gravity," gr-qc/0006067, Int. J. Mod. Phys. D10 (2001) 325-338.
32. Rodolfo Gambini, Jorge Pullin, "Nonstandard optics from quantum spacetime", Phys. Rev. D59 (1999) 124021, gr-qc/9809038;
33. Jorge Alfaro, Hugo A. Morales-Tžcotl, Luis F. Urrutia, "Loop quantum gravity and light propagation," Phys. Rev. D65 (2002) 103509, hep-th/0108061; "Quantum gravity and spin 1/2 particles effective dynamics," hep-th/0208192, Phys. Rev. D66:124006,2002.
34. Hanno Sahlmann, Thomas Thiemann, "Towards the QFT on Curved Spacetime Limit of QGR. I: A General Scheme," gr-qc/0207030; "Towards the QFT on Curved Spacetime Limit of QGR. II: A Concrete Implementation," gr-qc/0207031.
35. H. Kodama, Prog. Theor. Phys. 80, 1024(1988); Phys. Rev. D42(1990)2548.
36. L. Smolin and C. Soo, "The Chern-Simons Invariant as the Natural Time Variable for Classical and Quantum Cosmology," Nucl. Phys. B449 (1995) 289, gr-qc/9405015.
37. A. Ashtekar, private communication.
38. A. Sen, "On the existence of neutrino zero modes in vacuum spacetime," J. Math. Phys. 22 (1981) 1781, "Gravity as a spin system," Phys. Lett. B11 (1982) 89.
39. Abhay Ashtekar, "New variables for classical and quantum gravity," Phys. Rev. Lett. 57(18), 2244–2247 (1986).
40. C. Soo, "Wave function of the Universe and Chern-Simons Perturbation Theory," gr-qc/0109046.
41. P. A. M. Dirac, Lectures on Quantum Mechanics Belfer Graduate School of Science Monographs, no. 2 (Yeshiva University Press, New York,1964).
42. J. Stachel, "Einstein's search for general covariance, 1912-15" in *Einstein and the History of General Relativity* vol 1 of *Einstein Studies* eds. D. Howard and J. Stachel. (Birkhauser, Boston, 1989).
43. L. Smolin *Three Roads to Quantum Gravity* (Weidenfeld and Nicolson and Basic Books, London and New York, 2001)
44. Y. Ling and L. Smolin, "Supersymmetric spin networks and quantum supergravity," Phys. Rev. D61, 044008(2000), hep-th/9904016; "Holographic Formulation of Quantum Supergravity," hep-th/0009018, Phys. Rev. D63 (2001) 064010.
45. M. Atiyah, "Topological quantum field theory" Publ. Math. IHES 68 (1989) 175; The Geometry and Physics of Knots, Lezion Lincee (Cambridge University Press, Cambridge, 1990); G. Segal, Conformal field theory Oxford preprint (1988).
46. R. Floreanini and R. Percacci, Phys. Lett. B224 (1989) 291-294; B231:119-124, 1989. V.V. Fock, N.A. Nekrasov, A.A. Rosly, K.G. Selivanov "What we think

about the higher dimensional Chern-Simons theories" (Moscow, ITEP). ITEP-91-70, July 1991. 7pp. in Sakharov Conf.1991:465-472 (QC20:I475:1991)
47. M. Banados, M. Henneaux, C. Iannuzzo and C. M. Viallet, "A note on the gauge symmetries of pure Chern-Simons theory with p-form gauge fields" gr-qc/9703061; Max Banados, Luis J. Garay and Marc Henneaux, Nucl. Phys. B476:611-635,1996, hep-th/9605159; Phys. Rev. D53:593-596,1996, hep-th/9506187.
48. Y. Ling and L. Smolin, "Eleven dimensional supergravity as a constrained topological field theory," hep-th/0003285, Nucl. Phys. B601 (2001) 191-208.
49. J. Ambjorn, A. Dasgupta, J. Jurkiewiczcy and R. Loll, "A Lorentzian cure for Euclidean troubles", hep-th/0201104; J. Ambjorn and R. Loll, Nucl. Phys. B536 (1998) 407 [hep-th/9805108]; J. Ambjorn, J. Jurkiewicz and R. Loll, Phys. Rev. Lett. 85 (2000) 924 [hepth/ 0002050]; Nucl. Phys. B610 (2001) 347 [hep-th/0105267]; R. Loll, Nucl. Phys. B (Proc. Suppl.) 94 (2001) 96 [hep-th/0011194]; J. Ambjorn, J. Jurkiewicz and R. Loll, Phys. Rev. D64 (2001) 044011 [hep-th/0011276]; JHEP 09 (2001) 022 [hep-th/0106082].
50. T. Jacobson and L. Smolin, Phys. Lett. B 196 (1987) 39; Class. and Quant. Grav. 5 (1988) 583; J. Samuel, Pramana-J Phys. 28 (1987) L429.
51. J.F. Plebanski. "On the separation of Einsteinian" J. Math. Phys., 18:2511, 1977.
52. R. Capovilla, J. Dell and T. Jacobson, Phys. Rev. Lett. 21, 2325(1989); Class. Quant. Grav. 8, 59(1991); R. Capovilla, J. Dell, T. Jacobson and L. Mason, Class. and Quant. Grav. 8, 41(1991).
53. J. Barbero, "Real Ashtekar variables for Lorentzian signature spacetime," Phys. Rev. D51 (1995) 5507.
54. T. Thiemann, "Quantum Spin Dynamics (QSD) I & II," Class. Quant. Grav. 15 (1998) 839–905, gr-qc/9606089, gr-qc/9606090.
55. K. Krasnov, "On Quantum Statistical Mechanics of a Schwarzschild Black Hole," grqc/9605047, Gen. Rel. Grav. 30 (1998) 53–68; C. Rovelli, "Black hole entropy from loop quantum gravity," grqc/9603063.
56. A. Ashtekar, J. Baez, K. Krasnov, "Quantum Geometry of Isolated Horizons and Black Hole Entropy," gr-qc/0005126; A. Ashtekar, J. Baez, A. Corichi, K. Krasnov, "Quantum geometry and black hole entropy," gr-qc/9710007, Phys. Rev. Lett. 80 (1998) 904–907.
57. T. Jacobson and L. Smolin, Nucl. Phys. B 299 (1988) 295.
58. C. Rovelli and L. Smolin, "Knot theory and quantum theory," Phys. Rev. Lett 61(1988)1155; "Loop representation of quantum general relativity," Nucl. Phys. B331(1990)80-152.
59. R. Gambini and A. Trias, Phys. Rev. D23 (1981) 553, Lett. al Nuovo Cimento 38 (1983) 497; Phys. Rev. Lett. 53 (1984) 2359; Nucl. Phys. B278 (1986) 436; R. Gambini, L. Leal and A. Trias, Phys. Rev. D39 (1989) 3127.
60. M. P. Reisenberger. "Worldsheet formulations of gauge theories and gravity," in Proceedings of the 7th Marcel Grossman Meeting, ed. by R. Jantzen and G. MacKeiser, World Scientific, 1996; gr-qc/9412035; "A lattice worldsheet sum for 4-d Euclidean general relativity," gr-qc/9711052.
61. M. P. Reisenberger and C. Rovelli. "Sum-over-surface form of loop quantum gravity," gr-qc/9612035, Phys. Rev. D 56 (1997) 3490; "Spacetime as a Feynman diagram: the connection formulation," Class. Quant. Grav., 18:121140, 2001; "Spin foams as Feynman diagrams," gr-qc/0002083.

62. J. Barrett and L. Crane, "Relativistic spin networks and quantum gravity," J. Math. Phys. 39 (1998) 3296–3302, gr-qc/9709028.
63. J. Baez, "Spin foam models," Class. Quant. Grav. 15 (1998) 1827–1858, gr-qc/9709052; "An introduction to spin foam models of quantum gravity and BF theory," Lect. Notes Phys., 543:2594, 2000.
64. Fotini Markopoulou, "Dual formulation of spin network evolution," gr-qc/9704013.
65. Fotini Markopoulou, Lee Smolin, "Quantum geometry with intrinsic local causality," Phys. Rev. D58 (1998) 084032, gr-qc/9712067.
66. J. Iwasaki, "A reformulation of the Ponzano-Regge quantum gravity model in terms of surfaces," gr-qc/9410010; "A definition of the Ponzano-Regge quantum gravity model in terms of surfaces," gr-qc/9505043, J. Math. Phys. 36 (1995) 6288; L. Freidel and K. Krasnov, "Spin foam models and the classical action principle," Adv. Theor. Math. Phys., 2:11831247, 1999; R. De Pietri, L. Freidel, K. Krasnov, and C. Rovelli, "Barrett-Crane model from a Boulatov-Ooguri field theory over a homogeneous space," Nucl. Phys. B, 574:785806, 2000.
67. B. Bruegmann, R. Gambini and J. Pullin, Phys. Rev. Lett. 68 (1992) 431–434; Rodolfo Gambini, Jorge Griego, Jorge Pullin, "Chern-Simons states in spin-network quantum gravity," gr-qc/9703042, Phys. Lett. B413 (1997) 260–266; C. Di Bartolo, R. Gambini, J. Griego, J. Pullin, "Consistent canonical quantization of general relativity in the space of Vassiliev knot invariants," gr-qc/9909063, Phys. Rev. Lett. 84 (2000) 2314–2317; "Canonical quantum gravity in the Vassiliev invariants arena: I. Kinematical structure," gr-qc/9911009, Class. Quant. Grav. 17 (2000) 3211-3238.
68. R. Jackiw, Topological Investigations In Quantized Gauge Theories, p. 258, exercise 3.7, in S. B. Treiman et al. Current Algebra And Anomalies (World Scientific, 1985).
69. E. Witten, "A Note On The Chern-Simons And Kodama Wavefunctions," gr-qc/0306083.
70. L. Freidel and L. Smolin, "The linearization of the Kodama state," hep-th/0310224.
71. G. W. Gibbons and S. W. Hawking, "Cosmological Event Horizons, Thermodynamics, and Particle Creation," Phys. Rev. D 15, 2738 (1977).
72. L. N. Chang and C. Soo, "Ashtekar's variables and the topological phase of quantum gravity," Proceedings of the XXth. Conference on Differential Geometric Methods in Physics, (Baruch College, New York, 1991), edited by S. Catto and A. Rocha (World Scientific, 1992); Phys. Rev. D46 (1992) 4257; C. Soo and L. N. Chang, Int. J. Mod. Phys. D3 (1994) 529.
73. T. Jacobson, "New Variables for canonical supergravity," Class. Quant. Grav.5(1988)923; D. Armand-Ugon, R. Gambini, O. Obregon, J. Pullin, "Towards a loop representation for quantum canonical supergravity," hep-th/9508036, Nucl. Phys. B460 (1996) 615; L. F. Urrutia "Towards a loop representation of connection theories defined over a super-lie algebra," hep-th/9609010; H. Kunitomo and T. Sano "The Ashtekar formulation for canonical N=2 supergravity," Prog. Theor. Phys. suppl. (1993) 31; Takashi Sano and J. Shiraishi, "The Nonperturbative Canonical Quantization of the N=1 Supergravity," Nucl. Phys. B410 (1993) 423, hep-th/9211104; "The Ashtekar Formalism and WKB Wave Functions of N = 1,2 Supergravities," hep-th/9211103; T. Kadoyoshi and S. Nojiri, "N = 3 and N = 4 two form supergravities," Mod.

Phys. Lett. A12:1165-1174,1997, hep-th/9703149; K. Ezawa, "Ashtekar's formulation for N = 1, N = 2 supergravities as constrained BF theories," Prog. Theor. Phys. 95:863-882, 1996, hep-th/9511047.

74. Yi Ling, "Introduction to supersymmetric spin networks", hep-th/0009020, J. Math. Phys. 43 (2002) 154–169
75. T. Banks, "T C P, Quantum Gravity, The Cosmological Constant And All That", Nucl. Phys. B249 (1985) 332.
76. L. Smolin, in preparation.
77. A. Ashtekar, C. Rovelli and L. Smolin "Gravitons and Loops", Phys. Rev. D 44 (1991) 1740-1755; J. Iwasaki, C. Rovelli, "Gravitons as embroidery on the weave," Int. J. Mod. Phys. D 1 (1993) 533; "Gravitons from loops: non-perturbative loop-space quantum gravity contains the graviton-physics approximation," Class. Quantum Grav. 11 (1994) 1653.
78. M. Spradlin, A. Strominger, A. Volovich, "Les Houches Lectures on de Sitter Space," hep-th/0110007
79. G. Horowitz, "Exactly Soluble Diffeomorphism Invariant Theories", Commun. Math. Phys. 125 (1989) 417; V. Husain, "Topological Quantum Mechanics", Phys. Rev. D43 (1991) 1803.

Index

Lecture Notes in Physics

For information about earlier volumes please contact your bookseller or Springer LNP Online archive: springerlink.com

Vol.622: F. Benatti, R. Floreanini (Eds.), Irreversible Quantum Dynamics

Vol.623: M. Falcke, D. Malchow (Eds.), Understanding Calcium Dynamics, Experiments and Theory

Vol.624: T. Pöschel (Ed.), Granular Gas Dynamics

Vol.625: R. Pastor-Satorras, M. Rubi, A. Diaz-Guilera (Eds.), Statistical Mechanics of Complex Networks

Vol.626: G. Contopoulos, N. Voglis (Eds.), Galaxies and Chaos

Vol.627: S.G. Karshenboim, V.B. Smirnov (Eds.), Precision Physics of Simple Atomic Systems

Vol.628: R. Narayanan, D. Schwabe (Eds.), Interfacial Fluid Dynamics and Transport Processes

Vol.629: U.-G. Meißner, W. Plessas (Eds.), Lectures on Flavor Physics

Vol.630: T. Brandes, S. Kettemann (Eds.), Anderson Localization and Its Ramifications

Vol.631: D. J. W. Giulini, C. Kiefer, C. Lämmerzahl (Eds.), Quantum Gravity, From Theory to Experimental Search

Vol.632: A. M. Greco (Ed.), Direct and Inverse Methods in Nonlinear Evolution Equations

Vol.633: H.-T. Elze (Ed.), Decoherence and Entropy in Complex Systems, Based on Selected Lectures from DICE 2002

Vol.634: R. Haberlandt, D. Michel, A. Pöppl, R. Stannarius (Eds.), Molecules in Interaction with Surfaces and Interfaces

Vol.635: D. Alloin, W. Gieren (Eds.), Stellar Candles for the Extragalactic Distance Scale

Vol.636: R. Livi, A. Vulpiani (Eds.), The Kolmogorov Legacy in Physics, A Century of Turbulence and Complexity

Vol.637: I. Müller, P. Strehlow, Rubber and Rubber Balloons, Paradigms of Thermodynamics

Vol.638: Y. Kosmann-Schwarzbach, B. Grammaticos, K.M. Tamizhmani (Eds.), Integrability of Nonlinear Systems

Vol.639: G. Ripka, Dual Superconductor Models of Color Confinement

Vol.640: M. Karttunen, I. Vattulainen, A. Lukkarinen (Eds.), Novel Methods in Soft Matter Simulations

Vol.641: A. Lalazissis, P. Ring, D. Vretenar (Eds.), Extended Density Functionals in Nuclear Structure Physics

Vol.642: W. Hergert, A. Ernst, M. Däne (Eds.), Computational Materials Science

Vol.643: F. Strocchi, Symmetry Breaking

Vol.644: B. Grammaticos, Y. Kosmann-Schwarzbach, T. Tamizhmani (Eds.) Discrete Integrable Systems

Vol.645: U. Schollwöck, J. Richter, D.J.J. Farnell, R.F. Bishop (Eds.), Quantum Magnetism

Vol.646: N. Bretón, J. L. Cervantes-Cota, M. Salgado (Eds.), The Early Universe and Observational Cosmology

Vol.647: D. Blaschke, M. A. Ivanov, T. Mannel (Eds.), Heavy Quark Physics

Vol.648: S. G. Karshenboim, E. Peik (Eds.), Astrophysics, Clocks and Fundamental Constants

Vol.649: M. Paris, J. Rehacek (Eds.), Quantum State Estimation

Vol.650: E. Ben-Naim, H. Frauenfelder, Z. Toroczkai (Eds.), Complex Networks

Vol.651: J.S. Al-Khalili, E. Roeckl (Eds.), The Euroschool Lectures of Physics with Exotic Beams, Vol.I

Vol.652: J. Arias, M. Lozano (Eds.), Exotic Nuclear Physics

Vol.653: E. Papantonoupoulos (Ed.), The Physics of the Early Universe

Vol.654: G. Cassinelli, A. Levrero, E. de Vito, P. J. Lahti (Eds.), Theory and Appplication to the Galileo Group

Vol.655: M. Shillor, M. Sofonea, J.J. Telega, Models and Analysis of Quasistatic Contact

Vol.656: K. Scherer, H. Fichtner, B. Heber, U. Mall (Eds.), Space Weather

Vol.657: J. Gemmer, M. Michel, G. Mahler (Eds.), Quantum Thermodynamics

Vol.658: K. Busch, A. Powell, C. Röthig, G. Schön, J. Weissmüller (Eds.), Functional Nanostructures

Vol.659: E. Bick, F.D. Steffen (Eds.), Topology and Geometry in Physics

Vol.660: A.N. Gorban, I.V. Karlin, Invariant Manifolds for Physical and Chemical Kinetics

Vol.661: N. Akhmediev, A. Ankiewicz (Eds.) Dissipative Solitons

Vol.662: U. Carow-Watamura, Y. Maeda, S. Watamura (Eds.), Quantum Field Theory and Noncommutative Geometry

Vol.663: A. Kalloniatis, D. Leinweber, A. Williams (Eds.), Lattice Hadron Physics

Vol.664: R. Wielebinski, R. Beck (Eds.), Cosmic Magnetic Fields

Vol.665: V. Martinez (Ed.), Data Analysis in Cosmology

Vol.666: D. Britz, Digital Simulation in Electrochemistry

Vol.667: W. Dieter Heiss (Ed.), Quantum Dots: a Doorway to Nanoscale Physics

Vol.668: H. Ocampo, S. Paycha, A. Vargas (Eds.), Geometric and Topological Methods for Quantum Field Theory

Vol.669: G. Amelino-Camelia, J. Kowalski-Glikman (Eds.), Planck Scale Effects in Astrophysics and Cosmology

MIX
Papier aus verantwortungsvollen Quellen
Paper from responsible sources
FSC® C105338

If you have any concerns about our products,
you can contact us on
ProductSafety@springernature.com

In case Publisher is established outside the EU,
the EU authorized representative is:
Springer Nature Customer Service Center GmbH
Europaplatz 3, 69115 Heidelberg, Germany

Printed by Libri Plureos GmbH
in Hamburg, Germany